T0201960

Statistics for Biology and Health

Statistics for Biology and Health

Odd O. Aalen • Ørnulf Borgan • Håkon K. Gjessing

Survival and Event History Analysis

A Process Point of View

 Springer

Odd O. Aalen
Department of Biostatistics
Institute of Basic Medical Sciences
University of Oslo
Oslo, Norway
o.o.aalen@medisin.uio.no

Ørnulf Borgan
Department of Mathematics
University of Oslo
Oslo, Norway
borgan@math.uio.no

Håkon K. Gjessing
Norwegian Institute of Public Health
Oslo, Norway
and
Section for Medical Statistics
University of Bergen
Bergen, Norway
hakon.gjessing@fhi.no

Series Editors

M. Gail
National Cancer Institute
Rockville, MD 20892
USA

A. Tsiatis
Department of Statistics
North Carolina State University
Raleigh, NC 27695
USA

K. Krickeberg
Le Chatelet
F-63270 Manglieu
France

W. Wong
Department of Statistics
Stanford University
Stanford, CA 94305-4065
USA

J. Samet
Department of Epidemiology
School of Public Health
Johns Hopkins University
615 Wolfe Street
Baltimore, MD 21205-2103
USA

ISBN 978-0-387-20287-7   e-ISBN 978-0-387-68560-1
DOI: 10.1007/978-0-387-68560-1

Library of Congress Control Number: 2008927364

To Marit, John, and Margrete

To Tone, Eldrid, and Yngve

To Liên, Eli, Einar, Wenche, and Richard

Preface

Survival and event history analysis have developed into one of the major areas of biostatistics, with important applications in other fields as well, including reliability theory, actuarial science, demography, epidemiology, sociology, and econometrics. This has resulted in a number of substantial textbooks in the field. However, rapidly developing theoretical models, combined with an ever-increasing amount of high-quality data with complex structures, have left a gap in the literature. It has been our wish to provide a book that exposes the rich interplay between theory and applications. Without being unnecessarily technical, we have wanted to show how theoretical aspects of statistical models have direct, intuitive implications for applied work. And conversely, how apparently disparate and complex features of data can be set in a theoretical framework that enables comprehensive modeling.

Most textbooks in survival analysis focus on the occurrence of single events. In actual fact, much event history data consist of occurrences that are repeated over time or related among individuals. The reason for this is, in part, the development toward increased registration and monitoring of individual life histories in clinical medicine, epidemiology, and other fields. The material for research and analysis is therefore more extensive and complex than it used to be, and standard approaches in survival analysis are insufficient to handle this.

It is natural to view such life histories as stochastic processes, and this is the basic idea behind our book. We start with the now classical counting process theory and give a detailed introduction to the topic. Leaving out much mathematical detail, we focus on understanding the important ideas. Next, we introduce standard survival analysis methods, including Kaplan-Meier and Nelson-Aalen plots and Cox regression. We also give a careful discussion of the additive hazards model. Then we extend the use of the counting process framework by counting several events for each individual. This yields very fruitful models, especially when combined with the additive hazards model. We further include time-dependent covariates, or marker processes, to define what we term dynamic path analysis, an extension of classical path analysis to include time. This allows us to explicitly analyze how various processes influence one another. Most of this is new, or very recently developed, material.

Another new aspect of the book is the explicit connection drawn between event history analysis and statistical causality. The focus is on causal formulations where time is explicitly present, including ideas like Granger causality, local independence, and dynamic path analysis.

Unique to this book is the emphasis on models that give insight into the rather elusive concept of hazard rate, and the various shapes the hazard rate can have. The effect of unobserved heterogeneity, or frailty, is broadly discussed with focus on a number of artifacts implied by the frailty structure as well as on applying these models to multivariate survival data. A new class of process-based frailty models is introduced, derived from processes with independent increments. We also study models of underlying processes that, when hitting a barrier, lead to the event in question. It is emphasized that frailty is an essential concept in event history analysis, and wrong conclusions may be drawn if frailty is ignored.

The applied aspect of the book is supported by a range of real-life data sets. Most of the data are well known to us through active collaboration with research groups in other fields. By necessity, some of the examples based on them have been simplified to achieve a more pedagogical exposition. However, we have made a strong effort to keep close to the relevant research questions; in particular, our three case studies are intended as more or less full-blown analyses of data of current interest.

Hence, the book contains a lot of new material for both the practicing statistician and those who are more interested in the theory of the subject. The book is intended primarily for researchers working in biostatistics, but it should also be found interesting by statisticians in other fields where event history analysis is of importance. The reader should have some theoretical background in statistics, although the mathematics is kept at a "user" level, not going into the finer theoretical details.

The book should be well suited as a textbook for graduate courses in survival and event history analysis, in particular for courses on the analysis of multivariate or complex event history data. Some exercises are provided at the end of each chapter, and supplementary exercises, as well as some of the datasets used as examples, may be found on the book's Web page at www.springer.com/978-0-387-20287-7.

A number of friends and colleagues have read and commented on parts of this book, have participated in discussion on themes from the book, or have provided us with data used as examples, and we want to thank them all: Per Kragh Andersen, Vanessa Didelez, Ludwig Fahrmeir, Johan Fosen, Axel Gandy, Jon Michael Gran, Nina Gunnes, Robin Henderson, Nils Lid Hjort, Niels Keiding, Bryan Langholz, Stein Atle Lie, Bo Lindqvist, Torben Martinussen, Sven Ove Samuelsen, Thomas Scheike, Finn Skårderud, Anders Skrondal, Halvor Sommerfelt, Hans Steinsland, Hans van Houwelingen, and Stein Emil Vollset.

Hege Marie Bøvelstad and Marion Haugen did an invaluable job making graphs for a number of the examples. They also fearlessly took on the daunting task of preparing all our references, saving us long hours of tedious work.

Preliminary versions of the book have been used as the text for a graduate course in survival and event history analysis at the University of Oslo and for a Nordic course organized as part of a Nordplus program for master-level courses in biostatistics. We thank the students at these courses, who gave valuable feedback and pointed out a number of typos.

Most of the work on the book has been done as part of our regular appointments at the University of Oslo (Odd O. Aalen and Ørnulf Borgan) and the Norwegian Institute of Public Health (Håkon K. Gjessing), and we thank our employers for giving us time to work on the book as part of our regular duties. In addition, important parts of the work were done while the authors were participating in an international research group at the Centre for Advanced Study at the Norwegian Academy of Science and Letters in Oslo during the academic year 2005–2006. We express our gratitude to the staff of the center for providing such good working facilities and a pleasant social environment.

Oslo   *Odd O. Aalen*
January 2008   *Ørnulf Borgan*
                                                                                        Håkon K. Gjessing

Contents

Chapter 1

An introduction to survival and event history analysis

This book is about survival and event history analysis. This is a statistical methodology used in many different settings where one is interested in the occurrence of events. By events we mean occurrences in the lives of individuals that are of interest in scientific studies in medicine, demography, biology, sociology, econometrics, etc. Examples of such events are: death, myocardial infarction, falling in love, wedding, divorce, birth of a child, getting the first tooth, graduation from school, cancer diagnosis, falling asleep, and waking up. All of these may be subject to scientific interest where one tries to understand their cause or establish risk factors. In classical survival analysis one focuses on a single event for each individual, describing the occurrence of the event by means of survival curves and hazard rates and analyzing the dependence on covariates by means of regression models.

The connecting together of several events for an individual as they occur over time yields event histories. One might, for instance, be interested in studying how people pass through diseases. Clearly, the disease might have a number of stages. In parallel with the development, there will typically be a number of blood tests and other diagnostics, different treatment options may be attempted, and so on. The statistical analysis of such data, trying to understand how factors influence each other, is a great challenge.

Event histories are not restricted to humans. A sequence of events could also happen to animals, plants, cells, amalgam fillings, hip prostheses, light bulbs, cars – anything that changes, develops, or decays. Although a piece of technical equipment is very different from a human being, that does not prevent statistical methods for analyzing event history data to be useful, for example, in demography and medicine as well as in technical reliability. Motivated by our own research interests, the focus of this book is on applications of event history methodology to medicine and demography. But the methodology we present also should be of interest to researchers in biology, technical reliability, econometrics, sociology, etc., who want to apply survival and event history methods in their own fields.

Survival and event history analysis is used as a tool in many different settings, some of which are:

- Proving or disproving the value of medical treatments for diseases.

- Understanding risk factors, and thereby preventing diseases.
- Evaluating reliability of technical equipment.
- Understanding the mechanisms of biological phenomena.
- Monitoring social phenomena like divorce and unemployment.

Even though the purpose of a statistical analysis may vary from one situation to another, the ambitious aim of most statistical analyses is to help understand causality.

The purpose of this introductory chapter is twofold. Our main purpose is to introduce the reader to some basic concepts and ideas in survival and event history analysis. But we will also take the opportunity to indicate what lies ahead in the remaining chapters of the book. In Section 1.1 we first consider some aspects of classical survival analysis where the focus is on the time to a single event. Sometimes the event in question may occur more than once for an individual, or more than one type of event is of interest. In Section 1.2 such event history data are considered, and we discuss some methodological issues they give rise to, while we in Section 1.3 briefly discuss why survival analysis methods may be useful also for data that do not involve time. When events occur, a natural approach for a statistician would be to count them. In fact, counting processes, a special kind of stochastic process, play a major role in this book, and in Section 1.4 we provide a brief introduction to counting processes and their associated intensity processes and martingales. Finally, in Section 1.5, we give an overview of some modeling issues for event history data.

The counting process point of view adopted in this book is the most natural way to look at many issues in survival and event history analysis. Therefore the mathematical tools of counting processes and martingales should not be considered pure technicalities, but the reader should make an effort to understand the concepts and ideas they express. Pure technicalities, like regularity assumptions, will not be emphasized much in this book; our aim is to explain the concepts involved.

1.1 Survival analysis: basic concepts and examples

We shall start by considering classical survival analysis which focuses on the time to a single event for each individual, or more precisely the time elapsed from an initiating event to an event, or endpoint, of interest. Some examples:

- Time from birth to death.
- Time from birth to cancer diagnosis.
- Time from disease onset to death.
- Time from entry to a study to relapse.
- Time from marriage to divorce.

As a generic term, the time from the initiating event to the event of interest will be denoted a *survival time*, even when the endpoint is something different from death.

1.1.1 What makes survival special: censoring and truncation

Superficially, one might think that a survival time is just a measurement on a scale, and that an analysis of the survival times for a sample of individuals could be handled by the well-developed statistical methods for analyzing continuous, or possibly discrete, data. So, why not use ordinary linear regression and other standard statistical methods? The reason this will not usually work is a fundamental problem that one almost always meets when studying survival times. The point is that one actually has to wait for the event to happen, and when the study ends and the analysis begins, one will typically find that the event in question has occurred for some individuals but not for others. Some men had a testicular cancer during the period of observation, but most, luckily, had no such cancer. Those people might develop it later, but that is something we will not have knowledge about. Or, in a study of divorce, some couples were divorced during the time of the study, while others were not. Again, they may divorce later, but that is unknown to us when we want to analyze the data. Hence, the data come as a mixture of complete and incomplete observations. This constitutes a big difference compared to most other statistical data. If a doctor measures your blood pressure, it is done in one sitting in a couple of minutes. Most measurements are like that: they are made on a single occasion. But measuring survival times is altogether a different story, and this is what requires a different statistical theory.

The incomplete observations are termed *censored* survival times. An illustration of how censored survival times may arise is given in Figure 1.1. The figure illustrates a hypothetical clinical study where 10 patients are observed over a time period to see whether some specific event occurs. This event, or endpoint, could be death, remission of disease, relapse of disease, etc. The left panel shows the observations as they occur in calendar time. The patients enter the study at different times and are then followed until the event occurs or until the closure of the study after 10 years. For statistical analysis one often focuses on the time from entry to the event of interest. Each individual will then have his or her own starting point, with time zero being the time of entrance into the study. The right panel of Figure 1.1 shows the observations in this *study time* scale. For patients number 1, 2, 4, 5, 6, and 9 the event was observed within the period of the study, and we have complete observation of their survival times. Patients 3, 7, 8, and 10 had not yet experienced the event when the study closed, so their survival times are censored. Notice that the censoring present in these data cut off intervals on the right-hand side, and so one talks about *right-censoring*. In the simple example of Figure 1.1, closure of the study was the only reason for censoring. However, in real-life clinical studies, right-censored observations will also occur when an individual withdraws from the study or is lost to follow-up.

As indicated in Figure 1.1, the study time scale used in a statistical analysis will usually not be calendar time. There are commonly many possible different time scales that may be used. In a clinical study the initiating event, which corresponds to time zero in the study time scale, could be time of diagnosis, time of entry into the study, time of admission to hospital, time of remission, etc. The choice of time

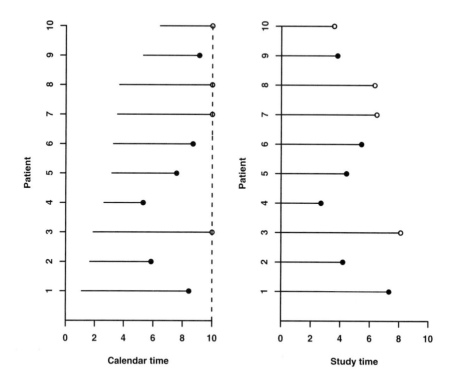

Fig. 1.1 *Patient follow up in a hypothetical clinical study with 10 patients. Left panel shows the actual calendar time. Right panel shows the same observations in the study time scale, where time 0 for each individual is his or her entry into the study. A filled circle indicates occurrence of the event, while an open circle indicates censoring. In the left panel the dotted vertical line indicates the closing date of the study.*

scale to use in a statistical analysis is usually a pragmatic one: what will make the analysis most relevant and clear with respect to the issues we study.

The set of individuals for which the event of interest has not happened before a given time t (in the chosen study time scale), and who have not been censored before time t, is termed the *risk set* at time t. In the right-hand panel of Figure 1.1 the risk set starts with 10 individuals, and then gradually declines to one and finally zero individuals.

A concept related to right-censoring is that of *left-truncation*. In a clinical study it may be that the patients come under observation some time after the initiating event. For instance, in a study of myocardial infarction only those who survive the initial phase and reach the hospital will be included in the study. Time zero for an individual patient may be the time of infarction, and if the patient reaches the hospital and is included in the study it may happen at different times for different patients. If the patient dies prior to reaching the hospital, he or she will never be entered in the study. The data arising here are left-truncated, and one may also use the term *delayed entry* about such data. This implies that the risk set will not only decline over time, but will also increase when new individuals enter the study. Originally, many methods of survival analysis were developed only for right-censored survival data. But the counting process approach, which we shall focus on in this book, clearly shows that most methods are equally valid for data with a combination of left-truncation and right-censoring.

As a simple example of survival data with right-censoring, we may consider the data on the 10 patients from the hypothetical clinical study of Figure 1.1. The data from the right-hand panel of the figure are, with an asterisk indicating a right-censored survival time,

$$7.32, 4.19, 8.11^*, 2.70, 4.42, 5.43, 6.46^*, 6.32^*, 3.80, 3.50^*.$$

This example illustrates well that right-censored survival times cannot be handled by ordinary statistical methods. In fact, even a simple mean cannot be calculated due to the censoring. If we cannot calculate the mean, then we cannot find a standard deviation or perform a t-test, a regression analysis, or almost anything else.

1.1.2 Survival function and hazard rate

Even though ordinary statistical methods cannot handle right-censored (and left-truncated) survival data, it is in fact quite simple to analyze such data. What one needs is the right concepts; and there are two basic ones that pervade the whole theory of survival analysis, namely the *survival function* and the *hazard rate*. One starts with a set of individuals at time zero and waits for an event that might happen. The survival function, $S(t)$, which one would usually like to plot as a *survival curve*, gives the expected proportion of individuals for which the event has not yet happened by time t. If the random variable T denotes the survival time, one may write more formally:

$$S(t) = P(T > t). \tag{1.1}$$

Remember that we use the term survival time in a quite general sense, and the same applies to the term survival function. Thus the survival function gives the probability that the event of interest has not happened by time t (in the study time scale), and it does not have to relate to the study of death. Often the survival function will tend to zero as t increases because over time more and more individuals will experience the event of interest. However, since we use the terms survival time and survival function also for events that do not necessarily happen to all individuals, like divorce or getting testicular cancer, the random variable T may be infinite. For such situations, the survival function $S(t)$ will decrease toward a positive value as t goes to infinity.

The survival function (1.1) specifies the unconditional probability that the event of interest has not happened by time t. The hazard rate $\alpha(t)$, on the other hand, is defined by means of a conditional probability. Assuming that T is absolutely continuous, that is, that it has a probability density, one looks at those individuals who have not yet experienced the event of interest by time t and considers the probability of experiencing the event in the small time interval $[t, t+dt)$. Then this probability equals $\alpha(t)dt$. To be more precise, the hazard rate is defined as a limit in the following way:

$$\alpha(t) = \lim_{\Delta t \to 0} \frac{1}{\Delta t} P(t \leqslant T < t + \Delta t \mid T \geq t). \tag{1.2}$$

Notice that while the survival curve is a function that starts at 1 and declines over time, the hazard rate can be essentially any nonnegative function. Note also that the hazard rate is the model counterpart of the incidence rate that is commonly computed in epidemiological studies.

The concepts are illustrated in Figure 1.2. The left-hand panels show two typical hazard rates, one that reaches a maximum and then declines and one that increases all the time. The right-hand panels show the corresponding survival curves. Notice that it is not so easy to see from the survival curves that the hazard rates are actually very different. In fact, the shape of the hazard rate is an issue we will return to several times in the book. The hazard rate may seem like a simple concept, but in fact it is quite elusive and, as we will see in Chapters 6, 10, and 11, it hides a lot of complexities.

From censored survival data, we can, as described in Section 3.2, easily estimate a survival curve by the *Kaplan-Meier estimator*. The estimation of a hazard rate is more tricky. But, as explained in Section 3.1, what can easily be done is to estimate the cumulative hazard rate

$$A(t) = \int_0^t \alpha(s)ds \tag{1.3}$$

by the *Nelson-Aalen estimator*. The increments of a Nelson-Aalen estimate may then be smoothed to provide an estimate of the hazard rate itself.

There are two fundamental mathematical connections between the survival function and the (cumulative) hazard rate that should be mentioned here. First note that by (1.2) and (1.3), we have

$$A'(t) = \alpha(t) = \lim_{\Delta t \to 0} \frac{1}{\Delta t} \frac{S(t) - S(t + \Delta t)}{S(t)} = -\frac{S'(t)}{S(t)}. \tag{1.4}$$

Then by integration, using that $S(0) = 1$, one gets

$$-\log\{S(t)\} = \int_0^t \alpha(s)ds,$$

and it follows that

$$S(t) = \exp\left\{-\int_0^t \alpha(s)ds\right\}. \tag{1.5}$$

In Appendix A.1 we describe how we may define the cumulative hazard rate for arbitrary survival times, which need be neither absolutely continuous nor discrete,

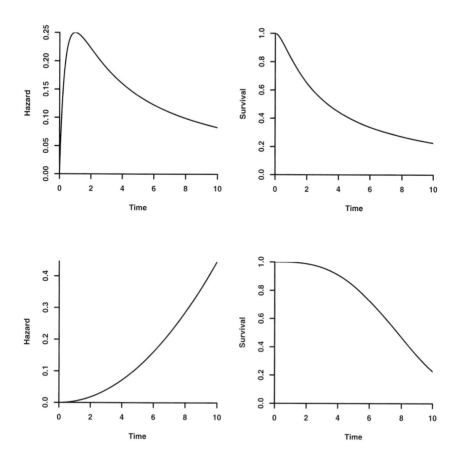

Fig. 1.2 *Illustrating hazard rates and survival curves. The hazard rates on the left correspond to the survival curves on the right.*

and we show how formula (1.5) is generalized to this situation. These results turn out to be useful when we study the properties of the Kaplan-Meier estimator and its relation to the Nelson-Aalen estimator; cf. Section 3.2.

1.1.3 Regression and frailty models

A main purpose of many studies is to assess the effect of one or more *covariates*, or explanatory variables, on survival. If there is only one categorical covariate, like gender or the stage of a cancer patient, the effect of the covariate may be assessed by estimating a survival curve for each level of the covariate and testing whether any observed differences are significant using the tests of Section 3.3. However, in most studies there will be a number of covariates, and some of them will be numeric. Then, as in other parts of statistics, regression modeling is called for. A number of regression models have been suggested for censored survival data, and we will consider some of them in Chapters 4 and 5.

The most used regression model for censored survival data is Cox's regression model. For this model it is assumed that the hazard rate of an individual with co-variates $x_1, \ldots, x_p$ takes the form

$$\alpha(t|x_1, \ldots, x_p) = \alpha_0(t) \exp\{\beta_1 x_1 + \cdots + \beta_p x_p\}. \tag{1.6}$$

Here $\alpha_0(t)$ is a *baseline hazard* that describes the shape of the hazard rate as a function of time, while $\exp\{\beta_1 x_1 + \cdots + \beta_p x_p\}$ is a *hazard ratio* (sometimes denoted a *relative risk*) that describes how the size of the hazard rate depends on covariates. Note that (1.6) implies that the hazard rates of two individuals are proportional (Exercise 1.5).

An alternative to Cox's model is the additive regression model due to Aalen, which assumes that the hazard rate of an individual with covariates $x_1, \ldots, x_p$ takes the form

$$\alpha(t|x_1, \ldots, x_p) = \beta_0(t) + \beta_1(t)x_1 + \cdots + \beta_p(t)x_p. \tag{1.7}$$

For this model $\beta_0(t)$ is the baseline hazard, while the *regression functions* $\beta_j(t)$ describe how the covariates affect the hazard rate at time t.

In (1.6) and (1.7) the covariates are assumed to be fixed over time. More generally, one may consider covariates that vary over time. In Sections 4.1 and 4.2 we will discuss more closely Cox's regression model and the additive regression model, including their extensions to time-varying covariates.

The regression models (1.6) and (1.7) may be used to model differences among individuals that may be ascribed to *observable* covariates. Sometimes one may want to model *unobservable* heterogeneity between individuals that may be due, for example, to unobserved genetic or environmental factors. One way of doing this is to assume that each individual has a *frailty* Z. Some individuals will have a high value of Z, meaning a high frailty, while others will have a low frailty. Then, conditional on the frailty, the hazard rate of an individual is assumed to take the form

$$\alpha(t|Z) = Z \cdot \alpha(t). \tag{1.8}$$

In Chapter 6 we will take a closer look at such proportional frailty models where each individual has its own unobserved frailty. Frailty models may also be used to model the dependence of survival times for individuals within a family (or some other cluster). Then, as discussed in Chapter 7, two or more individuals will share the same frailty or their frailties will be correlated.

1.1.4 The past

Another concept we shall introduce is that of a *past*. The definition of the hazard rate in (1.2) conditions with respect to the event of interest not having occurred *before* time *t*. It therefore makes an assumption about how information on the past influences the present and the future. This is a main reason why the hazard rate is a natural concept for analyzing events occurring in time. So even if we did not have the complication of censoring, the hazard rate would be a concept of interest. The idea of defining a past, and hence also a present and a future, is a fundamental aspect of the methods that shall be discussed in this book.

In (1.2) the only information we use on the past is that the event has not occurred before time *t*. When working with left-truncated and right-censored survival data, the information we use on the past will be more involved and will include information on all events, delayed entries, and censorings that occur *before* time *t* (in the study time scale), as well as information on the time-fixed covariates. (The handling of time-varying covariates requires some care; cf. the introduction to Chapter 4.) When working with event history data, where more than one event may occur for an individual, the information on the past becomes even more involved.

It turns out that it is extremely fruitful to introduce some ideas from the theory of stochastic processes, since parts of this theory are designed to precisely understand how the past influences the future development of a process. In particular counting processes and martingales turn out to be useful. We will give a brief introduction to these types of stochastic processes in Section 1.4, while a more thorough (although still fairly informal) treatment is given in Chapter 2.

As introduced in Section 1.1.1, right-censoring is a rather vague notion of lifetimes being incompletely observed. It is necessary to be more precise about the censoring idea to get mathematically valid results, and we will see in Sections 1.4 and 2.2.8 that the concept of a past and the use of counting processes help to provide a precise treatment of the censoring concept.

1.1.5 Some illustrative examples

The concepts and methods mentioned earlier play an increasingly important role in a number of applications, as the following examples indicate. Most of these examples, with associated data sets, will be used for illustrative purposes throughout the book. In connection with the examples, we will take the opportunity to discuss general problems and concepts that are the topics of later chapters in the book. We start with examples that are simple, seen from a methodological and conceptual point of view, and then move on to examples that require more complex concepts and methods.

Example 1.1. Time between births. The Medical Birth Registry of Norway was established in 1967; it contains information on all births in Norway since that time.

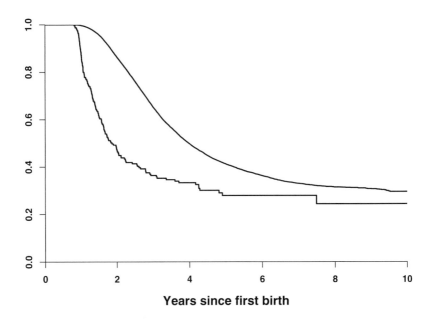

Fig. 1.3 *Empirical survival curves of the time between first and second birth. Upper curve: First child survived one year. Lower curve: First child died within one year. (Based on data from the Medical Birth Registry of Norway.)*

The registry is an invaluable source for epidemiological research of perinatal health problems as well as on demographic research related to fertility. Focusing on the latter, we will use information from the Birth Registry on the time between births for a woman. In particular, we will study the time between the first and second births of a woman (by the same father), and how this is affected if the first child dies within one year of its birth. Figure 1.3 shows Kaplan-Meier survival curves (cf. Section 3.2.1) for the time to the second birth for the 53 296 women for whom the first child survived one year and for the 262 women who lost their first child within one year of its birth. From the survival curves we see, for example, that it takes less than two years before 50% of the women who lost their first child will have another one, while it takes about four years before this is the case for the women who do not experience this traumatic event. We note that the survival curves give a very clear picture of the differences between the two group of women.

A more detailed description of the data on time between first and second births is given in Example 3.1. In addition to this example, the data will be used for illustration in Examples 3.8, 3.10, and 10.2. We will use data from the Birth Registry also to study the time between the second and third births of a woman and how this is affected by the genders of the two older children; cf. Examples 3.2, 3.9, and 3.14. □

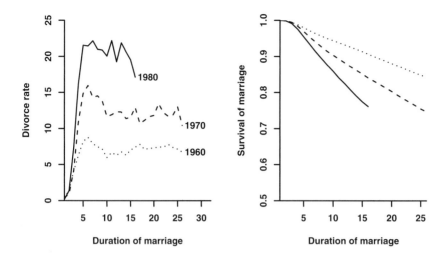

Fig. 1.4 *Rates of divorce per 1000 marriages per year (left panel) and empirical survival curves (right panel) for marriages contracted in 1960, 1970, and 1980. (Based on data from Statistics Norway.)*

Example 1.2. Divorce in Norway. The increase in divorce rates is a general feature of Western societies. The phenomenon is illustrated well by using the concepts of hazard rates and survival curves. In Figure 1.4 we show empirical hazard rates (rates of divorce) and empirical survival curves for marriages contracted in Norway in 1960, 1970, and 1980. The increase in divorce risk with marriage cohort is clearly seen. Furthermore, the hazard rates show an expected increase with duration of marriage until about five years where a slight decline and then leveling out occurs. The survival curves show how the proportions still married are decreasing to various degrees in the different cohorts. The concepts of survival curve and hazard rate are very suitable for describing the phenomenon, as opposed to the rather uninformative descriptions common in the popular press, comparing, for instance, divorces in a given year with the number of marriages contracted in the same year.

The rates of divorce and survival curves of Figure 1.4 are computed from data from Statistics Norway. The data are given in Example 5.4, where we also explain how the divorce rates and survival curves are computed. The divorce data are also used for illustrative purposes in Examples 11.1 and 11.2. □

Example 1.3. Myocardial infarction. Clinical medicine is probably the largest single area of application of the traditional methods of survival analysis. Duration is an important clinical parameter: to the severely ill patient it is of overriding importance how long he may expect to live or to stay in a relatively good state. An example

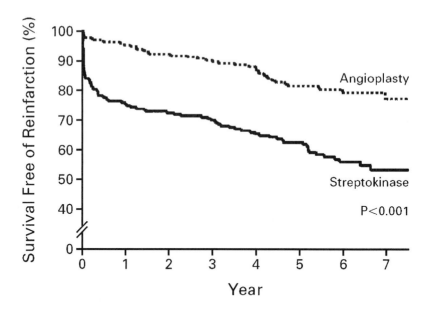

Fig. 1.5 *Survival curves for two treatments of myocardial infarction. Figure reproduced with permission from Zijlstra et al. (1999). Copyright The New England Journal of Medicine.*

is shown in Figure 1.5 comparing reinfarction-free survival of patients with myocardial infarction when randomized to treatment with angioplasty or streptokinase
(Zijlstra et al., 1999). The latter treatment is a medication, while angioplasty consists of inflating a tiny balloon in the blood vessel to restore the passage of blood.
The plots in the figure certainly give a clear conclusion: many lives would be spared
if all patients were given angioplasty.

A closer look tells us that the main difference between the groups comes very
early, in the first few weeks. A contentious issue in survival analysis is how to describe in the most fruitful way how survival depends on treatment and other covariates. Most commonly Cox's regression model (1.6) is used. This was done in the
paper by Zijlstra et al. (1999); adjusting for a number of factors the hazard ratio of
dying in the streptokinase group versus the angioplasty group was estimated to 2.31.
Cox's regression model assumes proportionality of hazards over time, and when this
assumption fails, the estimated hazard ratio will be an average measure that does not
consider the change in effect over time. If one considered a shorter time interval the
hazard ratio might be much larger. One focus in the current book is how to analyze
changes in effects over time, and it will be shown that ideas from frailty theory (cf.
Chapter 6) are essential to understand changes in hazard rates and hazard ratios. □

Example 1.4. Carcinoma of the oropharynx. Another example from clinical medicine is given by Kalbfleisch and Prentice (2002, section 1.1.2 and appendix A).

They present the results of a clinical trial on 195 patients with carcinoma of the oropharynx carried out by the Radiation Therapy Oncology Group in the United States. The patients were randomized into two treatment groups ("standard" and "test" treatment), and survival times were measured in days from diagnosis. A number of covariates were recorded for each patient at the entry to the study:

- x_1 = sex (1 = male, 2 = female),
- x_2 = treatment group (1 = standard, 2 = test),
- x_3 = grade (1 = well differentiated, 2 = moderately differentiated, 3 = poorly differentiated),
- x_4 = age in years at diagnosis,
- x_5 = condition (1 = no disability, 2 = restricted work, 3 = requires assistance with self-care, 4 = confined to bed),
- x_6 = T-stage (an index of size and infiltration of tumor ranging from 1 to 4, with 1 indicating a small tumor and 4 a massive invasive tumor),
- x_7 = N-stage (an index of lymph node metastasis ranging from 0 to 3, with 0 indicating no evidence of metastases and 3 indicating multiple positive nodes or fixed positive nodes).

The main purpose of such a clinical trial is to assess the effect of treatment. In addition, one would like to study the effects of the other covariates on survival. As these covariates are measured at entry and do not change over time, they are *fixed* covariates.

We will use the oropharynx data for illustration in Examples 4.6, 4.7, 4.8, 4.10, 4.13, and 10.1. In all these examples, we will exclude the two patients (numbers 136 and 159) who are missing information on some of the covariates. □

Example 1.5. Hip replacements. The Norwegian Arthroplasty Registry was started in 1987. At first only total hip replacements were registered, but from 1994 it includes replacements of other joints. By the end of 2005 the registry included information on more than 100 000 total hip replacements and about 30 000 replacements of other joints. The Arthroplasty Registry has provided valuable information on the epidemiology of joint replacements, in particular total hip replacements.

From September 1987 to February 1998 almost 40 000 patients had their first total hip replacement operation at a Norwegian hospital (Lie et al., 2000). In this book we will use a random sample of 5000 of these patients – 3503 females and 1497 males – to study the survival of patients who have had a total hip replacement operation. In particular we will compare the mortality among these patients to the mortality of the general Norwegian population. Figure 1.6 shows the Kaplan-Meier survival curve for the hip replacement patients (as a function of time since operation) together with the survival curve one would expect to get for a group of individuals from the general population with the same age and sex distribution as the hip replacement patients (see Section 3.2.5 for details). It is seen that the hip replacement patients have a slightly better survival than a comparable group from the general population. The main reason for this is that a patient needs to be in a fairly good health condition to be eligible for a hip replacement operation.

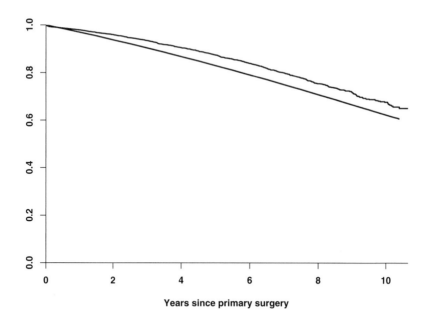

Fig. 1.6 *Empirical survival curve for patients who have had a total hip replacement (upper curve) and the expected survival curve for a hypothetical group of individuals from the general population with the same age and sex distribution as the hip replacement patients. (Based on data from The Norwegian Arthroplasty Registry.)*

The hip replacement data will be used for illustration in Examples 3.4, 3.11, 4.12, and 5.3. □

Example 1.6. Uranium miners. The Colorado Plateau uranium miners cohort was assembled to study the effects of radon exposure and smoking on lung cancer risk, and the data have been described in detail, for example, by Hornung and Meinhardt (1987). The cohort consists of 3347 male Caucasian miners who worked underground at least one month in the uranium mines of the four-state Colorado Plateau area and were examined at least once by Public Health Service physicians between 1950 and 1960. These miners were traced for mortality outcomes through December 31, 1982, by which time 258 lung cancer deaths and 1000 deaths from other causes were observed. Job histories were obtained from each of the mining companies and combined with available data on annual mine radon levels to construct each miner's radon exposure history. A smoking history was taken at the first examination and updated with each subsequent exam.

We will study how the risk of lung cancer death depends on cumulative radon and smoking exposures. These covariates are increasing with the age of a miner (which is the time scale we will use in our analyses), and they are hence *time-dependent* covariates. Since lung cancer victims survive about two years after being diagnosed and exposures after diagnosis have no effect on the course of the disease, the cumulative exposures are lagged by two years in our analyses. Due to this lagging, it is reasonable to consider the cumulative exposures for a miner to be *external* time-dependent covariates, that is, time-dependent covariates whose paths are *not* influenced by the (underlying) disease process of an individual. (A classification of various forms of time-dependent covariates is provided in the introduction to Chapter 4.)

The uranium miners' data will be used for illustration in Examples 4.2, 4.5, 4.14 and 4.16. □

Example 1.7. Liver cirrhosis. Between 1962 and 1969 about 500 patients diagnosed with liver cirrhosis at several Copenhagen hospitals were randomized to treatment with the synthetic hormone prednisone or placebo. The patients were followed until death or censoring at the closure of the study October 1, 1974, or until lost to follow-up alive before that date; see Andersen et al. (1993, pages 19–20) for a detailed description of the data and further references to the study.

Earlier analyses of the data suggest that there is an interaction between ascites (excess fluid in the abdomen) and treatment. In this book we will restrict attention to the larger group of patients with no ascites, and then we have 386 patients, of whom 191 received prednisone and 195 placebo. The number of deaths observed was 211, of which 94 occurred in the prednisone group and 117 in the placebo group.

A number of demographic, clinical, and biochemical covariates were recorded at entry to the study. In addition, some biochemical covariates were recorded at follow-up visits to a physician. Of these time-dependent covariates, we shall concentrate on the prothrombin index (defined as percentage of normal value of a blood test of some coagulation factors produced in the liver). In contrast to the time-dependent covariates of Example 1.6, the prothrombin index is an *internal* time-dependent covariate whose path is a marker for the (underlying) disease process of the individual.

An internal time-dependent covariate may be on the causal pathway between a time-fixed covariate (like treatment) and the occurrence of the event of interest. Therefore, as discussed in Section 9.2, special care needs to be exercised when including internal time-dependent covariates in a regression analysis.

The liver cirrhosis data will be used for illustration in Example 9.1. □

Example 1.8. Testicular cancer. Testicular cancer is a disease on the increase, becoming steadily more frequent. It is also mainly a disease of young men with a peak in the incidence (or hazard rate) at about the age of 30 years; cf. Figure 1.7.

We will use data from the Norwegian Cancer Registry on occurrences of testicular cancer in Norway between 1953 and 1993 to study how the incidence of testicular cancer varies with age and over time. In Section 6.9, a frailty model of the form (1.8) will be used to analyze the data. The analysis will indicate that a strong unobserved heterogeneity between individuals regarding the risk of acquiring the disease may be an explanation for the form of the incidence seen in Figure 1.7. Further analysis, which also takes the dependence within families into account, is provided in Section 7.5. □

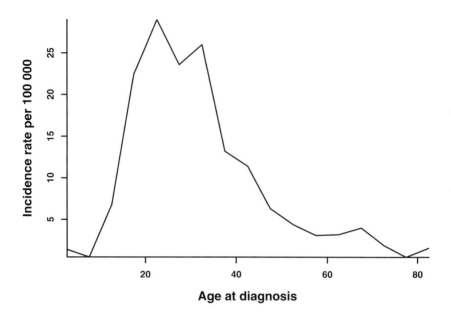

Fig. 1.7 *Age-specific incidence rates for testicular cancer. (Based on data from The Norwegian Cancer Registry 1997-2001.)*

Example 1.9. Amalgam fillings. When studying the duration of amalgam fillings in teeth, one may include several patients who each have a number of fillings. A study of this kind, including 32 patients with from 4 to 38 fillings for each patient, was performed by Aalen et al. (1995). Focusing on the duration (or survival time) of each amalgam filling, this is an example of *clustered survival data*, where the survival times of the fillings from the same patient are dependent. When studying the duration of amalgam fillings and how this depends on patient properties, one has to take this dependence into account. One way of doing this is to use a frailty model where all teeth from a patient share the same frailty.

The amalgam data will be used for illustrations in Examples 7.2 and 8.1. □

1.2 Event history analysis: models and examples

In classical survival analysis one focuses on the time to the occurrence of a single event for each individual. This may, however, be too simplistic in a number of situations. Sometimes the event in question may occur more than once for an individual. Examples of such *recurrent events* are myocardial infarctions, seizures in epileptic patients, and successive tumors in cancer studies. In other situations, more than one type of event are of interest. For example, an individual's civil status history may be as follows: the individual starts out being unmarried, then he may marry, get divorced, remarry and finally die. Another example is a cancer patient who may obtain remission, then have a relapse, after which she dies. Such situations, with more than one possible type of event for each individual, may conveniently be described by *multistate models*.

1.2.1 Recurrent event data

We now present two examples of recurrent event data that will be used for illustrative purposes later in the book, and we discuss some methodological issues such data give rise to.

Example 1.10. Bladder cancer. One example of recurrent event data is provided by a bladder cancer study conducted in the United States by the Veterans Administration Cooperative Urological Research Group (Byar, 1980). In this study, patients with superficial bladder tumors were included in a trial. The tumors were removed, and the patients were randomly assigned to one of three treatments: placebo, thiotepa and pyridoxine. The patients were followed up, and the recurrences of new tumors were recorded. In this book we will use the version of the bladder cancer data given by Kalbfleisch and Prentice (2002, section 9.4.3 and table 9.2). Here only the 86 patients receiving placebo or thiotepa are considered.

The main aim of the analysis of the data is to assess whether thiotepa helps to reduce the recurrence of new tumors. We are, however, also interested in studying the effect of covariates recorded at entry, like the number of initial tumors and the size of the largest initial tumor.

The bladder cancer data will be used for illustration in Examples 7.3, 8.3 and 8.6. □

Example 1.11. Movements of the small bowel. Aalen and Husebye (1991) describe an experiment in which one measured the cyclic pattern of motility (spontaneous movements) of the small bowel, called the migrating motor complex (MMC), in 19 healthy individuals. Phase III of MMC consists of a sequence of regular contractions migrating down the small bowel at irregular intervals; these intervals last from minutes to several hours. This cyclic pattern is invariably present in healthy individuals, and the absence is associated with severe disturbance of intestinal digestion and clearance. For each individual, the times of occurrence of MMC are recorded. Interest focuses on understanding various aspects of the MMC process for healthy individuals, for example, whether there is a variation in frequency between individuals, how the frequency changes over time, and what governs the duration between MMCs.

The data are given in Example 7.1, and are used for illustration in this example as well as in Examples 7.4 and 8.4. □

There are a number of methodological issues related to the analysis of recurrent event data. These will be considered in detail in Chapters 7 and 8.

The first of these issues is the choice of time scale. Should one use "global time" t measured from an initializing event (like randomization in Example 1.10 and start of the study in Example 1.11), or should one use "recurrence time" u, where time is measured from the last occurrence of an event (start of the study if there are no prior events)? The simplest example of a model with global time is a Poisson process. For a Poisson process with intensity $\alpha(t)$ (a nonnegative function), future occurrences of the event are not influenced by past occurrences, and the probability of an event occurring in the (global) time interval $[t, t+dt)$ is $\alpha(t) dt$. The simplest example of a model using recurrence time is a renewal process. For a renewal process, the probability of an event only depends on the time elapsed since the last event, and the probability that an event occurs in the time interval $[u, u+du)$ *after* the previous event equals $h(u) du$ for a hazard rate $h(u)$.

The second issue is how one should handle unobserved heterogeneity between individuals. For recurrent event data, it is commonly the case that some individuals will experience several events, while others experience only a few, and that this variation between individuals cannot be fully explained by the observed differences between the individuals. In some situations it makes sense to regard the variation between individuals as a nuisance and to use a *marginal* approach, where one focuses on the effect of the fixed covariates (e.g., treatment) averaged over the variation between individuals. We will discuss this approach more closely in Chapter 8. An alternative to the marginal approach is to assume, in a similar manner as in (1.8), that each individual has an unobserved *frailty* that influences the rate of the occurrence of events. Conditional on the frailty (and covariates) one may then model the occurrences of the events for an individual either as a Poisson process (using "global time") or as a renewal process (using "renewal time"). This approach will be further discussed and exemplified in Chapter 7. A third option is to use information on previous events for an individual as time-dependent covariates explaining the occurrence of future events. We will discuss this *dynamic* approach to recurrent event data in Chapter 8.

1.2.2 Multistate models

The simplest example of a multistate model is the model for *competing risks*. Competing risks occur, for example, in studies of cause-specific mortality where two or more causes of death "compete" for the life of an individual. We may model competing risks by a process with one transient state 0, corresponding to "alive," and k absorbing states, say, corresponding to "dead by cause h," $h = 1, 2, \ldots, k$. Figure 1.8 gives an illustration for $k = 3$ competing causes of death.

The concept of a hazard rate may be generalized to competing risks. For each cause h, the *cause-specific hazard* $\alpha_{0h}(t)$ describes the instantaneous risk of dying from that cause. Thus $\alpha_{0h}(t) dt$ is the probability that an individual will die of cause

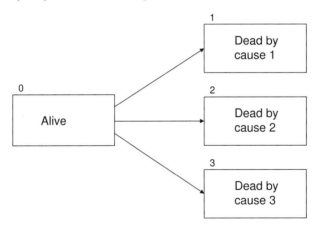

Fig. 1.8 *A model for k = 3 competing causes of death.*

h in the small time interval $[t, t+dt)$ given that the individual is still alive just prior to t; see (1.10) for a formal definition.

In mortality studies, the survival function $S(t)$ gives the probability that an individual will survive beyond time t, while $1 - S(t)$ gives the probability that the individual will be dead by time t. For competing risks we may in a similar manner consider the probability that an individual will be dead *due to cause h* by time t. We write $P_{0h}(0,t)$ for this probability, which is often denoted a *cumulative incidence function*. In epidemiology it is common to use "incidence rate" as synonymous for (empirical) hazard rate. Nevertheless the cumulative incidence function $P_{0h}(0,t)$ is different from the *cumulative cause-specific hazard* $A_{0h}(t) = \int_0^t \alpha_{0h}(u)du$, and the two should not be confused. Further discussion and comparison of the cumulative incidence function and the cumulative cause-specific hazard are provided in Section 3.4.1.

The notation $\alpha_{0h}(t)$ for the cause-specific hazards and $P_{0h}(0,t)$ for the cumulative incidence functions is chosen in accordance with common notation for transition intensities and transition probabilities for *Markov chains*; cf. (1.9) and (1.10). The relation between the cause-specific hazards and the cumulative incidence functions are derived in Appendix A.2.5 and summarized in Section 3.4.1.

Example 1.12. Causes of death in Norway. During the years 1974–78 all men and women aged 35–49 years who were residing in three rural Norwegian counties (Oppland, Sogn og Fjordane, and Finmark) were invited to a cardiovascular health screening exam. More than 90% participated in the screening and gave a self-report on their past and current smoking habits. Mortality in the resulting cohort of about 50 000 individuals was followed-up to the end of the year 2000 by record linkage with the cause of death registry at Statistics Norway using the Norwegian 11-digit birth number. In this book we will use a subset of 4000 individuals (2086 males and 1914 females) selected at random from this cohort to study the mortality for the four broad classes of cause of death:

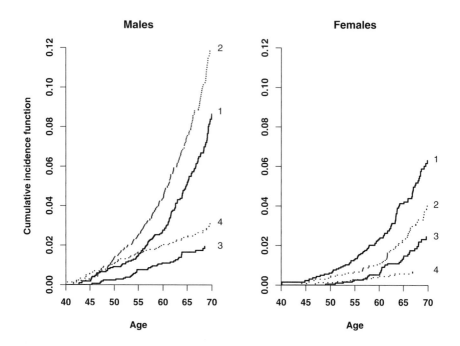

Fig. 1.9 *Empirical cumulative incidence functions for four causes of death among middle-aged Norwegian males (left) and females (right). 1) Cancer; 2) cardiovascular disease including sudden death; 3) other medical causes; 4) alcohol abuse, chronic liver disease, and accidents and violence.*

1) Cancer
2) Cardiovascular disease including sudden death
3) Other medical causes
4) Alcohol abuse, chronic liver disease, and accidents and violence

Figure 1.9 gives empirical cumulative incidence functions for these causes of death for males and females (computed as described in Example 3.15). For example, we see from the figure that a 40-year-old male has a 12% risk of cardiovascular death before the age of 70, while this risk is only 4% for females. For cancer death, the difference between the sexes is smaller; the estimated risks are 9% and 6%, respectively, for males and females. For further details about the data collection and results for the full cohort, see Vollset et al. (2006).

We will use the data for illustration in Examples 3.3, 3.15, 4.1, 4.4, and 5.5. □

To study the occurrence of a chronic disease as well as death, we may adopt an *illness-death model* with states 0, 1, and 2 corresponding to "healthy," "diseased," and "dead," respectively, and where no recovery (i.e., transition from state 1 to state 0) is possible; cf. Figure 1.10. If the probability for a transition from state 1 to

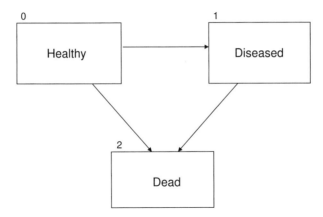

Fig. 1.10 *An illness-death model without recovery.*

state 2 does *not* depend on the duration in state 1, the process is Markov. Then the *transition intensities* $\alpha_{01}(t)$, $\alpha_{02}(t)$, and $\alpha_{12}(t)$ describe the instantaneous risks of transitions between the states. Thus $\alpha_{01}(t)dt$ is the probability that an individual who is healthy just prior to time t will get diseased in the small time interval $[t, t+dt)$, while $\alpha_{02}(t)dt$ and $\alpha_{12}(t)dt$ are the probabilities that an individual who is disease-free, respectively diseased, just before time t will die in the small time interval $[t, t+dt)$; see (1.10) for a formal definition of the transition intensities. We may also be interested in the probabilities of transitions between the states. We denote by $P_{gh}(s,t)$ the probability that an individual who is in state g at time s will be in state h at a later time t. The relations between the transition intensities and the transition probabilities for the Markov illness-death model are derived in Appendix A.2.5 and summarized in Section 3.4.2.

Other interpretations of the states than the ones given earlier are possible. In particular, in a study involving treatment of cancer, state 0 could correspond to "no response to treatment," state 1 to "response to treatment," and state 2 to "relapsed or dead." The transition probability $P_{01}(0,t)$ is then the *probability of being in response function* suggested by Temkin (1978) and sometimes used as an outcome measure when studying the efficacy of cancer treatment.

Example 1.13. Bone marrow transplantation. Klein and Moeschberger (2003, section 1.3) describe a study where 137 patients with acute leukemia received a bone marrow transplant during the period March 1, 1984, to June 30, 1989. The patients were followed for a maximum of seven years, and times to relapse and death were recorded. It was also recorded if and when the platelet count of a patient returned to a self-sustaining level. Such platelet recovery is a beneficial event: the rate at which patients are relapsing or dying is smaller after their platelets recover. The possible events for a patient may be described by a model of the form shown in Figure 1.10, but with the states 0, 1, and 2 corresponding to "transplanted," "platelet recovery," and "relapse or death," respectively.

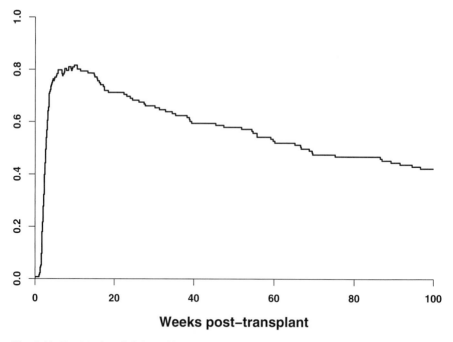

Fig. 1.11 *Empirical probability of being in response function for the bone marrow transplant patients.*

A patient starts out in state 0 at time $t = 0$ when he gets the bone marrow transplant. If his platelets recover, the patient moves to state 1, and if he then relapses or dies, he moves on to state 2. If the patient relapses or dies without the platelets returning to a normal level, he moves directly from state 0 to state 2. Platelet recovery is considered a response to the bone marrow transplantation, and Figure 1.11 shows the empirical probability of being in response function (computed as described in Example 3.16). This is an estimate of the transition probability $P_{01}(0,t)$ from state 0 to state 1 as a function of time t.

We will use the data for illustration in Examples 3.16 and 4.15. □

We have considered two simple multistate models; the competing risks model and the illness-death model. In general, a multistate model may be given by a *stochastic process X* in continuous time with finite state space $\mathscr{I} = \{0,1,2,\ldots,k\}$. The states in the state space correspond to the possible states of an individual, and $X(t) = g$ if the individual is in state g at time t. In general, the future state transitions of a multistate model may depend in a complicated way on its past. However, for the special case of a *Markov chain* the past and future are independent given its present state. Thus the future state transitions of a Markov chain depend only on its present state as described by the *transition probabilities:*

$$P_{gh}(s,t) = P(X(t) = h \mid X(s) = g) \qquad s < t, \quad g,h \in \mathscr{I}, \quad g \neq h. \qquad (1.9)$$

Corresponding to the hazard rate for a survival time, we may for a Markov chain define the *transition intensities:*

$$\alpha_{gh}(t) = \lim_{\Delta t \to 0} \frac{1}{\Delta t} P(X(t+\Delta t) = h \,|\, X(t-) = g), \qquad (1.10)$$

where $X(t-)$ denotes the value of X "just before" time t (formally the limit of $X(s)$ when s goes to t from below). Note that $\alpha_{gh}(t)\,dt$ is the probability that an individual who is in state g "just before" time t, will make a transition to state h in the small time interval $[t, t+dt)$, as exemplified earlier for the competing risks and illness-death models.

Appendix A.2 provides a summary of basic results for Markov chains, including the relation between the transition intensities and the transition probabilities. In Sections 3.1 and 3.4 we will show how one may estimate the cumulative transition intensities $A_{gh}(t) = \int_0^t \alpha_{gh}(u)\,du$ and the transition probabilities $P_{gh}(s,t)$ when there are no covariates; the extension to the case with covariates is studied in Section 4.2.9.

We end this section with two examples of more complicated multistate models.

Example 1.14. Sleep analysis. Yassouridis et al. (1999) studied the occurrence of various sleep patterns, including REM (rapid eye movement) sleep. Sleep was monitored continuously for a number of individuals, and events were defined as falling asleep, waking up, going into REM sleep, etc. Each individual may experience a number of such events during a night, and one may want to analyze the pattern. One question asked is the relation of sleep to measurement of the stress hormone cortisol, which was monitored every 20 minutes throughout the night.

The sleep data will be used for illustration in Examples 8.2, 8.5, and 9.3. The data were made available by the Max Planck institute for psychiatry, Munich, Germany. □

Example 1.15. Transition from HIV to AIDS. The model in Figure 1.12 presents states in the transition from HIV infection to AIDS (Aalen et al., 1997, 1999; Sweeting et al., 2005). The severity of infection is measured by the CD4 count, and whether the infection was known by a positive HIV test is also included. The latter information is of importance since a positive HIV test gives the possibility for treatment and may also induce behavior changes that influence the risk of further transmission of disease. The model has been used for prediction of incidence of new HIV and AIDS cases.

An interesting aspect of this model is that it is partly hidden – it may be considered a special case of the *Hidden Markov models.* What is observed may be the transition to the second level when the infection becomes known and then the further development until AIDS or death. Often the main interest of such a model will be on the infection rate, that is, the number of new cases into state 1. This, however, is not directly observed and may only be indirectly inferred from other information. □

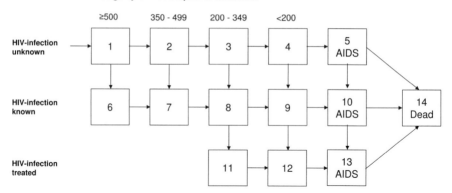

Fig. 1.12 *Model for progression of HIV disease, according to level of CD4 count and whether the infection was known (by positive HIV test) or not.*

1.3 Data that do not involve time

Basically, the methods to be discussed in this book have been developed to handle data on time to events, especially when censoring is involved. However, the same methods may also be of value for other types of data. The survival function, for instance, is nothing but a cumulative distribution function subtracted from 1. Hence, estimating a survival function is of general interest. One should note that ordinary two-sample t-tests and two-sample Wilcoxon tests both rest on an assumption that the survival function in one group has a parallel displacement compared to that of the other group. Therefore it is of interest routinely to check the survival functions whenever using these methods, something which is rarely done. For example, the Wilcoxon test is often recommended as a way out in situations where distributions are not normal, but a simple plot of survival functions would reveal that the assumption of parallel displacement is often not fulfilled.

There is, however, a deeper issue here. When considering time, the concept of a past is essential. However, a similar concept may be relevant for nontime data as well. As an example, consider measurements on "drive for thinness" which is a scale used in studies of anorexia nervosa. Measurements in a sample of individuals usually have an extremely skewed distribution, with some individuals (especially young girls) having a great drive to thinness, while many others may have no such drive. When analyzing such data, it may be natural to consider separately those that have a value above some limit, and to possibly move this limit gradually upward. This would correspond to studying individuals with more and more pathology. This idea of conditioning with respect to measurements being above certain limits is analogous to conditioning with respect to the past in time data.

In fact, there is a general problem how to handle severely skewed data where the extreme values indicate pathology in a medical sense. The common advice to take a logarithmic transformation to remove skewness may not be a good idea because one is interested in the extreme measurements precisely because they are extreme, and their particular impact may be lost after transformation. In addition, one may sometimes have a considerable group that is collected in the lowest point on the scale (e.g., at value zero), and a transformation does not solve this problem.

An alternative would be to use logistic regression, dividing the scale at some point that divides between a presumed pathological and a nonpathological group. However, the point of division of the scale may be rather arbitrary, and information is lost when using a dichotomized scale. A third alternative is to use classical nonparametric methods, but they are generally limited to simple comparisons and do not handle general regression. We shall suggest a fourth alternative, namely to use methods from survival analysis on severely skewed data and, as we will illustrate in Example 4.3, Cox regression and other survival analysis methods may be of considerable relevance for such data, even though they are not measurements of time.

1.4 Counting processes

As is seen from the previous sections, the focus in survival and event history analysis is on observing the occurrence of events over time. Such occurrences constitute point processes. These processes may be described by counting the number of events as they come along, which leads to the term *counting process*. Counting processes may arise in different ways. For instance, one may count the number of times a person wakes up during the night. Or the counting may include a number of individuals, for example, counting the deaths in a patient group in a clinical trial.

There is a very neat mathematical theory for counting processes, which turns out to be extremely useful for the statistical analysis of survival and event history data. In this and the following section we give a brief introduction to this theory and indicate how it may be useful for statistical modeling. A more detailed account is provided in Chapter 2, where we also discuss more closely concepts like martingales and stochastic integrals and give some useful rules for handling these.

1.4.1 What is a counting process?

We start by considering a single type of event. For a given time t, let $N(t)$ be the number of events that have occurred up to, and including, this time. Then $N(t)$ is a counting process. As an illustration, consider the data on the 10 patients from the hypothetical clinical study of Figure 1.1. Written in increasing order, the data of the right-hand panel of the figure are:

$$2.70, \ 3.50^*, \ 3.80, \ 4.19, \ 4.42, \ 5.43, \ 6.32^*, \ 6.46^*, \ 7.32, \ 8.11^*.$$

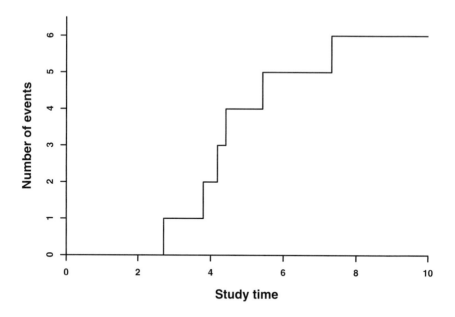

Fig. 1.13 *Illustration of a counting process.*

The counting process corresponding to these data is illustrated in Figure 1.13. The process jumps one unit at the time of each observed event and is constant between events. Even though it is not clear from the way we have represented $N(t)$ in the figure, it should be noted that a counting process by definition is *continuous from the right*.

A very well-known example of a counting process is the homogeneous Poisson process, where the jumps occur randomly and independently of each other. A homogeneous Poisson process is usefully described by its intensity λ, which is the probability of occurrence of an event in a small time interval divided by the length of the interval. When modeling a counting process, we shall extend this idea of an intensity. We consider a small time interval $[t, t + dt)$ and assume that at most a single event may occur in the interval. So either there is one event or none. Then the *intensity process* $\lambda(t)$ is the conditional probability that an event occurs in $[t, t + dt)$, given all that has been observed prior to this interval, divided by the length of the interval. Thus

$$\lambda(t)dt = P(dN(t) = 1 \mid \text{past}), \qquad (1.11)$$

where $dN(t)$ denotes the number of jumps of the process in $[t, t + dt)$. Since $dN(t)$ is a binary variable, we may alternatively write

$$\lambda(t)dt = \mathrm{E}(dN(t)\,|\,\text{past}). \qquad (1.12)$$

A reformulation of this relation gives

$$\mathrm{E}(dN(t) - \lambda(t)dt\,|\,\text{past}) = 0,$$

where $\lambda(t)dt$ can be moved inside the conditional expectation since it is a function of the past. Thus, if we introduce the process

$$M(t) = N(t) - \int_0^t \lambda(s)ds, \qquad (1.13)$$

we have that

$$\mathrm{E}(dM(t)\,|\,\text{past}) = 0. \qquad (1.14)$$

Formula (1.14) is very interesting because it is precisely the intuitive definition of a *martingale*. Hence the intuitive definition of an intensity process (1.11) is equivalent to asserting that the counting process minus the *cumulative intensity process*

$$\Lambda(t) = \int_0^t \lambda(s)ds \qquad (1.15)$$

is a martingale.

The left-hand panel of Figure 1.14 shows the counting process of Figure 1.13 together with its cumulative intensity process (computed as described in Example 1.18), while the right-hand panel of Figure 1.14 shows the accompanying martingale given as the difference of the two. Note that the martingale has a quite irregular behavior.

By (1.13) the increment $dN(t)$ of the counting process may be written

$$dN(t) = \lambda(t)dt + dM(t). \qquad (1.16)$$

This is a relation of the form "observation = signal + noise," with $dN(t)$ being the observation, $\lambda(t)dt$ the signal, and $dM(t)$ the noise term. For statistical modeling in Section 1.5 and later chapters, we will focus on the intensity process, while properties of the martingale $M(t)$ are crucial for studying the variability of statistical methods.

By (1.16), the interpretation of the martingale $M(t) = \int_0^t dM(s)$ is as cumulative noise. Thus the martingale (1.13) has a function similar to the sum of random errors in standard statistics. Martingales are a fundamental kind of stochastic processes with a large body of theory, and the connection opens up this wealth of knowledge to us, to pick and choose what we can use. In fact, some of the theory is very useful; in particular we shall need results about variation processes, stochastic integrals, and central limit theorems for martingales; cf. Chapter 2.

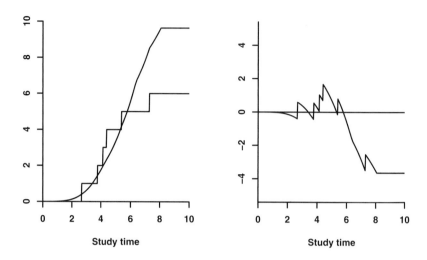

Fig. 1.14 *Illustration of a counting process and its cumulative intensity process (left panel) and the corresponding martingale (right panel).*

1.4.2 Survival times and counting processes

In this and the following subsection we give some examples of counting processes and their intensity processes. Here we consider the situation where the counting processes are derived from survival times, leaving more complicated situations to the next subsection. We start by considering one and several uncensored survival times (i.e., complete observation) and then indicate how censoring may be imposed on the uncensored data without changing the form of the intensity processes. In this connection the concept of *independent censoring* is crucial, a concept discussed more in depth in Section 2.2.8.

Example 1.16. One uncensored survival time. Let T be a survival time with hazard rate $\alpha(t)$. To this nonnegative random variable, we may associate the counting process $N^c(t) = I\{T \leq t\}$ that takes the value 1 if $T \leq t$ and the value 0 otherwise (superscript c for complete observation). To determine the intensity process, note that definition (1.2) of the hazard rate gives

$$P(dN^c(t) = 1 \,|\, \text{past}) = P(t \leq T < t + dt \,|\, \text{past}) = \begin{cases} \alpha(t)dt & \text{for } T \geq t, \\ 0 & \text{for } T < t, \end{cases}$$

since from the information on the past we know whether $T \geq t$ or $T < t$; cf. Section 1.1.4. Thus the intensity process takes the form $\lambda^c(t) = \alpha(t)I\{T \geq t\}$. $\square$

Example 1.17. Uncensored survival times. We then consider independent nonnegative random variables $T_1, T_2, \ldots, T_n$ corresponding to the survival times of n individuals and denote by $\alpha_i(t)$ the hazard rate of T_i. The survival times give rise to the counting processes $N_i^c(t) = I\{T_i \leq t\}$; $i = 1, 2, \ldots, n$. From the information on the past we here know *for all i* whether $T_i \geq t$ or $T_i < t$. Due to the independence of the survival times, information on the T_j for $j \neq i$ carries no information on T_i, and it follows from the previous example that the $N_i^c(t)$ have intensity processes

$$\lambda_i^c(t) = \alpha_i(t)I\{T_i \geq t\}; \qquad i = 1, \ldots, n. \tag{1.17}$$

From the *individual* counting processes $N_1^c(t), N_2^c(t), \ldots, N_n^c(t)$ we may obtain an *aggregated* process $N^c(t)$ by adding together the individual processes:

$$N^c(t) = \sum_{i=1}^n N_i^c(t).$$

This process counts how many individuals have experienced the event in question by time t. For the absolutely continuous case considered here, no two T_i are equal. Thus the aggregated process jumps one unit at a time, and hence it is a counting process. Now, by the linearity of expected values,

$$\mathrm{E}(dN^c(t) \,|\, \text{past}) = \sum_{i=1}^n \mathrm{E}\left(dN_i^c(t) \,|\, \text{past}\right), \tag{1.18}$$

and it follows by (1.12) that the intensity process $\lambda^c(t)$ of the aggregated counting process is given by

$$\lambda^c(t) = \sum_{i=1}^n \lambda_i^c(t). \tag{1.19}$$

For the particular case where $\alpha_i(t) = \alpha(t)$ for all i, that is, when the survival times are *iid* with hazard rate $\alpha(t)$, it follows by (1.17) and (1.19) that the aggregated counting process has intensity process of the multiplicative form

$$\lambda^c(t) = \alpha(t)Y^c(t).$$

Here $Y^c(t) = \sum_{i=1}^n I\{T_i \geq t\} = n - N(t-)$ is the number of individuals at risk "just before" time t. [$N(t-)$ denotes the limit of $N(s)$ when s goes to t from below, that is, the value of the aggregated counting process "just before" time t.] □

As discussed in Section 1.1.1, right-censoring is inevitable in most survival studies. In the survival analysis literature a number of different forms of right-censoring schemes have been introduced; see, for example, Klein and Moeschberger (2003, section 3.2) for a review. Some examples are:

- For *Type I censoring*, a survival time T_i is observed if it is no larger than a prespecified censoring time c_i, otherwise we only know that T_i exceeds c_i.
- For *Type II censoring*, observation is continued until a given number of events are observed, and then observation ceases.

- *Random censoring* is similar to Type I censoring, but the censoring times c_i are the observed values of random variables C_i that are independent of the T_i.

For the counting process framework adopted in this book, we do not need to assume that the censoring follows a specific scheme. Rather we will make the weakest possible assumption on the censoring that still allows valid statistical inference. We will here give a brief description of this general right-censoring scheme, called *independent censoring*, and refer the reader to Section 2.2.8 for a more thorough discussion.

As in Example 1.17, we assume that the uncensored survival times $T_1, \ldots, T_n$ are independent, and we denote by $\alpha_i(t)$ the hazard rate of T_i. However, we no longer observe (all) the uncensored survival times. Rather, for each individual i, we observe a (possibly) right-censored survival time $\widetilde{T}_i$ together with an indicator D_i that takes the value 1 when $\widetilde{T}_i = T_i$ (i.e., when we observe the actual survival time) and the value 0 when $\widetilde{T}_i < T_i$ (i.e., when we only observe a right-censoring time). *We then have independent right-censoring if an individual who is still at risk at time t* (i.e., has not experienced the event in question or been censored) *has the same risk of experiencing the event in the small time interval $[t, t + dt)$ as would have been the case in the situation without censoring.* Written a bit more formally, the independent censoring assumption takes the form:

$$P(t \leq \widetilde{T}_i < t + dt, D_i = 1 \mid \widetilde{T}_i \geq t, \text{past}) = P(t \leq T_i < t + dt \mid T_i \geq t). \qquad (1.20)$$

Note that from the information on the past at time t, we know *for all i* whether $\widetilde{T}_i \geq t$ or $\widetilde{T}_i < t$, and in the latter case we also know the value of D_i.

Situations where the independent censoring assumption (1.20) is *not* fulfilled occur when there is an underlying (unobserved) process that influences both the likelihood of the event and of censoring. One such example is a study of motion sickness described in the book by Altman (1991, section 13.2.1). Test persons were placed in a cubical cabin mounted on a hydraulic piston and subjected to vertical motion for up to two hours. The survival time of interest is the time when a person first vomited. Some persons requested an early stop of the experiment although they had not vomited, yielding censored survival times, while others successfully survived for two hours without vomiting and were censored by the end of the experiment. A person who gets motion sick is more likely to request an early stop of the experiment than a person who feels well. Therefore an individual who has not yet vomited or been censored before time t will be less likely to vomit in $[t, t + dt)$ than is the case for a person who has not yet vomited in the hypothetical situation where an early stop of the experiment is not allowed. This implies that the independent censoring assumption (1.20) does *not* hold for the motion sickness experiment.

The independent censoring assumption (1.20) may conveniently be expressed by means of the intensity processes of the counting processes

$$N_i(t) = I\{\widetilde{T}_i \leq t, D_i = 1\}; \qquad i = 1, \ldots n, \qquad (1.21)$$

registering the number of *observed* events (0 or 1) for the individuals. By (1.11) the intensity process $\lambda_i(t)$ of $N_i(t)$ is given by

$$\lambda_i(t)dt = P(dN_i(t) = 1 \,|\, \text{past}) = P(t \leq \widetilde{T}_i < t + dt, D_i = 1 \,|\, \text{past}).$$

Obviously $\lambda_i(t) = 0$ when $\widetilde{T}_i < t$. Therefore, by (1.2) and (1.20), we have independent censoring provided the intensity process for $N_i(t)$ takes the form

$$\lambda_i(t) = \alpha_i(t)Y_i(t), \qquad (1.22)$$

where

$$Y_i(t) = I\{\widetilde{T}_i \geq t\} \qquad (1.23)$$

is an at risk indicator for individual i. The intensity process (1.22) equals the hazard rate times the at risk indicator, as was also the case in the situation without censoring [cf. (1.17)]. In this sense, independent censoring preserves the form of the intensity processes.

As was the case for uncensored survival data (Example 1.17), one is often interested in the *aggregated* process

$$N(t) = \sum_{i=1}^{n} N_i(t) = \sum_{i=1}^{n} I\{\widetilde{T}_i \leq t, D_i = 1\} \qquad (1.24)$$

counting the number of *observed* events for all the n individuals. Formulas (1.18) and (1.19) remain valid for this aggregated process, so its intensity process takes the form

$$\lambda(t) = \sum_{i=1}^{n} \lambda_i(t) = \sum_{i=1}^{n} \alpha_i(t)Y_i(t). \qquad (1.25)$$

In particular, when $\alpha_i(t) = \alpha(t)$ for all i, the intensity process takes the multiplicative form

$$\lambda(t) = \alpha(t)Y(t), \qquad (1.26)$$

where $Y(t) = \sum_{i=1}^{n} Y_i(t)$ is the number of individuals at risk "just before" time t.

Example 1.18. A counting process derived from Weibull distributed censored survival times. Figure 1.14 shows a counting process together with its cumulative intensity process and accompanying martingale. The figure was made as follows. First we generated 10 survival times, T_i, from a Weibull distribution with hazard rate $\alpha(t) = 0.0031t^3$ and 10 censoring times, C_i, from a uniform distribution over the interval $(0, 10)$ to obtain 10 censored survival times $\widetilde{T}_i = \min(T_i, C_i)$ and censoring indicators $D_i = I(\widetilde{T}_i = T_i)$. We then computed the individual counting processes $N_i(t)$ and at risk indicators $Y_i(t)$ by (1.21) and (1.23) and the aggregated counting process $N(t)$ and its intensity process $\lambda(t)$ by (1.24) and (1.26). The left-hand panel of Figure 1.14 shows the aggregated counting process $N(t)$ and its cumulative intensity process $\Lambda(t) = \int_0^t \lambda(u)du$, while the right-hand panel shows the martingale $M(t) = N(t) - \Lambda(t)$. $\square$

So far we have only considered right-censored survival data. However, the results generalize almost immediately to survival data that are subject to both left-truncation and right-censoring. Our observations for the ith individual are then $(V_i, \widetilde{T}_i, D_i)$, where V_i is the left-truncation time (i.e., the time of entry) for the individual, and $\widetilde{T}_i$ and D_i are given as before. We then have *independent left-truncation and right-censoring* provided that the counting processes $N_i(t) = I\{\widetilde{T}_i \leq t, D_i = 1\}$ have intensity processes given by (1.22), where the at risk indicator now takes the form $Y_i(t) = I\{V_i < t \leq \widetilde{T}_i\}$ (cf. Exercise 1.9). Thus the form of the intensity processes is preserved under independent left-truncation and right-censoring.

1.4.3 Event histories and counting processes

In Section 1.2 we had a look at a number of event history models. We will indicate how these give rise to counting processes. As in Section 1.4.2 we start by considering counting processes for the situation where the event histories are completely observed and then discuss how censoring and truncation may be imposed without changing the form of the intensity processes.

We first consider multistate models (Section 1.2.2). Focusing on one specific transition between two states in a multistate model, say from state g to state h, we may consider the process $N_i^c(t)$ that counts the number of transitions from state g to state h for individual i in the time interval $[0, t]$, assuming complete observation; $i = 1, \ldots, n$. The intensity process $\lambda_i^c(t)$ of $N_i^c(t)$ has the form

$$\lambda_i^c(t) = \alpha_i(t) Y_i^c(t), \tag{1.27}$$

where $Y_i^c(t) = 1$ if the individual is in state g "just before" time t, and $Y_i^c(t) = 0$ otherwise, while $\alpha_i(t)dt$ is the probability that individual i makes a transition from state g to state h in the small time interval $[t, t + dt)$ given the past and given that the individual is in state g "just before" time t. If the multistate model is Markov, $\alpha_i(t)$ is the transition intensity from state g to state h, but in general it may depend in a complicated way on the past.

In a similar manner we may for recurrent event data (Section 1.2.1) consider the process $N_i^c(t)$ that counts the number of occurrences of the event in question for individual i in $[0, t]$, again assuming complete observation; $i = 1, \ldots, n$. The intensity process of $N_i^c(t)$ may be given by (1.27), where $Y_i^c(t) = 1$ for all t, and $\alpha_i(t)dt$ is the probability that individual i experiences an event in the small time interval $[t, t + dt)$ given the past. If the recurrent event data for the individuals come from independent Poisson process with common intensity $\alpha(t)$, we have $\alpha_i(t) = \alpha(t)$ for all i. If they come from independent renewal processes with hazard $h(u)$ for the time between events, we have $\alpha_i(t) = h(U_i(t))$, where $U_i(t)$ is the time elapsed since the last event for individual i (Exercise 1.8). In general, however, $\alpha_i(t)$ may depend in a complicated manner on covariates and past events.

The $\alpha_i(t)$ in (1.27) are our key model parameters and, as described in later chapters, a main aim of a statistical analysis is to infer how these (and derived quantities) depend on covariates and vary over time. However, such an analysis will usually be complicated by incomplete observation of the event histories.

Earlier, we considered left-truncation and right-censoring for survival data. We may also have left-truncation and right-censoring for general event histories, meaning that the individuals are not all followed from time zero (due to left-truncation) and that the follow-up time may differ between individuals (due to right-censoring). It is then crucial for valid inference on the $\alpha_i(t)$ that the truncation and censoring are *independent*. In Section 2.2.8 we will give a thorough discussion of the concept of independent censoring and truncation. Here we only note that we have independent censoring and truncation provided an individual who is at risk for the event in question at time t has the same risk of experiencing the event in the small time interval $[t, t + dt)$ as would have been the case in the situation without censoring and truncation. Then the intensity processes $\lambda_i(t)$ of the counting processes $N_i(t)$ registering the *observed* number of events have the same form as for the case with no truncation and censoring, that is,

$$\lambda_i(t) = \alpha_i(t)Y_i(t), \qquad (1.28)$$

where $\alpha_i(t)$ is the same as in (1.27), while $Y_i(t) = 1$ if individual i is "at risk" for experiencing the event "just before" time t; $Y_i(t) = 0$ otherwise. [For a multistate model, $Y_i(t) = 1$ if the individual is under observation *and* in state g "just before" time t, while for recurrent event data $Y_i(t) = 1$ if the individual is under observation "just before" time t.]

Assuming no tied event times, we may also for event history data consider the aggregated counting process $N(t) = \sum_{i=1}^{n} N_i(t)$ counting the *observed* number of events for all individuals. This aggregated counting process has the intensity process

$$\lambda(t) = \sum_{i=1}^{n} \alpha_i(t)Y_i(t). \qquad (1.29)$$

In particular, when $\alpha_i(t) = \alpha(t)$ for all i, the intensity process takes the form

$$\lambda(t) = \alpha(t)Y(t),$$

where $Y(t) = \sum_{i=1}^{n} Y_i(t)$ is the number of individuals at risk "just before" time t.

1.5 Modeling event history data

In Sections 1.4.2 and 1.4.3 we described how survival and event history data may be represented by *individual counting processes* $N_1(t), \ldots, N_n(t)$ registering the occurrences of an event of interest for each of n individuals, and we noted that their intensity processes typically take the form

$$\lambda_i(t) = \alpha_i(t)Y_i(t); \qquad i = 1,\ldots,n, \qquad (1.30)$$

where the $\alpha_i(t)$ are the quantities of interest, while the $Y_i(t)$ are "at risk indicators." In this section we will briefly discuss some important modeling issues for event history data. These issues will be elaborated in Chapters 3–11.

1.5.1 The multiplicative intensity model

When all the $\alpha_i(t)$ in (1.30) equal $\alpha(t)$, the *aggregated counting process* $N(t) = \sum_{i=1}^{n} N_i(t)$, registering the observed occurrences of the event of interest for a group of individuals, has an intensity process of the form

$$\lambda(t) = \alpha(t)Y(t) \qquad (1.31)$$

with $Y(t) = \sum_{i=1}^{n} Y_i(t)$ the number at risk "just before" time t.

In Section 3.1.2 we show that also a number of other situations give rise to a counting process $N(t)$ that has an intensity process $\lambda(t)$ of the multiplicative form (1.31) with $Y(t)$ an observable left-continuous process and $\alpha(t)$ a nonnegative (deterministic) function. The class of statistical models given by (1.31) is denoted the *multiplicative intensity model* for counting processes.

The basic nonparametric quantity to be estimated directly from a multiplicative intensity model would be the integral $A(t) = \int_0^t \alpha(s)ds$. This is estimated by the Nelson-Aalen estimator (Section 3.1). This estimator turns out to be a martingale after subtraction of the true $A(t)$, and its properties are simply deduced from this basic observation (Sections 3.1.5 and 3.1.6). The Kaplan-Meier estimator (Section 3.2) is important in the special case of survival data but would in other cases not be a useful estimator for the models we consider here.

Nonparametric comparison of the α-functions of two different groups can be done by considering the difference between the corresponding Nelson-Aalen estimates and computing a stochastic integral with a suitable weight function. This formulation encompasses almost all two-sample tests for censored data as special cases (Section 3.3).

1.5.2 Regression models

We will often want to study the effect of a number of covariates. We then focus on the individual counting processes $N_1(t),\ldots,N_n(t)$ and use a regression model for the $\alpha_i(t)$ in (1.30). This can be done in both a parametric and nonparametric framework. The standard nonparametric, or rather semiparametric choice, is the famous *Cox regression model* (Cox, 1972):

$$\alpha_i(t) = \alpha_0(t)\exp(\boldsymbol{\beta}^T \mathbf{x}_i(t)), \qquad (1.32)$$

where $\mathbf{x}_i(t) = (x_{i1}(t), \ldots, x_{ip}(t))^T$ is a vector of covariates for individual i that may vary over time, and $\boldsymbol{\beta} = (\beta_1, \ldots, \beta_p)^T$ is a vector of regression coefficients. The counting process approach has been of great importance in developing the theory for the Cox model. The basic connection was made by Andersen and Gill (1982). In addition to giving asymptotic theory, they extended the model beyond the survival data situation allowing for individual processes with recurrent events. We will study the Cox model for counting process models in Section 4.1.

A more pure nonparametric choice is the *additive regression model* suggested from counting process considerations by Aalen (1980). This specifies an additive model

$$\alpha_i(t) = \boldsymbol{\beta}(t)^T \check{\mathbf{x}}_i(t),$$

where $\check{\mathbf{x}}_i(t) = (1, x_{i1}(t), \ldots, x_{ip}(t))^T$. This model naturally accommodates arbitrarily varying regression coefficients $\boldsymbol{\beta}(t) = (\beta_0(t), \beta_1(t), \ldots, \beta_p(t))^T$, it gives simple estimation producing informative plots, and exact martingales arise in statistically useful ways (Section 4.2). We shall also apply the additive regression model to recurrent event data and show, among other things, interesting connections to the path analysis well known from classical linear models (Chapters 8 and 9).

Sometimes one may be interested in putting more structure on $\alpha(t)$ in (1.31) or $\alpha_0(t)$ in (1.32). This could be in the form of various parametric models with the Weibull model, bt^{k-1} for $k > 0$, being one important case (Chapter 5). We shall show how other parametric forms are generated from frailty models (Chapter 6) and first-passage-time models (Chapter 10). Parametric models are important and should be used more extensively in biomedical survival analysis. They are particularly useful in complex setups, for example, with interval censored or multivariate data.

1.5.3 Frailty models and first passage time models

The models discussed earlier are all in the framework of fixed effects models, with covariates being considered as fixed quantities. However, as in other areas of statistics, random effect models also play a major role. In fact, the time aspect of event history analysis introduces a number of special issues, which we shall address in detail, such as the occurrence of surprising artifacts in the comparison of hazard rates. The first thing to notice is that once we introduce random elements into the $\alpha_i(t)$, then they are no longer functions, but stochastic processes. One aim of this book is to point out how several kinds of stochastic processes may be put to use here (Chapters 10 and 11).

The simplest possibility is to let $\alpha_i(t)$ be the product of a single random variable, Z_i, and a function $\alpha(t)$:

$$\alpha_i(t) = Z_i \alpha(t).$$

This is the classical *proportional frailty* model of survival analysis [cf. (1.8)]. In spite of its simplicity, the proportional frailty model has played a large role in understanding the impact of random effects in survival models. It is also the basis of a very fruitful approach to analyzing multivariate survival data and forms, for instance, the backbone of the book by Hougaard (2000). We shall discuss this model in detail (Chapter 6); notice its close relationship to the regression model (1.32). The difference is that for frailty models we assume that (some of) the covariates are not observed fixed quantities, but rather unobserved random variables.

The proportional frailty model was introduced by Vaupel et al. (1979) in the late 1970s. At about the same time Woodbury and Manton (1977) proposed modeling $\alpha_i(t)$ as a quadratic function of an Ornstein-Uhlenbeck process. This important and elegant development has gone rather unnoticed. It is, however, conceptually important because one could easily imagine individual hazard rates, or individual frailty, to be stochastically varying over time. We shall therefore study this model, as well as a version where $\alpha_i(t)$ is modeled as a Lévy process (Chapters 10 and 11). The Lévy process is a natural model when one imagines that individual risk is advancing through the accumulation of independent shocks. There is a close relationship to the classical frailty models.

The idea of an underlying stochastic process determining the occurrence of events may also be introduced in another fashion. One then imagines that an event occurs for an individual when some underlying process crosses a limit (Chapter 10). Such models have been studied extensively by first-passage theory for Wiener processes by Whitmore and others; see, for instance, Lee and Whitmore (2006) and references therein. The inverse Gaussian distribution arises naturally in this context. Extensions of such models are discussed by Aalen and Gjessing (2001). The models are of both practical and conceptual value. For instance, they throw light on why the hazard rates assume specific shapes. The notion of *quasi-stationarity* is a central concept in this theory.

Finally, there is the emerging field of *joint models*, where one makes a stochastic model both for the covariate processes $\mathbf{x}_i(t)$, and for how these influence the hazard rates $\alpha_i(t)$. Such models have been proposed by Tsiatis et al. (1995), who suggest using various Gaussian processes for $\mathbf{x}_i(t)$ and combining this with a Cox model. It should be noted that all of the stochastic process models mentioned earlier can in principle be used to make joint models. For instance, Woodbury and Manton (1977), Myers (1981), and Manton and Stallard (1998) applied the Ornstein-Uhlenbeck model to carry out joint modeling. Their work predated the modern development in a more classical survival analysis setting by several years.

1.5.4 Independent or dependent data?

The typical assumption in event history analysis will be that one makes observations of a number of independent individuals. In medicine, this is connected with the use of these methods in the study of chronic diseases like cancer, where the independence assumption is highly reasonable. However, in a number of other cases one would expect dependence between individuals. This might be the case in infectious diseases, but also in the study of social phenomena, like divorce, where people may influence one another and where changes over time may be cultural trends in sub-groups or society as a whole. The methods based on counting process models are still of relevance. The counting process approach does not require independence. The intensity process governing the risk for an individual may be arbitrarily dependent on what has happened to other individuals.

In fact, the basic assumption in this book will be that certain processes are martingales. This is a weakening of the independence assumption. A very interesting result called the Doob, or more generally, Doob-Meyer decomposition (see Section 2.2.3), guarantees the existence of martingale components in almost any kind of process. So we are not really imposing a structure by the martingale assumption, but the task is rather to find the appropriate martingale.

1.6 Exercises

1.1 A survival time T is exponentially distributed if its survival function takes the form $S(t) = e^{-\gamma t}$ for $t > 0$.

a) Find the density function $f(t) = -S'(t)$ and the hazard rate $\alpha(t)$ for the exponential distribution.

A survival time T follows a Weibull distribution if its hazard rate takes the form $\alpha(t) = bt^{k-1}$ for $t > 0$.

b) Find the survival function $S(t)$ and the density function $f(t)$ for a Weibull distribution.

A survival time T is gamma distributed if its density function takes the form $f(t) = (\gamma^k/\Gamma(k))t^{k-1}e^{-\gamma t}$ for $t > 0$, where $\Gamma(k) = \int_0^\infty u^{k-1}e^{-u}\,du$ is the gamma function.

c) Find expressions for the survival function $S(t)$ and the hazard rate $\alpha(t)$ of the gamma distribution using the (upper) incomplete gamma function $\Gamma(k,x) = \int_x^\infty u^{k-1}e^{-u}\,du$.

1.2 Consider a survival time T with survival function $S(t)$ and cumulative distribution function $F(t) = 1 - S(t)$, and assume that the cumulative distribution function is continuous and strictly increasing. Then the pth fractile ξ_p of the distribution is uniquely determined by the relation $F(\xi_p) = p$. Of particular interest are the *median* $\xi_{0.50}$ and the *lower and upper quartiles* $\xi_{0.25}$ and $\xi_{0.75}$.

a) Show that ξ_p is uniquely determined by the relation $A(\xi_p) = -\log(1-p)$, where $A(t)$ is the cumulative hazard rate (1.3).
b) Find expressions for ξ_p for the exponential distribution and the Weibull distribution (cf. Exercise 1.1).

1.3 Consider a survival time T with survival function $S(t)$, and assume that T is a proper random variable, that is, $S(\infty) = 0$.

a) Show that the *mean survival time* may be given as $E(T) = \int_0^\infty S(u)du$.
   [*Hint:* Write $T = \int_0^\infty I(T > u)\,du$, where $I(\cdot)$ is the indicator function.]
b) Find expressions for the mean survival time for the exponential distribution and the Weibull distribution (cf. Exercise 1.1).

1.4 Figure 1.3 shows empirical survival curves (Kaplan-Meier estimates) for the time between first and second births for the women whose first child survived one year and the women whose first child died within one year of its birth.

a) Use the figure to give a crude estimate of the probability that a woman whose first child survived one year will have another child within two years of the birth of the first child. What is the estimate of this probability for a woman whose first child died within one year?
b) Estimate the lower quartile and the median in the survival distribution for the two groups of women.

1.5 Consider two individuals (individual 1 and individual 2) with covariates $x_{i1}, \ldots x_{ip}$; $i = 1, 2$; and assume that the hazard rates of their survival times are given by Cox's regression model (1.6):

$$\alpha_i(t) = \alpha(t|x_{i1}, \ldots, x_{ip}) = \alpha_0(t)\exp\{\beta_1 x_{i1} + \cdots + \beta_p x_{ip}\}; \qquad i = 1, 2.$$

a) Show that the *hazard ratio* $\alpha_2(t)/\alpha_1(t)$ does not depend on time t, that is, that the hazards are proportional.
b) Determine the hazard ratio $\alpha_2(t)/\alpha_1(t)$ for the special case where $x_{2j} = x_{1j} + 1$ and $x_{2l} = x_{1l}$ for all $l \neq j$. Use the result to give an interpretation of the quantities e^{β_j}; $j = 1, \ldots, p$.

1.6 Let T be an exponentially distributed survival time with hazard rate $\alpha(t) = 2$ for $t > 0$. Define $\tilde{T} = \min\{T, 1\}$ and $D = I(T \leq 1)$, and introduce the counting process $N(t) = I(\tilde{T} \leq t, D = 1)$.

a) Sketch $N(t)$ as a function of t when $D = 1$ and when $D = 0$.
b) Give an expression for the intensity process $\lambda(t)$ of $N(t)$, and sketch $\Lambda(t) = \int_0^t \lambda(u)du$ as a function of t when $D = 1$ and when $D = 0$.
c) Let $M(t) = N(t) - \Lambda(t)$, and sketch $M(t)$ as a function of t when $D = 1$ and when $D = 0$.

1.7 Let X be a birth- and death-process with $X(0) = x_0 > 0$ and with birth intensity ϕ and death intensity μ, that is,

$$P(X(t+dt) = k \,|\, X(t-) = j) = \begin{cases} j\phi dt & k = j+1, \\ 1 - j(\phi + \mu)dt & k = j, \\ j\mu dt & k = j-1. \end{cases}$$

Let $N_b(t)$ and $N_d(t)$ be the number of births and deaths, respectively, in $[0,t]$. (A transition from state j to state $j+1$ is a birth, while a transition from state j to state

$j-1$ is a death.) Use (1.11) to determine the intensity processes of the two counting processes.

1.8 Let $X_1, X_2, \ldots$ be independent and identically distributed random variables with hazard rate $h(x)$. Introduce the *renewal process* $T_n = X_1 + X_2 \cdots + X_n; n = 1, 2, \ldots;$ and let $N(t) = \sum_{n \geq 1} I(T_n \leq t)$. Use (1.11) to determine the intensity process of $N(t)$. [*Hint:* At time t the time elapsed since the last occurrence of an event equals the backward recurrence time $U(t) = t - T_{N(t-)}$.]

1.9 Let T be a survival time with hazard rate $\alpha(t)$, and let $v > 0$ be a constant. If we consider the survival time T conditional on $T > v$, we say that the survival time is *left-truncated*.

a) Show that the left-truncated survival time has a hazard rate that equals zero on the interval $[0, v]$ and equals $\alpha(t)$ on (v, ∞), and determine the intensity process of the counting process $N(t) = I(v < T \leq t)$. [*Hint:* Derive the conditional survival function $S(t|v) = P(T > t | T > v)$, and use this to find the hazard rate of the truncated survival time.]

Let $0 < v < u$ be constants, and consider the *left-truncated and right-censored* survival time $\widetilde{T} = T \wedge u$ obtained by censoring the left-truncated survival time at $u > v$, and let $D = I(\widetilde{T} = T)$.

b) Determine the intensity process of the counting process

$$N(t) = I(v < \widetilde{T} \leq t, D = 1).$$

[*Hint:* Consider the conditional distribution of T given $T > v$.]

1.10 Let $T_1, T_2, \ldots, T_n$ be independent survival times with hazard rate $\alpha_i(t)$ for T_i, and introduce the counting processes $N_i(t) = I(T_i \leq t); i = 1, \ldots, n$.

a) Let $\mu_i(t)$ for $i = 1, \ldots, n$ be known hazard functions (e.g., corresponding to the mortality in the general population). Find the intensity process of the aggregated process $N(t) = \sum_{i=1}^{n} N_i(t)$ in each of the three situations:

(i) $\alpha_i(t) = \alpha(t)$
(ii) $\alpha_i(t) = \mu_i(t)\alpha(t)$
(iii) $\alpha_i(t) = \mu_i(t) + \alpha(t)$

b) Which of the three situations in question (a) satisfy the multiplicative intensity model (1.31)?

Chapter 2
Stochastic processes in event history analysis

Event histories unfold in time. Therefore, one would expect that tools from the theory of stochastic processes would be of considerable use in event history analysis. This is indeed the case, and in the present chapter we will review some basic concepts and results for stochastic processes that will be used in later chapters of the book.

Event histories consist of discrete events occurring over time in a number of individuals. One can think of events as being counted as they happen. Therefore, as indicated in Section 1.4, counting processes constitute a natural framework for analyzing survival and event history data. We shall in this chapter develop this idea further, and in particular elaborate the fundamental martingale concept that makes counting processes such an elegant tool. In this book the focus is on models in continuous time. However, as some concepts and results for martingales and other stochastic processes are more easily understood in discrete time, we first, in Section 2.1, consider the time-discrete case. Then, in Section 2.2, we discuss how the concepts and results carry over to continuous time. To keep the presentation fairly simple, we restrict attention to univariate counting processes and martingales in this chapter. Extensions to the multivariate case are summarized in Appendix B.

With processes unfolding over time, one will also naturally come across processes that do not consist of discrete jumps of unit size, like counting processes do, where each jump corresponds to the occurrence of an event. For instance, one may imagine that an event is really just a manifestation of some underlying process that could be continuous; for example, a heart attack may occur when a blood clot grows beyond a certain size. In general there may be the idea of some continuous underlying process crossing a threshold and producing an event (cf. Chapter 10). This way of thinking is very natural and useful, and we shall apply some continuous stochastic processes. The most basic continuous stochastic processes is the Wiener process (or Brownian motion), which has independent increments that are normally distributed with mean zero and variance proportional to the length of the time interval. A review of the basic properties of the Wiener process is provided in Section 2.3.1, while a more extensive review and discussion of the more general diffusion processes is provided in Appendix A.4.

Transformations of Wiener processes arise as limits of martingales associated with counting processes when the number of individuals increases. Such approximations, which are given by the martingale central limit theorem, play an essential role in the statistical analysis of models based on counting processes. In Sections 2.3.2 and 2.3.3 conditions under which the martingale central limit theorem hold are discussed and formally stated.

The idea of independent increments is fundamental in stochastic process theory. From statistical inference one is well acquainted with the independent identically distributed random variables that form the basis of many statistical models. In stochastic process theory we have the Poisson and Wiener processes with their stationary and independent increments (cf. Sections 2.2.4 and 2.3.1). More general processes with stationary and independent increments also exist and are denoted Lévy processes. These have nice and important properties that are particularly useful in the theory of frailty and generalized frailty (cf. Chapters 6 and 11). An introduction to Lévy processes is given in Appendix A.5.

When processes do not have independent increments, the fundamental view taken in this book is that of the "French school" of stochastic processes. One then seeks to explain the future development of a process by means of what has happened previously. This is a dynamic point of view, connecting the past, present and future. This differs fundamentally from the theory of stationary processes that plays such a large role in time series analysis. The connection between the past and the future is given by means of *local characteristics* that describe how the past influences the changes taking place. The intensity process of a counting process is an example of a local characteristic. The mathematical foundation underlying this theory is given by a theorem called the *Doob-Meyer decomposition*. Although this is in its general form a quite heavy mathematical result, the intuitive content is simple and not hard to grasp (cf. Section 2.2.3).

The stochastic process concepts and results we present in this chapter and in Appendices A and B are based on quite heavy mathematics if one wants to go into every detail, and a number of regularity assumptions must be made to be mathematically precise. We shall not state these assumptions, but refer to Andersen et al. (1993) for the theory of counting processes with associated martingales and stochastic integrals and to the references provided in Appendix A for the theory of Wiener processes, diffusions, and Lévy processes. [For example, we will not state integrability conditions in this book, and thus we will not worry about whether a stochastic process is a (local) martingale or a (local) square integrable martingale.] It should be pointed out, however, that the basic ideas and results for stochastic processes are mostly relatively simple, and that they can be understood at an intuitive "working technical" level without going into mathematical details. This is the level of presentation we aim at in this book. Very often, the intuitive content in stochastic processes tends to drown in complex mathematical presentations, which is probably the reason this material is not so much used in applied statistics. We want to contribute to a "demystification" of martingales, stochastic integrals, and other stochastic process concepts.

2.1 Stochastic processes in discrete time

Although in this book we operate in continuous time, where events can occur at any time, it will be useful in this section to consider time-discrete processes. Mathematically, such processes are much simpler than the time-continuous ones, which may often be derived as limits of the time-discrete processes. So if we understand the basic ideas in a discrete context, it will also give us the required insight in the continuous setting. The results for time-discrete processes are also of interest in their own right, for example, for studying longitudinal data in discrete time (Borgan et al., 2007; Diggle et al., 2007).

2.1.1 Martingales in discrete time

Let $M = \{M_0, M_1, M_2, \ldots\}$ be a stochastic process in discrete time. The process M is a *martingale* if

$$E(M_n \mid M_0, M_1, \ldots, M_{n-1}) = M_{n-1} \qquad (2.1)$$

for each $n \geq 1$. Hence, the martingale property simply consists in asserting that the conditional expectation of a random variable in the process given the past equals the previous value. This innocent-looking assumption has a far greater depth than would appear at first look. The martingale property may be seen as a requirement to a fair game. If M_{n-1} is the collected gain after $n - 1$ games, then the expected gain after the next game should not change.

For the applications we have in mind, it will always be the case that $M_0 = 0$. As some of the following formulas become simpler when this is the case, we will tacitly assume that $M_0 = 0$ throughout.

In Chapter 1 we talked about the past in a rather unspecified fashion. The past could be just what is generated by the observed process as typified in formula (2.1). However, the past could also be defined as a *wider* amount of information. Often the past includes some external information (e.g., covariates) in addition to the previous values of the process itself. It is important to note that a process may be a martingale for some definitions of the past and not for others. In stochastic process theory the past is usually formulated as a σ-algebra of events. We will not give a formal definition of this σ-algebra. For our purpose it suffices to see it as the family of events that can be decided to have happened or not happened by observing the past. Such a σ-algebra is often termed $\mathscr{F}_n$ and is a formal way of representing what is known at time n. We will denote $\mathscr{F}_n$ as the *history at time n*, so that the entire history is represented by the increasing family of σ-algebras $\{\mathscr{F}_n\}$. The family has to increase, since our past knowledge will increase as time passes.

We shall now give a more general formulation of (2.1). Assume that $\mathscr{F}_n$, for each n, is generated by $M_1, \ldots, M_n$ plus possibly some external information. A technical formulation would be that the process $M = \{M_0, M_1, M_2, \ldots\}$ is *adapted* to the history $\{\mathscr{F}_n\}$. This means that, for each n, the random variables $M_1, \ldots, M_n$

are measurable with respect to the σ-algebra $\mathscr{F}_n$. A practical implication is that $M_1, \ldots, M_n$ may be considered as constants given the history $\mathscr{F}_n$; in particular:

$$E(M_m \,|\, \mathscr{F}_n) = M_m \qquad \text{for all } m \leq n. \tag{2.2}$$

The process $M = \{M_0, M_1, M_2, \ldots\}$ is a martingale with respect to the history $\{\mathscr{F}_n\}$ if

$$E(M_n \,|\, \mathscr{F}_{n-1}) = M_{n-1} \qquad \text{for all } n \geq 1, \tag{2.3}$$

cf. (2.1). This is equivalent to the more general statement

$$E(M_n \,|\, \mathscr{F}_m) = M_m \qquad \text{for all } n > m \tag{2.4}$$

(Exercise 2.1). Note the difference between (2.2) and (2.4). While (2.2) states that we know the past and present of the process M, (2.4) states that the expected value of the process in the future equals its present value.

As a consequence of (2.4) we get, using double expectations and the assumption $M_0 = 0$,

$$E(M_n) = E\{E(M_n \,|\, \mathscr{F}_0)\} = E(M_0) = 0. \tag{2.5}$$

Since the martingale has mean zero for all n, we say that it is a *mean zero martingale*. By a similar argument one may show that

$$\mathrm{Cov}(M_m, M_n - M_m) = 0 \qquad \text{for all } n > m$$

(Exercise 2.2), that is, the martingale has *uncorrelated increments*.

By (2.2), a reformulation of (2.3) is as follows:

$$E(M_n - M_{n-1} \,|\, \mathscr{F}_{n-1}) = 0 \qquad \text{for all } n \geq 1. \tag{2.6}$$

Here $\triangle M_n = M_n - M_{n-1}$, $n = 1, 2, \ldots$, are denoted martingale differences. Notice that (2.6) would also hold if the process had independent zero-mean increments. In this sense the concept of martingale differences is a weakening of the independent increment concept. We could also say that any sum of independent zero-mean random variables is a martingale (Exercise 2.3). The assumption of independence pervades statistics, but in many cases a martingale-type assumption would be sufficient for demonstrating unbiasedness, asymptotic normality, and so on.

2.1.2 Variation processes

Two processes describe the variation of a martingale $M = \{M_0, M_1, \ldots\}$. The *predictable variation process* is denoted $\langle M \rangle$ and for $n \geq 1$ is defined as the sum of conditional variances of the martingale differences:

$$\langle M \rangle_n = \sum_{i=1}^{n} E\{(M_i - M_{i-1})^2 \mid \mathscr{F}_{i-1}\} = \sum_{i=1}^{n} \text{Var}(\triangle M_i \mid \mathscr{F}_{i-1}), \tag{2.7}$$

while $\langle M \rangle_0 = 0$. The *optional variation process* $[M]$ is defined by

$$[M]_n = \sum_{i=1}^{n} (M_i - M_{i-1})^2 = \sum_{i=1}^{n} (\triangle M_i)^2 \tag{2.8}$$

for $n \geq 1$, and $[M]_0 = 0$. The following statements can be proved by simple calculations:

$$M^2 - \langle M \rangle \text{ is a mean zero martingale,} \tag{2.9}$$
$$M^2 - [M] \text{ is a mean zero martingale.} \tag{2.10}$$

We shall prove the second of these statements and leave (2.9) as an exercise for the reader (Exercise 2.4). To prove (2.10), we first note that $M_0^2 - [M]_0 = 0$. Then writing

$$M_n^2 = (M_{n-1} + M_n - M_{n-1})^2, \quad \text{and} \quad [M]_n = [M]_{n-1} + (M_n - M_{n-1})^2,$$

we get

$$\begin{aligned} &E(M_n^2 - [M]_n \mid \mathscr{F}_{n-1}) \\ &= E(M_{n-1}^2 + 2M_{n-1}(M_n - M_{n-1}) - [M]_{n-1} \mid \mathscr{F}_{n-1}) \\ &= M_{n-1}^2 - [M]_{n-1} + 2M_{n-1} E(M_n - M_{n-1} \mid \mathscr{F}_{n-1}) \\ &= M_{n-1}^2 - [M]_{n-1}, \end{aligned}$$

which gives exactly the martingale property.

2.1.3 Stopping times and transformations

One major advantage of the martingale assumption is that one can make certain manipulations of the process without destroying the martingale property. The independence property, on the other hand, would not survive these manipulations.

Our first example of this is the concept of an *optional stopping time* (or just stopping time). An example could be the first time M passes above a certain limit. We denote such a time by T and the value of the process by M_T. In general, a time T is called an optional stopping time if the event $\{T = t\}$ is only dependent on what has been observed up to and including time t. If T is defined as a first passage time, then it is an optional stopping time. It is easy to construct times that are not optional stopping times; for instance, let T be equal to the last time the process passes above a certain limit. The optional stopping time property can be decided from the past and present observations, while in the last example we would have to look ahead.

Stopping a fair game at an optional stopping time T preserves the fairness of the game. This is connected to a preservation of the martingale property under optional stopping. For a martingale M, define M *stopped at* T as follows:

$$M_n^T = M_{n \wedge T}. \tag{2.11}$$

(Here $n \wedge T$ denotes the minimum of n and T.) It can be proved that M^T is a martingale (Exercise 2.6). The idea of *fairness of a game* is strongly connected to the idea of *unbiasedness* in statistics, and this is a major reason for the statistical usefulness of martingales. Stopping will commonly occur in event history data in the form of censoring, and preserving the martingale property ensures that estimates and tests remain essentially unbiased.

A more general formulation of (2.11), which is of great use in our context, comes through defining a *transformation* of a process as follows. Let $X = \{X_0, X_1, X_2, \ldots\}$ be some general process with a history $\{\mathscr{F}_n\}$, and let $H = \{H_0, H_1, H_2, \ldots\}$ be a *predictable* process, that is, a sequence of random variables where each H_n is measurable with respect to $\mathscr{F}_{n-1}$ and hence is known one step ahead of time. The process Z defined by

$$Z_n = H_0 X_0 + H_1 (X_1 - X_0) + \cdots + H_n (X_n - X_{n-1}) \tag{2.12}$$

is denoted the *transformation* of X by H and written $Z = H \bullet X$.

Using (2.6) and the predictability of H, the following simple calculation shows that if M is a martingale, then so is $Z = H \bullet M$:

$$\begin{aligned}
\mathrm{E}(Z_n - Z_{n-1} \mid \mathscr{F}_{n-1}) &= \mathrm{E}(H_n(M_n - M_{n-1}) \mid \mathscr{F}_{n-1}) \\
&= H_n \mathrm{E}(M_n - M_{n-1} \mid \mathscr{F}_{n-1}) \\
&= 0.
\end{aligned}$$

Hence *a transformation preserves the martingale property.* Moreover, since $Z_0 = H_0 M_0 = 0$, the transformation is a mean zero martingale. Considering games, we see again that the fairness of the game is preserved; one often says that there is no betting system that can beat a fair game.

Note that for $n \geq 1$ we may write

$$Z_n = (H \bullet M)_n - \sum_{s=1}^{n} H_s \triangle M_s, \tag{2.13}$$

where $\triangle M_s = M_s - M_{s-1}$. In Section 2.2.2 we shall meet transformations under the more sophisticated disguise of stochastic integrals, and the formulation there will be seen to be very similar to that of (2.13). Most of the properties we will need for stochastic integrals are easily derived for transformations.

The variation processes obey the following rules under transformation:

$$\langle H \bullet M \rangle = H^2 \bullet \langle M \rangle \quad \text{and} \quad [H \bullet M] = H^2 \bullet [M],$$

or formulated as sums:

$$\langle H \bullet M \rangle_n = \sum_{s=1}^{n} H_s^2 \, \triangle \langle M \rangle_s, \tag{2.14}$$

$$[H \bullet M]_n = \sum_{s=1}^{n} H_s^2 \, \triangle [M]_s. \tag{2.15}$$

We will prove the first of these statements. By (2.13) and (2.7) we have:

$$\triangle(H \bullet M)_s = H_s \triangle M_s,$$
$$\triangle \langle M \rangle_s = \mathrm{Var}(\triangle M_s \,|\, \mathscr{F}_{s-1}).$$

Then using (2.7) and the predictability of H, the predictable variation process of the transformation $H \bullet M$ becomes

$$\begin{aligned}
\langle H \bullet M \rangle_n &= \sum_{s=1}^{n} \mathrm{Var}(H_s \triangle M_s \,|\, \mathscr{F}_{s-1}) \\
&= \sum_{s=1}^{n} H_s^2 \, \mathrm{Var}(\triangle M_s \,|\, \mathscr{F}_{s-1}) \\
&= \sum_{s=1}^{n} H_s^2 \, \triangle \langle M \rangle_s.
\end{aligned}$$

This proves (2.14). The proof of (2.15) is similar and is left as an exercise (Exercise 2.7).

2.1.4 The Doob decomposition

Martingales arise naturally whenever we try to explain the developments in a stochastic process as a function of its previous development and other observations of the past. It is possible to decompose an arbitrary stochastic process into a sequence of martingale differences and a predictable process. Let $X = \{X_0, X_1, X_2, \ldots\}$ be some general process, with $X_0 = 0$, with respect to a history $\{\mathscr{F}_n\}$, and define a process $M = \{M_0, M_1, M_2, \ldots\}$ by

$$M_0 = X_0, \quad M_n - M_{n-1} = X_n - E(X_n \,|\, \mathscr{F}_{n-1}).$$

It is immediately clear that the $\triangle M_n = M_n - M_{n-1}$ are martingale differences, since the expectation given the past $\mathscr{F}_{n-1}$ is zero. We can therefore write

$$X_n = E(X_n \,|\, \mathscr{F}_{n-1}) + \triangle M_n; \tag{2.16}$$

this is the Doob decomposition. The quantity $E(X_n \,|\, \mathscr{F}_{n-1})$ is a function of the past only, and hence the process taking these values is predictable. The martingale differences $\triangle M_n$ are often termed *innovations,* since they represent what is new and unexpected compared to past experience. Hence, formula (2.16) decomposes a process into what can be predicted from the past and what is new and "surprising," the innovations.

A time-continuous generalization of the Doob decomposition, called the Doob-Meyer decomposition, is in fact the key to the counting process approach in this book; cf. Section 2.2.3.

2.2 Processes in continuous time

We will now discuss stochastic processes in continuous time. We first consider time-continuous martingales and stochastic integrals and indicate how the results of the previous section carry over to the time-continuous case. We also discuss the Doob-Meyer decomposition for time-continuous stochastic processes, generalizing the Doob decomposition of Section 2.1.4. Then we briefly discuss the well-known Poisson process and show how this in a natural way gives rise to a time-continuous martingale. Finally, we consider counting processes and martingales derived from counting processes, and we review a number of results that will be of great use in later chapters of the book. To keep the presentation fairly simple, we do not here consider vector-valued counting processes, martingales, and stochastic integrals. The relevant results for such processes, which are multivariate extensions of the results presented in this chapter, are collected in Appendix B.

In practical applications, stochastic processes will be observed over a finite time interval. Unless otherwise stated, we will assume throughout the book that the time-continuous stochastic processes we consider are defined on the finite interval $[0, \tau]$.

Formally, we say that a stochastic process $X = \{X(t); t \in [0, \tau]\}$ is *adapted* to a history $\{\mathscr{F}_t\}$ (an increasing family of σ-algebras) if $X(t)$ is $\mathscr{F}_t$-measurable for each t. This means that at time t we know the value of $X(s)$ for all $s \leq t$ (possibly apart from unknown parameters). A realization of X is a function of t and is called a *sample path.* If the sample paths of a stochastic process are right-continuous and have left-hand limits, we say that the process is *cadlag* (continue à droite, limité à gauche). Unless otherwise stated, all time-continuous processes we encounter in the book are assumed to be cadlag.

Also in continuous time will we need the concept of a *stopping time.* We say that T is a stopping time if the event $\{T \leq t\}$ is $\mathscr{F}_t$-measurable for each t. This means that at time t we know whether $T \leq t$ or $T > t$.

2.2.1 Martingales in continuous time

A stochastic process $M = \{M(t); t \in [0, \tau]\}$ is a martingale relative to the history $\{\mathscr{F}_t\}$ if it is *adapted* to the history and satisfies the *martingale property:*

$$E(M(t) \mid \mathscr{F}_s) = M(s) \text{ for all } t > s. \qquad (2.17)$$

Note that the martingale property (2.17) corresponds to (2.4) for the time-discrete case. A heuristic way of formulating the martingale property, corresponding to (2.6), is to say that M is a martingale provided that

$$E(dM(t) \mid \mathscr{F}_{t-}) = 0.$$

Here $dM(t)$ is the increment of M over the small time interval $[t, t+dt)$, and $\mathscr{F}_{t-}$ means the history until *just before* time t.

As for the time-discrete case, we will tacitly assume throughout that $M(0) = 0$. This will cover all our applications. Then, by the argument used in (2.5), we have $EM(t) = 0$ for all t, that is, M is a *mean zero martingale*. Similarily we may show that a martingale has *uncorrelated increments*, that is,

$$\text{Cov}(M(t) - M(s), M(v) - M(u)) = 0 \qquad (2.18)$$

for all $0 \le s < t < u < v \le \tau$ (cf. Exercise 2.2).

We also for a time-continuous martingale M introduce the *predictable variation process* $\langle M \rangle$ and the *optional variation process* $[M]$. These are defined as the appropriate limits (in probability) of their time-discrete counterparts:

$$\langle M \rangle (t) = \lim_{n \to \infty} \sum_{k=1}^{n} \text{Var}(\triangle M_k \mid \mathscr{F}_{(k-1)t/n}), \qquad (2.19)$$

and

$$[M](t) = \lim_{n \to \infty} \sum_{k=1}^{n} (\triangle M_k)^2, \qquad (2.20)$$

where the time interval $[0, t]$ is partitioned into n subintervals each of length t/n, and $\triangle M_k = M(kt/n) - M((k-1)t/n)$ is the increment of the martingale over the kth of these subintervals. Informally, we have from (2.19) that

$$d\langle M \rangle(t) = \text{Var}(dM(t) \mid \mathscr{F}_{t-}), \qquad (2.21)$$

that is, the increment $d\langle M \rangle(t)$ of the predictable variation process over the small time interval $[t, t+dt)$ is the conditional variance of the increment of the martingale.

In a similar manner as for a discrete-time martingale, the following results hold [cf. (2.9) and (2.10)]:

$$M^2 - \langle M \rangle \text{ is a mean zero martingale}, \qquad (2.22)$$

$$M^2 - [M] \text{ is a mean zero martingale}. \qquad (2.23)$$

By (2.22) and (2.23), $M^2(t) - \langle M \rangle (t)$ and $M^2(t) - [M](t)$ have mean zero for all t. Therefore, since $M(t)$ has mean zero:

$$\text{Var}(M(t)) = E\left(M(t)^2\right) = E\langle M \rangle(t) = E[M](t). \qquad (2.24)$$

This shows how the variation processes of a martingale are closely linked to its variance, a fact that will be useful in later chapters when deriving estimators for the variances of statistical estimators and test statistics.

We will often encounter situations with several martingales, and it is then fruitful to define covariation processes for pairs of martingales M_1 and M_2. Corresponding to (2.19) and (2.20) we may define the *predictable covariation process* $\langle M_1, M_2 \rangle$ as the limit (in probability) of the sum of conditional covariances $\mathrm{Cov}(\triangle M_{1k}, \triangle M_{2k} | \mathscr{F}_{(k-1)t/n})$ and the *optional covariation process* $[M_1, M_2]$ as the limit of the sum of the products $\triangle M_{1k} \triangle M_{2k}$. Informally we may write

$$d\langle M_1, M_2 \rangle(t) = \mathrm{Cov}(dM_1(t), dM_2(t) | \mathscr{F}_{t-}).$$

Note that by the preceding definitions, $\langle M, M \rangle = \langle M \rangle$ and $[M, M] = [M]$. In a similar manner as (2.22) and (2.23) we have that:

$$M_1 M_2 - \langle M_1, M_2 \rangle \text{ is a mean zero martingale}, \qquad (2.25)$$

$$M_1 M_2 - [M_1, M_2] \text{ is a mean zero martingale}. \qquad (2.26)$$

As a consequence of these results

$$\mathrm{Cov}\,(M_1(t), M_2(t)) = \mathrm{E}\,(M_1(t)\,M_2(t)) = \mathrm{E}\langle M_1, M_2 \rangle(t) = \mathrm{E}[M_1, M_2](t) \qquad (2.27)$$

for all t; cf. (2.24).

The rules for evaluating the (co)variation processes of linear combinations of martingales are similar to the rules for evaluating (co)variances of linear combinations of ordinary random variables. As an example, the predictable variation processes of a sum of martingales may be written

$$\langle M_1 + M_2 \rangle = \langle M_1 \rangle + \langle M_2 \rangle + 2\langle M_1, M_2 \rangle, \qquad (2.28)$$

and a similar relation holds for the optional variation processes.

2.2.2 Stochastic integrals

We will now introduce the stochastic integral as an analog to the transformation for discrete-time martingales. Let $H = \{H(t); t \in [0, \tau]\}$ be a stochastic process that is *predictable*. Intuitively this means that for any time t, the value of $H(t)$ is *known just before* t (possibly apart from unknown parameters). A formal definition of predictability in continuous time is a bit intricate, and we will not go into details about this. But we note that sufficient conditions for H to be predictable are:

- H is *adapted* to the history $\{\mathscr{F}_t\}$.
- The sample paths of H are *left-continuous*.

Predictability may sound like an uninteresting technical assumption. It turns out, however, to be very important in practical calculations when defining test statistics and estimators.

We can now introduce the stochastic integral:

$$I(t) = \int_0^t H(s)\,dM(s).$$

This is a general concept valid for martingales from a much broader setting than counting processes. However, in our context most martingales will arise from counting processes (Section 2.2.5). Stochastic integration is the exact analogue of transformation of discrete-time martingales as defined in (2.12). Analogously to formula (2.13), the stochastic integral can be defined as a limit of such a transformation in the following sense:

$$I(t) = \lim_{n \to \infty} \sum_{k=1}^n H_k \triangle M_k,$$

where we have partitioned the time interval $[0,t]$ into n subintervals of length t/n, and we let $H_k = H((k-1)t/n)$ and $\triangle M_k = M(kt/n) - M((k-1)t/n)$. (In the general theory of stochastic integrals, which include integrals with respect to Wiener process martingales, this limiting definition is not valid and one has to introduce the concept of an Itô integral. In this book, Itô integrals are only used briefly in connection with stochastic differential equations in Section 10.4 and Appendix A.4.)

As for a transformation, the major interesting fact about a stochastic integral is that $I(t)$ is a mean zero martingale with respect to $\{\mathscr{F}_t\}$. Hence, *the martingale property is preserved under stochastic integration*. This follows from the corresponding fact for transformations (Section 2.1.3), since the stochastic integral can be seen as a limit of discrete-time versions.

In analogy with formulas (2.14) and (2.15), the following rules hold for evaluating the variation processes of a stochastic integral:

$$\left\langle \int H\,dM \right\rangle = \int H^2\,d\langle M \rangle, \tag{2.29}$$

$$\left[\int H\,dM \right] = \int H^2\,d[M], \tag{2.30}$$

while the following rules hold for the covariation processes:

$$\left\langle \int H_1\,dM_1, \int H_2\,dM_2 \right\rangle = \int H_1 H_2\,d\langle M_1, M_2 \rangle, \tag{2.31}$$

$$\left[\int H_1\,dM_1, \int H_2\,dM_2 \right] = \int H_1 H_2\,d[M_1, M_2]; \tag{2.32}$$

cf. Exercise 2.8.

2.2.3 The Doob-Meyer decomposition

In Section 2.1.4 we saw how a discrete-time stochastic process can be decomposed into a predictable process and a sequence of martingale differences. A similar result holds for processes in continuous time and is know as the *Doob-Meyer decomposition*.

To state the content of this decomposition, we first need to consider a specific class of processes. An adapted process $X = \{X(t); t \in [0, \tau]\}$ is called a *submartingale* if it satisfies

$$\mathrm{E}(X(t) \,|\, \mathscr{F}_s) \geq X(s) \quad \text{for all } t > s. \tag{2.33}$$

[Note the similarity with the martingale property (2.17).] Thus a submartingale is a process that tends to increase as time passes. In particular, any nondecreasing process, like a counting process, is a submartingale. The Doob-Meyer decomposition states that any submartingale X can be decomposed *uniquely* as

$$X = X^* + M, \tag{2.34}$$

where X^* is a nondecreasing predictable process, often denoted the *compensator* of X, and M is a mean zero martingale. Heuristically we have that

$$dX^*(t) = \mathrm{E}(dX(t) \,|\, \mathscr{F}_{t-})$$

and

$$dM(t) = dX(t) - \mathrm{E}(dX(t) \,|\, \mathscr{F}_{t-}).$$

Like the Doob decomposition in discrete time (Section 2.1.4), the Doob-Meyer decomposition (2.34) therefore tells us what can be predicted from the past, $dX^*(t)$, and what is the innovation, or surprising element, $dM(t)$.

In (2.22) we noted that if M is a martingale, so is $M^2 - \langle M \rangle$. Here M^2 is a submartingale (by Jensen's inequality) and $\langle M \rangle$ is a nondecreasing predictable process (by construction). This shows that the predictable variation process $\langle M \rangle$ is the compensator of M^2, and this offers an alternative definition of the predictable variation process.

The Doob-Meyer decomposition (2.34) extends immediately to a *special semimartingale*, that is, a process X that is a difference of two submartingales. But then the compensator X^* is no longer a nondecreasing predictable process, but a finite variation predictable process (i.e., a difference of two nondecreasing predictable processes).

2.2.4 The Poisson process

A homogeneous Poisson process describes the distribution of events that occur entirely independently of one another. One imagines a basic rate of occurrence, or intensity, denoted λ, such that the probability of an event occurring in the time interval $[t, t+dt)$ is $\lambda\,dt$. A homogeneous Poisson process has a number of well known properties:

- The time between events is exponentially distributed with probability density $\lambda e^{-\lambda t}$.
- The expected value and the variance of the number of events in a time interval of length h are both equal to λh.
- The number of events in a time interval of length h is Poisson distributed; the probability of exactly k events occurring is $(\lambda h)^k e^{-\lambda h}/k!$.
- The process has independent increments, that is, the number of events in nonoverlapping intervals are independent.

We let $N(t)$ be the number of events in $[0, t]$ and introduce the process

$$M(t) = N(t) - \lambda t, \qquad (2.35)$$

obtained by centering the Poisson process (by subtracting its mean). Further we denote by $\mathscr{F}_t$ the information about all events that happen in the time interval $[0, t]$. Due to the independent increments of a Poisson process, we have for all $t > s$:

$$\mathrm{E}\{M(t) - M(s) \,|\, \mathscr{F}_s\} = \mathrm{E}\{M(t) - M(s)\} = \mathrm{E}\{N(t) - N(s)\} - \lambda(t - s) = 0,$$

which yields

$$\mathrm{E}\{M(t) \,|\, \mathscr{F}_s\} = M(s). \qquad (2.36)$$

This is the martingale property (2.17), and hence the process (2.35) is a martingale. It follows that λt is the compensator of the Poisson process $N(t)$ (cf. Section 2.2.3). By a similar argument:

$$\mathrm{E}\{M^2(t) - \lambda t \,|\, \mathscr{F}_s\} = M^2(s) - \lambda s \qquad (2.37)$$

(Exercise 2.9), which shows that the process $M^2(t) - \lambda t$ is a martingale. Thus λt is also the compensator of $M^2(t)$, and it follows by the comment at the end of Section 2.2.3 that the martingale (2.35) has a predictable variation process

$$\langle M \rangle(t) = \lambda t. \qquad (2.38)$$

In the next subsection we will see that relations similar to (2.35), (2.36), and (2.38) are in general valid for counting processes.

2.2.5 Counting processes

As introduced in Section 1.4, a counting process $N = \{N(t); t \in [0, \tau]\}$ is a right-continuous process with jumps of size 1 at event times and constant in between. We assume that the counting process is adapted to the history $\{\mathscr{F}_t\}$, which is just a technical way of saying that the history is generated by N and possibly some external information as well.

The intensity process $\lambda(t)$ of a counting process (w.r.t. the history $\{\mathscr{F}_t\}$) is heuristically defined by

$$\lambda(t)dt = P(dN(t) = 1 \mid \mathscr{F}_{t-}) = E(dN(t) \mid \mathscr{F}_{t-}), \tag{2.39}$$

cf. (1.11) and (1.12). To give a precise mathematical definition of an intensity process, first note that since the counting process is nondecreasing, it is a submartingale (Section 2.2.3). Hence by the Doob-Meyer decomposition (2.34), there exist a unique predictable process $\Lambda(t)$, called the *cumulative intensity process*, such that $M(t) = N(t) - \Lambda(t)$ is a mean zero martingale.

Throughout the book we will consider the case where the cumulative intensity process is absolutely continuous. Then there exists a *predictable process* $\lambda(t)$ such that

$$\Lambda(t) = \int_0^t \lambda(s)ds, \tag{2.40}$$

and this gives a formal definition of the intensity process $\lambda(t)$ of the counting process. Further for the absolute continuous case, we have that

$$M(t) = N(t) - \int_0^t \lambda(s)ds \tag{2.41}$$

is a mean zero martingale. As indicated in Section 1.4, this is a key relation that we will use over and over again.

The predictable and optional variation processes of M are defined as the limits in (2.19) and (2.20), respectively. We first look at the latter of the two. To this end, note that the martingale (2.41) has jumps of size 1 at the jump times of N, and that it is continuous between the jump times. When the limit is approached in (2.20), only the jumps will remain. Thus the optional variation process becomes

$$[M](t) = N(t). \tag{2.42}$$

As for the predictable variation process, we will be content with a heuristic argument. By (2.21) and (2.41) we have that

$$\begin{aligned}
d\langle M\rangle(t) &= \mathrm{Var}(dM(t) \mid \mathscr{F}_{t-}) \\
&= \mathrm{Var}(dN(t) - \lambda(t)dt \mid \mathscr{F}_{t-}) \\
&= \mathrm{Var}(dN(t) \mid \mathscr{F}_{t-}),
\end{aligned}$$

since $\lambda(t)$ is predictable, and hence a fixed quantity given $\mathscr{F}_{t-}$. Now $dN(t)$ may only take the value 0 or 1, and it follows using (2.39) that

$$d\langle M\rangle(t) \approx \lambda(t)dt \{1 - \lambda(t)dt\} \approx \lambda(t)dt.$$

This motivates the relation

$$\langle M \rangle (t) = \int_0^t \lambda(s)ds, \tag{2.43}$$

which is another key result that will be used a number of times in later chapters.

Note that (2.41) and (2.43) imply that the cumulative intensity process (2.40) is the compensator of both the counting process, $N(t)$, and the square of the martingale, $M^2(t)$. Note also that (2.41) and (2.43) are similar to the relations (2.35) and (2.38) we derived for the Poisson process in Section 2.2.4. Thus one may say that a counting process has the same "local behavior" as a Poisson process. It is this "local Poisson-ness" of counting processes that is the source of their nice properties.

So far we have considered just a single counting process. In practice we will often have several of them, for instance, corresponding to different individuals or to different groups we want to compare. We assume that the counting processes are adapted to the same history $\{\mathscr{F}_t\}$, so the history is generated by all the counting processes and possibly some external information as well.

Generally, we shall require that no two counting processes in continuous time can jump simultaneously. Consider a pair N_1 and N_2 of counting processes, with corresponding martingales M_1 and M_2. Since the counting processes do not jump simultaneously, the same applies for the martingales M_1 and M_2. From this it follows that

$$\langle M_1, M_2 \rangle (t) = 0 \quad \text{for all } t \tag{2.44}$$

$$[M_1, M_2](t) = 0 \quad \text{for all } t \tag{2.45}$$

(Exercise 2.10). We say that the martingales are *orthogonal*. By (2.25) and (2.26) the orthogonality of M_1 and M_2 is equivalent to the fact that the product $M_1 M_2$ is a martingale.

2.2.6 Stochastic integrals for counting process martingales

In the context of counting processes the stochastic integral

$$I(t) = \int_0^t H(s)\,dM(s)$$

is simple to understand. Using formula (2.41) one simply splits the integral in two as follows:

$$I(t) = \int_0^t H(s)\,dN(s) - \int_0^t H(s)\,\lambda(s)\,ds.$$

For given sample paths of the processes, the last integral is simply an ordinary (Riemann) integral. The first integral, however, is to be understood as a sum of the values of H at every jump time of the counting process. Thus

$$\int_0^t H(s)\,dN(s) = \sum_{T_j \le t} H(T_j),$$

where $T_1 < T_2 < \cdots$ are the ordered jump times of N.

Using (2.29), (2.30), (2.42), and (2.43), we get the following expressions for the predictable and optional variation processes of a stochastic integral of a counting process martingale:

$$\left\langle \int H\,dM \right\rangle (t) = \int_0^t H^2(s)\lambda(s)\,ds, \tag{2.46}$$

$$\left[\int H\,dM \right] (t) = \int_0^t H^2(s)\,dN(s). \tag{2.47}$$

Consider counting processes $N_1, N_2, \ldots, N_k$ with no simultaneous jumps and with intensity processes $\lambda_1, \lambda_2, \ldots, \lambda_k$ (w.r.t. the same history). Then the corresponding martingales $M_1, M_2, \ldots, M_k$ are orthogonal [cf. (2.44)], and (2.31) implies that $\langle \int H_j\,dM_j, \int H_l\,dM_l \rangle(t) = 0$ for all t when $j \ne l$. Using (2.28), we then get the important relation:

$$\left\langle \sum_{j=1}^{k} \int H_j\,dM_j \right\rangle (t) = \sum_{j=1}^{k} \int_0^t H_j^2(s)\lambda_j(s)\,ds. \tag{2.48}$$

In a similar manner, the following result holds for the optional variation process:

$$\left[\sum_{j=1}^{k} \int H_j\,dM_j \right] (t) = \sum_{j=1}^{k} \int_0^t H_j^2(s)\,dN_j(s). \tag{2.49}$$

2.2.7 The innovation theorem

The intensity process of a counting process N *relative to a history* $\{\mathscr{F}_t\}$ is given informally by

$$\lambda^{\mathscr{F}}(t)\,dt = \mathrm{E}(dN(t)\,|\,\mathscr{F}_{t-}), \tag{2.50}$$

cf. (2.39). We here make the dependence on the history $\{\mathscr{F}_t\}$ explicit in the notation to point out that the intensity process depends on the history, and that if the history is changed, the intensity process *may* change as well. We will now have a closer look at this.

Consider a counting process N, and let $\{\mathscr{N}_t\}$ be the history (or filtration) generated by the counting process (denoted the *self-exiting filtration*). In Sections 2.2.5 and 2.2.6 it is a key assumption that a counting process is adapted to the history. This means that $\{\mathscr{N}_t\}$ is the smallest history we may consider if the results of Sections 2.2.5 and 2.2.6 are to hold true.

Usually, however, we will consider histories that are not only generated by N, but are generated by N as well as by other counting processes, censoring processes, covariates, etc. that are observed in parallel with N. Consider two such histories, $\{\mathscr{F}_t\}$ and $\{\mathscr{G}_t\}$, and assume that they are *nested*, that is, that $\mathscr{F}_t \subseteq \mathscr{G}_t$ for all t. Thus, at any time t, all information contained in $\mathscr{F}_t$ is also contained in $\mathscr{G}_t$, but $\mathscr{G}_t$ may contain information that is not contained in $\mathscr{F}_t$. Using double expectations, we then have

$$\mathrm{E}(dN(t) \mid \mathscr{F}_{t-}) = \mathrm{E}\{\mathrm{E}(dN(t) \mid \mathscr{G}_{t-}) \mid \mathscr{F}_{t-}\}.$$

By this and (2.50) it follows that the intensity processes of N with respect to the two histories are related as follows:

$$\lambda^{\mathscr{F}}(t) = \mathrm{E}(\lambda^{\mathscr{G}}(t) \mid \mathscr{F}_{t-}). \tag{2.51}$$

This result is called the *innovation theorem*.

It is important to note that the innovation theorem applies only to histories that are nested. Further, in order for N to be adapted to the histories, both of these need to contain $\{\mathscr{N}_t\}$. Thus the innovation theorem holds provided that $\mathscr{N}_t \subseteq \mathscr{F}_t \subseteq \mathscr{G}_t$.

When considering more than one history, it is important to use notation for intensity processes that points out their dependence on the history, as we have done in this subsection. In later chapters, however, it will usually be clear from the context how the history is defined. Then we will just write $\lambda(t)$ for the intensity process without explicitly mentioning the history.

2.2.8 Independent censoring

In Sections 1.4.2 and 1.4.3 we gave an informal discussion of the concept of *independent censoring*; the main point being that independent censoring preserves the form of the intensity processes of the counting processes at hand. We will now discuss more formally the concept of independent censoring.

In order to do that we have to operate with three different models:

(i) a model for the (hypothetical) situation *without censoring*, that is, where all occurrences of the event of interest are observed

(ii) a *joint model* for the (hypothetical) situation where all occurrences of the event of interest *as well as* the censoring processes are observed

(iii) a model for the situation *with censoring*, that is, for the occurrences of the event actually observed

The parameters of interest are defined according to model (i), the concept of independent censoring is defined by means of model (ii), while model (iii) is the one used in the actual statistical inference; cf. Figure 2.1.

We start by considering model (i), that is, the (hypothetical) situation where all occurrences of the event of interest are observed. Let $N_1^c(t), \ldots, N_n^c(t)$ be the counting processes registering these occurrences for each of n individuals, *assuming*

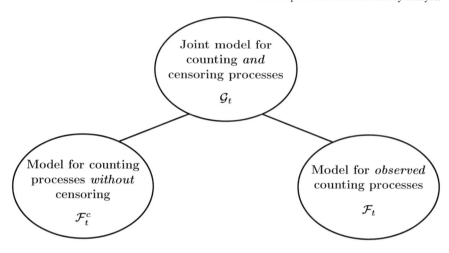

Fig. 2.1 *The three models that are involved in the definition of independent censoring. The model parameters are defined for the model without censoring, the conditions on the censoring are formulated for the joint model, while the statistical methods are derived and studied for the model for the observed counting processes.*

complete observation, and denote by $\mathscr{F}_t^c$ the information that would then have been available to the researcher by time t. The history $(\mathscr{F}_t^c)$ is generated by *all* the counting processes $N_i^c(t)$ and possibly also by covariate processes that run in parallel with the counting processes. We assume that the $(\mathscr{F}_t^c)$-intensity processes of the counting processes take the form

$$\lambda_i^{\mathscr{F}^c}(t) = Y_i^c(t)\alpha_i(t); \qquad i = 1,\ldots,n. \tag{2.52}$$

Here $Y_i^c(t)$ is a left-continuous $(\mathscr{F}_t^c)$-adapted indicator process that takes the value 1 if individual i *may* experience an event at time t, and is equal to 0 otherwise. (For example, we will have $Y_i^c(t) = 0$ if individual i has died by time t or if, in a multistate model, the individual is in a state from which the event of interest cannot take place.) The $\alpha_i(t)$ in (2.52) are our key model parameters and, as we will see in later chapters, a main aim of a statistical analysis is to infer how these (and derived quantities) depend on covariates and vary over time. It is important to note that the $\alpha_i(t)$ may be random and depend on covariates as well as previous occurrences of the event (through *the past* $\mathscr{F}_{t-}^c$).

The study of the $\alpha_i(t)$ is complicated by incomplete observation of the counting processes. To handle this, for each $i = 1,\ldots,n$ we introduce a left-continuous binary censoring process $Y_i^o(t)$ that takes the value 1 if individual i is under observation "just before" time t, and the value 0 otherwise. For the special case of right-censoring, $Y_i^o(t) = I\{t \leq C_i\}$ for a right-censoring time C_i. The observed counting processes are then given by

$$N_i(t) = \int_0^t Y_i^o(u)\,dN_i^c(u); \qquad i = 1, \ldots, n. \tag{2.53}$$

The censoring processes will often create some extra randomness, causing the observed counting processes $N_i(t)$ not to be adapted to the history $(\mathscr{F}_t^c)$. Then we can *not* define the intensity processes for the observed counting processes relative to the complete history $(\mathscr{F}_t^c)$.

To handle this problem, we have to consider the larger model (ii), that is, the *joint model* for the completely observed counting processes $N_i^c(t)$ *and* the censoring processes $Y_i^o(t)$. To this end we consider the larger history $(\mathscr{G}_t)$ generated by the complete history $(\mathscr{F}_t^c)$ *as well as* by the censoring processes. This history corresponds to the (hypothetical) situation where all occurrences of the event of interest are observed, and we *in addition* observe the censoring processes. We may then consider the $(\mathscr{G}_t)$-intensity processes $\lambda_i^{\mathscr{G}}(t)$ of the completely observed counting processes $N_i^c(t)$. If the censoring processes carry information on the likelihood of occurrence of the event, implying that individuals under observation have a different risk of experiencing the event than similar individuals that are not observed, the $\lambda_i^{\mathscr{G}}(t)$ will differ from the $(\mathscr{F}_t^c)$-intensity processes (2.52). We will, however, assume that this is not the case, so that the intensity processes relative to the two histories are the same:

$$\lambda_i^{\mathscr{G}}(t) = \lambda_i^{\mathscr{F}^c}(t); \qquad i = 1, \ldots, n. \tag{2.54}$$

When (2.54) is fulfilled, we say that *censoring is independent*.

Before we discuss the consequences of the independent censoring assumption (2.54), we will take a closer look at the assumption itself. To this end we concentrate on the situation with right-censoring, where n individuals are followed until observation stops at censoring or death (or, more generally, at the entry into an absorbing state). For this situation we have $Y_i^o(t) = I\{t \le C_i\}$ for (potential) right-censoring times C_i; $i = 1, \ldots, n$. A number of different right-censoring schemes are possible. For example:

- Censoring at fixed times that may differ between the individuals, that is, $C_i = c_i$ for given constants c_i (type I censoring).
- Censoring of all individuals at the time T when a specified number of occurrences have taken place, that is, $C_i = T$ for $i = 1, \ldots, n$ (type II censoring).
- Censoring of all individuals when the event has not occurred in a certain time interval.
- Censoring at random times C_i that may differ between the individuals and that are independent of the completely observed counting processes $N_i^c(t)$ (random censoring).

For the first three censoring schemes, the censoring times C_i are *stopping times* relative to the complete history $(\mathscr{F}_t^c)$. Therefore no additional randomness is introduced by the censoring, and there is no need to enlarge the history. Thus the histories $(\mathscr{G}_t)$ and $(\mathscr{F}_t^c)$ are the same, and the independent censoring assumption (2.54) is automatically fulfilled. For the last censoring scheme mentioned, additional randomness is introduced by the censoring. But as the (potential) censoring times C_i are assumed to be independent of the completely observed counting processes $N_i^c(t)$, the independent censoring assumption (2.54) holds for this censoring scheme as well.

We then consider the data actually observed, and we denote the history corresponding to the observed data by $(\mathscr{F}_t)$. The counting processes $N_i(t)$ given by (2.53) are observed, so these are adapted to $(\mathscr{F}_t)$. However, we do not necessarily observe the censoring processes $Y_i^o(t)$; e.g. we do not observe censoring after death. What is observed for each $i = 1, \ldots, n$ is the left-continuous process

$$Y_i(t) = Y_i^c(t) Y_i^o(t), \tag{2.55}$$

taking the value 1 if individual i is at risk for the event of interest "just before" time t and the value 0 otherwise. The processes $Y_i(t)$ are therefore adapted to $(\mathscr{F}_t)$ and, due to their left-continuity, they are in fact $(\mathscr{F}_t)$-predictable. We may then adopt an argument similar to the one used to derive the innovation theorem in the previous subsection, to find the $(\mathscr{F}_t)$-intensity processes $\lambda_i^{\mathscr{F}}(t)$ of the observed counting processes. Using (2.52), (2.53), (2.54) and (2.55) we obtain

$$\begin{aligned}
\lambda_i^{\mathscr{F}}(t)\,dt &= \mathrm{E}(\,dN_i(t)\,|\,\mathscr{F}_{t-}) = \mathrm{E}(\,Y_i^o(t)\,dN_i^c(t)\,|\,\mathscr{F}_{t-}) \\
&= \mathrm{E}\{\mathrm{E}(\,Y_i^o(t)\,dN_i^c(t)\,|\,\mathscr{G}_{t-})\,|\,\mathscr{F}_{t-}\} = \mathrm{E}\{Y_i^o(t)\,\mathrm{E}(\,dN_i^c(t)\,|\,\mathscr{G}_{t-})\,|\,\mathscr{F}_{t-}\} \\
&= \mathrm{E}\{Y_i^o(t)\,\lambda_i^{\mathscr{G}}(t)\,dt\,|\,\mathscr{F}_{t-}\} = \mathrm{E}\{Y_i^o(t)Y_i^c(t)\,\alpha_i(t)\,dt\,|\,\mathscr{F}_{t-}\} \\
&= Y_i(t)\,\mathrm{E}\{\alpha_i(t)\,|\,\mathscr{F}_{t-}\}\,dt
\end{aligned}$$

In order to progress further, an assumption on the observation of covariates is needed. Typically, the complete history $(\mathscr{F}_t^c)$ will be generated by the fully observed counting processes $N_i^c(t)$ as well as by covariate processes running in parallel with the counting processes, and the $\alpha_i(t)$ of (2.52) may depend on these covariates. For statistical inference on the $\alpha_i(t)$ to be feasible, we must assume that the covariates that enter into the specification of the $\alpha_i(t)$ are available to the researcher. A technical way of formulating this assumption is to assume that the $\alpha_i(t)$ are $(\mathscr{F}_t)$-predictable. Then the intensity processes of the observed counting processes take the form

$$\lambda_i^{\mathscr{F}}(t) = Y_i(t)\alpha_i(t); \qquad i = 1, \ldots, n. \tag{2.56}$$

Comparing this with (2.52), we see that *the form of the intensity processes is preserved under independent censoring.*

There is a close connection between drop-outs in longitudinal data and censoring for survival and event history data. In fact, *independent censoring* in survival and event history analysis is essentially the same as *sequential missingness at random* in longitudinal data analysis (e.g., Hogan et al., 2004).

2.3 Processes with continuous sample paths

In Section 2.1 we considered processes in discrete time, while the counting processes that were the focus of Section 2.2 have discrete state space. We will also consider processes where both time and state space are continuous. Examples of such processes are Wiener processes and Gaussian martingales. These have applications as models for underlying, unobserved processes and as limiting processes of stochastic integrals of counting process martingales.

2.3.1 The Wiener process and Gaussian martingales

The Wiener process, also called Brownian motion, has a similar fundamental character as the Poisson process. While the latter is the model of completely random events, the Wiener process is the model of completely random noise. In fact, the so-called white noise is a kind of derivative of the Wiener process. (This goes beyond the ordinary derivative, which is not valid here.)

Let $W(t)$ denote the value of the Wiener process at time t, and consider a time interval $(s,t]$. Then the increment $W(t) - W(s)$ over this interval is normally distributed with

$$\mathrm{E}\{W(t) - W(s)\} = 0 \quad \text{and} \quad \mathrm{Var}\{W(t) - W(s)\} = t - s.$$

Further, the Wiener process has continuous sample paths, and the increment of the Wiener process over a time interval is independent of its increments over all nonoverlapping intervals. Figure 2.2 shows one realization of the Wiener process.

The Wiener process may be modified in a number of ways; a drift may be introduced such that the process preferentially moves in a certain direction. Further extension of the Wiener process yields the diffusion processes. These extensions of the Wiener process are reviewed in Appendix A.4.

One role of the Wiener process and its allies is as models for underlying processes. By this we mean that an observed event reflects something that occurs on a deeper level. A divorce does not just happen; it is the end result of a long process of deterioration of a marriage. A myocardial infarction is the result of some disease process. Such underlying processes may be seen as diffusions. One would not usually observe them directly but could still use them as the basis for statistical models, which we shall show in Chapters 10 and 11.

Another role of the Wiener process is that time-transformations of the Wiener process arise as limits in a number of applications. In particular, this will be the case for stochastic integrals of counting process martingales, and this provides the basis for the necessary asymptotic theory of estimators and test statistics.

Let $V(t)$ be a strictly increasing continuous function with $V(0) = 0$, and consider the process $U(t) = W(V(t))$. The process U inherits the following properties from the Wiener process:

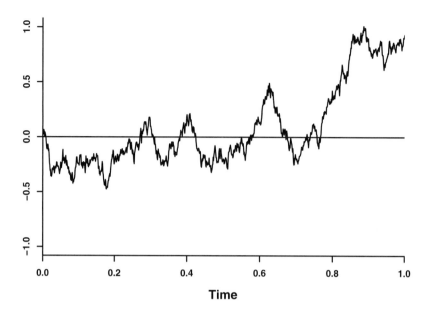

Fig. 2.2 *Simulation of a sample path of the Wiener process.*

- The sample paths are continuous.
- The increments over nonoverlapping intervals are independent.
- The increment over an interval $(s,t]$ is normally distributed with mean zero and variance $V(t) - V(s)$.

From these properties one may show that U is a mean zero martingale with predictable variation process $\langle U \rangle(t) = V(t)$ (Exercise 2.12). It is common to denote U a *Gaussian martingale*.

2.3.2 Asymptotic theory for martingales: intuitive discussion

We have pointed out a number of times that martingales are to be considered as processes of noise, or error, containing random deviations from the expected. In other parts of statistics, one is used to errors being approximately normally distributed. This holds for martingales as well. In fact there are central limit theorems for martingales that are closely analogous to those known for sums of independent random variables.

If we have a sequence of counting processes, where the number of jumps increases and gets more and more dense, the properly normalized associated martingales (or stochastic integrals with respect to these martingales) will converge to a limiting martingale with continuous sample path. The limiting martingale is closely connected to a Wiener process, or Brownian motion. In fact, if the predictable variation process of the limiting martingale equals a deterministic function $V(t)$, then the limiting martingale is exactly the Gaussian martingale discussed at the end of the previous subsection. This follows from the nice fact that a martingale with continuous sample paths is uniquely determined by its variation process.

So there are two things to be taken care of to ensure that a sequence of martingales converges to a Gaussian martingale:

(i) The predictable variation processes of the martingales shall converge to a deterministic function.

(ii) The sizes of the jumps of the martingales shall go to zero.

It is important to note that these two assumptions concern entirely different aspects of the sequence of processes. The first assumption implies a stabilization on the sample space of processes, while the second one is a requirement on the sample paths of the processes.

Figure 2.3 illustrates the convergence of a sequence of normalized counting process martingales to a Gaussian martingale. More specifically the figure shows $n^{-1/2}M(t)$ for $n = 10, 50, 250$, and 1250, where $M(t) = N(t) - \Lambda(t)$ is derived from censored Weibull survival data as described in Example 1.18 (for $n = 10$). In fact, the upper left-most panel of Figure 2.3 shows a normalized version of the martingale in the right-hand panel of Figure 1.14.

2.3.3 Asymptotic theory for martingales: mathematical formulation

There exist several versions of the central limit theorem for martingales that formalize requirements (i) and (ii) of the previous subsection. A very general and elegant theorem was formulated by Rebolledo (1980). The version of Rebolledo's theorem we present here is taken from Andersen et al. (1993, section II.5), where more details, and other versions of the conditions, can be found. Helland (1982) showed how Rebolledo's theorem in continuous time can be deduced from the simpler central limit theorem for discrete time martingales.

Let $\widetilde{M}^{(n)}$, $n \geq 1$, be a sequence of mean zero martingales defined on $[0, \tau]$, and let $\widetilde{M}_\varepsilon^{(n)}$ be the martingale containing all the jumps of $\widetilde{M}^{(n)}$ larger than a given $\varepsilon > 0$. Let $\xrightarrow{P}$ denote convergence in probability, and consider the conditions:

(i) $\langle \widetilde{M}^{(n)} \rangle(t) \xrightarrow{P} V(t)$ for all $t \in [0, \tau]$ as $n \to \infty$, where V is a strictly increasing continuous function with $V(0) = 0$.

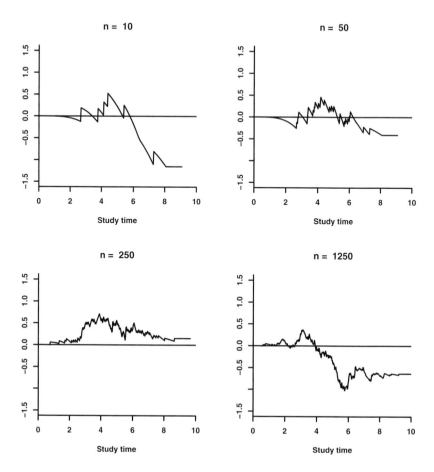

Fig. 2.3 *Illustration that a sequence of normalized counting process martingales converges to a Gaussian martingale (see text for details).*

(ii) $\langle \widetilde{M}_{\varepsilon}^{(n)} \rangle(t) \xrightarrow{P} 0$ for all $t \in [0, \tau]$ and all $\varepsilon > 0$ as $n \to \infty$.

Then, as $n \to \infty$, the sequence of martingales $\widetilde{M}^{(n)}$ converges in distribution to the mean zero Gaussian martingale U given by $U(t) = W(V(t))$.

Hence, under quite general assumptions there will be convergence in distribution to a limiting Gaussian martingale. Since the relevant statistics may also be functionals of the processes, many different probability distributions may arise from the theory. Note in particular that $\widetilde{M}^{(n)}(t)$ converges in distribution to a normally distributed random variable with mean zero and variance $V(t)$ for any given value of t.

We will use the martingale central limit theorem to derive the limiting behavior of (sequences of) stochastic integrals of the form $\int_0^t H^{(n)}(s)\, dM^{(n)}(s)$, where $H^{(n)}(t)$

is a predictable process and $M^{(n)}(t) = N^{(n)}(t) - \int_0^t \lambda^{(n)}(s)ds$ is a counting process martingale. More generally, we will consider sums of stochastic integrals

$$\sum_{j=1}^{k} \int_0^t H_j^{(n)}(s)\,dM_j^{(n)}(s),\tag{2.57}$$

where $H_j^{(n)}(t)$ is a predictable process for each n and

$$M_j^{(n)}(t) = N_j^{(n)}(t) - \int_0^t \lambda_j^{(n)}(s)ds$$

is a counting process martingale; $j = 1,\ldots,k$. For this situation, conditions (i) and (ii) take the form

$$\sum_{j=1}^{k} \int_0^t (H_j^{(n)}(s))^2 \lambda_j^{(n)}(s)\,ds \xrightarrow{P} V(t) \text{ for all } t \in [0, \tau],\tag{2.58}$$

$$\sum_{j=1}^{k} \int_0^t (H_j^{(n)}(s))^2 I\{|H_j^{(n)}(s)| > \varepsilon\}\lambda_j^{(n)}(s)\,ds \xrightarrow{P} 0 \text{ for all } t \in [0, \tau].\tag{2.59}$$

When we are going to use the martingale central limit theorem to show that an estimator or a test statistic converges in distribution, we have to check that conditions (2.58) and (2.59) hold. As we do not focus on regularity conditions in this book, we will not check the two conditions in detail in later chapters.

Assume that we may write $V(t) = \int_0^t v(s)ds$. Then, apart from some regularity conditions, a sufficient condition for (2.58) is that

$$\sum_{j=1}^{k} (H_j^{(n)}(s))^2 \lambda_j^{(n)}(s) \xrightarrow{P} v(s) > 0 \text{ for all } s \in [0, \tau], \text{ as } n \to \infty.\tag{2.60}$$

Furthermore, if k is fixed, a sufficient condition for (2.59) is that

$$H_j^{(n)}(s) \xrightarrow{P} 0 \text{ for all } j = 1,\ldots,k \text{ and } s \in [0, \tau], \text{ as } n \to \infty.\tag{2.61}$$

When $k = k_n$ is increasing with n (as is the case when $k = n$), it may be more involved to check condition (2.59), and we will not go into details on how this can be done.

To summarize: if (2.58) and (2.59) hold, then the sequence of sums of stochastic integrals $\sum_{j=1}^{k} \int_0^t H_j^{(n)}(s)\,dM_j^{(n)}(s)$, $n \geq 1$, converges in distribution to a mean zero Gaussian martingale with variance function $V(t) = \int_0^t v(s)ds$. In particular, for any given value of t, it converges in distribution to a normally distributed random variable with mean zero and variance $V(t)$. Under some regularity conditions, a sufficient condition for (2.58) is (2.60), and when k is fixed, a sufficient condition for (2.59) is (2.61).

In some applications in later chapters, we will need a multivariate version of the martingale central limit theorem. This is given in Appendix B.3.

2.4 Exercises

2.1 Show that (2.3) and (2.4) are equivalent. [*Hint:* By a general result for conditional expectations we have that $E(M_n \mid \mathscr{F}_{m_1}) = E\{E(M_n \mid \mathscr{F}_{m_2}) \mid \mathscr{F}_{m_1}\}$ for all $0 \le m_1 < m_2 < n$.]

2.2 Let $M = \{M_0, M_1, M_2, \ldots\}$ be a martingale. Show that $\mathrm{Cov}(M_m, M_n - M_m) = 0$ for all $n > m$. [*Hint:* Use the rule of double expectation to show that $\mathrm{Cov}(M_m, M_n - M_m) = E\{E(M_m(M_n - M_m) \mid \mathscr{F}_m)\}$.]

2.3 Let $M_n = \sum_{k=0}^{n} X_k$, where the $X_1, X_2, \ldots$ are independent random variables with zero mean and variance σ^2, and $X_0 = 0$.

a) Show that $M = \{M_0, M_1, M_2, \ldots\}$ is a martingale (w.r.t. the history generated by the process itself).
b) Use (2.7) to find the predictable variation process $\langle M \rangle$.
c) Use (2.8) to find the optional variation process $[M]$.

2.4 Prove (2.9), that is, that $M_0^2 - \langle M \rangle_0 = 0$ and that $E(M_n^2 - \langle M \rangle_n \mid \mathscr{F}_{n-1}) = M_{n-1}^2 - \langle M \rangle_{n-1}$.

2.5 Assume that the processes $M_1 = \{M_{10}, M_{11}, \ldots\}$ and $M_2 = \{M_{20}, M_{21}, \ldots\}$, with $M_{10} = M_{20} = 0$, are martingales with respect to the history $\{\mathscr{F}_n\}$. The *predictable covariation process* $\langle M_1, M_2 \rangle$ is for $n \ge 1$ defined by

$$\langle M_1, M_2 \rangle_n = \sum_{i=1}^{n} E\{\triangle M_{1i} \triangle M_{2i} \mid \mathscr{F}_{i-1}\} \tag{2.62}$$

$$= \sum_{i=1}^{n} \mathrm{Cov}(\triangle M_{1i}, \triangle M_{2i} \mid \mathscr{F}_{i-1}),$$

while $\langle M_1, M_2 \rangle_0 = 0$. The *optional covariation process* $[M_1, M_2]$ is defined by

$$[M_1, M_2]_n = \sum_{i=1}^{n} \triangle M_{1i} \triangle M_{2i} \tag{2.63}$$

for $n \ge 1$ and $[M_1, M_2]_0 = 0$.

a) Show that $M_1 M_2 - \langle M_1, M_2 \rangle$ and $M_1 M_2 - [M_1, M_2]_n$ are mean zero martingales.
b) Show that $\mathrm{Cov}(M_{1n}, M_{2n}) = E\langle M_1, M_2 \rangle_n = E[M_1, M_2]_n$ for all n.

2.6 Show that the process M^T defined by (2.11) is a martingale. [*Hint:* Use the martingale preservation property of the transformation $Z = H \bullet M$ with an appropriate choice of the predictable process H.]

2.7 Prove the statement (2.15). [*Hint:* Use the definition (2.8).]

2.8 Assume that the processes $M_1 = \{M_{10}, M_{11}, \dots\}$ and $M_2 = \{M_{20}, M_{21}, \dots\}$, with $M_{10} = M_{20} = 0$, are martingales with respect to the history $\{\mathscr{F}_n\}$, and let $H_1 = \{H_{10}, H_{11}, \dots\}$ and $H_2 = \{H_{20}, H_{21}, \dots\}$ be predictable processes. Show that

$$\langle H_1 \bullet M_1, H_2 \bullet M_2 \rangle_n = \sum_{s=1}^{n} H_{1s} H_{2s} \triangle \langle M_1, M_2 \rangle_s$$

$$[H_1 \bullet M_1, H_2 \bullet M_2]_n = \sum_{s=1}^{n} H_{1s} H_{2s} \triangle [M_1, M_2]_s.$$

[*Hint:* Use the definitions (2.62) and (2.63) in Exercise 2.5.]

2.9 Prove (2.37). [*Hint:* Use that $M^2(t) = N^2(t) - 2\lambda t N(t) + (\lambda t)^2$ and the properties of the Poisson process $N(t)$.]

2.10 Assume that the counting processes N_1 and N_2 do not jump simultaneously. Prove that the covariation processes $\langle M_1, M_2 \rangle$ and $[M_1, M_2]$ of the corresponding martingales are both identically equal to zero. [*Hint:* $N_1 + N_2$ is a counting process with $[M_1 + M_2] = N_1 + N_2$. The result follows from the analog to (2.28).]

2.11 Let $N(t)$ be an inhomogeneous Poisson process with intensity $\lambda(t)$. Then the number of events $N(t) - N(s)$ in the time interval $(s, t]$ is Poisson distributed with parameter $\int_s^t \lambda(u) \, du$, and the number of events in disjoint time intervals are independent. Let $\mathscr{F}_t$ be generated by $N(s)$ for $s \leq t$, and let $M(t) = N(t) - \int_0^t \lambda(u) \, du$.

a) Prove that $\mathrm{E}(M(t) | \mathscr{F}_s) = M(s)$ for all $s \leq t$, that is, that $M(t)$ is a martingale.
b) Prove that $\mathrm{E}\left(M^2(t) - \int_0^t \lambda(u) \, du \,|\, \mathscr{F}_s\right) = M^2(s) - \int_0^s \lambda(u) \, du$, that is, that $M^2(t) - \int_0^t \lambda(u) \, du$ is a martingale. Note that this shows that $\langle M \rangle(t) = \int_0^t \lambda(u) du$.

2.12 Let $W(t)$ be the Wiener process. Then the increment $W(t) - W(s)$ over the time interval $(s, t]$ is normally distributed with mean zero and variance $t - s$, and the increments over disjoint time intervals are independent. Let $V(t)$ be a strictly increasing continuous function with $V(0) = 0$, and introduce the stochastic process $U(t) = W(V(t))$. Finally let $\mathscr{F}_t$ be generated by $U(s)$ for $s \leq t$.

a) Prove that $\mathrm{E}(U(t) | \mathscr{F}_s) = U(s)$ for all $s \leq t$, that is, that $U(t)$ is a martingale.
b) Prove that $\mathrm{E}\left(U^2(t) - V(t) | \mathscr{F}_s\right) = U^2(s) - V(s)$, that is, that $U^2(t) - V(t)$ is a martingale. Note that this shows that $\langle U \rangle(t) = V(t)$.

Chapter 3
Nonparametric analysis of survival and event history data

In this chapter we study situations in which the data may be summarized by a counting process registering the occurrences of a specific event of interest or a few counting processes registering the occurrences of a few such events. One important example is the survival data situation, where one is interested in the time to a single event for each individual and the counting process is counting the number of occurrences of this event for a group of individuals. The event in question will depend on the study at hand; it may be the death of a laboratory animal, the relapse of a cancer patient, or the birth of a woman's second child. In order to emphasize that events other than death are often of interest in the survival data situation, we will use the term *event of interest* to denote the event under study. However, as mentioned in Section 1.1, we will use the terms *survival time* and *survival function* also when the event of interest is something different from death.

Another important example of the type of situations considered in this chapter is competing risks, where one wants to study two or more causes of death for a group of individuals (cf. Section 1.2.2). Here we have a counting process registering the total number of deaths from each cause. The competing risks model is a simple example of a Markov chain. More generally, we may use a Markov chain with a finite number of states to model the life histories for a group of individuals. The data may then be summarized by counting processes registering the total number of transitions between the states.

As discussed in Section 1.1.1, incomplete observation of survival times due to right-censoring, and possibly also left-truncation, are inevitable in most survival studies. Right-censoring occurs when for some individuals we only know that their survival times exceed certain censoring times, while left-truncation occurs when individuals are not under study from time zero (in the study time scale), but only from a later entry time. We may also have left truncation and right-censoring for Markov chain models, meaning that the individuals are not necessarily followed from time zero (due to left-truncation) and that the follow-up time may differ between individuals (due to right-censoring). Throughout, it is a crucial assumption that censoring and truncation are independent in the sense discussed in Sections 1.4.2, 1.4.3, and 2.2.8.

Most of the situations we encounter in this chapter fall into the following framework. For each of n individuals we have a counting process registering the number of occurrences of an event of interest; specifically $N_i(t)$ is the observed number of events for individual i in the time interval $[0,t]$. The intensity process of $N_i(t)$ is assumed to take the form

$$\lambda_i(t) = \alpha(t)Y_i(t), \tag{3.1}$$

where $\alpha(t)$ is a nonnegative function, while $Y_i(t) = 1$ if individual i is at risk "just before" time t and $Y_i(t) = 0$ otherwise. We summarize the data by aggregating the individual counting processes, that is, we consider the process $N(t) = \sum_{i=1}^{n} N_i(t)$ counting the total number of observed events. Assuming no tied event times, this aggregated counting process has the intensity process

$$\lambda(t) = \sum_{i=1}^{n} \lambda_i(t) = \alpha(t)Y(t), \tag{3.2}$$

where $Y(t) = \sum_{i=1}^{n} Y_i(t)$ is the number at risk "just before" time t. Counting processes obtained by counting the total number of events in the survival data situation, or the number of transitions between two states in a Markov chain, are of this form; cf. Sections 1.4.2 and 1.4.3.

In Section 3.1 we define and study the Nelson-Aalen estimator for $A(t) = \int_0^t \alpha(u)du$ and describe its use for a number of different situations. For the special case of survival data, $\alpha(t)$ is the hazard rate and $S(t) = \exp\{-A(t)\}$ is the survival function. In Section 3.2 we derive the Kaplan-Meier estimator for $S(t)$ and study its properties as well as its relation to the Nelson-Aalen estimator. Nonparametric tests for comparing the α-functions for two or more counting processes are considered in Section 3.3, and in Section 3.4 we consider an estimator for the matrix of transition probabilities in a finite Markov chain that is a generalization of the Kaplan-Meier estimator.

3.1 The Nelson-Aalen estimator

The Nelson-Aalen estimator is a nonparametric estimator that may (among other things) be used to estimate the cumulative hazard rate from censored survival data. Since no distributional assumptions are needed, one important use of the estimator is to check graphically the fit of parametric models, and this is why it was originally introduced for engineering applications by Nelson (1969, 1972). Independently of Nelson, Altshuler (1970) derived the Nelson-Aalen estimator in the context of competing risks animal experiments. Later, by adopting a counting process formulation, Aalen (1975, 1978b) extended the use of the estimator to Markov chains and other event history models, and he studied its small and large sample properties using martingale methods. The estimator is nowadays usually denoted the Nelson-Aalen estimator, although other names (the Nelson estimator, the Altshuler estimator, the empirical cumulative hazard estimator) are sometimes used.

In this section we first describe the Nelson-Aalen estimator and illustrate its use in the survival data situation. Then we consider the multiplicative intensity model for counting processes (cf. Section 1.5.1), which is the class of models for which the Nelson-Aalen estimator applies, and give a number of examples of use of the estimator. The counting process formulation is then used to study the small and large sample properties of the Nelson-Aalen estimator.

In most textbooks on survival analysis, the reader will encounter the Kaplan-Meier estimator before the Nelson-Aalen estimator. We have chosen to do it the other way around and to leave the Kaplan-Meier estimator until Section 3.2. The reason for this is that within the framework of this book, the Nelson-Aalen estimator is the most basic of the two. Moreover, the Nelson-Aalen estimator can be applied in a number of situations where the Kaplan-Meier estimator does not make sense, and as we will see in Section 3.2.4, the Kaplan-Meier estimator can be derived as a function of the Nelson-Aalen estimator.

3.1.1 The survival data situation

We first consider the survival data situation where one wants to study the time to occurrence of an event of interest for the individuals in a population. In this situation, the hazard rate $\alpha(t)$ specifies the instantaneous probability of the event, that is, for a randomly chosen individual, $\alpha(t)dt$ is the probability that the event will occur in the small time interval $[t, t+dt)$ given that it has not occurred earlier.

Assume that we have a sample of n individuals from the population. Our observation of the survival times for these individuals will typically be subject to right-censoring, and possibly also to left-truncation. In our general presentation, we will assume that there are no tied event times, leaving a brief discussion of the handling of ties to Section 3.1.3. Then the counting process N recording the occurrences of the event has intensity process of the multiplicative form (3.2) with $\alpha(t)$ the hazard rate and $Y(t)$ the number of individuals at risk at time t (or rather "just before" time t). Thus $N(t)$ is the number of occurrences of the event observed in $[0, t]$, while the jump times $T_1 < T_2 < \cdots$ of N are the ordered times when an occurrence of the event is observed.

Without any parametric assumptions, the hazard rate $\alpha(t)$ can be essentially any nonnegative function, and this makes it difficult to estimate. However, it turns out that it is easy to estimate the cumulative hazard

$$A(t) = \int_0^t \alpha(s)ds \qquad (3.3)$$

without assuming any structure on $\alpha(t)$. This is akin to estimating the cumulative distribution function, which is far easier than estimating a density function. The result is the Nelson-Aalen estimator, which is given as

$$\widehat{A}(t) = \sum_{T_j \leq t} \frac{1}{Y(T_j)}. \tag{3.4}$$

Note that the Nelson-Aalen estimator is an increasing right-continuous step function with increments $1/Y(T_j)$ at the observed event times.

In Section 3.1.5 we will give a formal derivation of the Nelson-Aalen estimator in the general setting of the multiplicative intensity model. Here we just give an intuitive justification of the estimator for censored survival data. The argument goes as follows. Partition the interval $[0,t]$ into small intervals, a typical interval being $[s, s+ds)$, and assume that each interval contains at most one observed event. The contribution to the cumulative hazard (3.3) over the interval $[s, s+ds)$ is $\alpha(s)ds$, which is the conditional probability of the occurrence of the event in the interval given that it has not happened before time s. If no event is observed in $[s, s+ds)$ then the natural estimate of $\alpha(s)ds$ will simply be zero. If one event is observed at time $T_j \in [s, s+ds)$, the natural estimate of $\alpha(s)ds$ will be one divided by the number of individuals at risk, that is $1/Y(s) = 1/Y(T_j)$. Aggregating these contributions over all time intervals, we arrive at $\widehat{A}(t)$ as a sensible estimator of $A(t)$.

In Section 3.1.5 we derive the estimator

$$\widehat{\sigma}^2(t) = \sum_{T_j \leq t} \frac{1}{Y(T_j)^2} \tag{3.5}$$

for the variance of the Nelson-Aalen estimator. Further, in Section 3.1.6 we show that the Nelson-Aalen estimator, evaluated at a given time t, is approximately normally distributed in large samples. Therefore, a standard $100(1-\alpha)\%$ confidence interval for $A(t)$ takes the form

$$\widehat{A}(t) \pm z_{1-\alpha/2}\widehat{\sigma}(t), \tag{3.6}$$

with $z_{1-\alpha/2}$ the $1-\alpha/2$ fractile of the standard normal distribution. The approximation to the normal distribution is improved by using a log-transformation giving the confidence interval

$$\widehat{A}(t) \exp\left\{\pm z_{1-\alpha/2}\widehat{\sigma}(t)/\widehat{A}(t)\right\} \tag{3.7}$$

(Exercise 3.3). Note that the two confidence intervals should be given a pointwise interpretation. The confidence interval (3.7) is satisfactory for quite small sample sizes (Bie et al., 1987). A confidence interval with small sample properties comparable to (3.7), or even slightly better, may be obtained by using an arcsine-transformation (Bie et al., op. cit.).

A Nelson-Aalen plot estimates the cumulative hazard, and hence its "slope" estimates the hazard rate itself. When interpreting a plot of the Nelson-Aalen estimator, we therefore focus on the "slope" of the plot; cf. the following examples. In this connection it is worth noting that the "slope" at one point in time can be interpreted independently of the "slope" at another time point. This is due to the fact, pointed

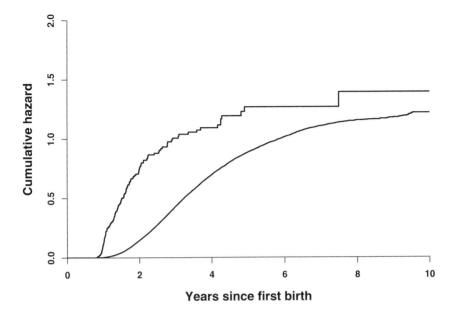

Fig. 3.1 *Nelson-Aalen estimates for the time between first and second births. Lower curve: first child survived one year; upper curve: first child died within one year.*

out in Section 3.1.5, that the increments of the Nelson-Aalen estimator over disjoint time intervals are uncorrelated.

Example 3.1. Time between births. We will use data from the Medical Birth Registry of Norway (cf. Example 1.1) to study the time between first and second births of a woman. In particular we will study how the time between the births is affected if the first child dies within one year of its birth. To this end we will use data on the 53 558 women who had their first birth in the period 1983–97, and where both parents were born after the establishment of the Birth Registry in 1967. Of these women, 262 lost their first child within one year of its birth.

Figure 3.1 gives Nelson-Aalen estimates for the time until second birth (by the same father) for the women whose first child survived one year and for the women who lost the child within one year, while Figure 3.2 gives the Nelson-Aalen estimate for the latter group of women with 95% log-transformed confidence intervals.

By considering the "slopes" of the Nelson-Aalen plots in Figure 3.1, we see that the hazard rate for the second birth is much larger between one and two years for the women who lost their first child than it is for the women who did not experience this traumatic event. More specifically, the average slopes of the Nelson-Aalen estimates between one and two years are 0.70 and 0.15, respectively, for the two groups of

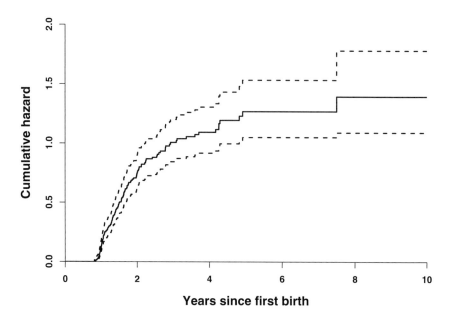

Fig. 3.2 *Nelson-Aalen estimates for the time between first and second births with log-transformed 95% confidence intervals for women who lost their first child within one year of its birth.*

women. Thus for the women who lost their first child, the hazard rate is about 0.70 per year between one and two years, while it is only about 0.15 per year for the women whose first child survived one year. After about two years, the Nelson-Aalen plots are fairly parallel, however, the average slope of both estimates being about 0.20 per year between two and four years after the birth of the first child.

The difference between the first two years and the period between two and four years after the first birth may be shown more clearly by making separate plots of the Nelson-Aalen estimates for the first two years and the increments of the Nelson-Aalen estimates between two and four years after the first birth. The resulting plots are shown in Figure 3.3. As mentioned before the example, these plots can be interpreted independently of each other, and they clearly show how the hazard for the women who lost their first child is increased until about two years after the first birth, but not thereafter.

In conclusion, therefore, our analysis shows that many women who lose their first child will quickly have another one. However, those who have not succeeded in about two years time are not more likely to have another child than the women whose first child survived one year. □

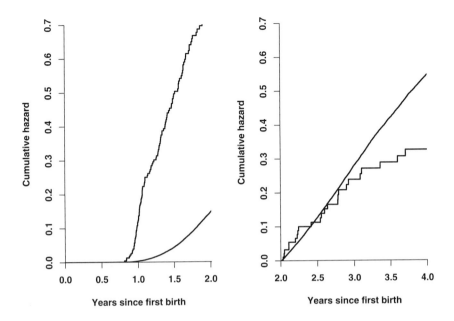

Fig. 3.3 *Nelson-Aalen estimates for the time between first and second births. Left panel: first two years after the first birth; right panel: two to four years after the first birth. Smooth curve: first child survived one year; irregular curve: first child died within one year.*

Example 3.2. Time between births. We consider the same data set as in the previous example. But here we restrict attention to the 16 116 women who had (at least) two births in the period 1983–97 and where both children survived at least one year. Of these 16 116 women, 4334 had two sons, 3759 had two daughters, 4067 had first a son and then a daughter, and 3956 had first a daughter and then a son. We will study the time until the third birth for these women (with the same father as the first two), and we will investigate whether the genders of the two older children affect the likelihood of a woman having a third birth.

   Figure 3.4 shows Nelson-Aalen plots for the cumulative hazard for the time between the second and third births depending on the genders of the two older children. For all four groups of women, the Nelson-Aalen plots are fairly linear corresponding to a constant hazard rate of the time to birth of the third child. Moreover, the plots roughly separate into two groups, with a hazard rate of about 0.12 per year for the women who have two children of the same gender and with a hazard rate of about 0.09 per year for those who have one child of each gender. Thus there are indications that the genders of the two older children affect the likelihood of a woman having a third birth. More specifically, there seems to be a slight preference among Norwegian couples to have (at least) one child of each gender. □

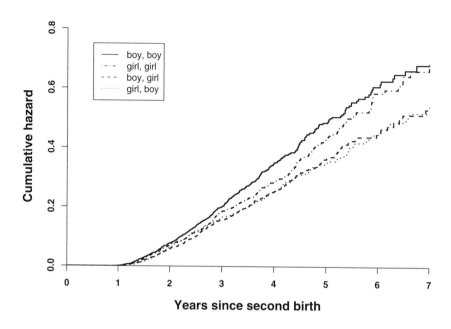

Fig. 3.4 *Nelson-Aalen estimates for the time between the second and third births depending on the gender of the two older children.*

The statistical properties of the Nelson-Aalen estimator (3.4) can be derived by classical methods (i.e., avoiding counting processes and martingales), but this is quite complicated, and unnecessarily strong conditions have to be made (Aalen, 1976). It is much simpler to use a counting process formulation of the model and to study the estimator using martingale methods (Sections 3.1.5 and 3.1.6). This not only has the benefit of making the derivations simpler; as we will see in the next subsection, it also extends the use of the Nelson-Aalen estimator beyond the survival data situation.

3.1.2 The multiplicative intensity model

We recall from Section 1.5.1 that a counting process $N(t)$ fulfills the multiplicative intensity model provided its intensity process $\lambda(t)$ takes the form

$$\lambda(t) = \alpha(t) Y(t) \tag{3.8}$$

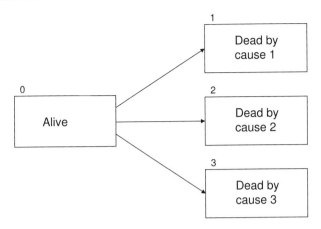

Fig. 3.5 *A model for competing risks with* $k = 3$.

with $\alpha(t)$ a nonnegative parameter function and $Y(t)$ a predictable process that does not depend on unknown parameters. In Section 3.1.5 we show that for the multiplicative intensity model we may use the Nelson-Aalen estimator (3.4), with $T_1 < T_2 < \cdots$ the ordered jump times of the counting process, to estimate $A(t) = \int_0^t \alpha(u)du$. Also the variance estimator (3.5) applies for the multiplicative intensity model.

The most typical example of the multiplicative intensity model is the standard model for survival data considered in the previous subsection. Here $Y(t)$ is the number of individuals at risk at time t, while $\alpha(t)$ is the hazard rate (cf. Section 1.4.2). There are also a number of other important examples of the multiplicative intensity model, and we now take a closer look at some such examples.

Example 3.3. Competing risks and causes of death in Norway. Assume that we want to study the time to death and the cause of death for the individuals in a population. As described in Section 1.2.2, this situation with competing causes of death may be modeled by a Markov chain with one transient state 0, corresponding to "alive," and k absorbing states corresponding to "dead by cause h," $h = 1, 2, \ldots, k$. Figure 3.5 gives an illustration for $k = 3$ competing causes of death. The transition intensity from state 0 to state h is denoted $\alpha_{0h}(t)$ and describes the instantaneous risk of dying from cause h, that is, $\alpha_{0h}(t)dt$ is the probability that an individual will die of cause h in the small time interval $[t, t+dt)$ given that the individual is still alive just prior to t. The $\alpha_{0h}(t)$ are also termed *cause-specific hazard rates*.

Based on a sample of n individuals from a population, let $N_{0h}(t)$ be the process counting the number of individuals who are observed to die from cause h (i.e., make a transition from state 0 to state h) in the interval $[0,t]$, and denote by $Y_0(t)$ the number of individuals at risk (i.e., in state 0) just prior to time t. Then $N_{0h}(t)$ has an intensity process of the multiplicative form $\alpha_{0h}(t)Y_0(t)$ (cf. Section 1.4.3), and we may use the Nelson-Aalen estimator [with $N(t) = N_{0h}(t)$ and $Y(t) = Y_0(t)$] to estimate the cumulative cause-specific hazard $A_{0h}(t) = \int_0^t \alpha_{0h}(u)du$.

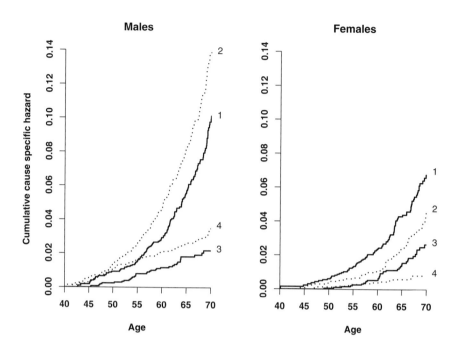

Fig. 3.6 *Nelson-Aalen estimates of the cumulative cause-specific hazard rates for four causes of death among middle-aged Norwegian males (left) and females (right). 1) Cancer; 2) cardiovascular disease including sudden death; 3) other medical causes; 4) alcohol abuse, chronic liver disease, and accidents and violence.*

As an illustration, we estimate cause-specific mortality among middle-aged Norwegian males and females in the period 1974–2000 using the data from three Norwegian counties described in Example 1.12. Because there are few individuals at risk below the age of 40 years, we have for our analysis left-truncated the data at 40 years. Moreover, as we focus on middle-aged death, we also right-censor all observations at the age of 70 years (if the individual is not already dead or censored by that age).

Figure 3.6 gives the Nelson-Aalen plots for the cause-specific hazards, with cause of death classified into the four broad groups:

1) Cancer
2) Cardiovascular disease including sudden death
3) Other medical causes
4) Alcohol abuse, chronic liver disease, and accidents and violence

For males, except for age below 50 years or so, cardiovascular disease is the leading cause of death, followed by cancer. Below about 50 years of age, the leading cause of death among males is alcohol abuse, chronic liver disease, and accidents and violence. Beyond this age, this cause of death has about the same cause specific hazard as other medical causes (the two Nelson-Aalen plots being fairly parallel). For females, the picture is quite different. Here, cancer is the leading cause of death, followed by cardiovascular disease and other medical causes, while the cause specific hazard for alcohol abuse, chronic liver disease, and accidents and violence is quite low in all ages. In conclusion, for the period 1974–2000, middle-aged Norwegian males have had a much higher mortality than females due to cardiovascular disease and alcohol abuse, chronic liver disease, and accidents and violence, while the difference between the two genders has been much smaller for deaths due to cancer and other medical causes. □

The competing risks model is an example of a simple Markov chain that finds applications in biostatistics. Another Markov chain of importance for biostatistical research is the illness-death model with states "healthy," "diseased," and "dead." More generally, we may consider any Markov chain with a finite number of states that may be used to model the life histories of the individuals in a population (cf. Section 1.2.2). For such a model, the transition intensity $\alpha_{gh}(t)$ describes the instantaneous risk of transition from state g to state h, that is, $\alpha_{gh}(t)dt$ is the probability that an individual who is in state g just before time t will make a transition to state h in the small time interval $[t, t + dt)$.

Suppose we have a sample of n individuals from the population under study. The individuals may be followed over different periods of time, so our observations of their life histories may be subject to left-truncation and right-censoring. We let $N_{gh}(t)$ count the number of observed transitions from state g to state h in $[0,t]$, and let $Y_g(t)$ be the number of individuals in state g just prior to time t. Then $N_{gh}(t)$ has intensity process of the multiplicative form $\alpha_{gh}(t)Y_g(t)$ (cf. Section 1.4.3), and we may use the Nelson-Aalen estimator [with $N(t) = N_{gh}(t)$ and $Y(t) = Y_g(t)$] to estimate the cumulative transition intensity $A_{gh}(t) = \int_0^t \alpha_{gh}(u)du$. In Section 3.4 we will see how such Nelson-Aalen estimators provide the building blocks for estimators of the transition probabilities (1.9).

For survival data and Markov chain models, the multiplicative intensity model (3.8) applies with $Y(t)$ the number at risk for the event in question. The multiplicative intensity model is, however, not restricted to such situations.

Example 3.4. Relative mortality and hip replacements. We consider the situation with right censored and/or left-truncated survival data with death as the event of interest. However, contrary to Section 3.1.1, we do not assume that all individuals share the same hazard rate $\alpha(t)$. Rather, we assume that the hazard rate $\alpha_i(t)$ of the ith individual may be written as the product

$$\alpha_i(t) = \alpha(t)\mu_i(t), \qquad (3.9)$$

where $\alpha(t)$ is a relative mortality common to all individuals, while $\mu_i(t)$ is a *known* mortality rate (hazard rate) at time t for a person from an external standard population corresponding to subject i (e.g., of the same sex and age as individual i). Then by (1.25) the process $N(t)$, counting the observed number of deaths in $[0,t]$, has

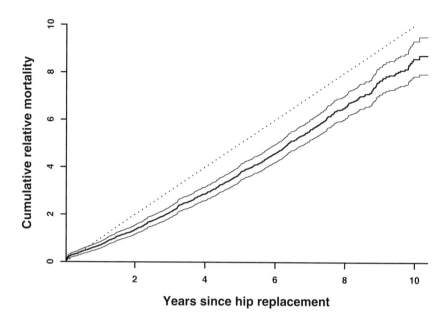

Fig. 3.7 *Nelson-Aalen estimate of the relative cumulative relative mortality with 95% standard confidence intervals for patients who have had a hip replacement in Norway in the period 1987– 97. A dotted line with unit slope is included for easy reference.*

an intensity process of the multiplicative form $\alpha(t)Y(t)$ with $Y(t) = \sum_{i=1}^{n} Y_i(t)\mu_i(t)$. Thus $Y(t)$ in the multiplicative intensity model (3.8) is no longer the number of individuals at risk, but rather the sum of the external rates for the individuals at risk. If we denote by $T_1 < T_2 < \cdots$ the times when deaths are observed, it follows that

$$\widehat{A}(t) = \sum_{T_j \le t} \frac{1}{\sum_{i=1}^{n} Y_i(T_j)\mu_i(T_j)} \tag{3.10}$$

is the Nelson-Aalen estimator for the cumulative relative mortality $A(t) = \int_0^t \alpha(u)du$.

As an illustration, we will use data from the Norwegian Arthroplasty Registry on 5000 patients – 3503 females and 1497 males – who had their first total hip replacement operation at a Norwegian hospital between September 1987 and February 1998; cf. Example 1.5. Denote by $\mu_m(a)$ and $\mu_f(a)$ the mortality at age a in the general Norwegian population, as it was reported by Statistics Norway for the year 1993. We will assume that the mortality rate (hazard rate) for the ith hip replacement patient t years after the operation may be given as

$$\alpha_i(t) = \alpha(t)\mu_{s_i}(a_i + t),$$

where s_i is the gender of the patient (f or m) and a_i is its age at operation. This is of the form (3.9) with $\mu_i(t) = \mu_{s_i}(a_i + t)$. We may therefore use (3.10) to estimate the cumulative relative mortality of the hip replacement patients.

The estimated cumulative mortality for the hip replacement patients is shown in Figure 3.7 with 95% standard confidence limits. For easy reference we have included a line with unit slope in the figure. For the first 5 to 6 years after the operation, the slope of the Nelson-Aalen estimate is about 0.75, clearly smaller than that of the unit line (except for a short period just after the operation). Thus the mortality of the hip replacement patients is smaller than that of the general population for the first 5 to 6 years after the operation. The main reason for this is that a patient needs to be in a fairly good health condition to be eligible for a hip replacement operation.

A careful look at Figure 3.7 reveals that the Nelson-Aalen estimate is above the unit line just after the operation. Figure 3.8 shows the estimate for the first six months after operation. The slope of the Nelson-Aalen estimate is about 2.7 per year the first month after operation, but thereafter it becomes smaller than 1. Thus the hip replacement patients have an increased mortality compared to the general population in the first month or so, but not thereafter. □

For the situations we have considered so far, the counting process $N(t)$ is obtained by aggregating individual counting processes, as described in the introduction to the chapter. We finally consider two situations in which this is not the case.

Example 3.5. An epidemic model. A simple model for the spread of an infectious disease in a community is the following. At the start of the epidemic, that is, at time $t = 0$, some individuals make contact with individuals from elsewhere and are thereby infected with the disease. There are no further infections from outside the community during the course of the epidemic. Let $N(t)$ count the number of new infections in $[0,t]$, and let $s(t)$ and $i(t)$ denote the number of susceptibles and infectives, respectively, just prior to time t. Assuming random mixing, the infection intensity in the community at time t, that is, the intensity process of $N(t)$, becomes $\lambda(t) = \alpha(t)s(t)i(t)$, where $\alpha(t)$ is the infection rate per possible contact. This intensity process is of the multiplicative form (3.8) with $Y(t) = s(t)i(t)$. See Becker (1989, section 7.6) for an illustration and further details. □

Example 3.6. Mating of Drosophila flies. In Table 3.1 we give the results of two (unpublished) mating experiment for Drosophila flies conducted by F. B. Christiansen in 1969. One of the purposes of the experiments was to investigate whether the mating pattern varies between different races of Drosophila flies. In each of the experiments, 30 female virgin flies and 40 virgin male flies were put in a plastic bowl covered with a transparent lid, called a "pornoscope." The flies were observed continuously, and times of initiatings of matings were recorded. In the two experiments considered here, both the female and male flies were of the same race: either a black ebony race or a yellow oregon race. We will study whether the mating pattern differs between these two races.

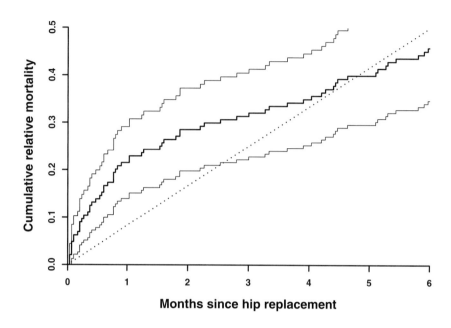

Fig. 3.8 *Nelson-Aalen estimate of the relative cumulative relative mortality with 95% standard confidence intervals for the first six months after the operation for patients who have had a hip replacement in Norway in the period 1987–97.*

Table 3.1 Times of initiation of matings in two pornoscope experiments (in seconds after insertion into the pornoscope).

Ebony	143	180	184	303	380	431	455	475	500	514
	521	552	558	606	650	667	683	782	799	849
	901	995	1131	1216	1591	1702	2212			
Oregon	555	742	746	795	934	967	982	1043	1055	1067
	1081	1296	1353	1361	1462	1731	1985	2051	2292	2335
	2514	2570	2970							

Consider one mating experiment, and let $N(t)$ be the number of matings initiated in the interval $[0,t]$. Then $f(t) = 30 - N(t-)$ and $m(t) = 40 - N(t-)$ are the numbers of virgin female and male flies, respectively, just before time t. If we assume random mating and that both female and male flies mate just once, the mating intensity in the experiment, that is, the intensity process of $N(t)$, becomes $\lambda(t) = \alpha(t)f(t)m(t)$, where $\alpha(t)$ is the mating rate per possible contact between a virgin female fly and a virgin male fly. This intensity process is of the multiplicative form (3.8) with $Y(t) = f(t)m(t)$. Therefore, if we denote by $T_1 < T_2 < \cdots$ the times when matings

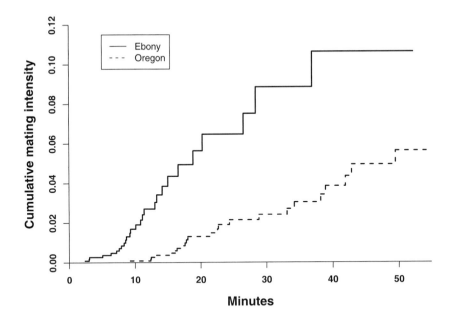

Fig. 3.9 *Nelson-Aalen estimates of the cumulative mating intensities for two races of Drosophila melanogaster.*

are initiated, we may estimate the cumulative mating intensity by the Nelson-Aalen estimator

$$\widehat{A}(t) = \sum_{T_j \leq t} \frac{1}{f(T_j)\, m(T_j)}.$$

Figure 3.9 shows Nelson-Aalen plots for the two races. From the plots we see that mating for the ebony flies starts earlier and takes place at a higher rate than is the case for the oregon flies.

The Drosophila flies mating data were earlier used for illustration by Aalen (1978b) and Andersen et al. (1993); see the latter reference for a more extensive data set and statistical analysis. □

3.1.3 Handling of ties

So far we have assumed that there are no tied event times, in agreement with our time-continuous model for the counting process N. However, in practice some tied event times are often present in a set of data, and we need to know how these should be handled. (Note that ties between censoring times and ties between an event time and a censoring time cause no problems. For such ties we adopt the common convention in survival and event history analysis: that event times precede censoring times.)

As we have seen, the multiplicative intensity model (3.8) covers a number of different counting process models. It is therefore difficult to give one recipe for the handling of ties that covers all possible situations. We will therefore concentrate on the situation described at the introduction to the chapter, where the process $N(t)$ is obtained by aggregating individual counting processes, and where $Y(t)$ is the number at risk "just before" time t. For this situation, we may deal with tied event times by two conceptually different approaches:

(i) According to the first approach, one assumes that the events actually happen in continuous time, and that in reality no two event times coincide. However, due to grouping and rounding, the recorded event times may be tied.

(ii) According to the second approach, one assumes that the data are genuinely discrete, so that tied event times are real and not due to grouping and rounding.

We will describe the tie corrections to which these two approaches give rise. To this end, we write the Nelson-Aalen estimator (3.4) as

$$\widehat{A}(t) = \sum_{T_j \le t} \triangle \widehat{A}(T_j). \tag{3.11}$$

For a time T_j when a single event is observed, we have $\triangle \widehat{A}(T_j) = 1/Y(T_j)$ in accordance with (3.4). However, when $d_j \ge 2$ events are observed at time T_j, a modification for ties is needed.

According to approach (i), the tied event times are due to rounding and grouping. It is therefore reasonable to use the estimate one would have gotten if they were (slightly) different. Thus approach (i) gives

$$\triangle \widehat{A}(T_j) = \sum_{l=0}^{d_j-1} \frac{1}{Y(T_j) - l}. \tag{3.12}$$

For approach (ii), we assume that the event times are genuinely discrete. Then it is reasonable to use the estimate

$$\triangle \widehat{A}(T_j) = \frac{d_j}{Y(T_j)}. \tag{3.13}$$

We also need to consider the estimator of the variance of the Nelson-Aalen estimator. This may be written

$$\widehat{\sigma}^2(t) = \sum_{T_j \le t} \triangle \widehat{\sigma}^2(T_j), \tag{3.14}$$

where $\triangle\widehat{\sigma}^2(T_j) = 1/Y(T_j)^2$ when a single event is observed at time T_j in accordance with (3.5). In the presence of tied event times, approach (i) gives

$$\triangle\widehat{\sigma}^2(T_j) = \sum_{l=0}^{d_j-1} \frac{1}{(Y(T_j)-l)^2}. \tag{3.15}$$

For approach (ii), a reasonable tie-corrected estimator takes the form

$$\triangle\widehat{\sigma}^2(T_j) = \frac{(Y(T_j)-d_j)\,d_j}{Y(T_j)^3}; \tag{3.16}$$

cf. the argument in Andersen et al. (1993, pages 180–181).

For practical purposes, the numerical difference between the two approaches to tie correction is usually quite small, and it is not very important which of the two one adopts.

3.1.4 Smoothing the Nelson-Aalen estimator

The Nelson-Aalen estimator (3.4) is an estimator of the *cumulative* α-function in the multiplicative intensity model (3.8). However, it is not the cumulative α-function, but $\alpha(t)$ itself that is the function of main interest. For many applications it is sufficient to estimate $\alpha(t)$ indirectly by looking at the slope the Nelson-Aalen plot as we do in Examples 3.1–3.6. But sometimes it may be useful to have an explicit estimate of $\alpha(t)$. Such an estimate may be obtained by smoothing the Nelson-Aalen estimator.

A number of methods may be applied to smooth the Nelson-Aalen estimator; see, e.g., Wang (2005) for a review. Here we will restrict our attention to kernel function smoothing. Then $\alpha(t)$ is estimated by a weighted average of the Nelson-Aalen increments $\triangle\widehat{A}(T_j)$ over an interval $[t-b, t+b]$:

$$\widehat{\alpha}(t) = \frac{1}{b}\sum_{T_j} K\left(\frac{t-T_j}{b}\right) \triangle\widehat{A}(T_j). \tag{3.17}$$

Here b is a *bandwidth*, while the *kernel function* $K(x)$ is a bounded function that vanishes outside $[-1, 1]$ and has integral 1. Examples of kernel functions are:

- The *uniform* kernel: $K(x) = 1/2$
- The *Epanechnikov* kernel: $K(x) = 3(1-x^2)/4$
- The *biweight* kernel: $K(x) = 15(1-x^2)^2/16$

Note that the formulas apply for $|x| \leq 1$ and that all the three kernel functions take the value 0 when $|x| > 1$.

Since the increments of the Nelson-Aalen estimator are uncorrelated, an estimator of the variance of $\widehat{\alpha}(t)$ is simply obtained as

$$\widehat{\mathrm{Var}}\left(\widehat{\alpha}(t)\right) = \frac{1}{b^2}\sum_{T_j} K\left(\frac{t-T_j}{b}\right)^2 \triangle\widehat{\sigma}^2(T_j), \tag{3.18}$$

where the $\triangle\widehat{\sigma}^2(T_j)$ are the increments of the variance estimator (3.5). Ramlau-Hansen (1983) was the first to study the kernel function estimator (3.17) within the framework of the multiplicative intensity model (3.8). He used counting process and martingale theory to derive the variance estimator (3.18) and to prove that $\widehat{\alpha}(t)$ is approximately normally distributed in large samples; see Andersen et al. (1993, section IV.2) for further details.

The choice of bandwidth b is critical for the smoothness of the kernel function estimate. There exist objective methods for making this choice (based on cross-validation and plug-in methods) that provide an optimal trade-off between bias and variablity (e.g., Wang, 2005). We will, however, not pursue this point here.

Another problem is how to estimate $\alpha(t)$ for the left-hand "tail," that is, for values of t in the time interval $[0,b]$. A solution is to adopt a boundary kernel $K_q(x)$ with support $[-1,q]$ and estimate $\alpha(t)$ for $t \in [0,b]$ by

$$\widehat{\alpha}(t) = \frac{1}{b}\sum_{T_j} K_q\left(\frac{t-T_j}{b}\right)\triangle\widehat{A}(T_j) \tag{3.19}$$

with $q = t/b$. Such boundary kernels were first considered by Gasser and Müller (1979) in the context of nonparametric regression, and they were used by Keiding and Andersen (1989) to smooth the Nelson-Aalen estimates for the cumulative transition intensities of a two-state Markov chain.

Müller and Wang (1994) derived improved boundary kernels for hazard rate estimation; see table 1 in their paper for the boundary kernels corresponding to the uniform kernel, the Epanechnikov kernel, and the biweight kernel. Müller and Wang (1994) also considered an extension of the kernel function estimator, where the bandwidth is allowed to depend on time, that is, where $b = b(t)$ in (3.17) and (3.19), thus allowing the degree of smoothing to adapt locally to the amount of data available.

Example 3.7. Time between births. This example is a continuation of Example 3.1. In that example we presented Nelson-Aalen plots for the time between first and second births for the 53 558 Norwegian women who gave birth to their first child in the period 1983–97 grouped according to whether the first child died within one year of its birth or not. Figure 3.10 shows estimated hazard rates for the two groups of women obtained by smoothing the increments of the Nelson-Aalen curves of Figure 3.1. The estimated hazards are obtained by the implementation `muhaz` in R (R Development Core Team, 2007) of the kernel function smoothing method of Müller and Wang (1994) using the Epanechnikov kernel with boundary correction and variable bandwidth. Figure 3.10 clearly shows that the hazard rate for the second birth is much larger between one and two years for the women who lost their first child than it is for the women who did not experience this traumatic event. Later

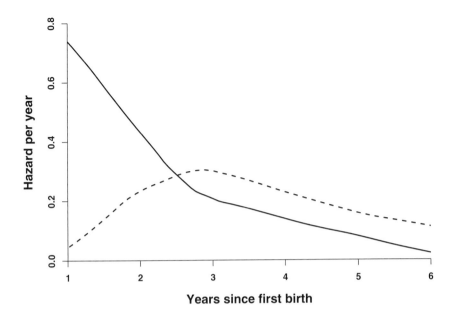

Fig. 3.10 *Estimated hazards for the time between first and second births. Drawn line: first child survived one year; dashed line: first child died within one year.*

on, the difference is much smaller; in fact, the latter group of women has the largest hazard rate after about 2.5 years. □

3.1.5 The estimator and its small sample properties

We consider a counting process $N(t)$ with an intensity process satisfying the multiplicative intensity model (3.8). We will derive the Nelson-Aalen estimator for $A(t) = \int_0^t \alpha(s)\,ds$ and study its statistical properties. Starting with formula (1.16) and inserting (3.8), we can write informally:

$$dN(t) = \alpha(t)Y(t)\,dt + dM(t). \qquad (3.20)$$

We want to divide both sides of the equation by $Y(t)$. However, this may not be possible since $Y(t)$ may equal zero. To avoid this problem, we introduce the indicator function $J(t) = I(Y(t) > 0)$. Multiplying (3.20) by $J(t)$ and dividing by $Y(t)$ yields

$$\frac{J(t)}{Y(t)}dN(t) = J(t)\alpha(t)dt + \frac{J(t)}{Y(t)}dM(t),$$

where $J(t)/Y(t)$ is interpreted as 0 whenever $Y(t)$ is zero. By integration we get

$$\int_0^t \frac{J(s)}{Y(s)}dN(s) = \int_0^t J(s)\alpha(s)\,ds + \int_0^t \frac{J(s)}{Y(s)}dM(s). \tag{3.21}$$

Note that the integral on the left-hand side is just the Nelson-Aalen estimator (3.4). This is because integrating over a counting process is the same as summing the integrand over the jump times $T_1 < T_2 < \cdots$ of the process. Hence the Nelson-Aalen estimator may be writen as

$$\widehat{A}(t) = \int_0^t \frac{J(s)}{Y(s)}dN(s). \tag{3.22}$$

Introducing the notation

$$A^*(t) = \int_0^t J(s)\alpha(s)\,ds,$$

formula (3.21) may be rewritten as

$$\widehat{A}(t) - A^*(t) = \int_0^t \frac{J(s)}{Y(s)}dM(s). \tag{3.23}$$

Since the right-hand side is a stochastic integral with respect to a martingale (cf. Section 2.2.2), it is itself a zero-mean martingale, which is a representation of noise in our context. In particular, it follows that $E\{\widehat{A}(t) - A^*(t)\} = 0$. Thus $\widehat{A}(t)$ is an unbiased estimator of $A^*(t)$. Ideally, we would want an unbiased estimator of $A(t)$, but this is impossible in this nonparametric framework since $\alpha(t)$ cannot be estimated when $Y(t) = 0$.

The martingale nature of the deviation in (3.23) is intimately connected to the nonparametric character of the estimator. In fact, we assume that the function $\alpha(t)$ can vary arbitrarily, so its estimate over some time interval should be unrelated to its estimate over some other nonoverlapping interval. Therefore, one would expect the increments of the Nelson-Aalen estimator over disjoint intervals to be uncorrelated, and this indeed is an implication of the martingale property (cf. Section 2.2.1).

We then use the rules for stochastic integrals to justify the variance estimator (3.5). The derivation is typical for the way counting process theory is used. The result could be derived without martingales, but that would be a much more lengthy argument. Using formula (2.47), we find the optional variation process of the martingale in (3.23) to be:

$$[\widehat{A} - A^*](t) = \int_0^t \frac{J(s)}{Y(s)^2}dN(s).$$

From (2.24), it further follows that

$$\mathrm{Var}(\widehat{A}(t) - A^*(t)) = \mathrm{E}[\widehat{A} - A^*](t).$$

Thus

$$\widehat{\sigma}^2(t) = \int_0^t \frac{J(s)}{Y(s)^2} dN(s), \qquad (3.24)$$

which is just the estimator (3.5), is an unbiased estimator for the variance of the Nelson-Aalen estimator.

3.1.6 Large sample properties

For all situations considered in Sections 3.1.1 and 3.1.2, except for Examples 3.5 and 3.6, the counting process N is obtained by aggregating n individual processes. To show that the Nelson-Aalen estimator is approximately normally distributed in such situations when n is large, we may apply the martingale central limit theorem to the martingale

$$\sqrt{n}(\widehat{A}(t) - A^*(t)) = \int_0^t \sqrt{n} \frac{J(s)}{Y(s)} dM(s).$$

To this end, we need to show that the two conditions of the martingale central limit theorem are fulfilled; cf. Section 2.3.3. We will be content with checking the versions (2.60) and (2.61) of these conditions (with $k = 1$), and we refer to Andersen et al. (1993, section IV.1.2) for more formal derivations.

We assume that $Y(t)/n$ becomes stable as n increases; more precisely that there exists a positive function y such that $Y(t)/n \xrightarrow{P} y(t)$ for all $t \in [0, \tau]$, as $n \to \infty$. Then, with $H(t) = \sqrt{n}J(t)/Y(t)$ and $\lambda(t) = Y(t)\alpha(t)$, we have

$$H(t)^2 \lambda(t) = \frac{J(t)\alpha(t)}{Y(t)/n} \xrightarrow{P} \frac{\alpha(t)}{y(t)}$$

$$H(t) = \frac{1}{\sqrt{n}} \frac{J(t)}{Y(t)/n} \xrightarrow{P} 0$$

for all $t \in [0, \tau]$ as $n \to \infty$. Thus (2.60) and (2.61) are fulfilled with $v(t) = \alpha(t)/y(t)$, and it follows by the martingale central limit theorem that $\sqrt{n}(\widehat{A} - A^*)$ converges in distribution to a mean zero Gaussian martingale with variance function $\sigma^2(t) = \int_0^t \{\alpha(s)/y(s)\} ds$.

Asymptotically there is no difference between $A(t)$ and $A^*(t)$, and it follows that also $\sqrt{n}(\widehat{A} - A)$ converges in distribution to a mean zero Gaussian martingale with variance function $\sigma^2(t)$. In particular, for a fixed value of t, the Nelson-Aalen estimator is asymptotically normally distributed, a fact that was used in connection with the confidence intervals (3.6) and (3.7).

3.2 The Kaplan-Meier estimator

To estimate the survival function from a sample of censored survival data, the
Kaplan-Meier estimator is the method of choice. The estimator may be obtained
as the limiting case of the classical actuarial estimator, and it seems to have been
first proposed by Böhmer (1912). It was, however, lost sight of by later researchers
and not further investigated until the important paper by Kaplan and Meier (1958)
appeared. The estimator is today usually named after these two authors, although it
is sometimes denoted the product-limit estimator.

    In this section, we first describe the Kaplan-Meier estimator, illustrate its use
in two particular situations, and discuss estimation of median and mean survival
times. Then we discuss the relation between a survival function and its cumulative
hazard, and we use this relation to express the Kaplan-Meier estimator as a product-
integral of the Nelson-Aalen estimator. Finally we indicate how this product-integral
representation is useful for the study of the statistical properties of the Kaplan-Meier
estimator.

3.2.1 The estimator and confidence intervals

We consider the survival data situation where one wants to study the time to occurrence of an event of interest. In this situation, the survival function $S(t)$ specifies the
probability that, for a randomly selected individual from the population under study,
the event will occur later than time t.

    To estimate the survival function we have at our disposal a sample of n individ-
uals from the population. Even though most results are valid under right-censoring
and left-truncation, we will at the outset only allow for right-censored survival times.
Modifications for the situation where we also have left-truncation (or delayed entry)
are briefly discussed at the end of this subsection. Moreover, in our general presentation we will assume that there are no tied survival times, leaving a brief discussion
of the handling of ties to Section 3.2.2.

    As in Section 3.1.1, we let $N(t)$ count the number of occurrences of the event in
$[0,t]$, while $Y(t)$ is the number of individuals at risk "just before" time t. As before,
we write $T_1 < T_2 < \cdots$ for the ordered times when an occurrence of the event is
observed, that is, for the jump times of N.

    To give an intuitive justification of the Kaplan-Meier estimator, we partition the
time interval $[0,t]$ into a number of small time intervals $0 = t_0 < t_1 < \cdots < t_K = t$
and use the multiplication rule for conditional probabilities to write

$$S(t) = \prod_{k=1}^{K} S(t_k \,|\, t_{k-1}). \qquad (3.25)$$

Here $S(v\,|\,u) = S(v)/S(u)$, for $v > u$, is the conditional probability that the event will
occur later than time v given that it has not yet occurred by time u. Since we assume

that there are no tied event times, we may make each of the time intervals so small that it contains at most one observed event, and that all censorings occur at the right-hand endpoint of an interval. Then, if no event is observed in $(t_{k-1}, t_k]$, we estimate $S(t_k \mid t_{k-1})$ by 1, where as if an event is observed at time $T_j \in (t_{k-1}, t_k]$, the natural estimate of $S(t_k \mid t_{k-1})$ is $1 - 1/Y(t_{k-1}) = 1 - 1/Y(T_j)$. Inserting these estimates into (3.25), we obtain

$$\widehat{S}(t) = \prod_{T_j \leq t} \left\{ 1 - \frac{1}{Y(T_j)} \right\}, \tag{3.26}$$

which is the Kaplan-Meier estimator. If there are no censored observations, the Kaplan-Meier estimator reduces to one minus the empirical cumulative distribution function (Exercise 3.5).

The variance of the Kaplan-Meier estimator may be estimated by

$$\widehat{\tau}^2(t) = \widehat{S}(t)^2 \sum_{T_j \leq t} \frac{1}{Y(T_j)^2}. \tag{3.27}$$

Note that this is just $\widehat{S}(t)^2$ times the variance estimator (3.5) for the Nelson-Aalen estimator. An argument for the variance estimator (3.27) is outlined in Section 3.2.6. It should be noted that (3.27) differs slightly from Greenwood's formula, which is commonly used to estimate the variance of the Kaplan-Meier estimator. Greenwood's formula takes the form

$$\widetilde{\tau}^2(t) = \widehat{S}(t)^2 \sum_{T_j \leq t} \frac{1}{Y(T_j)\{Y(T_j) - 1\}}. \tag{3.28}$$

If there are no censored observations, (3.28) reduces to $\widehat{S}(t)(1 - \widehat{S}(t))/n$, the standard binomial variance estimator (Exercise 3.5).

In Section 3.2.6 we also show that the Kaplan-Meier estimator, evaluated at a given time t, is approximately normally distributed in large samples. Therefore a standard $100(1 - \alpha)\%$ confidence interval for $S(t)$ takes the form

$$\widehat{S}(t) \pm z_{1-\alpha/2}\widehat{\tau}(t). \tag{3.29}$$

Alternatively we may replace $\widehat{\tau}(t)$ by $\widetilde{\tau}(t)$. The approximation to the normal distribution is improved by using the log-minus-log transformation giving the confidence interval

$$\widehat{S}(t)^{\exp\{\pm z_{1-\alpha/2}\widehat{\tau}(t)/(\widehat{S}(t)\log\widehat{S}(t))\}} \tag{3.30}$$

(Exercise 3.6). Note that the confidence intervals should be given a pointwise interpretation. The confidence interval (3.30) is satisfactory for quite small sample sizes (Borgan and Liestøl, 1990). Confidence intervals with small sample properties comparable to (3.30), or even slightly better, may be obtained by using the arcsine-square-root transformation (Borgan and Liestøl, op. cit.) or by basing the confidence interval on the likelihood ratio (Thomas and Grunkemeier, 1975; Cox and Oakes, 1984, section 4.3).

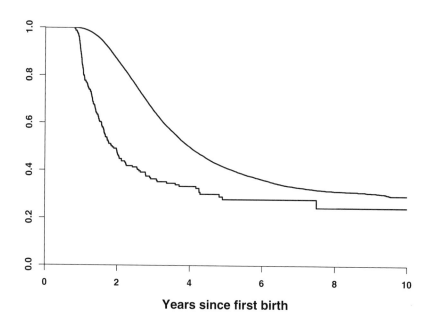

Fig. 3.11 *Kaplan-Meier estimates for the time between first and second birth. Upper curve: first child survived one year; lower curve: first child died within one year.*

Example 3.8. Time between births. This example is a continuation of Example 3.1. In that example we presented Nelson-Aalen plots for the time between first and second births for the 53 558 Norwegian women who gave birth to their first child in the period 1983–97 grouped according to whether the first child died within one year of its birth or not. Similar Kaplan-Meier plots are given in Figure 3.11. (A plot with pointwise confidence intervals for the women who lost their first child is provided in Figure 3.13 in Example 3.10.)

From the Kaplan-Meier plots we see, for example, that for the women for whom the first child survived one year, the estimated survival probabilities two and five years after the birth of the first child are, respectively, 86.0% and 40.5%. For the women who lost their first child within one year, the corresponding estimates are 43.4% and 24.8%. This shows (as also noted in Example 3.1) that a woman who loses her first child is more likely to have another child, and to have it earlier, than is the case for a woman who does not experience this traumatic event. □

Example 3.9. Time between births. This example is a continuation of Example 3.2. There we considered the 16 116 Norwegian women who had (at least) two births in the period 1983–97 and presented Nelson-Aalen plots for the time until the third birth depending on the genders of the two older children. Figure 3.12 gives the

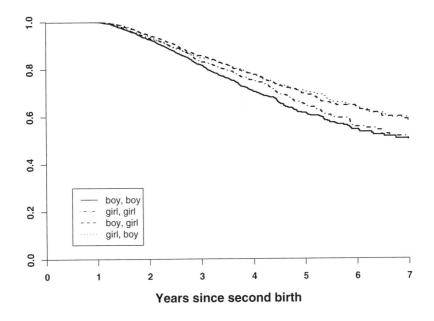

Fig. 3.12 *Kaplan-Meier estimates for the time between the second and third births depending on the gender of the two older children.*

corresponding Kaplan-Meier plots. The plots indicate that women who have one child of each gender are less likely to have a third child than the women who have two boys or two girls.

More specifically, we may consider the probability of having a third child within five years of the second one (estimated as one minus the five-year survival estimate). For a woman who has two boys, this probability is estimated to 38.5 % with a log-log-transformed confidence interval from 35.6% to 41.9%, while for a woman who has two girls the estimated probability of having a third one within five years is 35.1% (CI: 31.7% to 38.8%). For the women who first have a boy and then a girl and for the women who first have a girl and then a boy, the estimated probabilities are lower: 30.6% (CI: 27.5% to 34.0%) and 29.4% (CI: 26.5% to 32.6%), respectively. □

Right-censoring is not the only kind of data-incompleteness in survival analysis. Often, for example, in epidemiological applications, individuals are not followed from time zero (in the relevant time scale, typically age), but only from a later entry time (conditional on the event not having occurred by this entry time). Thus, in addition to right-censoring, the survival times are subject to left-truncation. For such data we may still use the Kaplan-Meier estimator (3.26) and estimate its variance by

(3.27) or (3.28). The number at risk $Y(t)$ is now the number of individuals who have entered the study before time t and are still in the study just prior to t. However, for left-truncated data the number at risk will often be low for small values of t. This will result in estimates $\widehat{S}(t)$ that have large sampling errors and therefore may be of little practical use. What can be usefully estimated in such situations is the conditional survival function $S(t\,|\,s)$ for a suitably chosen value of s. The estimation is performed as described earlier, the only modification being that the product in (3.26) and the sums in (3.27) and (3.28) are restricted to those T_j for which $s < T_j \le t$.

3.2.2 Handling tied survival times

In Section 3.1.3 we discussed two approaches for handling tied event times for the Nelson-Aalen estimator. The two approaches may be adopted to see how the Kaplan-Meier estimator should be modified for tied survival times. To this end, we note that when there are no tied survival times, the Kaplan-Meier estimator may be written

$$\widehat{S}(t) = \prod_{T_j \le t} \left\{ 1 - \triangle\widehat{A}(T_j) \right\}, \tag{3.31}$$

where $\triangle\widehat{A}(T_j) = 1/Y(T_j)$ is the increment of the Nelson-Aalen estimator (3.4); see also Section 3.2.4.

We then assume that there are d_j tied survival times at T_j. To obtain a tie-corrected version of the Kaplan-Meier estimator we may insert (3.12) or (3.13) in (3.31). In both cases, we get, after some simple algebra, that

$$\widehat{S}(t) = \prod_{T_j \le t} \left\{ 1 - \frac{d_j}{Y(T_j)} \right\}. \tag{3.32}$$

Thus the two approaches of Section 3.1.3 give the same tie-corrected Kaplan-Meier estimator.

When it comes to estimating the variance of the Kaplan-Meier estimator, the two approaches differ. Approach (i) gives the estimator $\widetilde{\tau}^2(t) = \widehat{S}(t)^2 \widehat{\sigma}^2(t)$, where $\widehat{\sigma}^2(t)$ is estimated by (3.14) and (3.15). On the other hand, approach (ii) gives Greenwood's formula:

$$\widetilde{\tau}^2(t) = \widehat{S}(t)^2 \sum_{T_j \le t} \frac{d_j}{Y(T_j)\{Y(T_j) - d_j\}}; \tag{3.33}$$

cf. the argument in Andersen et al. (1993, page 258). The difference between the two variance estimators is usually of little importance, but Greenwood's formula is the one commonly used.

3.2.3 Median and mean survival times

Use of the Kaplan-Meier estimator is not restricted to estimating survival probabilities for given times t. It may also be used to estimate fractiles such as the median survival time.

Consider the pth fractile ξ_p of the cumulative distribution function $F(t) = 1 - S(t)$, and assume that $F(t)$ has positive density $f(t) = F'(t) = -S'(t)$ in a neighborhood of ξ_p. Then ξ_p is uniquely determined by the relation $F(\xi_p) = p$, or equivalently, $S(\xi_p) = 1 - p$. The Kaplan-Meier estimator is a step function and hence does not necessarily attain the value $1 - p$. Therefore a similar relation cannot be used to define the estimator $\widehat{\xi}_p$ of pth fractile. Rather we define $\widehat{\xi}_p$ to be the smallest value of t for which $\widehat{S}(t) \leq 1 - p$, that is, the time t where $\widehat{S}(t)$ jumps from a value greater than $1 - p$ to a value less than or equal to $1 - p$.

In large samples the empirical fractile $\widehat{\xi}_p$ is approximately normally distributed with a standard error that may be estimated by

$$\mathrm{SE}(\widehat{\xi}_p) = \frac{\widehat{\tau}(\widehat{\xi}_p)}{\widehat{f}(\widehat{\xi}_p)} \tag{3.34}$$

(e.g. Andersen et al., 1993, example IV.3.6). Here $\widehat{f}(t)$ is an estimate of the density $f(t) = -S'(t)$. One may, for example, use

$$\widehat{f}(t) = \frac{1}{2b}\left(\widehat{S}(t-b) - \widehat{S}(t+b)\right) \tag{3.35}$$

for a suitable bandwidth b (corresponding to a kernel function estimator with uniform kernel; cf. Section 3.1.4).

This result may be used in the usual way to determine approximate confidence intervals, for example, for the median survival time $\xi_{0.50}$. For the purpose of determining a confidence interval for a fractile like the median, however, it is better to apply the approach of Brookmeyer and Crowley (1982). For the pth fractile, one then uses as a $100(1 - \alpha)\%$ confidence interval all hypothesized values ξ_p^0 of ξ_p that are not rejected when testing the null hypothesis $\xi_p = \xi_p^0$ against the alternative hypothesis $\xi_p \neq \xi_p^0$ at the α-level. As illustrated in the following example and further discussed in Exercise 3.8, such test-based confidence intervals can be read directly from the lower and upper confidence limits for the survival function in exactly the same manner that the empirical fractile $\widehat{\xi}_p$ can be read from the Kaplan-Meier curve itself.

Example 3.10. Time between births. In Examples 3.1 and 3.8 we studied the time between the births of the first and second child for women who lost the first child within one year of its birth. Figure 3.13 shows the Kaplan-Meier estimate with 95% log-log-transformed confidence intervals for these women. The estimate of the median time between the births is 1.86 years with a standard error estimate of 0.12 year computed according to (3.34) and (3.35) with bandwidth $b = 0.25$ year. Thus a

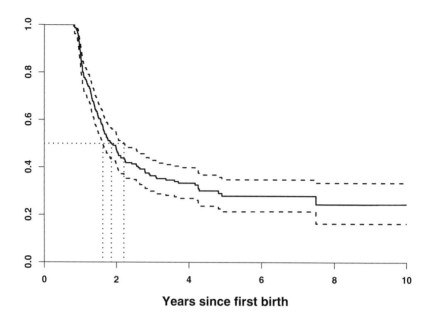

Years since first birth

Fig. 3.13 *Kaplan-Meier estimate with 95% log-log-transformed confidence intervals for the time between first and second birth for women who lost the first child within one year after birth. It is indicated at the figure how one may obtain the estimated median time with confidence limits.*

standard approximate 95% confidence interval is given as $1.86 \pm 1.96 \cdot 0.12$, that is, from 1.62 years to 2.10 years. A test-based log-log transformed confidence interval is from 1.63 years to 2.20 years; cf. Figure 3.13. The latter confidence interval is to be preferred to the standard one. □

The mean survival time $\mu = \int_0^\infty t f(t)dt = \int_0^\infty S(t)dt$ (cf. Exercise 1.3) depends heavily on the behavior of the right-hand tail of the distribution. In many survival studies, the longest survival times will be censored, resulting in little or no information in the data on which to base the estimation of the right-hand tail (unless one adopts a parametric model). This will usually make it impossible to get a reliable estimate of the mean survival time from censored data.

This is one important reason why in survival analysis the median is a more useful measure of location than the mean. What may be estimated from right-censored survival data is the expected survival time restricted to a fixed interval $[0,t]$, that is, $\mu_t = \int_0^t S(u)du$. This is estimated by

$$\widehat{\mu}_t = \int_0^t \widehat{S}(u)du,$$

the area under the Kaplan-Meier curve between 0 and t. Such an estimate may be of interest in its own right, or it may be compared with a similar population-based estimate to assess the expected number of years lost up to time t for a group of patients. In large samples $\widehat{\mu}_t$ is approximately normally distributed with a variance that may be estimated by

$$\widehat{\text{var}}(\widehat{\mu}_t) = \sum_{T_j \leq t} \frac{(\widehat{\mu}_t - \widehat{\mu}_{T_j})^2}{Y(T_j)^2}$$

(e.g., Andersen et al., 1993, example IV.3.8), a result that may be used to give approximate confidence limits for μ_t. By letting t tend to infinity, the result may be extended to estimation of the mean μ itself (Gill, 1983). However, the conditions (mainly on the censoring) needed for such an extension to be valid are usually not met in practice.

3.2.4 Product-integral representation

In order to study the statistical properties of the Kaplan-Meier estimator (in Section 3.2.6), we will exploit the relation between the Kaplan-Meier and Nelson-Aalen estimators. In this subsection we establish this relation, and more generally the relation between the cumulative hazard rate $A(t)$ and the survival function $S(t)$ for an arbitrary distribution.

It is common to assume that the survival function $S(t)$ is absolutely continuous (and we will usually assume this as well). Then

$$A'(t) = \alpha(t) = -\frac{S'(t)}{S(t)} \tag{3.36}$$

and

$$S(t) = e^{-A(t)}; \tag{3.37}$$

cf. (1.4) and (1.5). We will here describe how (3.36) and (3.37) may be generalized to an arbitrary distribution, which needs to be neither absolutely continuous nor discrete; see Appendix A.1 for a full account.

For an arbitrary distribution, the hazard rate $\alpha(t)$ is not defined, and the definition (1.3) of the cumulative hazard rate no longer applies. But given a survival time T from an arbitrary distribution with survival function $S(t) = P(T > t)$, we may in general define the cumulative hazard rate by

$$A(t) = -\int_0^t \frac{dS(u)}{S(u-)}. \tag{3.38}$$

For the absolutely continuous case, (3.38) specializes to

$$A(t) = -\int_0^t \frac{S'(u)}{S(u)}\,du$$

in agreement with (3.36). On the other hand, for a purely discrete distribution, (3.38) takes the form $A(t) = \sum_{u \leq t} \alpha_u$, where the discrete hazard

$$\alpha_u = P(T = u \mid T \geq u) = -\{S(u) - S(u-)\}/S(u-)$$

is the conditional probability that the event occurs exactly at time u given that it has not occurred earlier.

In order to express the survival function of an arbitrary distribution by means of its cumulative hazard rate, it is convenient to use the *product-integral*. Briefly the product-integral is defined as follows. Partition the time interval $[0,t]$ into a number of time intervals $0 = t_0 < t_1 < t_2 < \cdots < t_K = t$, and consider the finite product

$$\prod_{k=1}^{K} \{1 - (A(t_k) - A(t_{k-1}))\}.$$

If in this product we let the number K of time intervals increase while their lengths go to zero in a uniform way, the product will approach a limit, which is termed a product-integral. We write the product-integral as $\mathop{\pi}_{u \leq t} \{1 - dA(u)\}$. Here the product-integral notation π is used to suggest a limit of finite products $\prod$, just as the integral $\int$ is a limit of finite sums $\sum$.

By (3.38), we have

$$S(t_k \mid t_{k-1}) = \frac{S(t_k)}{S(t_{k-1})} = 1 + \frac{S(t_k) - S(t_{k-1})}{S(t_{k-1})} \approx 1 - (A(t_k) - A(t_{k-1}))$$

when t_{k-1} and t_k are close together. Comparing this with (3.25), it should come as no surprise that the product-integral equals the survival function:

$$S(t) = \mathop{\pi}_{u \leq t} \{1 - dA(u)\}. \tag{3.39}$$

For an absolutely continuous distribution, the product-integral (3.39) specializes to (3.37), while for a purely discrete distribution it takes the form $S(t) = \prod_{u \leq t}(1 - \alpha_u)$. More generally, we may decompose the cumulative hazard into a continuous and a discrete part, that is, $A(t) = A_c(t) + A_d(t)$, where $A_c(t)$ is continuous and $A_d(t) = \sum_{u \leq t} \alpha_u$ is a step function. Then the product-integral may be evaluated as

$$\mathop{\pi}_{u \leq t} \{1 - dA(u)\} = e^{-A_c(t)} \prod_{u \leq t} \{1 - \alpha_u\} \tag{3.40}$$

cf. (A.11) in Appendix A.

We now turn to the connection between the Kaplan-Meier and Nelson-Aalen estimators. To this end, note that the Nelson-Aalen estimator (3.4) may be given as

$$\widehat{A}(t) = \sum_{T_j \leq t} \triangle \widehat{A}(T_j),$$

where $\triangle \widehat{A}(T_j) = 1/Y(T_j)$. Thus the Nelson-Aalen estimator corresponds to the cumulative hazard of a purely discrete distribution. This empirical distribution has all its probability mass concentrated at the observed failure times, the discrete hazard at T_j being $\triangle \widehat{A}(T_j)$. The corresponding empirical survival function therefore takes the form

$$\widehat{S}(t) = \prod_{u \leq t} \left\{1 - d\widehat{A}(u)\right\} = \prod_{T_j \leq t} \left\{1 - \triangle \widehat{A}(T_j)\right\}, \tag{3.41}$$

that is, it is the Kaplan-Meier estimator. This shows that the Kaplan-Meier estimator is related to the Nelson-Aalen estimator in exactly the same way that the survival function is related to the cumulative hazard rate. This fact is lost sight of when one considers the relation (3.37) that is only valid for the continuous case.

3.2.5 Excess mortality and relative survival

In Example 3.4 we consider a model where the hazard rate of the ith individual takes the form $\alpha_i(t) = \alpha(t)\mu_i(t)$. Here $\mu_i(t)$ is a *known* mortality rate (hazard rate) for a person in the general population corresponding to subject i and $\alpha(t)$ is the *relative mortality* at time t.

An alternative is to assume that the hazard rate of the ith individual may be written as

$$\alpha_i(t) = \gamma(t) + \mu_i(t), \tag{3.42}$$

where $\gamma(t)$ is the *excess mortality* at time t. Note that the excess mortality will be negative if the mortality in the group under study is lower than in the general population.

For the excess mortality model (3.42) the processes $N_1(t), \ldots, N_n(t)$, counting observed deaths for the individuals in the group under study, have intensity processes of the form

$$\lambda_i(t) = \{\gamma(t) + \mu_i(t)\} Y_i(t); \qquad i = 1, \ldots, n.$$

Thus the intensity process of the aggregated process $N(t) = \sum_{i=1}^n N_i(t)$, recording the total number of observed deaths, becomes

$$\lambda(t) = \sum_{i=1}^n \lambda_i(t) = \{\gamma(t) + \overline{\mu}(t)\} Y(t).$$

Here $Y(t) = \sum_{i=1}^n Y_i(t)$ is the number of individuals at risk, and

$$\overline{\mu}(t) = \sum_{i=1}^n \mu_i(t) \frac{Y_i(t)}{Y(t)}$$

is the *average population mortality* among the individuals at risk at time t. We will derive an estimator for the cumulative excess mortality $\Gamma(t) = \int_0^t \gamma(s)\, ds$ and the *relative survival function*

$$R(t) = e^{-\Gamma(t)}. \tag{3.43}$$

As we will see, the relative survival function may be interpreted as the ratio of the survival function of the group under study and the survival function one would get for a similar group of individuals from the general population.

By the preceding results, we have

$$dN(t) = \{\gamma(t) + \overline{\mu}(t)\} Y(t)\, dt + dM(t),$$

where $M(t)$ is the counting process martingale corresponding to $N(t)$. This is a relation similar to (3.20) for the multiplicative intensity model. Following the arguments in Section 3.1.5, we therefore arrive at the estimator

$$\widehat{\Gamma}(t) = \int_0^t \frac{dN(s)}{Y(s)} - \int_0^t \overline{\mu}(s)\, ds \tag{3.44}$$

for the cumulative excess mortality. Note that (3.44) is the difference between the Nelson-Aalen estimator for the group under study and the cumultative average population mortality. By similar reasoning as for the Nelson-Aalen estimator (cf. Sections 3.1.5 and 3.1.6), one may show that $\widehat{\Gamma}(t)$ is approximately normally distributed with a variance that may be estimated by $\int_0^t Y(s)^{-2}\, dN(s)$.

By taking the product-integral of $\widehat{\Gamma}(t)$ and using (3.40) [which is valid for a wider class of functions than cumulative hazards], we get the following estimator for the relative survival function (3.43):

$$\widehat{R}(t) = \prod_{s \leq t} \left\{ 1 - d\widehat{\Gamma}(s) \right\} = \frac{\widehat{S}(t)}{\widehat{S}_e(t)}. \tag{3.45}$$

Here $\widehat{S}(t)$ is the Kaplan-Meier estimator for the group under study, and

$$\widehat{S}_e(t) = \exp\left\{ -\int_0^t \overline{\mu}(s)\, ds \right\} \tag{3.46}$$

is the *expected survival function*, that is, the survival function one would have had if the mortality in the group under study had been the same as in the general population (conditional on the observed numbers at risk). Note that the relative survival function will be below one if the group under study has a higher mortality than the general population, while it will be above one if the group has a mortality that is lower than the population mortality. By similar reasoning as for the Kaplan-Meier estimator (cf. Section 3.2.6), one may show that $\widehat{R}(t)$ is approximately normally distributed with a variance that may be estimated by $\widehat{R}(t)^2 \int_0^t Y(s)^{-2}\, dN(s)$.

The estimators (3.44) and (3.45) for the cumulative excess mortality and the relative mortality were proposed by Andersen and Væth (1989), who also studied the

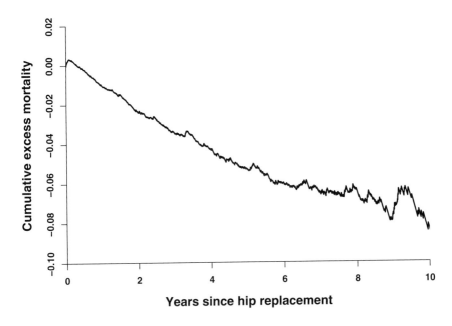

Fig. 3.14 *Estimate of the cumulative excess mortality for patients who have had a hip replacement in Norway in the period 1987–97.*

statistical properties of the estimators using counting processes and martingales. The estimator (3.45) is a continuous-time version of the relative survival rate commonly used in cancer epidemiology, while (3.46) is a continuous-time version of the expected survival rate (or expected survival curve). It should be noted that other definitions of expected survival curves are also available (e.g., Keiding, 2005).

Example 3.11 Hip replacements. We now consider the data from the Norwegian Arthroplasty Registry on 3503 females and 1497 males who had their first total hip replacement operation at a Norwegian hospital between September 1987 and February 1998; cf. Example 1.5. As in Example 3.4 we let $\mu_m(a)$ and $\mu_f(a)$ be the mortality at age a in the general Norwegian population, as it was reported by Statistics Norway for the year 1993. We here consider the excess mortality model where the mortality for the ith hip replacement patient t years after the operation is assumed to take the form

$$\alpha_i(t) = \gamma(t) + \mu_{s_i}(a_i + t),$$

where s_i is the gender of the patient (f or m) and a_i is its age at operation. This is of the form (3.42) with $\mu_i(t) = \mu_{s_i}(a_i + t)$. We may therefore use (3.44) to estimate the

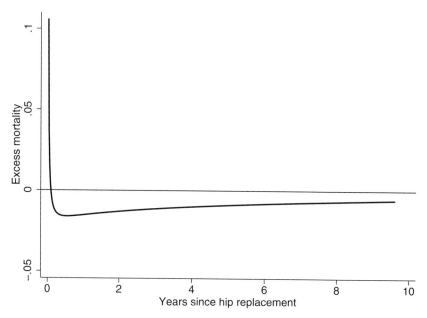

Fig. 3.15 *Estimate of the excess mortality for patients who have had a hip replacement in Norway in the period 1987–97.*

cumulative excess mortality of the hip replacement patients and (3.45) to estimate their relative survival.

Figure 3.14 gives the estimated cumulative excess mortality. Except for a short period just after the operation (and some random fluctuations later), the cumulative excess mortality is decreasing. This shows that the hip replacement patients have a lower mortality than the general population, except for a short period after the operation, in agreement with the results of Example 3.4. The same picture is revealed in Figure 3.15, where the cumulative rates have been smoothed by fractional polynomials (using Stata) to give an estimate of the excess mortality itself.

The relative survival curve of Figure 3.16 illustrates the same results in another way. We see that the relative survival is below one just after the operation. But later on it is above one, showing that hip replacement patients have a higher survival than the general population. The estimate of the relative survival function (3.45) is the ratio of the Kaplan-Meier estimate and the expected survival curve. The latter two are shown for the hip replacement patients in Figure 1.6 in Section 1.1.5. □

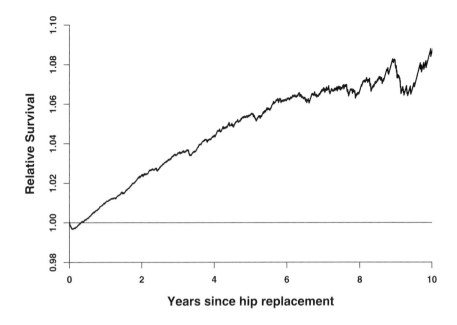

Fig. 3.16 *Relative survival curve for patients who have had a total hip replacement in Norway in the period 1987–97.*

3.2.6 *Martingale representation and statistical properties*

The product-integral formulation (3.41) of the Kaplan-Meier estimator shows its close relation to the Nelson-Aalen estimator, and it is also essential for the study of its statistical properties. We will outline the basic steps and refer to Andersen et al. (1993, section IV.3) for a detailed account.

As in Section 3.1.5 we let $J(t) = I\{Y(t) > 0\}$ and $A^*(t) = \int_0^t J(u)dA(u)$. We then introduce

$$S^*(t) = \prod_{u \le t}\{1 - dA^*(u)\}, \qquad (3.47)$$

and note that $S^*(t)$ is almost the same as $S(t)$, when there is only a small probability that there is no one at risk at times $u \le t$. A key to the study of the statistical properties of the Kaplan-Meier estimator is the relation

$$\frac{\widehat{S}(t)}{S^*(t)} - 1 = -\int_0^t \frac{\widehat{S}(u-)}{S^*(u)} d(\widehat{A} - A^*)(u), \qquad (3.48)$$

which is valid for arbitrary distributions; cf. formula (A.12) in Appendix A.1.

From Section 3.2.1 we know that $\widehat{A} - A^*$ is a martingale; cf. (3.23). Therefore, the right-hand side of (3.48) is a stochastic integral, and hence itself a mean zero martingale (cf. Section 2.2.2). Thus relation (3.48) provides a martingale representation of the Kaplan-Meier estimator. An immediate consequence of this martingale representation is that the expected value of $\widehat{S}(t)/S^*(t) - 1$ is equal to zero for all values of t. Thus $E\{\widehat{S}(t)/S^*(t)\} = 1$, which shows that the Kaplan-Meier estimator is almost unbiased when there is only a small probability that there is no one at risk at times $u \leq t$.

Using the martingale representation, one may further prove that the Kaplan-Meier estimator is uniformly consistent. Asymptotically we may therefore apply the approximations $\widehat{S}(u-)/S^*(u) \approx 1$, $S^*(t) \approx S(t)$, and $A^*(u) \approx A(u)$ in (3.48) to obtain

$$\frac{\widehat{S}(t)}{S(t)} - 1 \approx - \int_0^t d(\widehat{A} - A)(u)$$

or

$$\widehat{S}(t) - S(t) \approx -S(t)\left(\widehat{A}(t) - A(t)\right). \tag{3.49}$$

From this approximation we immediately get

$$\mathrm{Var}(\widehat{S}(t)) \approx S(t)^2 \, \mathrm{Var}(\widehat{A}(t)),$$

which shows that the variance of the Kaplan-Meier estimator may be estimated by $\widehat{\tau}^2(t) = \widehat{S}(t)^2 \, \widehat{\sigma}^2(t)$ [cf. (3.27)], where $\widehat{\sigma}^2(t)$ estimates the variance of the Nelson-Aalen estimator [cf. (3.5)].

To derive the asymptotic distribution of the Kaplan-Meier estimator, we assume that there exists a positive deterministic function y such that $Y(t)/n$ converges in probability to $y(t)$ for all $t \in [0, \tau]$ as $n \to \infty$, that is, that the proportion of the sample at risk at time t becomes stable as the sample size increases. Then, by the results of Section 3.1.6, $\sqrt{n}(\widehat{A} - A)$ converges in distribution to a mean zero Gaussian martingale. By the approximation (3.49), it therefore follows that $\sqrt{n}(\widehat{S} - S)$ asymptotically is distributed as a mean zero Gaussian process. In particular, for a fixed t, the Kaplan-Meier estimator (3.41) is approximately normally distributed, a fact that was used in connection with the confidence intervals (3.29) and (3.30). Also the asymptotic distributional results of the estimators for the median and mean survival times reviewed in Section 3.2.3 are consequences of this large sample result.

3.3 Nonparametric tests

In this section we consider the situation where we have two or more counting processes, each satisfying the multiplicative intensity model (3.8), and where the equality of the α-functions of their intensity processes is to be tested. One important special case is the survival data situation, where one wants to test the hypothesis that the hazard rates, or equivalently the survival functions, are the same in two or more populations. But, as illustrated in some of the following examples, the testing problem is also of importance for other applications of the multiplicative intensity model.

In most of the section we will, for the ease of exposition, concentrate on the two-sample case, that is, the situation with two counting processes. The extension to three or more counting processes will, however, also be briefly considered. Moreover, in our general presentation we will assume that there are no tied event times, leaving a brief discussion of the handling of ties to Section 3.3.4.

3.3.1 The two-sample case

Assume that we have counting processes N_1 and N_2 with intensity processes of the multiplicative form $\lambda_1(t) = \alpha_1(t)Y_1(t)$ and $\lambda_2(t) = \alpha_2(t)Y_2(t)$, respectively. We want to test whether their α-functions are the same over a time interval $[0, t_0]$. Usually we will chose t_0 as the upper limit τ of the study time interval, but other choices may sometimes be relevant. Thus we want to test the null hypothesis

$$H_0 : \alpha_1(t) = \alpha_2(t) \quad \text{for all } t \in [0, t_0].$$

The common value of $\alpha_1(t)$ and $\alpha_2(t)$ under the null hypothesis will be denoted $\alpha(t)$.

We will base a test for the null hypothesis on a comparison of the increments of the Nelson-Aalen estimators of $A_h(t) = \int_0^t \alpha_h(s)ds; h = 1,2$. These are given by

$$\widehat{A}_h(t) = \int_0^t \frac{J_h(s)}{Y_h(s)} dN_h(s), \tag{3.50}$$

where $J_h(t) = I\{Y_h(t) > 0\}$. Let $L(t)$ be a nonnegative predictable "weight process," assumed to be zero whenever $J_1(t)J_2(t) = 0$, that is, when at least one of $Y_1(t)$ and $Y_2(t)$ is zero. Different choices of the weight process are discussed later. Then

$$Z_1(t_0) = \int_0^{t_0} L(t) \left(d\widehat{A}_1(t) - d\widehat{A}_2(t) \right) \tag{3.51}$$

accumulates the weighted differences in the increments of the two Nelson-Aalen estimators for the time points $t \in [0, t_0]$ where a comparison is meaningful. Thus the statistic $Z_1(t_0)$ provides a reasonable basis for testing the null hypothesis versus noncrossing alternatives, that is, versus alternatives where $\alpha_1(t) \geq \alpha_2(t)$ for all $t \in [0, t_0]$ or $\alpha_1(t) \leq \alpha_2(t)$ for all $t \in [0, t_0]$.

To investigate the behavior of $Z_1(t_0)$, we first use (3.50) and the relation $J_1L = J_2L = L$ to write

$$Z_1(t_0) = \int_0^{t_0} \frac{L(t)}{Y_1(t)} dN_1(t) - \int_0^{t_0} \frac{L(t)}{Y_2(t)} dN_2(t). \tag{3.52}$$

Then we note that when $\alpha_1 = \alpha_2 = \alpha$, we have $dN_h(t) = \alpha(t)Y_h(t)dt + dM_h(t)$; for $h = 1, 2$; where the M_h are martingales. Inserting this into (3.52), we get

$$Z_1(t_0) = \int_0^{t_0} \frac{L(t)}{Y_1(t)} dM_1(t) - \int_0^{t_0} \frac{L(t)}{Y_2(t)} dM_2(t). \qquad (3.53)$$

Thus when the null hypothesis holds true, $Z_1(t_0)$ is a difference between two stochastic integrals (cf. Section 2.2.2), and hence itself a mean zero martingale (when considered as a process in t_0). In particular $EZ_1(t_0) = 0$.

To obtain an estimator for the variance of $Z_1(t_0)$, we first find the predictable variation process of the martingale (3.53). Now, using (2.48), we get

$$\langle Z_1 \rangle(t_0) = \int_0^{t_0} \left(\frac{L(t)}{Y_1(t)} \right)^2 \alpha(t)Y_1(t)dt + \int_0^{t_0} \left(\frac{L(t)}{Y_2(t)} \right)^2 \alpha(t)Y_2(t)dt$$

$$= \int_0^{t_0} \frac{L^2(t)Y_{\bullet}(t)}{Y_1(t)Y_2(t)} \alpha(t)dt, \qquad (3.54)$$

where the "dot" signifies summation over the relevant index, that is, $Y_{\bullet} = Y_1 + Y_2$. The variance of $Z_1(t_0)$ may be estimated by replacing $\alpha(t)dt$ in (3.54) by $d\widehat{A}(t) = dN_{\bullet}(t)/Y_{\bullet}(t)$, the increment of the Nelson-Aalen estimator based on the process $N_{\bullet} = N_1 + N_2$ counting observed events in both samples combined. The result is the estimator

$$V_{11}(t_0) = \int_0^{t_0} \frac{L^2(t)}{Y_1(t)Y_2(t)} dN_{\bullet}(t) \qquad (3.55)$$

for the variance of $Z_1(t_0)$. This variance estimator is unbiased under the null hypothesis (Exercise 3.10).

In Section 3.3.5 we show that $Z_1(t_0)$ is approximately normally distributed under the null. Thus a test for the null hypothesis may be based on the statistic

$$U(t_0) = \frac{Z_1(t_0)}{\sqrt{V_{11}(t_0)}}. \qquad (3.56)$$

The test statistic is approximately standard normally distribution when the null hypothesis holds true. Alternatively, we may use the test statistic

$$X^2(t_0) = \frac{Z_1(t_0)^2}{V_{11}(t_0)}, \qquad (3.57)$$

which is approximately chi-squared distributed with one degree of freedom under the null.

For the survival data situation, a number of nonparametric tests for comparing two survival functions have been suggested. Most of these are special cases of our general test statistic. Each test is obtained by choosing the weight process $L(t)$ in an appropriate way. The commonly used log-rank test corresponds to the choice $L(t) = Y_1(t)Y_2(t)/Y_{\bullet}(t)$. Another popular choice is the Harrington-Fleming family of

Table 3.2 Choice of weight process $L(t)$ for a number of two-sample tests

Test	Weight process[a]	Key references
Log-rank	$Y_1(t)Y_2(t)/Y_.(t)$	Mantel (1966), Peto and Peto (1972)
Gehan-Breslow	$Y_1(t)Y_2(t)$	Gehan (1965), Breslow (1970)
Efron[b]	$\widehat{S}_1(t-)\widehat{S}_2(t-)J_1(t)J_2(t)$	Efron (1967)
Tarone-Ware	$Y_1(t)Y_2(t)/\sqrt{Y_.(t)}$	Tarone and Ware (1977)
Peto-Prentice	$\widetilde{S}(t-)Y_1(t)Y_2(t)/(Y_.(t)+1)$	Peto and Peto (1972), Prentice (1978)
Harrington-Fleming	$\widehat{S}(t-)^\rho Y_1(t)Y_2(t)/Y_.(t)$	Harrington and Fleming (1982)

[a] $\widehat{S}_1(t)$ and $\widehat{S}_2(t)$ are the Kaplan-Meier estimators computed separately for each of the two samples, $\widehat{S}(t)$ is the Kaplan-Meier estimator based on the combined sample, and $\widetilde{S}(t) = \prod_{s \leq t}(1 - \frac{\Delta N_.(s)}{Y_.(s)+1})$ is a slightly modified version of the Kaplan-Meier estimator based on the combined sample.
[b] Modified as described by Andersen et al. (1993, page 352).

tests corresponding to $L(t) = \widehat{S}(t-)^\rho Y_1(t)Y_2(t)/Y_.(t)$. Here $\widehat{S}(t)$ is the Kaplan-Meier estimator based on the combined sample, and ρ is a constant specified by the user. The choice $\rho = 0$ gives the log-rank test, while $\rho = 1$ gives a Wilcoxon-type test, which puts more weight on differences for small values of t. An overview of various tests that have been suggested in the literature and the corresponding choices of the weight process $L(t)$ is given in Table 3.2; see Andersen et al. (1993, example V.2.1) for further details.

Example 3.12. Mating of Drosophila flies. In Example 3.6 we gave Nelson-Aalen estimates for the cumulative mating intensities among two races of Drosophila flies. Figure 3.9 indicates that the mating for the ebony flies starts earlier and takes place at a higher rate than is the case for oregon flies. We will now test if this difference is statistically significant.

Using the log-rank weights $L(t) = Y_1(t)Y_2(t)/Y_.(t)$, the statistic (3.52) takes the value $Z_1(t_0) = 15.56$ when t_0 is chosen as the duration of the experiment, and N_1 counts the matings for the ebony flies while N_2 counts the matings for the oregon flies. The corresponding variance estimate (3.55) is $V_{11}(t_0) = 8.03$, so that the test statistic (3.56) becomes

$$U(t_0) = \frac{Z_1(t_0)}{\sqrt{V_{11}(t_0)}} = \frac{15.56}{\sqrt{8.03}} = 5.49,$$

which is highly significant when compared with a standard normal distribution. □

We have derived $Z_1(t_0)$ as the accumulated difference of the Nelson-Aalen increments in the two samples; cf. (3.51). We will now rewrite $Z_1(t_0)$ in a way that offers an alternative interpretation, and which may more easily be generalized to more than two samples. Since $N_2 = N_. - N_1$, we obtain from (3.52) that

$$Z_1(t_0) = \int_0^{t_0} K(t)dN_1(t) - \int_0^{t_0} K(t)\frac{Y_1(t)}{Y_.(t)}dN_.(t), \tag{3.58}$$

where

$$K(t) = L(t)Y_.(t)\{Y_1(t)Y_2(t)\}^{-1} \tag{3.59}$$

is a nonnegative predictable "weight process." In particular, $K(t) = I\{Y_.(t) > 0\}$ for the log-rank test.

Thus for the log-rank test we have

$$Z_1(t_0) = N_1(t_0) - E_1(t_0),$$

where

$$E_1(t_0) = \int_0^{t_0} \frac{Y_1(t)}{Y_.(t)}dN_.(t) \tag{3.60}$$

can be interpreted as the number of events we would expect to observe in the first sample if the null hypothesis holds true (cf. Exercise 3.11). Thus the log-rank statistic is the difference between the observed and expected number of events in the first sample. Also the general statistic (3.58) can be given an "observed minus expected" interpretation, but here both the observed and expected number of events are weighted by the process K.

Example 3.13. Causes of death in Norway. In Example 3.3 we studied cause-specific mortality in Norway based on a sample of 2086 males and 1914 females. There we found that men had a much higher mortality than women due to cardiovascular disease and alcohol abuse, chronic liver disease, and accidents and violence, while the difference between the two genders is much smaller for deaths due to cancer and other medical causes. Here we will take a closer look at the mortality of the latter two causes.

We first look at cancer death, and let N_1 count the cancer deaths among males while N_2 counts the cancer deaths for females. We observe $N_1(t_0) = 129$ cancer deaths among the males between age 40 and 70 years, while we would expect $E_1(t_0) = 110.56$ if the cancer mortality is the same for the two genders (i.e., we use $t_0 = 70$). Thus by (3.58) we have $Z_1(t_0) = 129 - 110.56 = 18.44$. The variance estimate is $V_{11}(t_0) = 54.23$, so the test statistic (3.56) becomes

$$U(t_0) = \frac{Z_1(t_0)}{\sqrt{V_{11}(t_0)}} = \frac{18.44}{\sqrt{54.23}} = 2.50.$$

This gives a (two-sided) P-value of 1.2% when compared with a standard normal distribution. Thus, in the period 1974–2000, middle-aged Norwegian males had a significantly higher cancer mortality rate than females.

We then turn to deaths from other medical causes. Here we observe $N_1(t_0) = 34$ deaths among the males, which is close to what we would expect, $E_1(t_0) = 34.69$, if the mortality from other medical causes is the same for the two genders. The test statistic (3.56) here takes the nonsignificant value

$$U(t_0) = \frac{Z_1(t_0)}{\sqrt{V_{11}(t_0)}} = \frac{34 - 34.69}{\sqrt{16.99}} = -0.17,$$

corresponding to a (two-sided) P-value of 86.5%. $\square$

3.3.2 Extension to more than two samples

We will briefly indicate how the test statistic (3.57) may be extended to the situation where one has three or more counting processes, each satisfying the multiplicative intensity model, and the equality of the α-functions of their intensity processes is to be tested. More specifically, assume that we have the counting processes $N_1, \ldots, N_k$ with intensity process $\lambda_h(t) = \alpha_h(t)Y_h(t)$ for N_h; $h = 1, \ldots, k$. Based on these we want to test the null hypothesis

$$H_0 : \alpha_1(t) = \alpha_2(t) = \cdots = \alpha_k(t) \quad \text{for all } t \in [0, t_0].$$

For each h we define similar quantities as for the two-sample situation, and let $N_{\cdot} = \sum_{h=1}^{k} N_h$ and $Y_{\cdot} = \sum_{h=1}^{k} Y_h$. Further, we introduce a nonnegative predictable weight process $K(t)$, which is zero whenever $Y_{\cdot}(t)$ is zero. Generalizing the formulation (3.58) of the two-sample situation, we then define the processes

$$Z_h(t_0) = \int_0^{t_0} K(t) dN_h(t) - \int_0^{t_0} K(t) \frac{Y_h(t)}{Y_{\cdot}(t)} dN_{\cdot}(t) \tag{3.61}$$

for $h = 1, 2, \ldots, k$. One may show that the Z_h processes are mean zero martingales under the null hypothesis (when considered as a process in t_0). An unbiased estimator for the covariance between $Z_h(t_0)$ and $Z_j(t_0)$, which generalizes (3.55), is given by

$$V_{hj}(t_0) = \int_0^{t_0} K^2(t) \frac{Y_h(t)}{Y_{\cdot}(t)} \left(\delta_{hj} - \frac{Y_j(t)}{Y_{\cdot}(t)} \right) dN_{\cdot}(t), \tag{3.62}$$

where δ_{hj} is a Kronecker delta (i.e., equal to 1 when $h = j$ and zero otherwise).

Since $\sum_{h=1}^{k} Z_h(t_0) = 0$, we only consider the first $k-1$ of the $Z_h(t_0)$ when we form our test statistic. Thus we introduce the vector $\mathbf{Z}(t_0) = (Z_1(t_0), \ldots, Z_{k-1}(t_0))^T$ and the $(k-1) \times (k-1)$ matrix $\mathbf{V}(t_0)$ with entry (h, j) equal to $V_{hj}(t_0)$; for $h, j = 1, \ldots, k-1$. To test the null hypothesis, we may then use the test statistic

$$X^2(t_0) = \mathbf{Z}(t_0)^T \mathbf{V}(t_0)^{-1} \mathbf{Z}(t_0), \tag{3.63}$$

which is approximately chi-squared distributed with $k-1$ degrees of freedom when the null hypothesis holds true. This is a generalization of (3.57) to the situation with more than two samples.

Except for Efron's test, all of the tests for the survival data situation summarized in Table 3.2 can be extended to more than two samples. In order to obtain the weight process $K(t)$ for a given test, one just inserts the appropriate $L(t)$-process in (3.59).

Example 3.14. Time between births. In Examples 3.2 and 3.9 we gave Nelson-Aalen and Kaplan-Meier plots for the time between second and third births for 16 116 Norwegian women who had (at least) two births in the period 1983–97. The plots for the four groups of women indicate that the genders of the two older children affect the likelihood of a woman having a third birth. We will here investigate whether the observed differences are statistically significant.

We first test the null hypothesis that the hazard for the time between second and third births does not depend on the gender of the two older children. Using log-rank weights, the test statistic (3.63) takes the value $X^2(t_0) = 26.4$, with $t_0 = 7.5$ years corresponding to the largest observed time between second and third births. This is highly significant when compared to a chi-squared distribution with $4-1 = 3$ degrees of freedom, confirming that the hazard for the third birth depends on the gender of the two older children.

We then restrict attention to the women who have two children of the same gender and test the null hypothesis that the hazard for the third birth is the same for those women who have two boys and those who have two girls. Here the two-sample log-rank statistic (3.57) takes the value $X^2(t_0) = 3.90$, which is border line significant at conventional levels ($df = 1$, $P = 4.8\%$). Thus there is an indication that a woman who has two boys has a slightly increased hazard for a third birth compared to a woman who has two girls.

We finally look at the women who have one child of each gender and test the null hypothesis that the hazard for the third birth does not depend on the order of the boy and the girl. The two-sample log-rank statistic (3.57) now takes the insignificant value $X^2(t_0) = 0.08$ ($df = 1$, $P = 78\%$), so there is no indication of an effect of the order.

In conclusion, we have shown that the genders of the two older children significantly affect the likelihood of a woman having a third birth. This is mainly due to an increased hazard for a third birth for the women who have two children of the same gender compared to the women who have a girl and a boy. □

3.3.3 Stratified tests

In Sections 3.3.1 and 3.3.2 we consider two or more counting processes $N_h(t)$; $h = 1, \ldots, k$, satisfying the multiplicative intensity model $\lambda_h(t) = \alpha_h(t)Y_h(t)$, and we describe how we may test the hypothesis that the α_h-functions are equal.

An example of a situation covered by the results of Sections 3.3.1 and 3.3.2 is when the $N_h(t)$ count the number of deaths in each of k treatment groups (e.g., for

cancer patients), and we want to test the hypothesis that there is no difference between the treatments. When such a study is performed at m hospitals, say, there
will often be a variation in the results between the hospitals (e.g., due to differences in the patient populations). One would then like to test the hypothesis that the
effects of the treatments are the same at each hospital, allowing for heterogeneity
between hospitals. This is achieved by a simple extension of the test described in
Sections 3.3.1 and 3.3.2.

More generally, we will consider the situation in which we have counting pro-
cesses $N_{hs}(t)$; for $h = 1, \ldots, k$ and $s = 1, \ldots, m$; with intensity processes of the multiplicative form $\lambda_{hs}(t) = \alpha_{hs}(t) Y_{hs}(t)$. (For the example mentioned earlier, $N_{hs}(t)$
counts deaths for treatment h at hospital s.) We want to test the hypothesis that the
α_{hs}-functions are equal within each stratum s, while allowing them to vary across
strata. Thus we will test the hypothesis

$$H_0 : \alpha_{1s}(t) = \alpha_{2s}(t) = \cdots = \alpha_{ks}(t) \quad \text{for } t \in [0, t_0] \text{ for all } s = 1, \ldots, m.$$

In order to derive a test for H_0, we define for each stratum s the same quantities as
in Section 3.3.2, adding an index s to indicate the dependence on the actual stratum.
Thus we let $N_{\bullet s} = \sum_{h=1}^{k} N_{hs}$ and $Y_{\bullet s} = \sum_{h=1}^{k} Y_{hs}$, and introduce

$$Z_{hs}(t_0) = \int_0^{t_0} K_s(t) dN_{hs}(t) - \int_0^{t_0} K_s(t) \frac{Y_{hs}(t)}{Y_{\bullet s}(t)} dN_{\bullet s}(t)$$

and

$$V_{hjs}(t_0) = \int_0^{t_0} K_s^2(t) \frac{Y_{hs}(t)}{Y_{\bullet s}(t)} \left(\delta_{hj} - \frac{Y_{js}(t)}{Y_{\bullet s}(t)} \right) dN_{\bullet s}(t).$$

Further we introduce the vectors $\mathbf{Z}_s(t_0) = (Z_{1s}(t_0), \ldots, Z_{k-1,s}(t_0))^T$ and denote by
$\mathbf{V}_s(t_0)$ the $(k-1) \times (k-1)$ matrix with entry (h, j) equal to $V_{hjs}(t_0)$; $h, j = 1, \ldots, k-$
1. To test the null hypothesis, we may then aggregate information over the m strata
to obtain the test statistic

$$X^2(t_0) = \left(\sum_{s=1}^{m} \mathbf{Z}_s(t_0) \right)^T \left(\sum_{s=1}^{m} \mathbf{V}_s(t_0) \right)^{-1} \left(\sum_{s=1}^{m} \mathbf{Z}_s(t_0) \right),$$

which is approximately chi-squared distributed with $k-1$ degrees of freedom when
the null hypothesis holds true.

3.3.4 Handling of tied observations

In Section 3.1.3 we discuss how to handle tied event times for the Nelson-Aalen
estimator. Here we will describe the handling of ties for the k-sample tests of Section 3.3.2. In a similar manner as for the Nelson-Aalen estimator, we will concentrate on situations where the processes $N_1(t), \ldots, N_k(t)$ are obtained by aggregating

individual counting processes, and where $Y_h(t)$ is the number at risk "just before" time t for process h; $h = 1, 2, \ldots, k$. In addition, we now have to assume that no two $Y_h(t)$ are identically equal (as will be the case if we are comparing intensities in a Markov chain going from the same state).

In Section 3.1.3 we consider two approaches to the handling of ties for the Nelson-Aalen estimator. Approach (i) assumes that tied event times are due to grouping and rounding (of distinct event times), while approach (ii) assumes that the event times are genuinely discrete.

When it comes to testing, approach (i) is no longer straightforward to use, because we would have to take into account in which process an event occurs. For testing, approach (ii) is therefore the one commonly used. We let $T_1 < T_2 < \ldots < T_D$ be the distinct times when events are observed (in any process), denote by d_{hl} the number of observed events for process h at time T_l, and let $d_l = \sum_{h=1}^{k} d_{hl}$ be the total number of events at T_l. Then, corresponding to (3.61), we have the tie-corrected test statistic

$$Z_h(t_0) = \sum_{l=1}^{D} K(T_l)d_{hl} - \sum_{l=1}^{D} K(T_l)\frac{Y_h(T_l)}{Y.(T_l)}d_l, \qquad (3.64)$$

while the tie-corrected covariance estimator [cf. (3.62)] becomes:

$$V_{hj}(t_0) = \sum_{l=1}^{D} K^2(T_l)\frac{Y_h(T_l)}{Y.(T_l)}\left(\delta_{hj} - \frac{Y_j(T_l)}{Y.(T_l)}\right)\frac{Y.(T_l) - d_l}{Y.(T_l) - 1}d_l. \qquad (3.65)$$

Andersen et al. (1993, pages 352–354) show that (3.64) is a mean zero martingale in an extended model under the null hypothesis and that (3.65) is an unbiased estimator for the covariance between $Z_h(t_0)$ and $Z_j(t_0)$.

3.3.5 Asymptotics

In this subsection we give the main steps in the derivation of the large sample properties of the two-sample test statistics (3.56) for the situation without stratification. For a detailed account, also covering more than two samples, the reader is referred to Andersen et al. (1993, section V.2.2).

For ease of presentation, we concentrate on the log-rank test, where the weight process takes the form $L(t) = Y_1(t)Y_2(t)/Y.(t)$, and we only comment briefly at the end on the modifications needed for the general situation. We assume that the counting processes N_1 and N_2 are obtained by aggregating n_1 and n_2 individual processes, respectively, and let $n = n_1 + n_2$. For the log-rank test, we have under the null that

$$\frac{1}{\sqrt{n}}Z_1(t_0) = \int_0^{t_0} \frac{Y_2(t)}{\sqrt{n}Y.(t)}dM_1(t) - \int_0^{t_0} \frac{Y_1(t)}{\sqrt{n}Y.(t)}dM_2(t), \qquad (3.66)$$

is a martingale when considered as a process in t_0. This martingale is of the form (2.57) of Section 2.3.3 with $k = 2$,

$$H_1(t) = \frac{Y_2(t)}{\sqrt{n}Y_.(t)} \quad \text{and} \quad H_2(t) = -\frac{Y_1(t)}{\sqrt{n}Y_.(t)},$$

and $\lambda_h(t) = \alpha(t)Y_h(t)$ for $h = 1, 2$.

To show that (3.66) converges in distribution to a Gaussian martingale, we therefore need to check the conditions (2.60) and (2.61) of the martingale central limit theorem. To this end, we will assume that there exist positive deterministic functions y_h, $h = 1, 2$, such that $Y_h(t)/n$ converges in probability to $y_h(t)$ for all $t \in [0, t_0]$ as $n \to \infty$. Then

$$H_1(t)^2 \lambda_1(t) + H_2(t)^2 \lambda_2(t)$$

$$= \left(\frac{Y_2(t)}{\sqrt{n}Y_.(t)}\right)^2 \alpha(t)Y_1(t) + \left(-\frac{Y_1(t)}{\sqrt{n}Y_.(t)}\right)^2 \alpha(t)Y_2(t)$$

$$= \frac{Y_1(t)Y_2(t)}{nY_.(t)} \alpha(t) = \frac{(Y_1(t)/n) \cdot (Y_2(t)/n)}{Y_.(t)/n} \alpha(t)$$

$$\xrightarrow{P} \frac{y_1(t)y_2(t)}{y_.(t)} \alpha(t)$$

as $n \to \infty$, where $y_. = y_1 + y_2$. Thus condition (2.60) is fulfilled. Furthermore

$$H_1(t) = \frac{Y_2(t)}{\sqrt{n}Y_.(t)} = \frac{1}{\sqrt{n}} \frac{Y_2(t)/n}{Y_.(t)/n} \xrightarrow{P} 0$$

as $n \to \infty$, and similarly for $H_2(t)$, which shows that condition (2.61) is fulfilled as well.

By the martingale central limit theorem, it follows that (3.66) converges in distribution to a Gaussian martingale with variance function

$$v_{11}(t_0) = \int_0^{t_0} \frac{y_1(t)y_2(t)}{y_.(t)} \alpha(t)dt$$

when considered as a process in t_0. In particular for a given value of t_0, $n^{-1/2}Z_1(t_0)$ is asymptotically normally distributed with mean zero and variance $v_{11}(t_0)$. Further, one may show that n^{-1} times the variance estimator (3.55) of the log-rank test converges in probability to $v_{11}(t_0)$, and from this the asymptotic standard normality of the test statistic (3.56) follows.

With a general weight process, the proper normalization of Z_1 may no longer be $n^{-1/2}$ but will depend on the choice of $L(t)$. The normalization should be chosen such that condition (2.60) is fulfilled for the normalized martingale. With this modification, the arguments for the log-rank test may be repeated to show that (3.56) is asymptotically standard normally distributed in the general case.

3.4 The empirical transition matrix

The survival data situation may be described by a process with two states: "alive" and "dead." Splitting the state "dead" into two or more states, corresponding to different causes of death, the model for competing risks is obtained; cf. Section 1.2.2 and Example 3.3. The model for competing risks is a Markov chain. As discussed in Section 1.2.2, other Markov chains are also of importance in biomedical research; one such example being the three-state illness-death model.

For survival data, the probability of a transition from state "alive" to state "dead" may be estimated by one minus the Kaplan-Meier estimator. The Kaplan-Meier estimator may be generalized to arbitrary Markov chains with a finite number of states. Such a generalization was considered by Aalen (1978a) for the competing risks model, and by Aalen and Johansen (1978) and Fleming (1978a,b) for the general case. In particular, the product-integral formulation of Aalen and Johansen (1978) shows how the resulting estimator can be seen as a matrix version of the Kaplan-Meier estimator. We will call the estimator *the empirical transition matrix*, but note that it is often denoted the Aalen-Johansen estimator.

In this section, we first consider the competing risks model and the three-state Markov illness-death model for a chronic disease. This gives illustrations of the empirical transition matrix in two simple situations where its elements take an explicit form. We then present the estimator in general and show how it is obtained as the product-integral of the Nelson-Aalen estimators for the cumulative transition intensities. We also briefly indicate how the product-integral representation may be used to study the statistical properties of the empirical transition matrix, and how this provides estimators of variances and covariances.

3.4.1 Competing risks and cumulative incidence functions

In Section 1.2.2 we saw how the situation with k competing causes of death may be modeled by a Markov chain with one transient state 0, corresponding to "alive," and k absorbing states corresponding to "dead by cause h," $h = 1, 2, \ldots, k$. In this model for competing risks, the *cause-specific hazards* $\alpha_{0h}(t)$ describe the instantaneous risk of dying from each of the causes.

We denote by $P_{0h}(s,t)$ the probability that an individual who is in state 0 (i.e., alive) at time s will be in state h (i.e., dead from cause h) at a later time t; $h = 1, \ldots, k$, and remind the reader that the $P_{0h}(s,t)$ are often denoted *cumulative incidence functions*. The probability that an individual who is alive (i.e., in state 0) at time s will still be alive at a later time t is denoted $P_{00}(s,t)$. Thus $P_{00}(s,t) = 1 - \sum_{h=1}^{k} P_{0h}(s,t)$. One may show that (see Appendix A.2.5)

$$P_{00}(s,t) = \exp\left(-\int_s^t \sum_{h=1}^k \alpha_{0h}(u)du\right) \qquad (3.67)$$

and

$$P_{0h}(s,t) = \int_s^t P_{00}(s,u)\alpha_{0h}(u)du \qquad (3.68)$$

for $h = 1, 2, \ldots, k$.

Assume that we have a sample of n individuals from the population under study. Each individual is followed from an entry time to death or censoring. We denote by $T_1 < T_2 < \cdots$ the times when deaths *from any cause* are observed, and let $N_{0h}(t)$ be the process counting the number of individuals who are observed to die from cause h (i.e., make a transition from state 0 to state h) in the interval $[0,t]$. Further we write $N_{0.}(t) = \sum_{h=1}^k N_{0h}(t)$ for the total number of deaths in $[0,t]$, and let $Y_0(t)$ denote the number of individuals at risk (i.e., in state 0) just prior to time t. Then the survival probability (3.67) is estimated by the Kaplan-Meier estimator

$$\widehat{P}_{00}(s,t) = \prod_{s < T_j \le t} \left(1 - \frac{\triangle N_{0.}(T_j)}{Y_0(T_j)}\right), \qquad (3.69)$$

while the transition probabilities (3.68) are estimated by the *empirical cumulative incidence functions*

$$\widehat{P}_{0h}(s,t) = \sum_{s < T_j \le t} \widehat{P}_{00}(s, T_{j-1})\triangle\widehat{A}_{0h}(T_j). \qquad (3.70)$$

Here $\triangle\widehat{A}_{0h}(T_j) = \triangle N_{0h}(T_j)/Y_0(T_j)$ is the increment at time T_j of the Nelson-Aalen estimator $\widehat{A}_{0h}(t) = \int_0^t Y_0(u)^{-1}dN_{0h}(u)$ for the cumulative cause-specific hazard $A_{0h}(t) = \int_0^t \alpha_{0h}(u)du$; cf. Example 3.3. It should be noted that we obtain (3.70) from (3.68) by replacing $P_{00}(s,u) = P_{00}(s,u-)$ by $\widehat{P}_{00}(s,u-)$ and $\alpha_{0h}(u)du$ by $d\widehat{A}_{0h}(u) = dN_{0h}(s)/Y_0(s)$.

The variance of the Kaplan-Meier estimator $\widehat{P}_{00}(s,t)$ may be estimated as described in Section 3.2.1, while the variance of $\widehat{P}_{0h}(s,t)$ may be estimated by formula (3.89) given in Section 3.4.5. As the empirical cumulative incidence functions are approximately normally distributed in large samples (cf. Section 3.4.4), this may be used to provide confidence intervals for $P_{0h}(s,t)$ as illustrated in the following example.

Example 3.15. Causes of death in Norway. We consider the data of Example 1.12 on causes of death in a sample of 2086 males and 1914 females. Figure 3.17 shows the empirical cumulative incidence functions $\widehat{P}_{0h}(40,t)$ for the four causes of death:

1) Cancer
2) Cardiovascular disease including sudden death
3) Other medical causes
4) Alcohol abuse, chronic liver disease, and accidents and violence

The plots of the empirical cumulative incidence functions resemble the Nelson-Aalen plots of the cumulative cause-specific hazards given in Figure 3.6. However, the interpretations of the plots are different. While the *slope* of a Nelson-Aalen plot estimates the hazard for a given cause, the *value* of an empirical cumulative

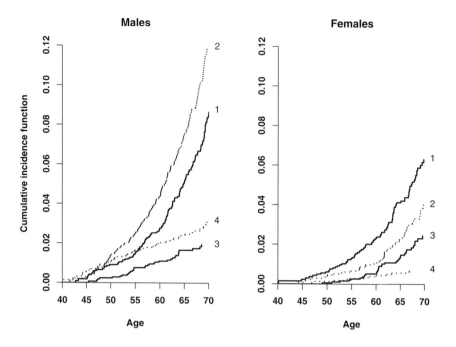

Fig. 3.17 *Empirical cumulative incidence functions for four causes of death among middle-aged Norwegian males (left) and females (right). 1) Cancer; 2) cardiovascular disease including sudden death; 3) other medical causes; 4) alcohol abuse, chronic liver disease, and accidents and violence.*

incidence function estimates the probability, or *absolute risk*, of death from a given cause, taking into account the risk of death from all other causes. For example, we see from Figure 3.17 that the estimated absolute risk of cardiovascular death between 40 and 70 years of age is 12.0% for males, while it is only 4.1% for females. For cancer death, the difference between the sexes is smaller; the estimated absolute risks are 8.6% and 6.3%, respectively, for males and females.

Figure 3.18 gives the empirical cumulative incidence function for cancer death for males and females with 95% confidence intervals. For comparison, the figure also shows the estimated absolute risk of cancer death disregarding the three competing causes of death (computed as one minus the Kaplan-Meier estimator treating deaths from the other causes as censorings). The latter is sometimes interpreted as estimating the probability of death due to cancer assuming this to be the only possible cause of death. Such an interpretation may be quite speculative, however; see the discussion in Kalbfleisch and Prentice (2002, chapter 8). The absolute risk estimates disregarding competing risks are of course larger than the estimates that take the competing causes of death into account. However, for females, the difference between the two is very small due to the fairly low risk of death from competing

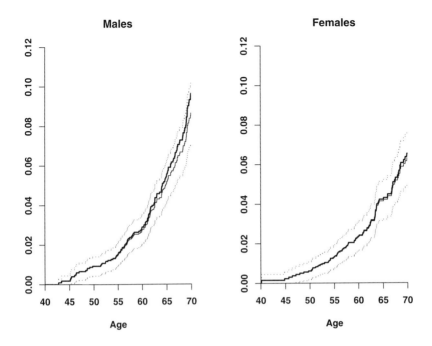

Fig. 3.18 *Empirical cumulative incidence functions for cancer death (thin drawn line) among middle-aged Norwegian males (left) and females (right) with 95% confidence intervals (dotted lines). Absolute risk estimates disregarding other causes of death are also given (thick drawn line).*

causes. For males, the difference is somewhat larger due in particular to the quite high risk of cardiovascular deaths. □

For the situation of Example 3.15, there were fairly small differences between the empirical cumulative incidence function for cancer death and the estimated absolute risk of cancer death disregarding the competing causes of death (cf. Figure 3.18). In other situations, this difference may be much more pronounced; see Putter et al. (2007) for one such example.

3.4.2 An illness-death model

In Section 1.2.2 we indicated that we may adopt the Markov illness-death model of Figure 3.19 to study the occurrence of a chronic disease as well as death. The transition intensities of the model are denoted $\alpha_{01}(t)$, $\alpha_{02}(t)$, and $\alpha_{12}(t)$, and they describe the instantaneous risks of transitions between the states. Further, for an individual who is healthy (i.e., in state 0) at time s, we write $P_{01}(s,t)$ for the probability

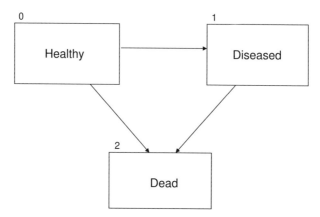

Fig. 3.19 *An illness-death model without recovery.*

that he is diseased (i.e., in state 1) at a later time t, while $P_{00}(s,t)$ is the probability that he is still healthy (i.e., in state 0) at that time. Similarly, for an individual who is diseased (i.e., in state 1) at time s, we let $P_{11}(s,t)$ denote the probability that he is still alive (i.e., in state 1) at time t. Then we have (see Appendix A.2.5)

$$P_{00}(s,t) = \exp\left\{-\int_s^t [\alpha_{01}(u) + \alpha_{02}(u)]\,du\right\},\tag{3.71}$$

$$P_{11}(s,t) = \exp\left(-\int_s^t \alpha_{12}(u)\,du\right),\tag{3.72}$$

$$P_{01}(s,t) = \int_s^t P_{00}(s,u)\alpha_{01}(u)P_{11}(u,t)\,du.\tag{3.73}$$

It is seen that (3.71) and (3.72) are of the same form as the survival probability in the survival data situation.

Assume that we have a sample of n individuals from the population under study and that each individual is followed from an entry time to death or censoring. Exact times of disease occurrences and deaths are recorded, and we denote by $T_1 < T_2 < T_3 < \cdots$ the times of *any observed event* (disease occurrence or death). Further, we let $N_{01}(t)$ count the number of individuals who get diseased (i.e., make a transition from state 0 to state 1) in $[0,t]$, while $N_{02}(t)$ and $N_{12}(t)$ count the numbers of disease-free, respectively diseased, individuals who are observed to die in this time interval. Finally, we write $N_{0.}(t) = N_{01}(t) + N_{02}(t)$, and let $Y_0(t)$ and $Y_1(t)$ be the number of healthy (i.e., in state 0) and diseased (i.e. in state 1) individuals, respectively, just prior to time t. Then (3.71) and (3.72) may be estimated by the Kaplan-Meier estimators

$$\widehat{P}_{00}(s,t) = \prod_{s < T_j \le t} \left(1 - \frac{\triangle N_{0.}(T_j)}{Y_0(T_j)}\right),\tag{3.74}$$

$$\widehat{P}_{11}(s,t) = \prod_{s<T_j\leq t}\left(1 - \frac{\triangle N_{12}(T_j)}{Y_1(T_j)}\right), \tag{3.75}$$

while an estimator for (3.73) is

$$\widehat{P}_{01}(s,t) = \sum_{s<T_j\leq t} \widehat{P}_{00}(s,T_{j-1})\triangle\widehat{A}_{01}(T_j)\widehat{P}_{11}(T_j,t). \tag{3.76}$$

Here $\triangle\widehat{A}_{01}(T_j) = \triangle N_{01}(T_j)/Y_0(T_j)$ is the increment of the Nelson-Aalen estimator $\widehat{A}_{01}(t) = \int_0^t Y_0(u)^{-1}dN_{01}(u)$. Note that (3.76) is obtained from (3.73) by replacing $P_{00}(s,u) = P_{00}(s,u-)$ by $\widehat{P}_{00}(s,u-)$, $P_{11}(u,t)$ by $\widehat{P}_{11}(u,t)$ and $\alpha_{01}(u)du$ by $d\widehat{A}_{01}(u) = dN_{01}(u)/Y_0(u)$.

The variances of the Kaplan-Meier estimators (3.74) and (3.75) may be estimated in the usual way (cf. Section 3.2), while the variance of $\widehat{P}_{01}(s,t)$ may be estimated by formula (3.90) given in Section 3.4.5. Used in conjunction with the approximate normality of the estimators in large samples (Section 3.4.4), this in the usual manner provides approximate confidence intervals for the transition probabilities.

Before we illustrate these results, let us mention that other interpretations of the states are possible than the ones given earlier. One interpretation arises in the study of complications to a disease. Here state 0 could correspond to "diseased with no complications," state 1 to "diseased with complications," and state 2 to "dead." Another interpretation is relevant in a study involving treatment of cancer. Here state 0 could correspond to "no response to treatment," state 1 to "response to treatment," and state 2 to "relapsed or dead." The probability $P_{01}(s,t)$ is then the *probability of being in response function* suggested by Temkin (1978) and sometimes used as an outcome measure when studying the efficacy of cancer treatment. In the following example, we adopt the latter interpretation of the states.

Example 3.16. Bone marrow transplantation. We consider the data of Example 1.13 on bone marrow transplantations for 137 patients with acute leukemia. The patients were followed for a maximum of seven years, and times to relapse and death were recorded. It was also recorded if and when the platelet count of a patient returned to a self-sustaining level.

The possible events for a patient may be described by a model of the form shown in Figure 3.19, but with the states 0, 1, and 2 corresponding to "transplanted," "platelet recovered," and "relapsed or dead," respectively. A patient starts out in state 0 at time $t = 0$ when he gets the bone marrow transplant. If his platelets recover, the patient moves to state 1, and if he then relapses or dies, he moves on to state 2. If the patient relapses or dies without the platelets returning to a normal level, he moves directly from state 0 to state 2.

Figure 3.20 shows Nelson-Aalen plots of the cumulative transition intensities $A_{01}(t) = \int_0^t \alpha_{01}(u)du$, $A_{02}(t) = \int_0^t \alpha_{02}(u)du$, and $A_{12}(t) = \int_0^t \alpha_{12}(u)du$. It is seen from the figure that the transition intensity for platelet recovery is very high the first few weeks after transplantation. We also see that the intensity for relapse or death is smaller after platelet recovery.

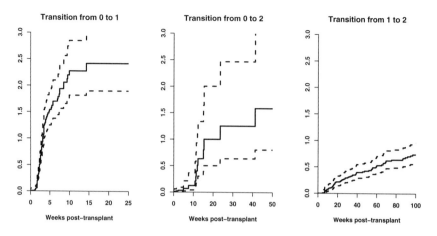

Fig. 3.20 *Nelson-Aalen estimates with log-transformed 95% confidence intervals of the cumulative transition intensities for the bone marrow transplant patients. The states are 0: "transplanted," 1: "platelet recovered," and 2: "relapsed or dead." Note that the time scale is not the same for the three estimates.*

We will consider platelet recovery as a response to the bone marrow transplantation. It is then of interest to estimate the probability of being in response function, that is, the probability of being alive t weeks after transplantation with platelet count on a self-sustaining level. The empirical probability of being in response function $\widehat{P}_{01}(0,t)$ is shown in Figure 3.21. We note that the empirical probability of being in response function increases steeply the first few weeks after transplantation as the platelets recover for most patients, and it reaches a maximum value of around 80% about 10 weeks after transplantation. Then the curve decreases due to relapse and death, and it is about 40% two years after transplantation. $\square$

3.4.3 The general case

We next consider a general Markov chain with a finite number of states that may be used to model the life histories of individuals from a population. Let $\mathscr{I} = \{0,1,\ldots,k\}$ be the state space of the Markov chain, and denote by $\alpha_{gh}(t)$ the transition intensity from state $g \in \mathscr{I}$ to state $h \in \mathscr{I}$, $g \neq h$. As described in Section 1.2.2, these transition intensities describe the instantaneous risks of transitions between the states. Further, for all $g,h \in \mathscr{I}$, we let $P_{gh}(s,t)$ denote the probability that an individual who is in state g at time s will be in state h at a later time t, and we write $\mathbf{P}(s,t)$ for the $(k+1) \times (k+1)$ matrix of these transition probabilities. Only for simple Markov chains, like the competing risks and illness-death models con-

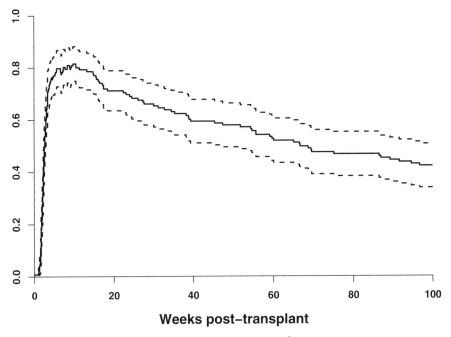

Fig. 3.21 *Empirical probability of being in response function $\widehat{P}_{01}(0,t)$ for the bone marrow transplant patients.*

sidered earlier, is it possible to give explicit expressions for the $P_{gh}(s,t)$ in terms of the transition intensities; cf. (3.67), (3.68) and (3.71)–(3.73).

One may, however, quite generally express the transition probability matrix $\mathbf{P}(s,t)$ in terms of the matrix of the transition intensities. To see how this can be done, we partition the time interval $(s,t]$ into a number of time intervals: $s = t_0 < t_1 < t_2 < \cdots < t_K = t$ and use the Markov property to write the transition probability matrix from time $s = t_0$ to $t = t_K$ as the following matrix product:

$$\mathbf{P}(s,t) = \mathbf{P}(t_0,t_1) \times \mathbf{P}(t_1,t_2) \times \cdots \times \mathbf{P}(t_{K-1},t_K) \tag{3.77}$$

If the number of time points increases, while the distance between them goes to zero in a uniform way, the product on the right-hand side of (3.77) approaches a limit termed a (matrix-valued) product-integral.

The product-integral is most usefully written in terms of the $(k+1) \times (k+1)$ matrix $\boldsymbol{\alpha}(u)$ of the transition intensities, that is the matrix where the (g,h)th off-diagonal elements equal the transition intensities $\alpha_{gh}(u)$; $g \neq h$; and the diagonal elements $\alpha_{gg}(u) = -\sum_{h \neq g} \alpha_{gh}(u)$ are chosen so that all row sums are zero. The transition intensities describe the instantaneous probabilities of transitions between the states. Therefore $\mathbf{P}(u, u+du) \approx \mathbf{I} + \boldsymbol{\alpha}(u)du$, where $\mathbf{I}$ is the $(k+1) \times (k+1)$ identity matrix. This explains why we may write the limit of equation (3.77) in product-integral form as $\mathbf{P}(s,t) = \prod_{(s,t]}\{\mathbf{I}+\boldsymbol{\alpha}(u)du\}$; see Appendix A.3.2 for further details. Alternatively, if we let $\mathbf{A}(t)$ denote the cumulative transition intensity matrix with elements $A_{gh}(t) = \int_0^t \alpha_{gh}(u)du$, we may write

$$\mathbf{P}(s,t) = \prod_{(s,t]}\{\mathbf{I}+d\mathbf{A}(u)\}. \tag{3.78}$$

This is a generalization of the product-integral representation (3.39) of the survival function. Note that the product-integral representation of the transition probability matrix is not restricted to the situation where transition intensities exist. In fact, (3.78) only assumes the existence of cumulative transition intensities $A_{gh}(t)$, which do not need to be absolutely continuous.

We may derive an estimator for $\mathbf{P}(s,t)$ by a heuristic argument paralleling the one we gave for the Kaplan-Meier estimator in Section 3.2.1. Suppose that we have a sample of n individuals from the population under study. The individuals may be followed over different periods of time, so our observations of their life histories may be subject to left-truncation and right-censoring. We assume that exact times for transitions between the states are recorded and denote by $T_1 < T_2 < \cdots$ the times when transitions *between any two states* are observed. Assuming that there are no tied transition times, we may make each of the time intervals $(t_{m-1},t_m]$ used in expression (3.77) so small that each of them contains at most one observed transition and that all left-truncations and right-censorings occur at the right-hand endpoint of an interval. Then, if no transition is observed in $(t_{m-1},t_m]$, we estimate $\mathbf{P}(t_{m-1},t_m)$ by $\mathbf{I}$. On the other hand, if a transition from state g to state h is observed at time $T_j \in (t_{m-1},t_m]$, we estimate $\mathbf{P}(t_{m-1},t_m)$ by a matrix of the form

$$\begin{pmatrix} 1 & 0 & 0 & 0 & 0 \\ 0 & 1 & 0 & 0 & 0 \\ 0 & 0 & 1-1/Y_g(T_j) & 0 & 1/Y_g(T_j) \\ 0 & 0 & 0 & 1 & 0 \\ 0 & 0 & 0 & 0 & 1 \end{pmatrix}, \tag{3.79}$$

where we show the case with five states (i.e., $k = 4$) and with a transition from state $g = 2$ to state $h = 4$. (Remember that the five states are labeled 0, 1, 2, 3, and 4.) This is an identity matrix except for row g where a part of the probability mass is moved to column h. By (3.77) we multiply such matrices for all transition times T_j with $s < T_j \leq t$ to obtain the *empirical transition matrix* $\widehat{\mathbf{P}}(s,t)$. We note that this empirical transition matrix in a very natural way generalizes the Kaplan-Meier estimator to finite-state Markov chains.

The empirical transition matrix may be expressed as a product-integral of the Nelson-Aalen estimators. To see how this can be done, for $g \neq h$, we introduce the process $N_{gh}(t)$ counting the number of individuals who are observed to experience a transition from state g to state h in $[0,t]$, and let $Y_g(t)$ be the number of individuals in state g just prior to time t. For $g \neq h$ we may then estimate the cumulative transition intensity $A_{gh}(t)$ by the Nelson-Aalen estimator $\widehat{A}_{gh}(t) = \int_0^t dN_{gh}(s)/Y_g(s)$. We further introduce $\widehat{A}_{gg}(t) = -\sum_{h \neq g}\widehat{A}_{gh}(t)$, and let $\widehat{\mathbf{A}}(t)$ be the $(k+1) \times (k+1)$ matrix with these elements. By (3.78) it is reasonable to estimate the transition probability matrix by $\widehat{\mathbf{P}}(s,t) = \prod_{(s,t]}\{\mathbf{I}+d\widehat{\mathbf{A}}(u)\}$. Now $\widehat{\mathbf{A}}(t)$ is a matrix of step functions with a finite number of jumps in $(s,t]$, and therefore this product-integral is just the finite matrix product

$$\widehat{\mathbf{P}}(s,t) = \prod_{s<T_j\leq t}\left(\mathbf{I}+\triangle\widehat{\mathbf{A}}(T_j)\right), \qquad (3.80)$$

where the matrix product is taken in the order of increasing T_j. We note that each of the factors in (3.80) is a matrix of the form (3.79). Thus (3.80) is the empirical transition matrix derived earlier by a heuristic argument.

For simple models like the competing risks model and the three-state illness-death model, we are able to give explicit expressions for the elements of (3.80); cf. (3.69), (3.70) and (3.74)–(3.76). In general, however, this is not possible. But in any case a direct implementation of (3.80) is simple, using software that can handle matrix multiplications.

We are often interested in estimating the probability, $p_h(t)$, that an individual is in state h at time t. One such situation is illustrated in Example 3.16. If all individuals start out in state 0 at time 0, we have $p_h(t) = P_{0h}(0,t)$. More generally, if we have no left-truncation but an individual starts out in state g with probability $p_g(0)$, we have

$$p_h(t) = \sum_{g=0}^{k} p_g(0)P_{gh}(0,t).$$

This state occupation probability may be estimated by

$$\widehat{p}_h(t) = \sum_{g=0}^{k} \widehat{p}_g(0)\widehat{P}_{gh}(0,t), \qquad (3.81)$$

where $\widehat{p}_g(0)$ is the proportion of the n individuals who start out in state g at time 0. Datta and Satten (2001) showed that the estimator (3.81) of the state occupation probability is valid also for non-Markov models provided that the censoring time for an individual is independent of the states occupied by the individual and his times of transition between the states. [Note that this is a stronger assumption than the common assumption of independent censoring (Sections 1.4 and 2.2.8), where censoring is allowed to depend on events in the past.] When censoring depends on the states occupied and the times of transition between the states, the estimator (3.81) may be modified using "inverse probability of censoring weighting" (cf. Section 4.2.7) to obtain a valid estimator for the state occupation probability (Datta and Satten, 2002).

3.4.4 Martingale representation and large sample properties

The product-integral formulation of the empirical transition matrix is useful for the study of its statistical properties. We will indicate a few main steps and refer to

Andersen et al. (1993, section IV.4) for a detailed account. For each $g \in \mathscr{I}$ we introduce an indicator $J_g(t)$ that is one if there is at least one individual in state g just before time t and zero otherwise. Further, for all $g, h \in \mathscr{I}$ we define $A_{gh}^*(t) = \int_0^t J_g(u) dA_{gh}(u)$ and let $\mathbf{A}^*(t)$ be the $(k+1) \times (k+1)$ matrix with these elements. Finally we introduce $\mathbf{P}^*(s,t) = \prod_{(s,t]} \{\mathbf{I} + d\mathbf{A}^*(u)\}$ and note that this is almost the same as $\mathbf{P}(s,t)$ [cf. (3.78)] when there is only a small probability that one or more states will be empty at times u between s and t. By Duhamel's equation [formula (A.25) in Appendix A.3.2], we may then write

$$\widehat{\mathbf{P}}(s,t)\mathbf{P}^*(s,t)^{-1} - \mathbf{I} = \int_{(s,t]} \widehat{\mathbf{P}}(s,u-)\mathrm{d}(\widehat{\mathbf{A}} - \mathbf{A}^*)(u)\mathbf{P}^*(s,u)^{-1}. \tag{3.82}$$

This is a matrix version of formula (3.48) for the Kaplan-Meier estimator. Now $\widehat{\mathbf{A}} - \mathbf{A}^*$ is a $(k+1) \times (k+1)$ matrix of martingales; cf. (3.23). It follows that the right-hand side of (3.82) is a matrix-valued stochastic integral and therefore itself a $(k+1) \times (k+1)$ matrix of mean zero martingales. As a consequence of this

$$\mathrm{E}\left\{\widehat{\mathbf{P}}(s,t)\mathbf{P}^*(s,t)^{-1}\right\} = \mathbf{I},$$

so the empirical transition matrix estimator is almost unbiased.

For large sample purposes, we may replace $\mathbf{P}^*$ by $\mathbf{P}$ in (3.82). Further, since the empirical transition matrix is uniformly consistent, we may also replace $\widehat{\mathbf{P}}$ by $\mathbf{P}$ on the right-hand side of (3.82) to obtain the approximation

$$\widehat{\mathbf{P}}(s,t)\mathbf{P}(s,t)^{-1} - \mathbf{I} \approx \int_{(s,t]} \mathbf{P}(s,u)\,\mathrm{d}(\widehat{\mathbf{A}} - \mathbf{A}^*)(u)\,\mathbf{P}(s,u)^{-1}.$$

Multiplication by $\mathbf{P}(s,t) = \mathbf{P}(s,u)\mathbf{P}(u,t)$ then gives

$$\widehat{\mathbf{P}}(s,t) - \mathbf{P}(s,t) \approx \int_{(s,t]} \mathbf{P}(s,u)\,\mathrm{d}(\widehat{\mathbf{A}} - \mathbf{A}^*)(u)\,\mathbf{P}(u,t). \tag{3.83}$$

Using (3.83) and the asymptotic normality of the Nelson-Aalen estimator (Section 3.1.6), one may show that the elements of the empirical transition matrix are approximately normally distributed in large samples.

3.4.5 Estimation of (co)variances

Based on the approximaton (3.83) we may obtain an estimator for the covariance matrix of the empirical transition matrix. The covariance matrix of this $(k+1) \times (k+1)$ *matrix-valued* estimator is defined as follows. First we arrange the columns of $\widehat{\mathbf{P}}(s,t)$ on top of each other into one vector of dimension $(k+1)^2$ denoted $\mathrm{vec}\{\widehat{\mathbf{P}}(s,t)\}$. Then the covariance matrix of $\widehat{\mathbf{P}}(s,t)$ is defined as the ordinary covariance matrix of the vector $\mathrm{vec}\{\widehat{\mathbf{P}}(s,t)\}$. We note that the covariance matrix

$\text{cov}\{\widehat{\mathbf{P}}(s,t)\}$ is a $(k+1)^2 \times (k+1)^2$ matrix and that the (g,h)th block of it, when written as a partitioned matrix in the obvious way, is the $(k+1) \times (k+1)$ matrix of covariances between the elements of the gth and hth columns of $\widehat{\mathbf{P}}(s,t)$.

We denote the covariance matrix of the mean zero matrix-valued martingale $\widehat{\mathbf{A}}(u) - \mathbf{A}^*(u)$ by $\mathbf{\Psi}(u)$. Using elementary properties of the vec-operator and Kronecker products of matrices (e.g., Mardia et al., 1979, appendix A.2.5), the covariance matrix of the integrand on the right-hand side of (3.83) may be written

$$\text{cov}\{\mathbf{P}(s,u)\,\text{d}(\widehat{\mathbf{A}} - \mathbf{A}^*)(u)\,\mathbf{P}(u,t)\} \tag{3.84}$$

$$= \{\mathbf{P}(u,t)^T \otimes \mathbf{P}(s,u)\}\,d\mathbf{\Psi}(u)\,\{\mathbf{P}(u,t) \otimes \mathbf{P}(s,u)^T\}.$$

We may then accumulate estimates of the right-hand side of (3.84) over all u between s and t (using the fact that the martingale $\widehat{\mathbf{A}}(u) - \mathbf{A}^*(u)$ has uncorrelated increments) to obtain the estimator

$$\widehat{\text{cov}}\{\widehat{\mathbf{P}}(s,t)\} = \int_s^t \{\widehat{\mathbf{P}}(u,t)^T \otimes \widehat{\mathbf{P}}(s,u)\}\,d\widehat{\mathbf{\Psi}}(u)\,\{\widehat{\mathbf{P}}(u,t) \otimes \widehat{\mathbf{P}}(s,u)^T\} \tag{3.85}$$

of the covariance matrix of the empirical transition matrix.

$\widehat{\mathbf{\Psi}}(u) = \{\widehat{\psi}_{hj,lm}(u)\}$ is a $(k+1)^2 \times (k+1)^2$ matrix with elements given as follows. For $h \neq j$, we let

$$\widehat{\sigma}_{hj}^2(u) = \int_0^u Y_h(v)^{-2} dN_{hj}(v) \tag{3.86}$$

be the estimated variance of the Nelson-Aalen estimator $\widehat{A}_{hj}(u)$; cf. (3.24). Then

$$\widehat{\psi}_{hj,lm}(u) = \begin{cases} \sum_{j \neq h} \widehat{\sigma}_{hj}^2(u) & \text{for } h = j = l = m, \\ -\widehat{\sigma}_{hm}^2(u) & \text{for } h = j = l \neq m, \\ \widehat{\sigma}_{hj}^2(u) & \text{for } h = l \neq j = m, \\ 0 & \text{otherwise.} \end{cases} \tag{3.87}$$

The estimator (3.85) is a generalization of the variance estimator (3.27) for the Kaplan-Meier estimator.

The $(k+1)^2 \times (k+1)^2$ matrix (3.85) is written in a very compact form. One may, however, derive explicit expressions for its elements. Specifically, one may show that for any $g,h,m,r \in \mathscr{I}$, the covariance between $\widehat{P}_{gh}(s,t)$ and $\widehat{P}_{mr}(s,t)$ may be estimated by

$$\widehat{\text{cov}}(\widehat{P}_{gh}(s,t), \widehat{P}_{mr}(s,t)) = \sum_{l=0}^k \sum_{q \neq l} \sum_{s < T_j \leq t} \left\{ \widehat{P}_{gq}(s,T_j)\widehat{P}_{mq}(s,T_j) \right.$$

$$\left. \times [\widehat{P}_{lh}(T_j,t) - \widehat{P}_{qh}(T_j,t)][\widehat{P}_{lr}(T_j,t) - \widehat{P}_{qr}(T_j,t)] \triangle \widehat{\sigma}_{ql}^2(T_j) \right\}, \tag{3.88}$$

where $\triangle \widehat{\sigma}_{ql}^2(T_j) = \triangle N_{ql}(T_j)/Y_q(T_j)^2$ is the increment of the variance estimator of the Nelson-Aalen estimator $\widehat{A}_{ql}(t)$.

This formula can be simplified in special cases. In particular, for competing risks (cf. Section 3.4.1), we get

$$\widehat{\text{var}}\,\widehat{P}_{0h}(s,t) = \sum_{s<T_j\leq t} [\widehat{P}_{00}(s,T_j)\,\widehat{P}_{0h}(T_j,t)]^2 \triangle \widehat{\sigma}_{0.}^2(T_j) \qquad (3.89)$$

$$+ \sum_{s<T_j\leq t} \widehat{P}_{00}(s,T_j)^2\,[1 - 2\widehat{P}_{0h}(T_j,t)]\,\triangle \widehat{\sigma}_{0h}^2(T_j),$$

where $\triangle \widehat{\sigma}_{0.}^2(T_j) = \sum_{h=1}^{k}\triangle \widehat{\sigma}_{0h}^2(T_j) = \triangle N_{0.}(T_j)/Y_0(T_j)^2$. For the illness-death model (cf. Section 3.4.2), we get the variance estimator:

$$\widehat{\text{var}}\,\widehat{P}_{01}(s,t) = \sum_{s<T_j\leq t} \widehat{P}_{00}(s,T_j)^2\,[\widehat{P}_{11}(T_j,t) - \widehat{P}_{01}(T_j,t)]^2 \triangle \widehat{\sigma}_{01}^2(T_j)$$

$$+ \sum_{s<T_j\leq t} [\widehat{P}_{00}(s,T_j)\widehat{P}_{01}(T_j,t)]^2 \triangle \widehat{\sigma}_{02}^2(T_j) \qquad (3.90)$$

$$+ \sum_{s<T_j\leq t} [\widehat{P}_{01}(s,T_j)\widehat{P}_{11}(T_j,t)]^2 \triangle \widehat{\sigma}_{12}^2(T_j).$$

Also the variance estimator (3.27) for the Kaplan-Meier estimator is a special case of (3.88).

3.5 Exercises

3.1 The data in the table are from Freireich et al. (1963) and show the result of a study where children with leukemia are treated with a drug (6-MP) to prevent relapse, and where this treatment is compared with placebo. The numbers in the table are remission lengths in weeks; a * indicates a censored observation.

Placebo	1	1	2	2	3	4	4	5	5	8	8
	8	8	11	11	12	12	15	17	22	23	
6-MP	6	6	6	6*	7	9*	10	10*	11*	13	16
	17*	19*	20*	22	23	25*	32*	32*	34*	35*	

a) Compute the Nelson-Aalen estimates for the 6-MP group and for the placebo group.
b) Plot both Nelson-Aalen estimates in the same figure. What can you learn from the plots?

3.2 Suggest at least one method based on the Nelson-Aalen estimator that may be used to check graphically whether two hazard rates are proportional. Illustrate the method(s) using the leukemia data from Exercise 3.1.

3.3 In Section 3.1.6 it is proved that, for a given value of t, the Nelson-Aalen estimator $\widehat{A}(t)$ is approximately normally distributed with mean $A(t)$ and a variance that may be estimated by $\widehat{\sigma}^2(t)$ given by (3.5).

a) Let g be a strictly increasing and continuously differentiable function. By a Taylor series expansion we have that

$$g(\widehat{A}(t)) \approx g(A(t)) + g'(A(t))\{\widehat{A}(t) - A(t)\}.$$

Use this to show informally that $g(\widehat{A}(t))$ is approximately normally distributed with mean $g(A(t))$ and a variance that may be estimated by $\{g'(\widehat{A}(t))\}^2\, \widehat{\sigma}^2(t)$. (This is a special case of the well-known delta method.)

b) Show that $g(\widehat{A}(t)) \pm z_{1-\alpha/2}\, g'(\widehat{A}(t))\, \widehat{\sigma}(t)$ is an approximate $100(1-\alpha)\%$ confidence interval for $g(A(t))$.

By inverting the confidence interval in b) (i.e., by taking the inverse g^{-1} of its lower and upper limits), we obtain an approximate $100(1-\alpha)\%$ confidence interval for $A(t)$. By a suitable choice of the transformation g, such a transformed confidence interval may have better properties for small and moderate sample sizes than the standard one given in (3.6).

c) Show that the transformation $g(x) = \log x$ yields the log-transformed confidence interval (3.7).

3.4 We consider the leukemia data from Exercise 3.1.

a) Compute the Kaplan-Meier estimates for the 6-MP group and for the placebo group.

b) Plot both Kaplan-Meier estimates in the same figure. What can you learn from the plots?

3.5 In this exercise we will take a look at the Kaplan-Meier estimator $\widehat{S}(t)$ and Greenwood's formula for the situation where there is no censoring. To this end assume that $X_1, X_2, \ldots X_n$ are *iid* nonnegative random variables with survival function $S(t) = P(X_i > t)$ and cumulative distribution function $F(t) = P(X_i \le t) = 1 - S(t)$.

a) Show that $\widehat{S}(t) = 1 - \widehat{F}(t)$, where $\widehat{F}(t) = n^{-1}\sum_{i=1}^{n} I(X_i \le t)$ is the empirical cumulative distribution function.

b) Let $\widetilde{\tau}^2(t)$ be given by (3.28). Show that $\widetilde{\tau}^2(t) = n^{-1}\{\widehat{S}(t)(1 - \widehat{S}(t))\}$. [*Hint:* Show that $\widetilde{\tau}^2(t)/\widehat{S}(t)^2$ and $(1 - \widehat{S}(t))/(n\widehat{S}(t))$ are equal by observing that they are constant between the observed survival times and by showing that their increments at the observed survival times are the same.]

3.6   In Section 3.2.6 it is proved that, for given value of t, the Kaplan-Meier estimator $\widehat{S}(t)$ is approximately normally distributed with mean $S(t)$ and a variance that may be estimated by $\widehat{\tau}^2(t)$ given by (3.27) [or alternatively by Greenwood's formula (3.28)]. In Exercise 3.3 we discussed (in the context of the Nelson-Aalen estimator) how one may derive confidence intervals based on transformations. Use this approach to show that the transformation $g(x) = -\log(-\log x)$ yields the log-minus-log transformed confidence interval (3.30) for the survival function.

3.7   We consider the leukemia data from Exercise 3.1.

a) Make a plot of the Kaplan-Meier estimate for the placebo group with 95% log-minus-log transformed confidence limits.
b) Use the plot to find an estimate and a 95% confidence interval for the median relapse time.

3.8   By the results of Section 3.2.6 we have that the standardized Kaplan-Meier estimator $\{\widehat{S}(t) - S(t)\}/\widehat{\tau}(t)$ is approximately standard normally distributed, for given value of t, where $\widehat{\tau}^2(t)$ is the variance estimator (3.27). [Alternatively we may estimate the variance by Greenwood's formula (3.28).]

a) Let ξ_p be the pth fractile of the survival distribution (cf. Section 3.2.3), and consider testing the hypothesis $H_0 : \xi_p = \xi_p^0$ versus the alternative $H_A : \xi_p \neq \xi_p^0$. Explain why we get a test with level α if we reject the hypothesis if $|\widehat{S}(\xi_p^0) - (1-p)|/\widehat{\tau}(\xi_p^0) > z_{1-\alpha/2}$.
b) A $100(1-\alpha)\%$ confidence interval for ξ_p is given as all values ξ_p^0 of ξ_p that are not rejected by the test in question a). Explain why this interval may be read directly from the lower and upper confidence limits (3.29) as illustrated in Figure 3.13.

For ease of presentation we consider the situation without transformations in this exercise. However, the arguments in questions a) and b) carry over with only minor modifications to the situation in which we use a transformation of the Kaplan-Meier estimator.

3.9   We consider the leukemia data from Exercise 3.1. Use the log-rank test to test the hypothesis that the relapse rate is the same in the two groups.

3.10   Show that the variance estimator (3.55) for the two sample tests is unbiased under the null hypothesis. [*Hint:* The aggregated counting process $N_{\textbf{.}}(t)$ has intensity process $\lambda_{\textbf{.}}(t) = \alpha(t)Y_{\textbf{.}}(t)$ under the null hypothesis.]

3.11   Consider $E_1(t_0)$ given by (3.60). Show that $E\{E_1(t_0)\} = E\{N_1(t_0)\}$ under the null hypothesis, that is, that $E_1(t_0)$ can be interpreted as the number of events we would expect to observe in the first sample if the null hypothesis holds true. [*Hint:* $Z_1(t) = N_1(t) - E_1(t)$ is a mean zero martingale under the null hypothesis.]

3.12 Let $N(t)$ be a counting process that satisfies the multiplicative intensity model $\lambda(t) = \alpha(t)Y(t)$, and consider the null hypothesis

$$H_0 : \alpha(t) = \alpha_0(t) \quad \text{for all } t \in [0, t_0],$$

where $\alpha_0(t)$ is a *known function*. In this exercise we derive a test for the null hypothesis that has good properties for alternatives of the form $\alpha(t) \geq \alpha_0(t)$ or $\alpha(t) \leq \alpha_0(t)$.

a) Let $J(t) = I\{Y(t) > 0\}$ and $\widehat{A}(t) = \int_0^t \{J(s)/Y(s)\}dN(s)$, and introduce $A_0^*(t) = \int_0^t J(s)\alpha_0(s)ds$. Show that $\widehat{A} - A_0^*$ is a mean zero martingale when the null hypothesis holds true.

b) Find an expression for the predictable variation process of $\widehat{A} - A_0^*$ when the null hypothesis holds true.

Consider the test statistic

$$Z(t_0) = \int_0^{t_0} L(t)\{d\widehat{A}(t) - dA_0^*(t)\},$$

where $L(t)$ is a nonnegative predictable "weight process" that takes the value 0 whenever $Y(t) = 0$.

c) Show that $Z(t_0)$ is a mean zero martingale when the null hypothesis holds true (when considered as a process in t_0), and explain why it is reasonable to use $Z(t_0)$ as a test statistic.

d) Show that

$$\langle Z \rangle(t_0) = \int_0^{t_0} \frac{L^2(t)}{Y(t)} \alpha_0(t)dt$$

when the null hypothesis holds true, and explain why this is an unbiased estimator for the variance of $Z(t_0)$ under the null hypothesis.

e) Explain briefly why the standardized test statistic $Z(t_0)/\sqrt{\langle Z \rangle(t_0)}$ is approximately standard normally distributed when the null hypothesis holds true.

One possible choice of weight process is $L(t) = Y(t)$. The test then obtained is usually denoted the one-sample log-rank test.

f) Show that, for the one-sample log-rank test, we have $Z(t_0) = N(t_0) - E(t_0)$, where $E(t_0) = \int_0^{t_0} Y(s)\alpha_0(s)ds$. Explain why $E(t_0)$ may be interpreted as (an estimate of) the expected number of events under the null hypothesis.

g) Show that the standardized version of the one-sample log-rank statistic takes the form $\{N(t_0) - E(t_0)\}/\sqrt{E(t_0)}$.

3.13 In this exercise we use data from a study of an intrauterine device (IUD) for prevention (Peterson, 1975). One hundred women participated in the study and were followed for a maximum of 300 days. For each woman, the number of days from the insertion of the device to the first of the following events were recorded:

1) The IUD is expelled
2) The IUD is removed (for personal or medical reasons)
3) The woman is lost to follow-up (censoring)

The recorded times of these events are given in the table.

IUD expelled	IUD removed		Censoring
2	14	166	86
8	21	178	203
10	27	183	207
25	40	272	
28	42	272	
28	92	288	
32	110	288	
63	147	288	
86	148	297	
158	165		

a) Describe the competing risks in this situation.
b) Compute and plot the Nelson-Aalen estimates $\widehat{A}_{0h}(t)$ for the cumulative cause-specific hazard rates for causes $h = 1, 2$ and interpret the plots.
c) Calculate and plot the cumulative incidence functions $\widehat{P}_{0h}(0,t)$ for causes $h = 1, 2$ and interpret the plots.
d) Discuss the interpretation of the estimates in b) and c).

Chapter 4
Regression models

In the previous chapter we considered situations where the data may be summarized into one or a few counting processes registering the occurrences of an event of interest. Such situations occur when the population in question is grouped into a few subpopulations according to the value of one or two categorical covariates. However, usually there are more than two covariates of interest in a study, and some of them may be numeric. Then, as in almost all parts of statistics, grouping is no longer a useful option, and regression models are called for.

For regression models we have to consider one counting process for each individual under study. Specifically, if we have data from n individuals, we consider the counting processes $N_1, N_2, \ldots, N_n$ with $N_i(t)$ counting the number of occurrences of the event of interest for individual i in $[0,t]$ (Section 1.5.2). For survival data $N_i(t) = 1$ if by time t the event has been observed to occur for individual i, otherwise $N_i(t) = 0$. More generally, we consider situations where the event in question may occur more than once for each individual. For such recurrent event data the process $N_i(t)$ may take integer values larger than 1.

Regression models for counting process data may be formulated in different ways. A common approach is to model the effect of the covariates on the intensity process of a counting process. To this end, assume that at time t for individual i we have available a covariate vector $\mathbf{x}_i(t) = (x_{i1}(t), \ldots, x_{ip}(t))^T$ whose components are fixed or time-varying, and that the intensity process of N_i may be written

$$\lambda_i(t) = Y_i(t)\alpha(t|\mathbf{x}_i). \tag{4.1}$$

Here $Y_i(t)$ is, as usual, an indicator taking the value 1 if individual i is at risk for the event of interest just before time t and the value 0 otherwise, while the intensity or hazard rate $\alpha(t|\mathbf{x}_i)$ of individual i is defined conditional on the values of the fixed covariates as well as on the observed paths of the time-varying ones. Note that we use "hazard rate" as a generic term for $\alpha(t|\mathbf{x}_i)$ also outside the survival data situation. Then, to obtain a regression model, we need to specify how $\alpha(t|\mathbf{x}_i)$ depends on $\mathbf{x}_i(t)$. In Section 4.1 we consider Cox's semiparametric regression model and related relative risk regression models, while the nonparametric additive regression model due to Aalen (1980, 1989) is considered in Section 4.2.

Standard use of the relative risk regression models and the additive regression model requires that we collect information on the covariate values for all individuals in a study even when only a few individuals experience the event of interest. This may be very expensive, or even logistically impossible, in large studies. In Section 4.3 we describe how case-control sampling techniques offer a useful alternative in such situations.

In model (4.1) there is an implicit assumption that censoring and truncation are independent in the sense discussed in Sections 1.4.2, 1.4.3, and 2.2.8. Thus $\alpha(t|\mathbf{x}_i)$ is the same as one would have in the situation without censoring and truncation. The independent censoring assumption implies that censoring may depend on information in "the past" but not on future events. As a consequence, censoring may depend on covariates that are included in the model but not on covariates that are not included.

Some comments on covariates are in order. Throughout the book we tacitly assume that all covariates are *predictable*. This implies that a *fixed* covariate should be measured in advance (i.e., at time zero) and remain fixed throughout the study, while the value at time t of a time-dependent covariate should be known just before time t.

In the following, it is important to distinguish between *external* and *internal* covariates (Kalbfleisch and Prentice, 2002, section 6.3). The most important type of external (or exogenous) covariates are the *fixed* covariates. External covariates may, however, also depend on time. There are two types of external time-dependent covariates. For a *defined* time-dependent covariate the complete path of the covariate is given at the outset of the study. One example of a defined time-dependent covariate is an individual's age at study time t. Another example is a covariate of the form $xg(t)$, where x is a fixed covariate and $g(t)$ is a given function of time, for example, $g(t) = \log t$. An *ancillary* time-dependent covariate is the observed path of a stochastic process whose development over time is *not* influenced by the occurrences of the event being studied. An example of an ancillary time-dependent covariate is the observed level of air pollution. For the purpose of statistical modeling, we may assume that the complete paths of external time-dependent covariates are given at the outset of the study (and hence are part of the available information on the past at any time t). Time-dependent covariates that are not external are called *internal* (or endogenous). Examples of internal time-dependent covariates are the values of biochemical markers measured on the individuals during follow-up.

Most of the results of this chapter are valid for both external and internal covariates. Exceptions are the results of Sections 4.1.2, 4.1.6, and 4.2.9 on estimation of survival probabilities and Markov transition probabilities that require covariates to be external. However, as the interpretation of regression analyses with internal time-dependent covariates is not at all straightforward, we will in this chapter only consider practical examples in which all covariates are external. The situation with internal time-dependent covariates is discussed in Chapters 8 and 9.

For ease of exposition, in the main body of the chapter we will assume that we are interested in only one type of event for each individual. However, most results carry over with only minor modifications to the situation where more than one type of event is of interest (e.g., deaths due to different causes), and this situation is considered briefly in Section 4.2.9.

4.1 Relative risk regression

The most common regression models in survival and event history analysis are the relative risk regression models. For these models it is assumed that the vector of covariates $\mathbf{x}_i(t)$ for individual i is related to the hazard rate $\alpha(t|\mathbf{x}_i)$ of the individual by the relation

$$\alpha(t\,|\,\mathbf{x}_i) = \alpha_0(t)\,r(\boldsymbol{\beta},\mathbf{x}_i(t)). \tag{4.2}$$

Here $r(\boldsymbol{\beta},\mathbf{x}_i(t))$ is a relative risk function, $\boldsymbol{\beta} = (\beta_1,\ldots,\beta_p)^T$ is a vector of regression coefficients describing the effect of the covariates, and $\alpha_0(t)$ is a baseline hazard rate that is left unspecified. Model (4.2) contains both a nonparametric part (the baseline hazard) and a parametric part (the relative risk function), and is therefore said to be semiparametric. We normalize the relative risk function by assuming $r(\boldsymbol{\beta},\mathbf{0}) = 1$. Thus $\alpha_0(t)$ corresponds to the hazard rate of an individual with all covariates identically equal to zero. For the exponential relative risk function $r(\boldsymbol{\beta},\mathbf{x}_i(t)) = \exp\{\boldsymbol{\beta}^T\mathbf{x}_i(t)\}$, formula (4.2) gives the usual Cox regression model (Cox, 1972). Other possibilities include the linear relative risk function $r(\boldsymbol{\beta},\mathbf{x}_i(t)) = 1 + \boldsymbol{\beta}^T\mathbf{x}_i(t)$ and the excess relative risk model $r(\boldsymbol{\beta},\mathbf{x}_i(t)) = \prod_{j=1}^{p}\{1 + \beta_j x_{ij}(t)\}$.

Consider two individuals, indexed 1 and 2, with vector of covariates $\mathbf{x}_1(t)$ and $\mathbf{x}_2(t)$, respectively. The ratio of their hazard rates is

$$\frac{\alpha(t\,|\,\mathbf{x}_2)}{\alpha(t\,|\,\mathbf{x}_1)} = \frac{r(\boldsymbol{\beta},\mathbf{x}_2(t))}{r(\boldsymbol{\beta},\mathbf{x}_1(t))}. \tag{4.3}$$

If all covariates are fixed, this ratio is constant over time, and the model (4.2) is called a proportional hazards model. [Note that if $\mathbf{x}_1(t) = \mathbf{0}$ the hazard rate ratio (4.3) becomes $r(\boldsymbol{\beta},\mathbf{x}_2(t))$. Thus it would be more appropriate to denote $r(\boldsymbol{\beta},\mathbf{x}(t))$ a hazard rate ratio function. Nevertheless, we will stick to the more commonly used term relative risk function.]

Assume that all components of $\mathbf{x}_1(t)$ and $\mathbf{x}_2(t)$ are equal, except the jth component, where $x_{2j}(t) = x_{1j}(t) + 1$, and consider the special case of Cox's model. Then the hazard ratio (4.3) becomes

$$\frac{\alpha(t\,|\,\mathbf{x}_2)}{\alpha(t\,|\,\mathbf{x}_1)} = \exp\left\{\boldsymbol{\beta}^T\left(\mathbf{x}_2(t) - \mathbf{x}_1(t)\right)\right\} = e^{\beta_j}.$$

Thus the effect of one unit's increase in the jth covariate, when all other covariates are kept the same, is to multiply the hazard rate by e^{β_j}. It is common to say that e^{β_j}

is the *relative risk* of the jth covariate (even though the term hazard rate ratio is more appropriate). Similar simple interpretations of the regression coefficients are not available for the linear relative risk model or the excess relative risk model; this is one reason why Cox's model is the one commonly used.

Cox's regression model has been given extensive treatment in a number of books. Andersen et al. (1993), Fleming and Harrington (1991), and Kalbfleisch and Prentice (2002) give detailed theoretical treatments using counting process theory. More applied presentations are given by Hosmer and Lemeshow (1999), Klein and Moeschberger (2003), and Therneau and Grambsch (2000); the last emphasizes computational aspects.

We do not aim at an exhaustive treatment of the Cox model and other relative risk regression models in this section. Rather, our goal is to present the basic properties together with some selected topics that are of particular interest for other parts of the book. One such topic is the martingale residual processes discussed in Section 4.1.3. We find these to be quite useful for model checking in a number of situations. However, the reader should be aware that many other methods for model checking are available for the Cox model; see, in particular, the last three books mentioned earlier.

4.1.1 Partial likelihood and inference for regression coefficients

The semiparametric nature of model (4.2) makes it impossible to use ordinary likelihood methods (to be considered in Chapter 5) to estimate the regression coefficients. Instead one has to resort to a partial likelihood. We will show how this can be derived, essentially following Cox's original argument (Cox, 1975).

First, note that, by combining (4.1) and (4.2), the intensity process of $N_i(t)$ may be written

$$\lambda_i(t) = Y_i(t)\,\alpha_0(t)\,r(\boldsymbol{\beta}, \mathbf{x}_i(t)). \tag{4.4}$$

Then introduce the aggregated counting process $N_\cdot(t) = \sum_{l=1}^n N_l(t)$, registering events among all individuals, and note that this has intensity process

$$\lambda_\cdot(t) = \sum_{l=1}^n \lambda_l(t) = \sum_{l=1}^n Y_l(t)\,\alpha_0(t)\,r(\boldsymbol{\beta}, \mathbf{x}_l(t)). \tag{4.5}$$

The intensity process of $N_i(t)$ may be factorized as $\lambda_i(t) = \lambda_\cdot(t)\,\pi(i|t)$, where

$$\pi(i|t) = \frac{\lambda_i(t)}{\lambda_\cdot(t)} = \frac{Y_i(t)\,r(\boldsymbol{\beta}, \mathbf{x}_i(t))}{\sum_{l=1}^n Y_l(t)\,r(\boldsymbol{\beta}, \mathbf{x}_l(t))} \tag{4.6}$$

is the conditional probability of observing an event for individual i at time t, given the past and that an event is observed at that time.

The partial likelihood for $\boldsymbol{\beta}$ is obtained by multiplying the conditional probabilities (4.6) over all observed event times, thereby disregarding the information on the regression coefficients contained in the aggregated counting process $N_\cdot(t)$. We assume that there are no tied event times and denote by $T_1 < T_2 < \cdots$ the times when events are observed. (A comment on the handling of ties is given at the end of this subsection.) Then, if i_j is the index of the individual who experiences an event at T_j, the partial likelihood becomes

$$L(\boldsymbol{\beta}) = \prod_{T_j} \pi(i_j \mid T_j) = \prod_{T_j} \frac{Y_{i_j}(T_j) \, r(\boldsymbol{\beta}, \mathbf{x}_{i_j}(T_j))}{\sum_{l=1}^{n} Y_l(T_j) \, r(\boldsymbol{\beta}, \mathbf{x}_l(T_j))}.$$

If we introduce the notation $\mathscr{R}_j = \{l \mid Y_l(T_j) = 1\}$ for the *risk set* at T_j, the partial likelihood may be written in the more familiar form

$$L(\boldsymbol{\beta}) = \prod_{T_j} \frac{r(\boldsymbol{\beta}, \mathbf{x}_{i_j}(T_j))}{\sum_{l \in \mathscr{R}_j} r(\boldsymbol{\beta}, \mathbf{x}_l(T_j))}. \tag{4.7}$$

The partial likelihood (4.7) may also be derived as a profile likelihood for $\boldsymbol{\beta}$ (Exercise 4.7).

The maximum partial likelihood estimator $\widehat{\boldsymbol{\beta}}$ is the value of $\boldsymbol{\beta}$ that maximizes (4.7). In Section 4.1.5 we show that the maximum partial likelihood estimator enjoys large sample properties similar to ordinary maximum likelihood estimators. In particular, in large samples, $\widehat{\boldsymbol{\beta}}$ is approximately multivariate normally distributed around the true value of $\boldsymbol{\beta}$ with a covariance matrix that may be estimated by $\mathscr{I}(\widehat{\boldsymbol{\beta}})^{-1}$, where $\mathscr{I}(\boldsymbol{\beta})$ is the expected information matrix given by formula (4.48) in Section 4.1.5. Alternatively, we may estimate the covariance matrix by $\mathbf{I}(\widehat{\boldsymbol{\beta}})^{-1}$, where $\mathbf{I}(\boldsymbol{\beta}) = \left\{ -\frac{\partial^2}{\partial \beta_h \partial \beta_j} \log L(\boldsymbol{\beta}) \right\}$ is the observed information matrix. When the relative risk function is of the exponential form of Cox's model, the observed information matrix coincides with the expected information matrix, but this is not the case for other relative risk functions. Of the two covariance matrix estimators, we recommend the one based on expected information for routine use (cf. Section 4.1.5).

In order to test the simple null hypothesis $\boldsymbol{\beta} = \boldsymbol{\beta}_0$, where $\boldsymbol{\beta}_0$ is known, one may apply the usual likelihood-based tests. Specifically, writing $\mathbf{U}(\boldsymbol{\beta}) = \frac{\partial}{\partial \beta} \log L(\boldsymbol{\beta})$ for the vector of score functions, one may adopt one of the test statistics:

- The *likelihood ratio test statistic:*

$$\chi_{LR}^2 = 2 \left\{ \log L(\widehat{\boldsymbol{\beta}}) - \log L(\boldsymbol{\beta}_0) \right\} \tag{4.8}$$

- The *score test statistic:*

$$\chi_{SC}^2 = \mathbf{U}(\boldsymbol{\beta}_0)^T \mathscr{I}(\boldsymbol{\beta}_0)^{-1} \mathbf{U}(\boldsymbol{\beta}_0) \tag{4.9}$$

- The *Wald test statistic:*

Table 4.1 Smoking habits in three Norwegian counties.

Smoking habit	Males	Females
Never smoked	445	947
Former smoker	676	333
1–9 cigarettes per day	199	225
10–19 cigarettes per day	457	341
20 or more cigarettes per day	241	65
Pipe or cigar smoker	68	3

$$\chi_W^2 = (\widehat{\boldsymbol{\beta}} - \boldsymbol{\beta}_0)^T \mathscr{I}(\widehat{\boldsymbol{\beta}})(\widehat{\boldsymbol{\beta}} - \boldsymbol{\beta}_0) \tag{4.10}$$

These three test statistics are asymptotically equivalent, and they are all approximately chi-squared distributed with p degrees of freedom under the null hypothesis. For the one-parameter case (i.e., $p = 1$) one may alternatively use the versions $Z_{SC} = U(\beta_0)\mathscr{I}(\beta_0)^{-1/2}$ and $Z_W = (\widehat{\beta} - \beta_0)\mathscr{I}(\widehat{\beta})^{1/2}$ for the score and Wald test statistics, respectively, both being approximately standard normally distributed when $\beta = \beta_0$. (Alternatively, one may use observed rather than expected information in the score and Wald test statistics.)

All three tests can be generalized to the situation in which one wants to test a composite null hypothesis that r of the p regression coefficients are equal to zero, or equivalently (after a reparameterization) that there are r linear restrictions among the β_j. In particular, if $\boldsymbol{\beta}^*$ is the maximum partial likelihood estimator under the null hypothesis, then the likelihood ratio statistic takes the form $\chi_{LR}^2 = 2\{\log L(\widehat{\boldsymbol{\beta}}) - \log L(\boldsymbol{\beta}^*)\}$ and is approximately chi-squared distributed with r degrees of freedom when the null hypothesis holds true.

We give two examples that illustrate these results. The first is an analysis of total mortality in three Norwegian counties (Example 1.12) using Cox's model, while the other is an analysis of lung cancer mortality in a cohort of uranium miners (Example 1.6) using the excess relative risk model.

Example 4.1. Mortality in three Norwegian counties. In Examples 3.3 and 3.15 we used the data on mortality in three Norwegian counties to study the cause-specific mortality according to four causes of death. In this example we concentrate on total mortality (i.e., mortality due to any cause) and use Cox's regression model to study the effects of sex and smoking habit.

Table 4.1 gives a summary of the smoking habits reported by the 4000 individuals in our sample. To study the effect of smoking habits, we concentrate on cigarette smoking and disregard the 71 individuals who smoked a pipe or cigar.

We fit a Cox regression model with the five covariates:

Table 4.2 Estimated regression coefficients with standard errors based on a Cox regression analysis of the total mortality in three Norwegian counties.

j	Covariate x_j	$\widehat{\beta}_j$	$\mathrm{se}(\widehat{\beta}_j)$
1	Sex	−0.541	0.094
2	Former smoker	0.315	0.134
3	1–9 cigarettes per day	0.891	0.146
4	10–19 cigarettes per day	0.895	0.127
5	20 or more cigarettes per day	1.086	0.154

$$x_{i1} = \begin{cases} 1 \text{ if individual } i \text{ is a female} \\ 0 \text{ if individual } i \text{ is a male} \end{cases}$$

$$x_{i2} = \begin{cases} 1 \text{ if individual } i \text{ is a former smoker} \\ 0 \text{ otherwise} \end{cases}$$

$$x_{i3} = \begin{cases} 1 \text{ if individual } i \text{ smokes 1–9 cigarettes per day} \\ 0 \text{ otherwise} \end{cases}$$

$$x_{i4} = \begin{cases} 1 \text{ if individual } i \text{ smokes 10–19 cigarettes per day} \\ 0 \text{ otherwise} \end{cases}$$

$$x_{i5} = \begin{cases} 1 \text{ if individual } i \text{ smokes 20 or more cigarettes per day} \\ 0 \text{ otherwise} \end{cases}$$

The Cox regression analysis gives the results summarized in Table 4.2. [Note that for most statistical software, one does not need to explicitly define the covariates x_{i2}, x_{i3}, x_{i4}, and x_{i5}. For example, in R (R Development Core Team, 2007) one just declares smoking habit to be a "factor" and the program defines the covariates internally.]

When interpreting the results of a Cox regression analysis, it is common to focus on the relative risks (or hazard rate ratios) e^{β_j}. From Table 4.2 we find that the estimated relative risk for sex is $e^{-0.541} = 0.582$. Thus the mortality rate (hazard rate) of a female is 58.2% of that of a male with the same smoking habit. We obtain a 95% confidence interval of the relative risk e^{β_j} by exponentiating the lower and upper limits of the standard 95% confidence interval for the regression coefficient [i.e., $\widehat{\beta}_j \pm 1.96\,\mathrm{se}(\widehat{\beta}_j)$]. Thus a 95% confidence interval for the relative risk for sex has lower limit $e^{-0.541-1.96\cdot0.094} = 0.484$ and upper limit $e^{-0.541+1.96\cdot0.094} = 0.700$.

Table 4.3 gives the estimated relative risks with 95% confidence intervals for all the covariates. For example, we see that a former smoker has a mortality rate that is 1.37 times that of an individual who has never smoked, while an individual who smokes 20 or more cigarettes per day has a mortality rate that is almost three times that of an individual who has never smoked. □

Table 4.3 Estimated relative risks (hazard rate ratios) with 95% confidence intervals (c.i.) based on a Cox regression analysis of the total mortality in three Norwegian counties.

Covariate	Hazard ratio	95% c.i.
Sex	0.58	0.48–0.70
Former smoker	1.37	1.05–1.78
1–9 cigarettes per day	2.44	1.83–3.25
10–19 cigarettes per day	2.45	1.91–3.14
20 or more cigarettes per day	2.96	2.19–4.00

Example 4.2. Uranium miners. The Colorado Plateau uranium miners' cohort was introduced in Example 1.6. Here we will estimate the effect of radon exposure and smoking on the risk of lung cancer death.

We consider age as the basic time scale and summarize radon and smoking data into cumulative exposures lagged by two years. Thus we consider the covariates $\mathbf{x}(t) = (x_{i1}(t), x_{i2}(t))^T$, where $x_{i1}(t)$ is cumulative radon exposure measured in working level months (WLM) up to two years prior to age t, and $x_{i2}(t)$ is cumulative smoking in number of packs smoked up to two years prior to t.

As has been the case in previous analyses of these data [see, e.g., Langholz and Goldstein (1996) and their references], we adopt the excess relative risk model. Thus the lung cancer mortality rate for miner i is assumed to take the form

$$\alpha(t \mid \mathbf{x}_i) = \alpha_0(t)\{1 + \beta_1 x_{i1}(t)\}\{1 + \beta_2 x_{i2}(t)\}. \tag{4.11}$$

The estimated radon excess relative risk (with standard error) is $\widehat{\beta}_1 = 0.40$ (0.12) per 100 WLMs cumulative radon exposure, while the smoking excess relative risk is $\widehat{\beta}_2 = 0.19$ (0.05) per 1000 packs of cigarettes smoked. The estimates are almost uncorrelated, the estimated correlation being 0.04. The standard errors are evaluated as the square roots of the diagonal elements of the inverse of the expected information matrix. The estimates are only slightly changed when we instead use the observed information matrix. Then the standard errors of $\widehat{\beta}_1$ and $\widehat{\beta}_2$ become 0.13 and 0.06, respectively.

Not surprisingly, both radon and smoking are seen to have a significant effect on the risk of lung cancer death when adjusted for the effect of the other. According to the estimated model, the effect of each 100 WLMs increase in the radon exposure is to increase the relative lung cancer risk by 40%, compared to an unexposed individual, while the relative risk increases by 19% for each additional 1000 packs of cigarettes smoked. □

As mentioned in Section 1.3, Cox regression and other survival analysis methods can also be useful in some situations that do not include time measurements. One such situation is described in the next example.

Example 4.3. Drive for thinness. In a study by Skårderud et al. (2005), a measure called "drive for thinness" was registered for 285 teenagers; 127 boys and 158 girls.

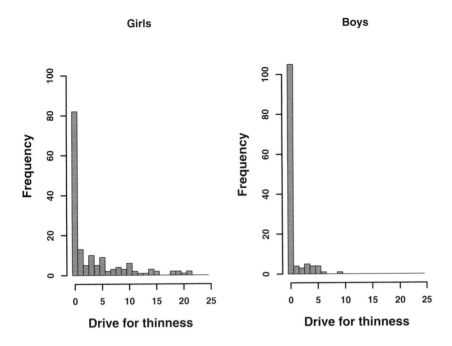

Fig. 4.1 *Distribution of "drive for thinness" for girls (left panel) and boys (right panel).*

As seen from the histograms in Figure 4.1 the distribution of this measure is extremely skewed, especially among the girls.

Certainly, normality-based methods cannot be used to study these data. Attempting to transform the data by a logarithmic transformation also does not work because of a clustering of values at zero and because the importance of the large values would be obscured. In order to judge whether a standard nonparametric approach is valid, one may look at the Kaplan-Meier plots (equivalent to considering the empirical cumulative distributions). The assumption of the standard Wilcoxon test is that the two distributions are the same, just that one is shifted to the right compared to the other. From the Kaplan-Meier curves shown in the left-hand panel of Figure 4.2 it is obvious that this is not true.

Looking at the Nelson-Aalen plots in the right-hand panel of Figure 4.2 there is an indication of a proportional hazards model, although with a large number of ties at zero. This points toward using a log-rank test, and then a Cox regression when covariates shall be judged. The log-rank test gives a chi-squared value of 40.2 with one degree of freedom, which is highly significant. A further Cox regression analysis (Table 4.4) discloses that much of the difference between the sexes is due to low self-esteem being related to "drive for thinness" (and low self-esteem being

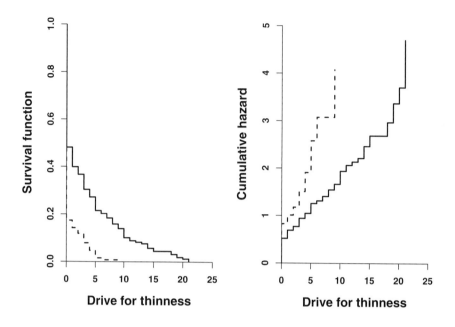

Fig. 4.2 *Kaplan-Meier plots (left panel) and Nelson-Aalen plots (right panel) of "drive for thinness." Girls: drawn line; boys: dashed line.*

much more prevalent among girls than among boys), while there is no clear effect of age. □

In the derivation of the partial likelihood (4.7), we assumed that there are no tied event times. In practice, a few ties may be broken at random. But there also exist modifications of the partial likelihood that can handle tied event times in a systematic manner. The most commonly used modifications are the ones proposed by Breslow and Efron; see Kalbfleisch and Prentice (2002, section 4.2.3) or Klein and Moeschberger (2003, section 8.4) for details.

Table 4.4 Cox regression analysis of "drive for thinness".

Covariate	$\widehat{\beta}_j$	$\text{se}(\widehat{\beta}_j)$	Hazard ratio	Wald statistic
Sex	−0.381	0.133	0.683	8.19
Age	−0.034	0.044	0.967	0.58
Low self-esteem	0.123	0.021	0.884	32.8

4.1.2 Estimation of cumulative hazards and survival probabilities

To derive an estimator for the cumulative baseline hazard $A_0(t) = \int_0^t \alpha_0(u)du$, we take the aggregated counting process $N_\bullet(t) = \sum_{l=1}^n N_l(t)$ as our starting point. By (4.5) its intensity process takes the form

$$\lambda_\bullet(t) = \left(\sum_{l=1}^n Y_l(t)\, r(\boldsymbol{\beta}, \mathbf{x}_l(t)) \right) \alpha_0(t).$$

If we had known the true value of $\boldsymbol{\beta}$, this would have been an example of the multiplicative intensity model (cf. Section 3.1.2), and we could have estimated $A_0(t)$ by the Nelson-Aalen estimator

$$\widehat{A}_0(t; \boldsymbol{\beta}) = \int_0^t \frac{dN_\bullet(u)}{\sum_{l=1}^n Y_l(u)\, r(\boldsymbol{\beta}, \mathbf{x}_l(u))}. \tag{4.12}$$

Since $\boldsymbol{\beta}$ is unknown, we replace it by $\widehat{\boldsymbol{\beta}}$ to obtain

$$\widehat{A}_0(t) = \int_0^t \frac{dN_\bullet(u)}{\sum_{l=1}^n Y_l(u)\, r(\widehat{\boldsymbol{\beta}}, \mathbf{x}_l(u))} = \sum_{T_j \le t} \frac{1}{\sum_{l \in \mathcal{R}_j} r(\widehat{\boldsymbol{\beta}}, \mathbf{x}_l(T_j))} \tag{4.13}$$

as an estimator for the cumulative baseline hazard. The estimator (4.13) is usually denoted the Breslow estimator.

If all the covariates are fixed, the (adjusted) cumulative hazard corresponding to an individual with a given covariate vector $\mathbf{x}_0$ is

$$A(t \mid \mathbf{x}_0) = \int_0^t \alpha(u \mid \mathbf{x}_0)\, du = r(\boldsymbol{\beta}, \mathbf{x}_0) A_0(u), \tag{4.14}$$

and this may be estimated by

$$\widehat{A}(t \mid \mathbf{x}_0) = r(\widehat{\boldsymbol{\beta}}, \mathbf{x}_0) \widehat{A}_0(t). \tag{4.15}$$

When some of the covariates are time-dependent, it is usually not meaningful to estimate the cumulative hazard corresponding to a fixed value of the vector of covariates. What can be meaningfully estimated in such situations is the cumulative hazard over an interval $[0,t]$ corresponding to a given covariate path $\mathbf{x}_0(s); 0 < s \le t$. The cumulative hazard corresponding to such a path is

$$A(t \mid \mathbf{x}_0) = \int_0^t r(\boldsymbol{\beta}, \mathbf{x}_0(u))\, \alpha_0(u)du, \tag{4.16}$$

and this may be estimated by

$$\widehat{A}(t \mid \mathbf{x}_0) = \int_0^t r(\widehat{\boldsymbol{\beta}}, \mathbf{x}_0(u))\, d\widehat{A}_0(u) = \sum_{T_j \le t} \frac{r(\widehat{\boldsymbol{\beta}}, \mathbf{x}_0(T_j))}{\sum_{l \in \mathcal{R}_j} r(\widehat{\boldsymbol{\beta}}, \mathbf{x}_l(T_j))}. \tag{4.17}$$

In Section 4.1.6 we show that (4.17), for a given value of t, is asymptotically normally distributed around its true value with a variance that can be estimated by (4.52). The estimator (4.15) for fixed covariates is a special case.

For survival data we may use (4.15) or (4.17) to derive an estimator for the adjusted survival function corresponding to given values of time-fixed covariates or given covariate paths of *external* time-dependent covariates. To this end we use (3.39) to write the corresponding survival function $S(t \,|\, \mathbf{x}_0)$ as the product-integral

$$S(t \,|\, \mathbf{x}_0) = \prod_{u \leq t} \{1 - dA(u \,|\, \mathbf{x}_0)\}.$$

Inserting $\widehat{A}(t \,|\, \mathbf{x}_0)$ in this expression, we get the Kaplan-Meier-type estimator

$$\widehat{S}(t \,|\, \mathbf{x}_0) = \prod_{u \leq t} \left\{1 - d\widehat{A}(u \,|\, \mathbf{x}_0)\right\} = \prod_{T_j \leq t} \left\{1 - \triangle\widehat{A}(T_j \,|\, \mathbf{x}_0)\right\}. \qquad (4.18)$$

This estimator is asymptotically normally distributed around its true value with variance that can be estimated by (4.53); cf Section 4.1.6.

Example 4.4. Mortality in three Norwegian counties. In Example 4.1 we studied the effect of sex and smoking habit on mortality due to any cause in three Norwegian counties. The estimated regression coefficients were given in Table 4.2. Based on the fitted model, we may use (4.18) to estimate the survival function for given values of the covariates. Figure 4.3 shows the estimated survival curves (given survival to 40 years) for each gender for individuals who have never smoked and for individuals who smoke 20 or more cigarettes per day. It is seen that a 40-years-old male who has never smoked has about an 85% probability of surviving to 70 years of age, while this probability is about 60% for a male who smokes 20 or more cigarettes per day. For a female the corresponding probabilities are, respectively, 90% and 75%. $\square$

For Markov chains, where the transition intensities are modeled by relative risk regression models, we may use (4.15) to estimate the cumulative transition intensities for given values (or paths) of the covariates. When all covariates are external, these estimates may be combined as described in Section 3.4 to obtain estimates of covariate dependent Markov transition probabilities; see Andersen et al. (1991) and Borgan (2002) for details.

4.1.3 Martingale residual processes and model check

The martingale residuals, first considered by Barlow and Prentice (1988), provide a useful tool for checking the fit of the relative risk regression model (4.2). For the special case of Cox's model (with exponential relative risk function), different plots and goodness-of-fit tests have been proposed based on these residuals; see Therneau and Grambsch (2000, section 5.7) for a review and further discussion. The martingale residuals are the difference between the observed and expected numbers of

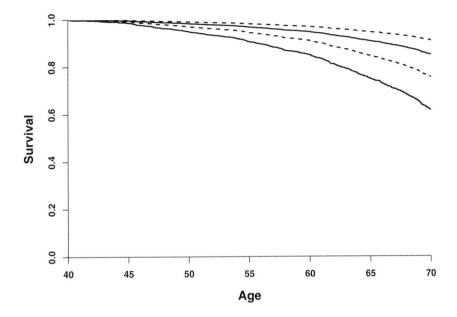

Fig. 4.3 *Estimated survival curves (given survival to 40 years) for some covariate values based on a Cox regression analysis of the total mortality in three Norwegian counties. Males: drawn lines; females: dashed lines. Upper lines: never smoked; lower lines: smokes 20 or more cigarettes per day.*

events for each individual over the full study time interval [cf. (4.21)], and they are therefore not well suited to detect time-dependent deviations from the model. In this subsection we will focus on plots and goodness-of-fit tests based on the martingale residual *processes*, which compare the observed and expected numbers of events for each individual as a function of time t [cf. (4.20)]. These were introduced by Aalen (1993) for the additive regression model of Section 4.2 and implemented for Cox's model by Grønnesby and Borgan (1996).

To define the martingale residual processes, we first introduce the cumulative intensity processes

$$\Lambda_i(t) = \int_0^t \lambda_i(u)\,du = \int_0^t Y_i(u)\,r(\boldsymbol{\beta}, \mathbf{x}_i(u))\,\alpha_0(u)\,du \qquad (4.19)$$

corresponding to (4.4). By (2.40), it follows that $M_i(t) = N_i(t) - \Lambda_i(t)$; $i = 1, 2, \ldots, n$; are martingales when model (4.2) holds true. If we now insert the maximum partial likelihood estimator $\widehat{\boldsymbol{\beta}}$ for $\boldsymbol{\beta}$ and the increment $d\widehat{A}_0(u)$ of the Breslow estimator for $\alpha_0(u)\,du$ in (4.19), we get the estimated cumulative intensity processes

$$\widehat{\Lambda}_i(t) = \int_0^t Y_i(u)\, r(\widehat{\boldsymbol{\beta}}, \mathbf{x}_i(u))\, d\widehat{A}_0(u) = \sum_{T_j \le t} \frac{Y_i(T_j)\, r(\widehat{\boldsymbol{\beta}}, \mathbf{x}_i(T_j))}{\sum_{l \in \mathscr{R}_j} r(\widehat{\boldsymbol{\beta}}, \mathbf{x}_l(T_j))}$$

and the *martingale residual processes*

$$\widehat{M}_i(t) = N_i(t) - \widehat{\Lambda}_i(t). \tag{4.20}$$

Evaluating the martingale residual processes at $t = \tau$, where τ is the upper time limit for the study, we arrive at the *martingale residuals*

$$\widehat{M}_i = \widehat{M}_i(\tau) = N_i(\tau) - \widehat{\Lambda}_i(\tau) \tag{4.21}$$

first considered by Barlow and Prentice (1988).

The individual martingale residual processes may be useful for recurrent event data, where each individual may experience a number of events (see Section 8.3.2). For survival data, however, each of the martingale residual processes contains too little information to be of much use on its own. However, by aggregating them over groups of individuals, useful plots and goodness-of-fit tests may be obtained.

Specifically, assume that we have some grouping of the individuals, typically based on the values of one or two covariates, and denote the groups by $J = 1, \ldots, G$. We will allow the grouping of the individuals to depend on time. Thus an individual may move from one group to another as time passes, as is the case when the grouping is performed on the basis of one or more time-dependent covariates. It is a prerequisite, however, that the information used for grouping at time t is available just before that time. Then, if we denote by $J(u)$ the set of all individuals who belong to group J at time u, the group J martingale residual process takes the form

$$\widehat{M}_J(t) = \int_0^t \sum_{i \in J(u)} d\widehat{M}_i(u) = N_J(t) - \sum_{T_j \le t} \frac{\sum_{i \in \mathscr{R}_j \cap J(T_j)} r(\widehat{\boldsymbol{\beta}}, \mathbf{x}_i(T_j))}{\sum_{l \in \mathscr{R}_j} r(\widehat{\boldsymbol{\beta}}, \mathbf{x}_l(T_j))}. \tag{4.22}$$

Here $N_J(t) = \int_0^t \sum_{i \in J(u)} dN_i(u)$ is the observed number of events in group J in $[0, t]$, while the last term on the right-hand side of (4.22) is an estimate of the expected number of events in the group when the relative risk regression model (4.2) holds true. In fact, if we could have used the true value $\boldsymbol{\beta}$ instead of its estimate $\widehat{\boldsymbol{\beta}}$ in (4.22), then the grouped martingale residual processes would have been martingales. However, since the regression coefficients have to be estimated, the grouped martingale residual processes are only approximately martingales.

Example 4.5 illustrates how we may use plots of the grouped martingale residual processes to check the fit of model (4.2). In addition to provide useful plots, the grouped martingale residual processes may be used to derive formal goodness-of-fit tests. In particular, one may use a chi-squared test based on a comparison of the observed and expected number of events in the G groups in K disjoint time intervals. To see how this can be done, let $0 = a_0 < a_1 < \cdots < a_{K-1} < a_K = \tau$ be a partitioning of the study time interval, and introduce (for $H = 1, 2, \ldots, K$ and $J = 1, 2, \ldots, G$)

$$\widehat{M}_{HJ} = \widehat{M}_J(a_H) - \widehat{M}_J(a_{H-1}) = O_{HJ} - E_{HJ}. \qquad (4.23)$$

Here $O_{HJ} = N_J(a_H) - N_J(a_{H-1})$ is the observed number of events in group J in time interval H, while

$$E_{HJ} = \sum_{a_{H-1} < T_j \le a_H} \frac{\sum_{i \in \mathscr{R}_j \cap J(T_j)} r(\widehat{\boldsymbol{\beta}}, \mathbf{x}_i(T_j))}{\sum_{l \in \mathscr{R}_j} r(\widehat{\boldsymbol{\beta}}, \mathbf{x}_l(T_j))}$$

is the corresponding expected number under model (4.2). The martingale residual processes (4.22) sum to zero at any given time t. To derive a chi-squared goodness-of-fit test, we therefore disregard the contribution from one of the groups, say the first group, and consider the $K(G-1)$-vector $\widehat{\mathbf{M}}$ with elements $\widehat{M}_{HJ}$ for $H = 1, 2, \ldots, K; J = 2, 3, \ldots, G$. The covariance matrix of $\widehat{\mathbf{M}}$ may be estimated by the matrix $\widehat{\boldsymbol{\Sigma}} = \{\widehat{\sigma}_{LI,HJ}\}$ with elements given by (4.29). A goodness-of-fit test may then be based on the statistic $\chi^2 = \widehat{\mathbf{M}}^T \widehat{\boldsymbol{\Sigma}}^{-1} \widehat{\mathbf{M}}$, which is approximately chi-squared distributed with $K(G-1)$ degrees of freedom in large samples when model (4.2) holds true.

Example 4.5. Uranium miners. We consider the uranium miners' data, and will show how the grouped martingale residual processes and the accompanying goodness-of-fit test can be used to check the fit of the excess relative risk model (4.11) of Example 4.2. For illustration, we concentrate on the modeling of radon exposure and group the individuals (at the time of each lung cancer death) into three groups defined by their cumulative radon exposure: group I: less than 500 WLMs; group II: 500–1500 WLMs; group III: more than 1500 WLMs.

Figure 4.4 shows the grouped martingale residual processes (4.22) for these groups. From the figure we see that more lung cancer deaths than expected occur in the high exposure group (group III) below the age of 60 years, while fewer cases than expected occur above this age. Thus the grouped martingale residual plot indicates that there is an interaction between cumulative radon exposure and age.

We will use the chi-squared test described earlier to investigate whether the interaction between age and radon exposure indicated by Figure 4.4 is statistically significant. Table 4.5 gives the observed and expected number of lung cancer death in the three cumulative radon exposure groups for ages below and above 60 years. The table summarizes and quantifies what we saw in Figure 4.4: below age 60 years we observe 81 lung cancer deaths in group III, while only 68.6 are expected under the model. Above 60 years it is the other way around; 36 deaths are observed in this group, while 47.5 are expected. The chi-squared goodness-of-fit statistic with $2(3-1) = 4$ degrees of freedom corresponding to the observed and expected numbers of Table 4.5 takes the value 10.2, corresponding to a P-value of 3.8%. Thus there are clear indications that model (4.11), with a common baseline for all individuals and where the effect of radon depends only on the cumulative radon exposure and not on the variation in the exposure over time, may be a bit too simplistic. □

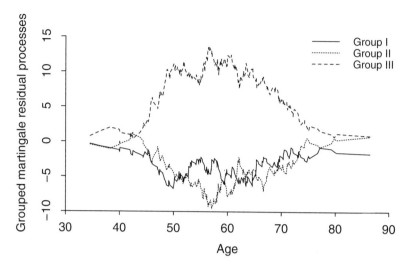

Fig. 4.4 *Grouped martingale residual processes for the uranium miners data. Grouping is done according to cumulative radon exposure: Group I: below 500 WLMs; group II: 500–1500 WLMs; group III: above 1500 WLMs.*

Table 4.5 Observed and expected numbers of lung cancer deaths in two age intervals and three radon exposure groups.

Exposure group[a]	Observed numbers	Expected numbers
Below 60 years of age		
Group I	30	34.5
Group II	39	46.9
Group III	81	68.6
Above 60 years of age		
Group I	27	24.2
Group II	45	36.3
Group III	36	47.5

a) Group I: below 500 WLMs; group II: 500–1500 WLMs, group III: above 1500 WLMs.

We conclude this subsection by studying the large sample distribution of the grouped martingale residual processes and by deriving the estimator for the co-variance matrix for the goodness-of-fit test. A comment on computation of the goodness-of-fit test for the special case of Cox's model is also given. The reader may omit these derivations and comments without loss of continuity.

One may show that in large samples, the grouped martingale residual processes are approximately normally distributed with mean zero when model (4.2) holds true

(Grønnesby and Borgan, 1996; Borgan and Langholz, 2007). Furthermore, for $s \leq t$, the covariance between $\widehat{M}_I(s)$ and $\widehat{M}_J(t)$ may be estimated by

$$\widehat{\sigma}_{IJ}(s,t) = \widehat{\phi}_{IJ}(0,s) - \widehat{\boldsymbol{\psi}}_I(0,s)^T \mathscr{I}(\widehat{\boldsymbol{\beta}})^{-1} \widehat{\boldsymbol{\psi}}_J(0,t). \tag{4.24}$$

Here

$$\widehat{\phi}_{IJ}(t_1,t_2) = \sum_{t_1 < T_j \leq t_2} \frac{S_I^{(0)}(\widehat{\boldsymbol{\beta}},T_j)}{S^{(0)}(\widehat{\boldsymbol{\beta}},T_j)} \left\{ \delta_{IJ} - \frac{S_J^{(0)}(\widehat{\boldsymbol{\beta}},T_j)}{S^{(0)}(\widehat{\boldsymbol{\beta}},T_j)} \right\} \tag{4.25}$$

and

$$\widehat{\boldsymbol{\psi}}_J(t_1,t_2) = \sum_{t_1 < T_j \leq t_2} \left\{ \frac{\mathbf{S}_J^{(1)}(\widehat{\boldsymbol{\beta}},T_j)}{S^{(0)}(\widehat{\boldsymbol{\beta}},T_j)} - \frac{S_J^{(0)}(\widehat{\boldsymbol{\beta}},T_j)\mathbf{S}^{(1)}(\widehat{\boldsymbol{\beta}},T_j)}{S^{(0)}(\widehat{\boldsymbol{\beta}},T_j)^2} \right\}, \tag{4.26}$$

where δ_{IJ} is a Kronecker delta (i.e., equal to 1 when $I = J$ and equal to 0 otherwise). Further,

$$S^{(0)}(\boldsymbol{\beta},t) = \sum_{l=1}^{n} Y_l(t)\, r(\boldsymbol{\beta}, \mathbf{x}_l(t)) \tag{4.27}$$

and

$$\mathbf{S}^{(1)}(\boldsymbol{\beta},t) = \sum_{l=1}^{n} Y_l(t)\, \dot{\mathbf{r}}(\boldsymbol{\beta}, \mathbf{x}_l(t)), \tag{4.28}$$

where $\dot{\mathbf{r}}(\boldsymbol{\beta}, \mathbf{x}_i(t)) = \frac{\partial}{\partial \boldsymbol{\beta}} r(\boldsymbol{\beta}, \mathbf{x}_i(t))$, while $S_J^{(0)}(\boldsymbol{\beta},t)$ and $\mathbf{S}_J^{(1)}(\boldsymbol{\beta},t)$ are defined similarly to $S^{(0)}(\boldsymbol{\beta},t)$ and $\mathbf{S}^{(1)}(\boldsymbol{\beta},t)$, but with the sums restricted to the individuals who belong to group J at time t.

By these approximate distributional results for the grouped martingale residual processes, it follows that the vector $\widehat{\mathbf{M}}$ with elements given by (4.23) is approximately mean zero multivariate normally distributed in large samples when model (4.2) holds true. Its covariance matrix may be estimated by the matrix $\widehat{\boldsymbol{\Sigma}} = \{\widehat{\sigma}_{LI,HJ}\}$ with elements

$$\widehat{\sigma}_{LI,HJ} = \widehat{\mathrm{Cov}}(\widehat{M}_{LI}, \widehat{M}_{HJ}) \tag{4.29}$$

$$= \delta_{LH}\, \widehat{\phi}_{IJ}(a_{H-1}, a_H) - \widehat{\boldsymbol{\psi}}_I(a_{L-1}, a_L)^T \mathscr{I}(\widehat{\boldsymbol{\beta}})^{-1} \widehat{\boldsymbol{\psi}}_J(a_{H-1}, a_H);$$

$H, L = 1, 2, \ldots, K$; $J, I = 2, 3, \ldots, G$; where δ_{LH} is a Kronecker delta. This shows that the goodness-of-fit statistic $\chi^2 = \widehat{\mathbf{M}}^T \widehat{\boldsymbol{\Sigma}}^{-1} \widehat{\mathbf{M}}$ is approximately chi-squared distributed with $K(G-1)$ degrees of freedom in large samples when model (4.2) holds true.

Finally, consider the extension of model (4.2) where an individual i who belongs to group J at time $t \in (a_{H-1}, a_H]$ has a hazard rate of the form

$$\alpha(t; \mathbf{x}_i) = \alpha_0(t)\, r(\boldsymbol{\beta}, \mathbf{x}_i(t))\, e^{\gamma_{HJ}}. \tag{4.30}$$

Then by some algebra along the lines of May and Hosmer (1998, appendix A) one may show that the goodness-of-fit statistic χ^2 is equivalent to the score test for the hypothesis that the additional $K(G-1)$ parameters γ_{HJ} in (4.30) are all equal to zero. For the special case of Cox's regression model, the extended model (4.30) becomes a Cox model as well. Thus for the Cox model, the chi-squared goodness-of-fit test can be computed as the score test for the addition of categorical grouping variables using standard software.

4.1.4 Stratified models

The relative risk regression model (4.2) assumes a common baseline hazard for all individuals. If this is not a realistic assumption, one may adopt a stratified version of the model. Then the study population is grouped into k strata, say, and for an individual i in stratum s, the hazard is assumed to take the form

$$\alpha(t\,|\,\mathbf{x}_i) = \alpha_{s0}(t)\,r(\boldsymbol{\beta},\mathbf{x}_i(t)). \tag{4.31}$$

Note that the effects of the covariates are assumed to be the same across strata, while the baseline hazard may vary between strata. [Andersen et al. (1993, section VII.1) describe how it is possible to relax the assumption that the effects of covariates are the same across strata.] Note also that, although we have not made it explicit in the notation, the grouping into strata may depend on time. It is a key requirement, however, that the information used to group the individuals into strata only depend on past information and not on future events.

For the stratified proportional hazards model (4.31), estimation of the vector of regression coefficients $\boldsymbol{\beta}$ is based on the partial likelihood

$$L(\boldsymbol{\beta}) = \prod_{s=1}^{k}\prod_{T_{sj}} \frac{r(\boldsymbol{\beta},\mathbf{x}_{i_j}(T_{sj}))}{\sum_{l\in\mathscr{R}_{sj}} r(\boldsymbol{\beta},\mathbf{x}_l(T_{sj}))}, \tag{4.32}$$

where $T_{s1} < T_{s2} < \cdots$ are the times when events are observed in stratum s, and $\mathscr{R}_{sj}$ is the risk set in this stratum at time T_{sj}. The statistical properties of the maximum partial likelihood estimator $\widehat{\boldsymbol{\beta}}$ are the same as for the model without stratification. Further, the Breslow estimator for the stratum-specific cumulative baseline hazard $A_{s0}(t) = \int_0^t \alpha_{s0}(u)du$ takes the form

$$\widehat{A}_{s0}(t) = \sum_{T_{sj}\leq t} \frac{1}{\sum_{l\in\mathscr{R}_{sj}} r(\widehat{\boldsymbol{\beta}},\mathbf{x}_l(T_{sj}))}, \tag{4.33}$$

while we may estimate the stratum specific-survival functions for given covariate values as in (4.18). Also the martingale residual processes are easily adapted to the stratified model; see Borgan and Langholz (2007) for details.

4.1.5 Large sample properties of $\widehat{\boldsymbol{\beta}}$

It took almost 10 years from Cox's seminal paper on relative risk regression models (Cox, 1972) for a rigorous proof of the large sample properties of the maximum partial likelihood estimator to be provided (Tsiatis, 1981). A main reason is that the partial likelihood (4.7) is a product of dependent factors, and this makes it difficult to handle within the classical framework of *iid* random variables. However, as shown by Andersen and Gill (1982), counting processes and martingales provide a very natural framework for studying Cox's regression model.

In this subsection, following Andersen and Gill (1982), we outline the main steps in the derivation of the large sample properties of the maximum partial likelihood estimator $\widehat{\boldsymbol{\beta}}$ for the (unstratified) relative risk regression model (4.2). In particular, we will show that $\widehat{\boldsymbol{\beta}}$ is asymptotically multivariate normally distributed around the true value of $\boldsymbol{\beta}$ with a covariance matrix that may be estimated by the inverse of the expected information matrix (4.48). For ease of presentation we concentrate on Cox's model, where the relative risk function takes the exponential form $r(\boldsymbol{\beta}, \mathbf{x}_i(t)) = \exp(\boldsymbol{\beta}^T \mathbf{x}_i(t))$, and we only briefly mention the relevant modifications for the general case. A detailed account for Cox's model, including the necessary regularity conditions and the extension to stratified models, is given by Andersen et al. (1993, section VII.2.2).

So far the notation $\boldsymbol{\beta}$ has been used to denote the true value of the regression coefficient [e.g., in (4.2)] as well as to denote the argument of the partial likelihood and other quantities depending on $\boldsymbol{\beta}$. When studying large sample properties, it is useful to distinguish more clearly between the two uses. Therefore, here and in Section 4.1.6, we will denote the true value of the regression coefficient by $\boldsymbol{\beta}_0$, while $\boldsymbol{\beta}$ is used for the argument of the partial likelihood and similar quantities.

We start out by deriving the (partial) score function and observed (partial) information matrix for Cox's model. To this end note that, when $r(\boldsymbol{\beta}, \mathbf{x}_i(t)) = \exp(\boldsymbol{\beta}^T \mathbf{x}_i(t))$, the logarithm of the partial likelihood (4.7) may be written

$$l_{\mathrm{cox}}(\boldsymbol{\beta}) = \sum_{i=1}^{n} \int_0^\tau \left\{ \boldsymbol{\beta}^T \mathbf{x}_i(u) - \log S_{\mathrm{cox}}^{(0)}(\boldsymbol{\beta}, u) \right\} dN_i(u),$$

where τ is the upper time limit for the study and

$$S_{\mathrm{cox}}^{(0)}(\boldsymbol{\beta}, t) = \sum_{l=1}^{n} Y_l(t) \exp\{\boldsymbol{\beta}^T \mathbf{x}_l(t)\}. \tag{4.34}$$

We also introduce the notation

$$\mathbf{S}_{\mathrm{cox}}^{(1)}(\boldsymbol{\beta}, t) = \sum_{l=1}^{n} Y_l(t) \mathbf{x}_l(t) \exp\{\boldsymbol{\beta}^T \mathbf{x}_l(t)\} \tag{4.35}$$

and

$$\mathbf{S}_{\text{cox}}^{(2)}(\boldsymbol{\beta},t) = \sum_{l=1}^{n} Y_l(t)\mathbf{x}_l(t)^{\otimes 2} \exp\{\boldsymbol{\beta}^T\mathbf{x}_l(t)\}, \tag{4.36}$$

where $\mathbf{v}^{\otimes 2}$ of a column vector $\mathbf{v}$ equals the matrix $\mathbf{v}\mathbf{v}^T$. Then the score function becomes

$$\mathbf{U}_{\text{cox}}(\boldsymbol{\beta}) = \frac{\partial}{\partial\boldsymbol{\beta}} l_{\text{cox}}(\boldsymbol{\beta}) = \sum_{i=1}^{n} \int_0^{\tau} \left\{ \mathbf{x}_i(u) - \frac{\mathbf{S}_{\text{cox}}^{(1)}(\boldsymbol{\beta},u)}{S_{\text{cox}}^{(0)}(\boldsymbol{\beta},u)} \right\} dN_i(u), \tag{4.37}$$

while the observed information matrix may be written

$$\mathbf{I}_{\text{cox}}(\boldsymbol{\beta}) = -\frac{\partial}{\partial\boldsymbol{\beta}^T} \mathbf{U}_{\text{cox}}(\boldsymbol{\beta}) = \int_0^{\tau} \mathbf{V}_{\text{cox}}(\boldsymbol{\beta},u)\, dN_{\bullet}(u), \tag{4.38}$$

where

$$\mathbf{V}_{\text{cox}}(\boldsymbol{\beta},t) = \frac{\mathbf{S}_{\text{cox}}^{(2)}(\boldsymbol{\beta},t)}{S_{\text{cox}}^{(0)}(\boldsymbol{\beta},t)} - \left(\frac{\mathbf{S}_{\text{cox}}^{(1)}(\boldsymbol{\beta},t)}{S_{\text{cox}}^{(0)}(\boldsymbol{\beta},t)}\right)^{\otimes 2}. \tag{4.39}$$

If we insert $\boldsymbol{\beta}_0$ in the score function (4.37) and use the decomposition

$$dN_i(u) = \lambda_i(u)du + dM_i(u) = Y_i(u)\alpha_0(u)\exp\{\boldsymbol{\beta}_0^T\mathbf{x}_i(u)\}du + dM_i(u),$$

we find, after some simple algebra,

$$\mathbf{U}_{\text{cox}}(\boldsymbol{\beta}_0) = \sum_{i=1}^{n} \int_0^{\tau} \left\{ \mathbf{x}_i(u) - \frac{\mathbf{S}_{\text{cox}}^{(1)}(\boldsymbol{\beta}_0,u)}{S_{\text{cox}}^{(0)}(\boldsymbol{\beta}_0,u)} \right\} dM_i(u). \tag{4.40}$$

Here the integrands are predictable processes. Thus the score function is a (vector-valued) stochastic integral when evaluated at the true value of the regression coefficient. In particular $E\mathbf{U}_{\text{cox}}(\boldsymbol{\beta}_0) = \mathbf{0}$. If on the right-hand side of (4.40) we replace the upper limit of integration by t, we get a stochastic process. This stochastic process is a martingale with a predictable variation process that, evaluated at τ, becomes [cf. (B.4) and (B.6) in Appendix B]

$$\langle\mathbf{U}_{\text{cox}}(\boldsymbol{\beta}_0)\rangle(\tau) = \int_0^{\tau} \mathbf{V}_{\text{cox}}(\boldsymbol{\beta}_0,u) S_{\text{cox}}^{(0)}(\boldsymbol{\beta}_0,u)\, \alpha_0(u)du. \tag{4.41}$$

Note that, since the aggregated counting process $N_{\bullet}(t)$ has intensity process $\lambda_{\bullet}(t) = S_{\text{cox}}^{(0)}(\boldsymbol{\beta}_0,t)\alpha_0(t)$, the observed information matrix (4.38) evaluated at $\boldsymbol{\beta}_0$ may be decomposed as

$$\mathbf{I}_{\text{cox}}(\boldsymbol{\beta}_0) = \langle\mathbf{U}_{\text{cox}}(\boldsymbol{\beta}_0)\rangle(\tau) + \int_0^{\tau} \mathbf{V}_{\text{cox}}(\boldsymbol{\beta}_0,u)\, dM_{\bullet}(u), \tag{4.42}$$

where $M_{\bullet} = \sum_{l=1}^{n} M_l$ is a martingale. Thus at the true value of the regression coefficient, the observed information matrix equals the predictable variation process of the score function plus a stochastic integral. In particular, by taking expectations, it follows that the expected information matrix equals the covariance matrix of the score function. This is known to be a key property in maximum likelihood estimation.

By the martingale central limit theorem (Section 2.3.3 and Appendix B.3), we may now show, under suitable regularity conditions, that $n^{-1/2}\mathbf{U}_{\text{cox}}(\boldsymbol{\beta}_0)$ converges in distribution to a multivariate normal distribution with mean zero and with a covariance matrix $\boldsymbol{\Sigma}_{\text{cox}}$ that is the limit in probability of $n^{-1}\langle \mathbf{U}_{\text{cox}}(\boldsymbol{\beta}_0)\rangle (\tau)$. Using (4.42) we may also show that both $n^{-1}\mathbf{I}_{\text{cox}}(\boldsymbol{\beta}_0)$ and $n^{-1}\mathbf{I}_{\text{cox}}(\widehat{\boldsymbol{\beta}})$ converge in probability to $\boldsymbol{\Sigma}_{\text{cox}}$.

From these results the large sample properties of $\widehat{\boldsymbol{\beta}}$ follow in the usual way. The main steps in the derivations are as follows. First note that $\widehat{\boldsymbol{\beta}}$ is the solution to the score equation $\mathbf{U}_{\text{cox}}(\widehat{\boldsymbol{\beta}}) = \mathbf{0}$. Then Taylor expand the score equation around $\boldsymbol{\beta}_0$, to get

$$\mathbf{0} = \mathbf{U}_{\text{cox}}(\widehat{\boldsymbol{\beta}}) \approx \mathbf{U}_{\text{cox}}(\boldsymbol{\beta}_0) - \mathbf{I}_{\text{cox}}(\boldsymbol{\beta}_0)(\widehat{\boldsymbol{\beta}} - \boldsymbol{\beta}_0). \tag{4.43}$$

From this we obtain

$$\sqrt{n}\left(\widehat{\boldsymbol{\beta}} - \boldsymbol{\beta}_0\right) \approx \left(n^{-1}\mathbf{I}_{\text{cox}}(\boldsymbol{\beta}_0)\right)^{-1} n^{-1/2}\mathbf{U}_{\text{cox}}(\boldsymbol{\beta}_0) \approx \boldsymbol{\Sigma}_{\text{cox}}^{-1} n^{-1/2}\mathbf{U}_{\text{cox}}(\boldsymbol{\beta}_0),$$

and it follows that $\sqrt{n}(\widehat{\boldsymbol{\beta}} - \boldsymbol{\beta}_0)$ converges in distribution to a multivariate normal distribution with mean zero and covariance matrix $\boldsymbol{\Sigma}_{\text{cox}}^{-1}\boldsymbol{\Sigma}_{\text{cox}}\boldsymbol{\Sigma}_{\text{cox}}^{-1} = \boldsymbol{\Sigma}_{\text{cox}}^{-1}$. Thus $\widehat{\boldsymbol{\beta}}$ is approximately multivariate normally distributed around $\boldsymbol{\beta}_0$ with a covariance matrix that may be estimated by $\mathbf{I}_{\text{cox}}(\widehat{\boldsymbol{\beta}})^{-1}$.

The arguments just given for Cox's model go through with only minor modifications for a general relative risk function $r(\boldsymbol{\beta}, \mathbf{x}_i(t))$, but the formulas become more involved. The score function now becomes

$$\mathbf{U}(\boldsymbol{\beta}) = \sum_{i=1}^{n} \int_0^\tau \left\{ \frac{\dot{\mathbf{r}}(\boldsymbol{\beta}, \mathbf{x}_i(u))}{r(\boldsymbol{\beta}, \mathbf{x}_i(u))} - \frac{\mathbf{S}^{(1)}(\boldsymbol{\beta}, u)}{S^{(0)}(\boldsymbol{\beta}, u)} \right\} dN_i(u), \tag{4.44}$$

where $\dot{\mathbf{r}}(\boldsymbol{\beta}, \mathbf{x}_i(t)) = \frac{\partial}{\partial \boldsymbol{\beta}} r(\boldsymbol{\beta}, \mathbf{x}_i(t))$ and $S^{(0)}(\boldsymbol{\beta}, t)$ and $\mathbf{S}^{(1)}(\boldsymbol{\beta}, t)$ are given by (4.27) and (4.28), respectively. If we introduce

$$\mathbf{S}^{(2)}(\boldsymbol{\beta}, t) = \sum_{l=1}^{n} Y_l(t) \frac{\dot{\mathbf{r}}(\boldsymbol{\beta}, \mathbf{x}_l)^{\otimes 2}}{r(\boldsymbol{\beta}, \mathbf{x}_l)} \tag{4.45}$$

and

$$\mathbf{V}(\boldsymbol{\beta}, t) = \frac{\mathbf{S}^{(2)}(\boldsymbol{\beta}, t)}{S^{(0)}(\boldsymbol{\beta}, t)} - \left(\frac{\mathbf{S}^{(1)}(\boldsymbol{\beta}, t)}{S^{(0)}(\boldsymbol{\beta}, t)} \right)^{\otimes 2}, \tag{4.46}$$

the predictable variation process of the score function, evaluated at $\boldsymbol{\beta}_0$, may be written

$$\langle \mathbf{U}(\boldsymbol{\beta}_0)\rangle (\tau) = \int_0^\tau \mathbf{V}(\boldsymbol{\beta}_0, u) S^{(0)}(\boldsymbol{\beta}_0, u) \alpha_0(u) du. \tag{4.47}$$

As for Cox's model we may show that the observed information matrix, evaluated at $\boldsymbol{\beta}_0$, may be written as a sum of the predictable variation process of the score and a stochastic integral. Thus by taking expectations, it again follows that the expected information matrix equals the covariance matrix of the score function. The martingale central limit theorem and a Taylor series expansion give, also in the general case, that $\widehat{\boldsymbol{\beta}}$ is approximately multivariate normally distributed around $\boldsymbol{\beta}_0$. In order to estimate the covariance matrix of $\widehat{\boldsymbol{\beta}}$ we may use the inverse of either the observed or the (estimated) expected information matrix. The latter is obtained by inserting the increment $d\widehat{A}_0(u;\boldsymbol{\beta}) = dN_{\boldsymbol{\cdot}}(u)/S^{(0)}(\boldsymbol{\beta},u)$ [cf. (4.12)] for $\alpha_0(u)du$ in (4.47). Thus the expected information matrix becomes

$$\mathscr{I}(\boldsymbol{\beta}) = \int_0^\tau \mathbf{V}(\boldsymbol{\beta},u)\,dN_{\boldsymbol{\cdot}}(u) = \sum_{T_j} \mathbf{V}(\boldsymbol{\beta},T_j), \qquad (4.48)$$

where $\mathbf{V}(\boldsymbol{\beta},t)$ is given by (4.46).

For the exponential relative risk function, it is seen that $S^{(0)}(\boldsymbol{\beta},t)$, $\mathbf{S}^{(1)}(\boldsymbol{\beta},t)$, $\mathbf{S}^{(2)}(\boldsymbol{\beta},t)$, and $\mathbf{V}(\boldsymbol{\beta},t)$ reduce to the similar quantities defined for Cox's model by (4.34), (4.35), (4.36), and (4.39). Thus, by comparing (4.38) and (4.48), it is seen that for Cox's model the observed and expected information matrices coincide. In general they will differ, but the difference between the two will often be of little practical importance. However, as the expected information matrix depends only on quantities that are aggregates over the risk sets, while the observed information matrix (formula not shown) depends specifically on the covariate values of the individuals who experience events, the expected information matrix tends to be the more stable of the two, and it is the one we recommend on a routine basis.

4.1.6 Large sample properties of estimators of cumulative hazards and survival functions

We now study the large sample properties of the estimator (4.17) for the cumulative hazard corresponding to a given covariate path $\mathbf{x}_0(s)$; $0 < s \le t$; and the estimator (4.18) for the corresponding survival probability. Again we concentrate on the main steps in the derivations for the (unstratified) relative risk regression model (4.2), and refer readers to Andersen et al. (1993, sections VII.2.2 and VII.2.3) for a detailed treatment of the fixed covariate case (including stratified models) for Cox's model.

First note that the difference between (4.17) and (4.16) may be written as

$$\begin{aligned}
\widehat{A}(t\,|\,\mathbf{x}_0) - A(t\,|\,\mathbf{x}_0) &= \int_0^t r(\widehat{\boldsymbol{\beta}}, \mathbf{x}_0(u))\left(d\widehat{A}_0(u;\widehat{\boldsymbol{\beta}}) - d\widehat{A}_0(u;\boldsymbol{\beta}_0)\right) \\
&\quad + \int_0^t r(\widehat{\boldsymbol{\beta}}, \mathbf{x}_0(u))\left(d\widehat{A}_0(u;\boldsymbol{\beta}_0) - \alpha_0(u)du\right) \qquad (4.49) \\
&\quad + \int_0^t \left(r(\widehat{\boldsymbol{\beta}}, \mathbf{x}_0(u)) - r(\boldsymbol{\beta}_0, \mathbf{x}_0(u))\right)\alpha_0(u)du,
\end{aligned}$$

with

$$\widehat{A}_0(t;\boldsymbol{\beta}) = \int_0^t \frac{dN_{\bullet}(u)}{S^{(0)}(\boldsymbol{\beta},u)};$$

cf. (4.12), (4.13), and (4.27).

Asymptotically we may replace $r(\widehat{\boldsymbol{\beta}}, \mathbf{x}_0(u))$ by $r(\boldsymbol{\beta}_0, \mathbf{x}_0(u))$ in the first two terms on the right-hand side of (4.49). Then by a Taylor series expansion, the first term may be approximated by

$$-\int_0^t r(\boldsymbol{\beta}_0, \mathbf{x}_0(u)) \frac{\mathbf{S}^{(1)}(\boldsymbol{\beta}_0, u)}{S^{(0)}(\boldsymbol{\beta}_0, u)^2} dN_{\bullet}(u) \left(\widehat{\boldsymbol{\beta}} - \boldsymbol{\beta}_0\right), \tag{4.50}$$

while for the second term, we may use the decomposition

$$dN_{\bullet}(u) = S^{(0)}(\boldsymbol{\beta}_0, u)\alpha_0(u)du + dM_{\bullet}(u)$$

to get the approximation

$$\int_0^t \frac{r(\boldsymbol{\beta}_0, \mathbf{x}_0(u))}{S^{(0)}(\boldsymbol{\beta}_0, u)} dM_{\bullet}(u).$$

Finally, by a Taylor series expansion, the last term on the right-hand side of (4.49) is approximately equal to

$$\int_0^t \dot{\mathbf{r}}(\boldsymbol{\beta}_0, \mathbf{x}_0(u))\alpha_0(u)du \left(\widehat{\boldsymbol{\beta}} - \boldsymbol{\beta}_0\right).$$

Further, we may, asymptotically, replace $dN_{\bullet}(u)$ in (4.50) by $S^{(0)}(\boldsymbol{\beta}_0, u)\alpha_0(u)du$. Then (4.49) gives the approximation

$$\widehat{A}(t \mid \mathbf{x}_0) - A(t \mid \mathbf{x}_0) \approx \int_0^t \frac{r(\boldsymbol{\beta}_0, \mathbf{x}_0(u))}{S^{(0)}(\boldsymbol{\beta}_0, u)} dM_{\bullet}(u) \tag{4.51}$$

$$+ \int_0^t \left\{ \dot{\mathbf{r}}(\boldsymbol{\beta}_0, \mathbf{x}_0(u)) - r(\boldsymbol{\beta}_0, \mathbf{x}_0(u)) \frac{\mathbf{S}^{(1)}(\boldsymbol{\beta}_0, u)}{S^{(0)}(\boldsymbol{\beta}_0, u)} \right\} \alpha_0(u)du \left(\widehat{\boldsymbol{\beta}} - \boldsymbol{\beta}_0\right).$$

We may now argue in a similar way as we did for the Nelson-Aalen estimator in Sections 3.1.5 and 3.1.6 to show that, for a given value of t, the leading term on the right-hand side of (4.51) is approximately normally distributed with mean zero and with a variance that may be estimated by

$$\widehat{\omega}^2(t \mid \mathbf{x}_0) = \int_0^t \left(\frac{r(\widehat{\boldsymbol{\beta}}, \mathbf{x}_0(u))}{S^{(0)}(\widehat{\boldsymbol{\beta}}, u)}\right)^2 dN_{\bullet}(u) = \sum_{T_j \le t} \left(\frac{r(\widehat{\boldsymbol{\beta}}, \mathbf{x}_0(T_j))}{\sum_{l \in \mathcal{R}_j} r(\widehat{\boldsymbol{\beta}}, \mathbf{x}_l(T_j))}\right)^2.$$

Further, using the approximate multivariate normality of $\widehat{\boldsymbol{\beta}}$ (cf. Section 4.1.5), it follows that the second term on the right-hand side of (4.51) is approximately normally distributed with mean zero and with a variance that may be estimated by $\widehat{\mathbf{G}}(t\,|\,\mathbf{x}_0)^T \mathscr{I}(\widehat{\boldsymbol{\beta}})^{-1}\widehat{\mathbf{G}}(t\,|\,\mathbf{x}_0)$, where

$$\widehat{\mathbf{G}}(t\,|\,\mathbf{x}_0) = \int_0^t \left\{ \dot{\mathbf{r}}(\widehat{\boldsymbol{\beta}}, \mathbf{x}_0(u)) - r(\widehat{\boldsymbol{\beta}}, \mathbf{x}_0(u)) \frac{\mathbf{S}^{(1)}(\widehat{\boldsymbol{\beta}}, u)}{S^{(0)}(\widehat{\boldsymbol{\beta}}, u)} \right\} d\widehat{A}_0(u)$$

$$= \sum_{T_j \leq t} \frac{\dot{\mathbf{r}}(\widehat{\boldsymbol{\beta}}, \mathbf{x}_0(T_j))}{\sum_{l \in \mathscr{R}_j} r(\widehat{\boldsymbol{\beta}}, \mathbf{x}_l(T_j))} - \sum_{T_j \leq t} r(\widehat{\boldsymbol{\beta}}, \mathbf{x}_0(T_j)) \frac{\sum_{l \in \mathscr{R}_j} \dot{\mathbf{r}}(\widehat{\boldsymbol{\beta}}, \mathbf{x}_l(T_j))}{\left(\sum_{l \in \mathscr{R}_j} r(\widehat{\boldsymbol{\beta}}, \mathbf{x}_l(T_j))\right)^2}.$$

Finally, one may show that the two terms on the right-hand side of (4.51) are asymptotically independent. Thus it follows that $\widehat{A}(t\,|\,\mathbf{x}_0)$ is approximately normally distributed around its true value with a variance that may be estimated by

$$\widehat{\sigma}^2(t\,|\,\mathbf{x}_0) = \widehat{\omega}^2(t\,|\,\mathbf{x}_0) + \widehat{\mathbf{G}}(t\,|\,\mathbf{x}_0)^T \mathscr{I}(\widehat{\boldsymbol{\beta}})^{-1}\widehat{\mathbf{G}}(t\,|\,\mathbf{x}_0) \tag{4.52}$$

as stated in Section 4.1.2.

Finally, let us briefly consider the large sample properties of the estimator $\widehat{S}(t\,|\,\mathbf{x}_0)$ for the (adjusted) survival function $S(t\,|\,\mathbf{x}_0)$ corresponding to given values of time-fixed covariates or given covariate paths of *external* time-dependent covariates. By an argument similar to the one giving (3.49), one may show that

$$\widehat{S}(t\,|\,\mathbf{x}_0) - S(t\,|\,\mathbf{x}_0) \approx -S(t\,|\,\mathbf{x}_0)\left(\widehat{A}(t\,|\,\mathbf{x}_0) - A(t\,|\,\mathbf{x}_0)\right).$$

From this it follows that $\widehat{S}(t\,|\,\mathbf{x}_0)$ is approximately normally distributed around $S(t\,|\,\mathbf{x}_0)$ with a variance that is $S(t\,|\,\mathbf{x}_0)^2$ times the variance of $\widehat{A}(t\,|\,\mathbf{x}_0)$. Thus an estimator for the variance of $\widehat{S}(t\,|\,\mathbf{x}_0)$ is given by

$$\widehat{\tau}^2(t\,|\,\mathbf{x}_0) = \widehat{S}(t\,|\,\mathbf{x}_0)^2 \widehat{\sigma}^2(t\,|\,\mathbf{x}_0), \tag{4.53}$$

where $\widehat{\sigma}^2(t\,|\,\mathbf{x}_0)$ is given by (4.52).

4.2 Additive regression models

For the relative risk regression models (4.2), the effect of the covariates is to act multiplicatively on the baseline hazard, and their effect is assumed to be constant over time. In this section we consider an alternative regression model, namely the additive nonparametric model proposed by Aalen (1980, 1989).

For Aalen's model the hazard rate at time t for an individual i with vector of covariates $\mathbf{x}_i(t) = (x_{i1}(t), x_{i2}(t), \ldots, x_{ip}(t))^T$ takes the form

$$\alpha(t\,|\,\mathbf{x}_i) = \beta_0(t) + \beta_1(t)x_{i1}(t) + \cdots + \beta_p(t)x_{ip}(t). \tag{4.54}$$

Here $\beta_0(t)$ is the baseline hazard corresponding to the hazard rate of an individual with all covariates identically equal to zero, while $\beta_j(t)$ is the increase in the hazard at time t corresponding to a unit's increase in the jth covariate. In this sense $\beta_j(t)$ is the *excess risk* at time t for the jth covariate. We note that the parameters $\beta_j(t)$ in (4.54) are arbitrary *regression functions*, allowing the effects of the covariates to change over time. Thus the model is fully nonparametric. The vector $(\beta_0(t), \beta_1(t), \ldots, \beta_p(t))^T$ shall be denoted $\boldsymbol{\beta}(t)$.

Semiparametric additive regression models, where more structure is put on the functions in (4.54) have been proposed by Lin and Ying (1994) and McKeague and Sasieni (1994). Lin and Ying considered the model where the baseline $\beta_0(t)$ may be any nonnegative function, while the regression functions $\beta_j(t); j = 1, 2, \ldots, p;$ are assumed to be constant over time (i.e., real-valued parameters). The model proposed by McKeague and Sasieni is intermediate between (4.54) and the Lin and Ying model in that some of the $\beta_j(t)$ are allowed to be arbitrary regression functions, while others are assumed to be constant over time. While these alternative excess risk regression models have their virtues, we find the nonparametric additive model to be particularily useful due to its simplicity, and since its nonparametric nature makes it a flexible model that allows the effects of covariates to change over time. In this section we therefore concentrate on the nonparametric additive regression model (4.54). For a thorough discussion of the model of McKeague and Sasieni the readers are referred to the book by Martinussen and Scheike (2006), where this model and the nonparametric model (4.54) play prominent roles.

The use of an additive model for hazard rates is a bit unconventional, since one would typically prefer models where the object of modeling (in this case the hazard rate) is forced to stay within its natural boundaries. In the Cox model the hazard rate is necessarily nonnegative, and in logistic regression the probability must stay within the interval (0,1). An additive model, on the other hand, does not have this natural restriction and can stray into negative values for the hazard rate; indeed this is occasionally seen in estimation, especially if the model fit is not very good. Therefore, the use of an additive model in this area is a kind of statistical heresy. Why then do we propose such a model? There are a number of reasons:

1. An additive model may in some cases be a more correct description of the actual relationship than a multiplicative model. Especially, Rothman (2002) has focused on the importance of additive models when evaluating independent risk factors. Still, the point is not really additivity, since one may have interaction effects, but rather the linear dependence on the parameter functions.
2. A criticism raised toward the ubiquity of relative risk measures is that, although generally useful, they may be misleading. A relative risk of, say, 2 may indicate that the risk factor is quite important. But if the condition is very rare, something not revealed by the relative risk, the medical doctor may still feel that the risk factor is not of great public health importance. Tang and Dickinson (1998) point out that relative risk measures may be misleading as to the actual impact of a risk factor. Difference measures (or excess risk estimates) may be better in many circumstances and also useful for comparing groups with different baseline rates.

It is also shown in Section 4.2.8 how relative survival functions, which come naturally out of the additive model, are a useful alternative to hazard ratios.

3. Interaction between risk factors is often of interest. It is well known that there could easily be a negative interaction in a multiplicative model and a positive interaction in an additive model for the same data; see Perneger (1998) for an interesting comment on this. It is not obvious which of the models correspond best to what one might understand as biological interaction.

4. The theory of the additive model given in this section allows very simple estimation of how the effect of covariates changes over time. No smoothing is needed, which is an advantage since smoothing may tend to blur effects if changes happen fast. Such a simple approach is not available for the Cox model, where smoothing may be required if effects change over time. But it is important to note that change in the additive effects may be different from change in the multiplicative effects, although they often seem surprisingly similar.

5. Additivity (or linearity) fits very nicely with the underlying martingale theory. Estimates and residual processes give exact martingales, something that is not the case for the Cox model. Also the additive model can be incorporated in a very natural way with the empirical transition matrix, yielding again exact martingale results; see Section 4.2.9.

6. Covariates measured with uncertainty, or covariates dropped from or added to the model, are easily accommodated due to linearity (Aalen, 1989). In nonlinear models, like the Cox model, the model structure will in principle not be preserved under such changes. This vulnerability is in stark contrast to the practice of entering and removing variables ad libitum, although in practice this limitation may not be so important.

7. Additivity (or linearity) allows the implementation of dynamic path analysis with decomposition in direct and indirect effects, something not possible for nonlinear models; see Chapters 8 and 9. In particular, Kaufman et al. (2004) point out how decomposition into direct and indirect effects can, under certain conditions, be performed for difference measures, but not for ratio measures.

8. Additivity is useful when analyzing dynamic covariates; see Chapter 8. There is a natural connection with frailty models and with stochastic processes of the Polya type. Putting dynamic covariates into the Cox model may give a stochastic process that explodes in a finite time interval (Section 8.6.3).

9. The additive model is useful when estimating the weights to be used in inverse probability of censoring weighting (IPCW), as demonstrated by Satten et al. (2001), Datta and Satten (2002), and in Section 4.2.7. The reason for the usefulness is the flexibility of the model.

10. Although negative hazard estimates may be seen, this is mainly relevant when studying predictions for individuals. The purpose of most regression analyses would be to study the size of effects of various covariates, not to make individual predictions. Estimated survival curves can be easily adjusted to become monotone; see Section 4.2.6. This sort of adjustment is in the spirit of other nonparametric regression, see, e.g., Hall and Müller (2003). The additive model may also fruitfully be used for analyzing excess hazard where negative estimates may

be quite appropriate. In fact, it has been pointed out as a disadvantage of the proportional excess hazards model that it is forced to give positive estimates, and an additive model has been presented as more suitable; see Section 4.2.8.

4.2.1 Estimation in the additive hazard model

Estimation in the additive nonparametric model (4.54) focuses on the cumulative regression functions $B_q(t) = \int_0^t \beta_q(u)du$. Roughly speaking, the estimation is performed at each event time by regressing the $dN_i(t)$ for the individuals at risk on their covariates. Although any single regression poorly estimates the increments $dB_q(t) = \beta_q(t)dt$, stability in the estimates for the cumulative regression functions $B_j(t)$ themselves is achieved by aggregating the estimated increments over time.

More specifically, we may proceed as follows. Using (4.1), we note that when model (4.54) applies, the counting process $N_i(t)$ has an intensity process of the form

$$\lambda_i(t) = Y_i(t)\left\{\beta_0(t) + \beta_1(t)x_{i1}(t) + \cdots + \beta_p(t)x_{ip}(t)\right\}. \qquad (4.55)$$

Now $dN_i(t) = \lambda_i(t)dt + dM_i(t)$, so we may write

$$dN_i(t) = Y_i(t)dB_0(t) + \sum_{j=1}^{p} Y_i(t)x_{ij}(t)dB_j(t) + dM_i(t); \qquad (4.56)$$

$i = 1,2,\ldots,n$. Note that, for a given value of t, this relation has the form of an ordinary linear regression model with the $dN_i(t)$ being the observations, the $Y_i(t)x_{ij}(t)$ the covariates, the $dB_j(t)$ the parameters to be estimated, and the $dM_i(t)$ the random errors.

It is convenient to rewrite (4.56) using vector and matrix notation. To this end, let $\mathbf{N}(t) = (N_1(t),\ldots,N_n(t))^T$ be the vector of counting processes, and let $\mathbf{M}(t) = (M_1(t),\ldots,M_n(t))^T$ be the corresponding vector of martingales. We also introduce the vector of cumulative regression functions $\mathbf{B}(t) = (B_0(t),B_1(t),\ldots,B_p(t))^T$ and the matrix $\mathbf{X}(t)$ whose ith row has the elements $Y_i(t),Y_i(t)x_{i1}(t),\ldots,Y_i(t)x_{ip}(t)$. Note that $\mathbf{X}(t)$ is a $n \times (p+1)$ matrix. Then (4.56) may be written as

$$d\mathbf{N}(t) = \mathbf{X}(t)d\mathbf{B}(t) + d\mathbf{M}(t), \qquad (4.57)$$

which has the form of a linear regression model on matrix form with design matrix $\mathbf{X}(t)$. Ordinary least squares regression therefore gives

$$d\widehat{\mathbf{B}}(t) = \left(\mathbf{X}(t)^T\mathbf{X}(t)\right)^{-1}\mathbf{X}(t)^T d\mathbf{N}(t)$$

when $\mathbf{X}(t)$ has full rank. We introduce $J(t)$ as the indicator of $\mathbf{X}(t)$ having full rank, and the least squares generalized inverse

$$\mathbf{X}^-(t) = \left(\mathbf{X}(t)^T\mathbf{X}(t)\right)^{-1}\mathbf{X}(t)^T. \tag{4.58}$$

Accumulating the increments $d\widehat{\mathbf{B}}(t)$ over the times $T_1 < T_2 < \cdots$ when an event occurs and $\mathbf{X}(t)$ has full rank, we obtain the estimator

$$\widehat{\mathbf{B}}(t) = \int_0^t J(u)\mathbf{X}^-(u)\,d\mathbf{N}(u) = \sum_{T_j \le t} J(T_j)\mathbf{X}^-(T_j)\triangle\mathbf{N}(T_j) \tag{4.59}$$

for the vector of cumulative regression functions. Here $\triangle\mathbf{N}(T_j)$ is a vector of zeros except for a one for the component corresponding to the individual who experiences the event at time T_j. Comparing (4.57) with (3.20) and (4.59) with (3.22), we see that the nonparametric additive regression model is a multivariate generalization of the multiplicative intensity model of Section 3.1.2, while the estimator of the vector of cumulative regression functions is a multivariate Nelson-Aalen estimator. In particular, if we have an additive model with only one binary covariate ($x_1 = 0$ or 1), then the estimated cumulative baseline $\widehat{B}_0(t)$ is the Nelson-Aalen estimator for the reference group ($x_1 = 0$) while $\widehat{B}_1(t)$ is the difference between the Nelson-Aalen estimators for the two groups (Exercise 4.4).

We shall use martingale methods to derive an estimator for the covariance matrix of $\widehat{\mathbf{B}}(t)$ and to study its statistical properties. To this end we introduce

$$\mathbf{B}^*(t) = \int_0^t J(u)\,d\mathbf{B}(u), \tag{4.60}$$

which is almost the same as $\mathbf{B}(t)$ when the probability of $\mathbf{X}(u)$ having full rank for all $u \in [0,t]$ is close to one. If we then insert (4.57) in (4.59), we obtain the key relation

$$\widehat{\mathbf{B}}(t) - \mathbf{B}^*(t) = \int_0^t J(u)\mathbf{X}^-(u)\,d\mathbf{M}(u). \tag{4.61}$$

The right-hand side of (4.61) is a vector-valued stochastic integral, and hence is itself a mean zero martingale. In particular, $E\{\widehat{\mathbf{B}}(t) - \mathbf{B}^*(t)\} = \mathbf{0}$, so $\widehat{\mathbf{B}}(t)$ is an unbiased estimator of $\mathbf{B}^*(t)$ and an almost unbiased estimator for $\mathbf{B}(t)$.

In order to derive an estimator of the covariance matrix of $\widehat{\mathbf{B}}(t)$, we introduce the vector of intensity processes $\boldsymbol{\lambda}(t) = (\lambda_1(t),\ldots,\lambda_n(t))^T$ and note that by (B.4) in Appendix B the predictable variation process of the martingale (4.61) takes the form

$$\langle\widehat{\mathbf{B}} - \mathbf{B}^*\rangle(t) = \int_0^t J(u)\mathbf{X}^-(u)\,\mathrm{diag}\{\boldsymbol{\lambda}(u)du\}\,\mathbf{X}^-(u)^T. \tag{4.62}$$

Here $\mathrm{diag}\{\mathbf{v}\}$ means the diagonal matrix with the vector $\mathbf{v}$ as diagonal. An estimator for the covariance matrix of $\widehat{\mathbf{B}}(t) - \mathbf{B}^*(t)$, and hence for $\widehat{\mathbf{B}}(t)$ as well, is obtained by inserting an estimate for $\boldsymbol{\lambda}(u)du$ in this expression. Here there are two options. One option is to estimate $\boldsymbol{\lambda}(u)du$ by $d\mathbf{N}(u)$ to obtain the covariance matrix estimator

$$\widehat{\boldsymbol{\Sigma}}(t) = \sum_{T_j \le t} J(T_j)\mathbf{X}^-(T_j)\,\mathrm{diag}\{\triangle\mathbf{N}(T_j)\}\,\mathbf{X}^-(T_j)^T. \tag{4.63}$$

Alternatively, we may use the estimate $\mathbf{X}(u)\,d\widehat{\mathbf{B}}(u)$ for $\boldsymbol{\lambda}(u)du$ to obtain the model-based covariance matrix estimator

$$\widehat{\boldsymbol{\Sigma}}_{\mathrm{mod}}(t) = \sum_{T_j \le t} J(T_j)\mathbf{X}^-(T_j)\,\mathrm{diag}\{\mathbf{X}(T_j)\triangle\widehat{\mathbf{B}}(T_j)\}\mathbf{X}^-(T_j)^T. \qquad (4.64)$$

Using the martingale central limit theorem, one may prove that $\sqrt{n}(\widehat{\mathbf{B}} - \mathbf{B}^*)$ converges in distribution to a mean zero multivariate Gaussian martingale with a covariance function that is the limit in probability of n times (4.62); see Andersen et al. (1993, section VII.4.2) for details and the necessary regularity conditions. Asymptotically there is no difference between $\mathbf{B}^*(t)$ and $\mathbf{B}(t)$, so $\sqrt{n}(\widehat{\mathbf{B}} - \mathbf{B})$ converges weakly to this Gaussian martingale as well. In particular, for a fixed value of t, the cumulative regression function estimator $\widehat{\mathbf{B}}(t)$ is approximately multivariate normally distributed around its true value $\mathbf{B}(t)$ with a covariance matrix that may be estimated as described earlier.

Thus an approximate $100(1 - \alpha)\%$ confidence interval for the qth cumulative regression function $B_q(t)$ is given as

$$\widehat{B}_q(t) \pm z_{1-\alpha/2}\sqrt{\widehat{\sigma}_{qq}(t)},$$

with $z_{1-\alpha/2}$ the $1 - \alpha/2$ fractile of the standard normal distribution and $\widehat{\sigma}_{qq}(t)$ the qth diagonal element of either $\widehat{\boldsymbol{\Sigma}}(t)$ or $\widehat{\boldsymbol{\Sigma}}_{\mathrm{mod}}(t)$.

In practice the cumulative regression function estimator is used for plotting, as will be illustrated in Example 4.6. Due to the cumulative character of a plot, one should focus on the slope of the estimated curve and judge how this changes over time. If the additive model is not true, the slope may give a wrong picture of the effect of a covariate. Note, however, that Henderson and Milner (1991) have proposed an interesting supplement to the cumulative regression plot that may extend its use even to the proportional hazards situation.

The cumulative plots may seem to be a rather indirect way of viewing the regression functions, and one might be interested in a more direct estimation of the regression functions themselves. This problem is parallel to that of estimating the hazard rate, rather than the cumulative hazard. We may therefore use kernel function smoothing as discussed in Section 3.1.4; all we have to do is to replace the increments of the Nelson-Aalen estimate by the increments of the estimated cumulative regression functions. In analogy with (3.17) this gives the estimator

$$\widehat{\boldsymbol{\beta}}(t) = \frac{1}{b}\sum_{T_j} K\left(\frac{t - T_j}{b}\right)\triangle\widehat{\mathbf{B}}(T_j), \qquad (4.65)$$

where $K(t)$ is a kernel function, for example, the Epanechnikov kernel. Since the increments of the cumulative regression function estimator are uncorrelated (as a consequence of the martingale property), an estimator of the covariance matrix of $\widehat{\boldsymbol{\beta}}(t)$ is simply obtained as the corresponding weighted sum (with squared weights) of the terms of the sum in (4.63), that is, by:

$$\hat{\boldsymbol{\Sigma}}_\beta(t) = \frac{1}{b^2} \sum_{T_j} K^2\left(\frac{t-T_j}{b}\right) J(T_j)\mathbf{X}^-(T_j)\operatorname{diag}\{\triangle \mathbf{N}(T_j)\}\mathbf{X}^-(T_j)^T.$$

This formula is analogous to (3.18) for the smoothed Nelson-Aalen estimate.

The choice of the bandwidth b is critical for the smoothness of the kernel estimate. There exist methods for making this choice that have certain optimality properties. We will not consider such methods here, and we will be content with choosing the bandwidth subjectively by what seems to be a reasonable degree of smoothing.

Another important problem arising in the application of kernel smoothing is how to take care of the tails. Depending on the size of the window $[t-b,t+b]$ there will be a tail at each end of the time interval in which no estimate is given. In Example 4.6 we apply the Epanechnikov boundary kernel given by Gasser and Müller (1979) [see also Keiding and Andersen (1989)] at the left-hand tail (that is, close to time 0). At the right-hand tail estimation will only be continued as long as the ordinary Epanechnikov kernel works, that is, as long as the window is wholly enclosed in the time interval where $\mathbf{X}(t)$ has full rank. The reason for this is that random variation becomes very great at the right-hand tail.

Example 4.6. Carcinoma of the oropharynx. In Example 1.4 we described a set of survival data concerning treatment of carcinoma of the oropharynx. Data were given for 195 patients, 2 of whom (numbers 136 and 159) are not included in the following analysis due to missing values. To resolve ties, a random number between 0 and 1 was added to all survival times (in days).

We perform an analysis of the data by the additive regression model, including simultaneously the seven covariates:

- x_1 = sex (1 = male, 2 = female),
- x_2 = treatment group (1 = standard, 2 = test),
- x_3 = grade (1 = well differentiated, 2 = moderately differentiated, 3 = poorly differentiated),
- x_4 = age (in years),
- x_5 = condition (1 = no disability, 2 = restricted work, 3 = requires assistance with self-care, 4 = confined to bed),
- x_6 = T-stage (an index of size and infiltration of tumor ranging from 1 to 4, with 1 indicating a small tumor and 4 a massive invasive tumor),
- x_7 = N-stage (an index of lymph node metastasis ranging from 0 to 3, with 0 indicating no evidence of metastases and 3 indicating multiple positive nodes or fixed positive nodes).

The cumulative regression plots for some of the covariates are shown in Figure 4.5. Sex does not seem to be important. But there are significant initial effects for the covariates condition, T-stage, and N-stage; these effects tend to disappear after about one year. A Cox analysis with the seven covariates yields the values 5.71, 2.29, and 1.73 of the Wald test statistic for condition, T-stage, and N-stage; to be compared with a standard normal distribution. The two-sided P-values are below 5%, except for N-stage, which gives a P-value of 8.4%. The regression plot shows that this is due to a strong initial positive effect being "watered down" by a lack of,

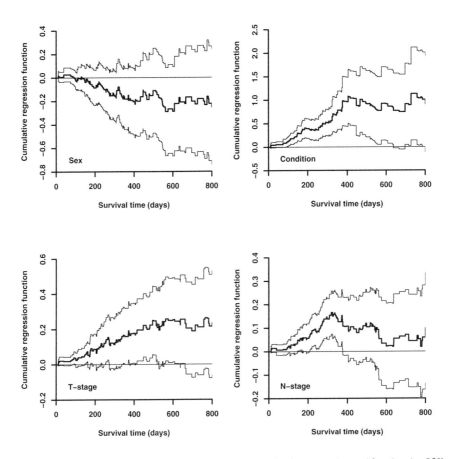

Fig. 4.5 *Oropharynx data: Cumulative regression curves for four covariates with pointwise 95% confidence limits*

or even a slightly negative effect after one year. Hence, not taking into consideration the change in effect over time may lead to missing significant effects. In fact, the lack of much effect after one year is confirmed in a Cox analysis when only survival beyond one year is analyzed (Aalen, 1989).

The TST test statistics (see Section 4.2.3) for the covariates condition, T-stage, and N-stage are 4.11, 2.52, and 1.99, respectively, that is, relatively similar to the Wald test statistics in a Cox setting.

Smoothed curves of the actual regression functions are presented in Figure 4.6, using bandwidth $b = 150$ days. The significant initial effects, especially for the covariates condition and N-stage, are confirmed and it is seen how they disappear

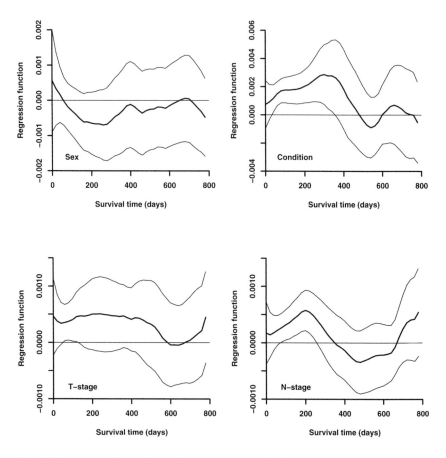

Fig. 4.6 *Oropharynx data: Smoothed regression curves for four covariates with pointwise 95% confidence limits*

later. Toward the end of the time interval, the estimates are rather uncertain due to few patients remaining in the risk set. □

We have used the least squares generalized inverse (4.58) for the estimator $\widehat{\mathbf{B}}(t)$ and its covariance matrix estimators (4.63) and (4.64). Other choices of generalized inverse are possible. In particular, a weighted least squares generalized inverse could have been used (Huffer and McKeague, 1991). However, this would have required a two-step procedure, where one first estimates the weights using a smoothed version of (4.59) and then estimates the cumulative regression functions using a weighted least squares generalized inverse. As this two-step procedure complicates the method, and the efficiency gain to our experience is modest, we have throughout restricted our attention to the use of the least squares generalized inverse (4.58).

4.2.2 Interpreting changes over time

The tendency in Figure 4.5 for covariates to lose their effect over time would be expected since the covariates are measured at time 0 and therefore the measurements "age" as time goes by.

In general, there are (at least) two forces that might lead to diminishing the effect of a covariate, namely ageing of measurements and frailty (see Chapter 6). Hence one should clearly be cautious in assuming a constant effect of a covariate over time as is done in the proportional hazards setting. In regression modeling for survival data, there is typically too little emphasis on changes in effects over time, and hence one gets an unnecessarily limited understanding of the process. There is in general no reason why effects should be constant over time. The hazards could converge, diverge, or even cross when two or more groups are compared. An interesting example of crossing hazards is presented in a comparison of surgery and medical treatment for prevention of stroke in a *Lancet* paper by Rothwell and Warlow (1999). An example of strongly converging hazards may be found in a paper comparing the medication streptokinase with angioplasty ("balloon" treatment) in patients with acute myocardial infarction (Zijlstra et al., 1999). An epidemiological example showing attenuation of effects over time is given by Nielsen et al. (2006).

Of course, some analysis of changes over time can easily be done within the proportional hazards framework as well, for example, by dividing up the time interval and performing separate analyses within each subinterval. One may also adopt more sophisticated smoothing methods or the dynamic landmarking approach proposed by van Houwelingen (2007). However, this is usually not done for at least two reasons. First, standard computer software often gives just the standard proportional hazards analysis and does not encourage the study of changes of effects over time. Second, the tradition in the medical setting, where these methods are used most, is to focus on the classical proportional hazards setting. The fact that this assumption is mainly due to mathematical convenience may not be appreciated by nonmathematical users.

It is important to realize that the causal meaning of changes over time is not obvious. As pointed out in Chapter 6, frailty may lead to converging hazards even if the individual effects are constant over time. So one should be careful with the interpretation.

Concerning the issue of ageing of measurements, one may ask why updated measurements are not used more. One reason is that measurements taken at or before time zero of a study are much easier to handle in the analysis. Updated measurements would mean time-dependent covariates, which could easily give complications in the causal interpretation. This is discussed more in Chapters 8 and 9, where one possible solution to the causal problem is also presented.

4.2.3 Martingale tests and a generalized log-rank test

Various procedures for testing influence of the covariates in the additive model have been studied, for example, by Martinussen and Scheike (2006). Here we shall focus on tests derived by integrating certain weight functions with respect to the estimated cumulative regression functions.

We consider first how one may test the null hypothesis that a covariate, say the qth, has no effect on the hazard rate. Thus we want to test the hypothesis

$$H_0 : \beta_q(t) = 0 \quad \text{for all } t \in [0, t_0]$$

for a suitably chosen time point t_0. Usually $[0, t_0]$ will be the full study time interval $[0, \tau]$, but we may also restrict attention to a smaller interval. We consider tests for the null hypothesis that have good properties versus alternatives of the form $\beta_q(t) > 0$ (or $\beta_q(t) < 0$).

If the null hypothesis holds true, the increments $\triangle \widehat{B}_q(T_j)$ will tend to fluctuate around zero, while they will tend to be positive (or negative) under the alternative. Thus a test for H_0 can be based on the stochastic integral

$$Z_q(t_0) = \int_0^{t_0} L_q(t) \, d\widehat{B}_q(t) = \sum_{T_j \le t_0} L_q(T_j) \triangle \widehat{B}_q(T_j), \qquad (4.66)$$

where $L_q(t)$ is a nonnegative predictable "weight process" assumed to be zero whenever $J(t) = 0$. Considered as a process in t_0, $Z_q(t_0)$ is a martingale under the null hypothesis with predictable variation process

$$\langle Z_q \rangle (t_0) = \int_0^{t_0} L_q^2(t) \, d\langle \widehat{B}_q \rangle (t).$$

It follows that $\mathrm{E}\{Z_q(t_0)\} = 0$ when the null hypothesis holds true. Moreover, we may estimate the variance of $Z_q(t_0)$ by replacing $d\langle \widehat{B}_q \rangle (t)$ in the expression for the predictable variation process by $d\widehat{\sigma}_{qq}(t)$, the increment of the qth diagonal element of (4.63). Hence the variance may be estimated by

$$V_{qq}(t_0) = \int_0^{t_0} L_q^2(t) \, d\widehat{\sigma}_{qq}(t) = \sum_{T_j \le t_0} L_q^2(T_j) \triangle \widehat{\sigma}_{qq}(T_j), \qquad (4.67)$$

where $\triangle \widehat{\sigma}_{qq}(T_j)$ is the increment at time T_j of qth diagonal element of (4.63). Thus a test for H_0 may be based on the standardized test statistic $Z_q(t_0) V_{qq}(t_0)^{-1/2}$, which by the martingale central limit theorem is asymptotically standard normal under the null.

The variance estimator (4.67) of the test statistic is based on the covariance matrix estimator (4.63), and it is valid under both the null and alternative hypotheses. Alternatively one may base the estimation of the variance on the model-based covariance matrix estimator in (4.64) evaluated under the null hypothesis.

In practice the weight process $L_q(t)$ can be chosen in many different ways to emphasize various effects that might arise, and there is generally no best way to make a choice. One tentative recommendation of Aalen (1989) is to use a weight function that may be derived from the matrix

$$\mathbf{K}(t) = \left\{ \text{diag} \left[\left(\mathbf{X}(t)^T \mathbf{X}(t)\right)^{-1} \right] \right\}^{-1}. \tag{4.68}$$

This suggestion was inspired by ordinary least squares regression where the variances of the estimators are proportional to the diagonal elements of the matrix $\left(\mathbf{X}^T \mathbf{X}\right)^{-1}$, with $\mathbf{X}$ being the design matrix, and where the weight attached to an estimate is in inverse proportion to its variance. When testing if a specific regression function $\beta_q(t)$ is identically equal to zero, one uses the weight function $L_q(t) = K_{qq}(t)$ from the diagonal of the matrix $\mathbf{K}(t)$. We denote this the *TST test*. If we have an additive model with only one binary covariate ($x_1 = 0$ or 1), then the TST test for the null hypothesis that $\beta_1(t)$ is identically equal to zero is the log-rank test (Exercise 4.5).

When it comes to testing several regression functions simultaneously, the situation is more complex. For instance, Aalen (1989) suggested a test for several regression functions simultaneously based on the matrix $\mathbf{K}(t)$. However, as was pointed out by Bhattacharyya and Klein (2005), tests based on $\mathbf{K}(t)$ are not suitable when performing an omnibus test for a number of regression functions. The reason is that they lead to an inconsistency, in the sense that when comparing several groups the results depend on which group is used as a baseline. In fact, most choices of weight functions would lead to this kind of inconsistency, and therefore the appropriate choice of weight functions is rather limited. We shall, however, give a general procedure that works well.

Let us first assume that all covariates are centered, that is, the mean of the covariates for individuals at risk is subtracted prior to every jump time. It was then noted by Gandy et al. (2008) that one can remove the diagonal operator from (4.68), using instead the weight matrix $\mathbf{K}_{\text{mod}}(t) = \mathbf{X}(t)^T \mathbf{X}(t)$. In the case that all regression functions except $\beta_0(t)$ are tested simultaneously, one then simply gets the Cox score test or the log-rank test when all covariates are categorical. Using this weighting principle, the test statistic can be written

$$\mathbf{U}_{\text{mod}}(t_0) = \int_0^{t_0} \mathbf{X}(t)^T \mathbf{X}(t) d\widehat{\mathbf{B}}(t)$$

$$= \int_0^{t_0} \mathbf{X}(t)^T \mathbf{X}(t) \left(\mathbf{X}(t)^T \mathbf{X}(t)\right)^{-1} \mathbf{X}(t)^T d\mathbf{N}(t)$$

$$= \int_0^{t_0} \mathbf{X}(t)^T d\mathbf{N}(t).$$

The actual log-rank test follows by a chi-square construction as in formula (4.70).

This idea can be generalized to the situation where just some of the regression functions are tested. Following Gandy et al. (2008) we define what we term *a generalized log-rank test* for the null hypothesis that a subset of the regression functions are identically equal to zero over some time interval. More precisely, for some integer $q < p$, consider the null hypothesis

$$H_0 : \beta_{q+1}(t) = \beta_{q+2}(t) = \cdots = \beta_p(t) = 0 \qquad \text{for } 0 \leq t \leq t_0. \qquad (4.69)$$

Let $\mathbf{X}_1(t)$ and $\mathbf{X}_2(t)$ be matrices consisting, respectively, of the $q+1$ first and $p-q$ last columns of $\mathbf{X}(t)$, that is, $\mathbf{X}_2(t)$ corresponds to the covariate functions that enter the null hypothesis. The idea is to test whether $\mathbf{X}_2(t)$ contains important predictors after adjustment for $\mathbf{X}_1(t)$. The first step in constructing the test is to orthogonalize $\mathbf{X}_2(t)$ with respect to $\mathbf{X}_1(t)$. This is achieved by the transformation

$$\widetilde{\mathbf{X}}_2(t) = \{\mathbf{I} - \mathbf{X}_1(t)\mathbf{X}_1^+(t)\}\mathbf{X}_2(t),$$

where the notation $\mathbf{A}^+$ for a matrix $\mathbf{A}$ denotes its Moore-Penrose inverse. When the matrix has full rank, the Moore-Penrose inverse coincides with the least squares generalized inverse $\mathbf{A}^-$ defined in (4.58), but the Moore-Penrose inverse is also defined for matrices that are not of full rank, which could be of relevance here. The Moore-Penrose inverse inverse can be calculated in computer packages, for instance, by means of the singular value decomposition.

Using the same principle as defining $\mathbf{U}_{\text{mod}}(t_0)$, we suggest the following test statistic:

$$\mathbf{U}(t_0) = \int_0^{t_0} \widetilde{\mathbf{X}}_2(t)^T d\mathbf{N}(t).$$

Note that if $q = 0$ we retrieve exactly the Cox score (log-rank) test $\mathbf{U}_{\text{mod}}(t_0)$. Since the test here is defined also for subsets of the regression functions, we denote it a generalized log-rank test.

With an appropriate covariance estimator $\widehat{\text{cov}}(\mathbf{U}(t_0))$ for the test statistic, the full test is defined by

$$\chi^2(\mathbf{U}) = \mathbf{U}(t_0)^T \, \widehat{\text{cov}}(\mathbf{U}(t_0))^{-1} \mathbf{U}(t_0), \qquad (4.70)$$

which is chi-squared distributed with $p - q$ degrees of freedom under the null hypothesis. There are two natural ways of estimating the covariance matrix, one using the assumed structure under the null hypothesis, and the other not using this structure. The first approach was recommended by Grønnesby (1997) and corresponds to what is used in the global log-rank test for the Cox model. However, it was shown by Gandy et al. (2008), based on simulation studies, that this is not always the best choice. Still, we feel that the general recommendation would be to use the covariance exploiting the null hypothesis structure, and this is defined by

$$\widehat{\text{cov}}_*(\mathbf{U}(t_0)) = \int_0^{t_0} \widetilde{\mathbf{X}}_2(t)^T \text{diag}\left\{\mathbf{X}_1(t)\mathbf{X}_1^+(t)d\mathbf{N}(t)\right\} \widetilde{\mathbf{X}}_2(t). \qquad (4.71)$$

Notice that the other estimator, not assuming the null hypothesis, is a slight modification, given as

$$\widehat{\text{cov}}_g(\mathbf{U}(t_0)) = \int_0^{t_0} \widetilde{\mathbf{X}}_2(t)^T \text{diag}\{d\mathbf{N}(t)\} \widetilde{\mathbf{X}}_2(t). \qquad (4.72)$$

The choice of covariance estimator can sometimes be very important. Gandy et al. (2008) analyze a data set where there is a large difference between the test as calculated with one or the other covariance estimator; the chi-square test statistics being, respectively, 22.35 with covariance from (4.71) and 9.91 with covariance from (4.72), each with three degrees of freedom.

A limitation of the cumulative regression function estimator $\widehat{\mathbf{B}}(t)$ is that it can only be computed as long as $\mathbf{X}(t)$ has full rank. It is therefore important to use a parsimonious model in order to be able to preserve the full rank as long as possible. It will then be useful to perform log-rank tests on covariates in order to exclude those that are clearly insignificant (with the caveat that significance may change over time). This is particularly relevant when nonlinear terms or interaction terms are included as part of model checking.

Note that the TST test defined earlier still has a role, being analogous to the Wald test of single parameters in a Cox model. The generalized log-rank test, on the other hand, gives a test for additional information in a set of covariates adjusted for other covariates that are already in the model. Hence this is a perfect tool for deciding how large the model should be and thus for performing the kind of analysis that would usually be done by the likelihood ratio test in parametric models.

Example 4.7. Carcinoma of the oropharynx. This example is a continuation of Example 4.6. We apply the generalized log-rank test with covariance estimator (4.71). First including only the covariate condition yields a chi-square value of 40.83 with one degree of freedom, giving a P-value of 1.66×10^{-10}. Then adding the two covariates T-stage and N-stage gives a chi-square value of 9.84 with two degrees of freedom, giving a P-value of 0.007.

It is of interest to compute these test statistics with the covariance estimator (4.72). The chi-square value for the covariate condition is now only 18.28. Adding T-stage and N-stage yields the chi-square value 8.44. Again, this demonstrates a considerable effect of the choice of covariance matrix. $\square$

4.2.4 Martingale residual processes and model check

Model check can be approached in several ways. For goodness-of-fit testing, see Gandy and Jensen (2005a,b). Here we shall focus on studying the residuals in various graphical manners.

Similar to the results for the relative risk regression models in Section 4.1.3, we shall consider martingale residual processes by comparing the counting processes with their estimated cumulative intensity processes. The comparison can only be performed when the model is estimable, that is, $\mathbf{X}(t)$ is nonsingular. Hence, the vector of martingale residual processes is defined by

$$\mathbf{M}_{\text{res}}(t) = \int_0^t J(u) \, d\mathbf{N}(u) - \int_0^t J(u)\mathbf{X}(u)d\widehat{\mathbf{B}}(u). \tag{4.73}$$

Renaming the two components on the right-hand side, we may write

$$\mathbf{M}_{\text{res}}(t) = \mathbf{N}^*(t) - \widehat{\mathbf{\Lambda}}^*(t), \tag{4.74}$$

where $\mathbf{N}^*$ and $\widehat{\mathbf{\Lambda}}^*$ are the vectors of counting processes and estimated cumulative intensity processes apart from those time intervals where $\mathbf{X}(t)$ is singular. The formula is analogous to the corresponding one for the relative risk regression models; see (4.21).

We will prove that $\mathbf{M}_{\text{res}}(t)$ is an exact martingale. We can write:

$$\mathbf{M}_{\text{res}}(t) = \int_0^t J(u)(\mathbf{I} - \mathbf{X}(u)\mathbf{X}^-(u)) \, d\mathbf{N}(u).$$

Since $\int_0^t J(u)(\mathbf{I} - \mathbf{X}(u)\mathbf{X}^-(u))\mathbf{X}(u)\, d\mathbf{B}(u)$ is identically zero [by the definition (4.58) of the least squares generalized inverse], we may use formula (4.57) to write:

$$\mathbf{M}_{\text{res}}(t) = \int_0^t J(u)(\mathbf{I} - \mathbf{X}(u)\mathbf{X}^-(u)) \, d\mathbf{M}(u). \tag{4.75}$$

If the additive model is true, $\mathbf{M}(t)$ is the vector of counting process martingales, and the integral in (4.75) is a stochastic integral of a predictable process with respect to a vector-valued martingale. Hence $\mathbf{M}_{\text{res}}(t)$ is itself a (vector-valued) martingale, which justifies the name *martingale residual process* in an exact martingale sense. A further study of the residual process for recurrent event data is given in Section 8.3.2.

The standard approaches in the Cox model for checking residuals can all be applied here. Basically there are four ways of using the residuals to assess model fit:

1. *Martingale residuals:* We can evaluate the martingale residual processes at $t = \tau$, where τ is the upper time limit for the study, and get the martingale residuals

$$\widehat{M}_i = M_{\text{res},i}(\tau). \tag{4.76}$$

These can be plotted in a similar way as for the Cox model (Therneau and Grambsch, 2000). Specifically, assume that we have right-censored survival data and that all covariates are time-fixed. Then it can be proven that $\mathbf{X}(0)^T \mathbf{M}_{\text{res}}(\tau) = 0$ (Exercise 4.6). The result means that, if the model is correctly specified, there should be no linear trends in the martingale residuals with respect to the covariates. Hence, a useful way of checking the model is to analyze the residuals with respect to nonlinear functions of the covariates, for example, square values or products of the covariates; see Aalen (1993) for an application of this.

2. *Martingale residual process plots:* One may consider the residual processes as functions of time. It may then be useful to aggregate the individual residual processes over a few subgroups as demonstrated for the relative risk regression models in Section 4.1.3; see also Aalen (1993) and Grønnesby and Borgan (1996). If the model fits well, the suggested plots would be expected to fluctuate around

the zero line. The method is a special case of the more general procedure of considering linear combinations (contrasts) of the martingale residual processes. Consider a $q \times n$ matrix $\mathbf{V}$ of real numbers and define

$$\mathbf{M}_{\text{res}}^{\mathbf{V}}(t) = \mathbf{V}\mathbf{M}_{\text{res}}(t).$$

The quadratic variation process provides an estimator of the covariance matrix of $\mathbf{M}_{\text{res}}^{\mathbf{V}}(t)$. It may be derived from the theory of stochastic integrals [cf. (B.5) in Appendix B], to yield

$$[\mathbf{M}_{\text{res}}^{\mathbf{V}}(t)](t) \tag{4.77}$$

$$= \int_0^t J(s)\mathbf{V}(\mathbf{I} - \mathbf{X}(s)\mathbf{X}^-(s))\, \text{diag}(d\mathbf{N}(s))(\mathbf{I} - \mathbf{X}(s)\mathbf{X}^-(s))^T\mathbf{V}^T.$$

For statistical testing, one may appeal to the central limit theorem for martingales in order to use asymptotic normal distributions.

3. *Arjas plots:* Instead of considering the residual processes with respect to time, one may consider the two parts on the right-hand side of (4.74) separately and compare $N_i^*(t)$ aggregated over suitable subgroups of individuals with $\widehat{\Lambda}_i^*(t)$ aggregated over the same subgroups. This produces a so-called Arjas plot (Arjas, 1988) and corresponds to comparing observed and expected occurrences.

4. *Cox-Snell residuals:* Finally, the classical approach of Cox-Snell residuals for right-censored survival data could be applied. Then the set of $\widehat{\Lambda}_i^*(\tau)$ is considered as a censored sample from a unit exponential distribution when the underlying model is true, with the random variable being censored if $N_i^*(\tau)$ is 0 and uncensored otherwise. The distribution of the residuals can be simply checked by a Nelson-Aalen plot.

Martingale residuals, as defined in (4.76), do not behave in the way one might expect by analogy with the ordinary linear models of statistics. For instance, the distribution of martingale residuals cannot be expected to bear any similarity to the normal distribution. Indeed, for both the relative risk regression models and the additive model, the distribution may be rather skewed, or even bimodal, especially if there is much censoring. Hence, the interpretation of residuals is certainly not as straightforward as in the classic linear regression models. One possibility might be to transform the residuals, but much of the problem may remain.

Methods of plotting martingale residuals have been studied extensively for the Cox model. These plots may be difficult to judge due to the problems just mentioned, and, in any case, they depend on ad hoc smoothing techniques. We therefore recommend plotting martingale residual processes over time or making Arjas plots. One should try to pick groups of individuals that might be expected to show deviation, for instance, those with extreme covariate values. The advantage of the martingale residual process plot is that it explicitly involves the time scale, indicating for which time intervals the fit might be bad, and that one might make pointwise confidence limits (or confidence bands) to judge the significance of deviations. The Arjas

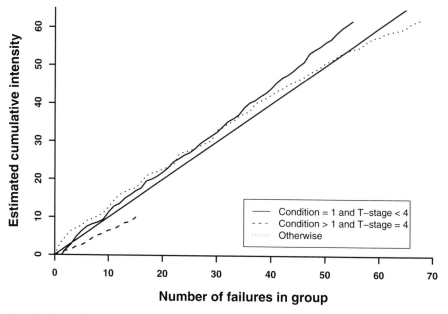

Fig. 4.7 *Arjas plot for the oropharynx survival data divided into a high-risk, an intermediate-risk, and a low-risk group.*

plot, on the other hand, may give a clearer picture of how important the deviations really are.

Example 4.8. Carcinoma of the oropharynx. This example is a continuation of Example 4.6. Figure 4.7 shows an Arjas plot for a division of the oropharynx data into a high-risk, an intermediate-risk, and a low-risk group. Clearly, the deviations are very small, indicating a good model fit. Other plots and analyses give the same result and are not shown. □

Example 4.9. Graft survival times in renal transplantation. In order to understand the value of the various residual plots, it is of interest to see an example where the model does not fit well. Henderson and Milner (1991) present data on graft survival times (in months) of 148 renal patients. The data come from the Leicester and Newcastle transplant centers, and were collected between 1983 and 1987. Only one covariate was considered, namely the total number of HLA-B or DR antigen mismatches between donor and recipient, a number ranging from 0 to 4. This is denoted a mismatch score. Only 31 failures were observed; the rest were censored. Henderson and Milner assert that the additive model does not fit the data when the covariate is used as it is given, and this will be checked here (see also Aalen, 1993).

First, an Arjas plot is shown. The individuals are stratified into three groups: mismatch score equal to 0 or 1, mismatch score equal to 2, and mismatch score equal to 3 or 4. The Arjas plot for this stratification is shown in Figure 4.8. Clearly the curves

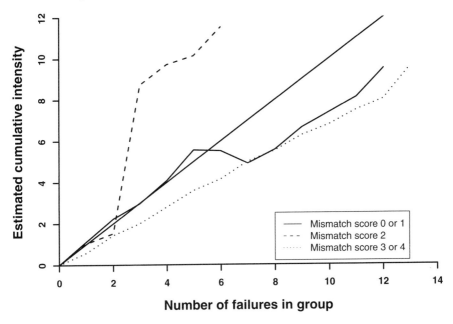

Fig. 4.8 *Arjas plot for the graft survival data.*

differ from the identity straight line, which indicates that the model does not fit well. We have also plotted martingale residual processes for the three groups, presenting the results in Figure 4.9. If the model fit well, the graphs would be expected to stay close to 0. The impression of a deviation is therefore present, for instance, one sees that the deviation is mainly limited to the first four months. In Figure 4.10 the martingale residual process for the group with mismatch score 2 is plotted together with curves indicating deviation of two standard errors. The upper pointwise confidence bound is seen to cross the zero line, indicating that the apparent lack of fit of the model is not entirely due to random variation.

In view of these results it seems natural to include a quadratic term in the original nonparametric regression analysis. This has been attempted, and the TST test value for the quadratic term is found to be equal to 1.97. Further analysis, including residual analysis, shows that this extended model fits the data well. □

4.2.5 Combining the Cox and the additive models

Cox regression analysis is the standard tool for regression analysis of survival data. However, sometimes the model does not fit sufficiently well, and one possibility is then to combine it with the additive model. The idea is that the combination of two very different models may be advantageous. Extensive work in this direction

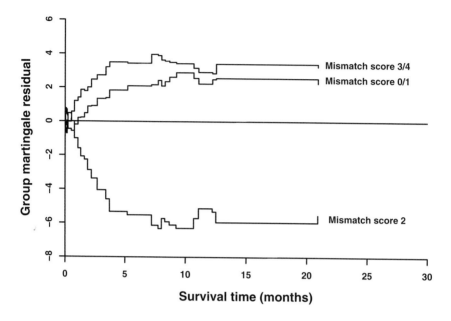

Fig. 4.9 *Martingale residual plot for the graft survival data.*

has been carried out by Martinussen and Scheike, as presented in their book (Martinussen and Scheike, 2006), which also describes an implementation of the method in R (R Development Core Team, 2007). In Scheike and Zhang (2003) it is demonstrated how the combined model yields a much better prediction of survival for subgroups. A closely related procedure was suggested by Zahl (2003).

The additive model may also be useful for analyzing the residuals of the Cox model. This follows the principle of the Mizon-Richard encompassing procedure; for details see Martinussen et al. (2008).

4.2.6 Adjusted monotone survival curves for comparing groups

Kaplan-Meier survival curves are commonly presented as summary plots in survival analysis. However, quite often the groups to be compared differ in many respects and one might like to see adjusted survival curves. A common procedure is to use stratified Cox analysis for the groups (Section 4.1.4), and then to enter the mean of the covariates in the resulting estimate of the survival curves. This probably works

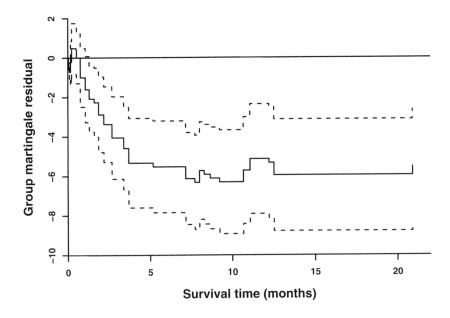

Fig. 4.10 *Martingale residual plot for the graft survival data. Mismatch score 2 with confidence intervals*

well in many cases, but it has also been reported that it can give misleading results (e.g., Ghali et al., 2001).

The additive model can be applied to this situation and gives a flexible adjustment. The cumulative hazard for a set of fixed covariates $\mathbf{x}_0 = (x_{01}, x_{02}, \ldots, x_{0p})^T$ is

$$A(t\,|\,\mathbf{x}_0) = \int_0^t \alpha(u\,|\,\mathbf{x}_0)du = B_0(t) + B_1(t)x_{01} + \cdots + B_p(t)x_{0p},$$

and it may be estimated by

$$\widehat{A}(t\,|\,\mathbf{x}_0) = \widehat{B}_0(t) + \widehat{B}_1(t)x_{01} + \cdots + \widehat{B}_p(t)x_{0p} = \check{\mathbf{x}}_0^T \widehat{\mathbf{B}}(t). \qquad (4.78)$$

Here $\check{\mathbf{x}}_0 = (1, x_{01}, x_{02}, \ldots, x_{0p})^T$, that is, the vector of covariates with a leading 1 added. The adjusted survival function $S(t\,|\,\mathbf{x}_0) = \exp\{-A(t\,|\,\mathbf{x}_0)\}$ may then be estimated by the Kaplan-Meier-type estimator

$$\widehat{S}(t\,|\,\mathbf{x}_0) = \prod_{T_j \le t} \left(1 - \triangle\widehat{A}(T_j\,|\,\mathbf{x}_0)\right). \qquad (4.79)$$

This is similar to (4.18) for the relative risk regression models.

By arguments similar to those of Section 3.2.6, we may show that (4.79) is approximately normally distributed around $S(t \,|\, \mathbf{x}_0)$ with a variance that may be estimated by $\widehat{S}(t \,|\, \mathbf{x}_0)^2 \, \widehat{\sigma}^2(t \,|\, \mathbf{x}_0)$. Here $\widehat{\sigma}^2(t \,|\, \mathbf{x}_0) = \check{\mathbf{x}}_0^T \widehat{\boldsymbol{\Sigma}}(t) \check{\mathbf{x}}_0$ is an estimator for the variance of $\widehat{A}(t \,|\, \mathbf{x}_0)$, and $\widehat{\boldsymbol{\Sigma}}(t)$ is one of the covariance matrix estimators (4.63) and (4.64) for $\widehat{\mathbf{B}}(t)$.

An alternative to the estimator (4.79) is

$$\widetilde{S}(t \,|\, \mathbf{x}_0) = \exp\left(-\widehat{A}(t \,|\, \mathbf{x}_0)\right).$$

In practice, the two estimators will be approximately equal, and it does not matter much which of the two one adopts. But from a theoretical point of view, the estimator (4.79) is the "canonical" one; cf. the discussion in Section 3.2.4.

Occasionally, the estimated survival curve (4.79) could have increasing bits due to negative increments of the estimated cumulative hazard (4.78). A simple way of taking care of this problem is to fit a monotone curve to the survival estimator as used in isotonic regression (Barlow et al., 1972). Such a monotone version is given recursively for increasing event times $T_1 < T_2 < \cdots$ as follows:

$$\widehat{S}_{\mathrm{mon}}(T_j \,|\, \mathbf{x}_0) = \begin{cases} \widehat{S}(T_j \,|\, \mathbf{x}_0) & \text{if } \widehat{S}(T_j \,|\, \mathbf{x}_0) < \widehat{S}_{\mathrm{mon}}(T_{j-1} \,|\, \mathbf{x}_0) \\ \widehat{S}_{\mathrm{mon}}(T_{j-1} \,|\, \mathbf{x}_0) & \text{otherwise.} \end{cases} \tag{4.80}$$

Clearly, the adjusted cumulative hazard estimate (4.78) can be made monotone in a similar fashion. The variance estimate of the monotone survival curve is not likely to deviate much from that of the unadjusted curve. In case of doubt, one can easily perform bootstrap estimation.

In fact, the issue of monotone estimation of survival curves with the additive model is very similar to the problem of order-preserving nonparametric regression, discussed, for instance, by Hall and Müller (2003). In this case one also has a series of linear regressions, and the question is whether the results satisfy a monotonicity property. Just like for the additive model there may be diffculty with this; especially at "boundaries of the distribution of the explanatory variables" (Hall and Müller, 2003, page 598). The approach in (4.80) is similar to a simple initial suggestion by these authors, described as "filling in the valleys" in their figure 1. They also have a far more elaborate solution that might be of use here as well.

Example 4.10. Carcinoma of the oropharynx. An illustration of adjusted survival curves, compared to Kaplan-Meier curves, is given in Figure 4.11, based on the oropharynx data (Example 1.4). We divide the data set according to whether the condition equals 1 or is larger than 1. We calculate the survival curves for the mean values (from the whole sample) of the other covariates by using formula (4.79). The lower left curve in the figure is originally slightly nonmonotone and has then been monotonized by the procedure (4.80). The corrections are shown by small deviations in the figure.

One sees that the adjusted survival curves are closer together than the Kaplan-Meier curves. Adjusted survival curves would seem a useful supplement to the

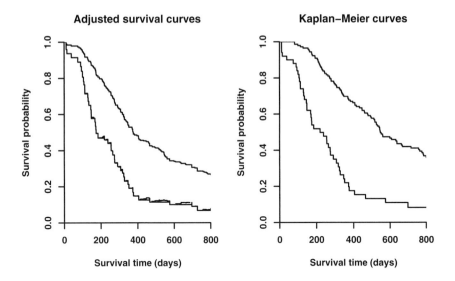

Fig. 4.11 *Survival curves for oropharynx data divided into condition = 1 (upper curves) and condition > 1 (lower curves). The right curves are ordinary Kaplan-Meier plots. The left curves are
adjusted by additive regression for other covariates, inputing the mean values of these covariates
for the whole sample. Monotone adjustment to the lower-left survival curve has been made as
indicated on the curve.*

standard presentation of unadjusted Kaplan-Meier curves. In papers, one often sees
unadjusted survival curves together with adjusted hazard rate analysis (e.g., by the
Cox model), and this is clearly inconsistent. □

4.2.7 Adjusted Kaplan-Meier curves under dependent censoring

The independent censoring assumption, which underlies almost all methods discussed in this book, implies that censoring may depend on information in "the past,"
but not on future events. Thus censoring may depend on the covariates that are included in the model, but not on covariates that are not included.

When computing a Kaplan-Meier estimate, we adopt a model with no covariates.
Thus when censoring depends on the covariates, the ordinary Kaplan-Meier estima-
tor may be biased. Two approaches may be adopted to give an adjusted Kaplan-
Meier estimator in such situations.

The first approach is as follows. Assume that we have a sample of n individ-
uals with time-fixed covariates $\mathbf{x}_1, \mathbf{x}_2, \ldots, \mathbf{x}_n$. We fit an additive model with these
covariates. Then, if censoring only depends on the covariates included in the model
(and the additive model gives a reasonable model fit), we get unbiased estimates of

the cumulative regression functions. An adjusted survival function estimate is then given by:

$$\widehat{S}_{\text{adj},1}(t) = \frac{1}{n} \sum_{i=1}^{n} \widehat{S}(t \mid \mathbf{x}_i), \qquad (4.81)$$

where $\widehat{S}(t \mid \mathbf{x}_i)$ is given by (4.79) with $\mathbf{x}_0$ replaced by $\mathbf{x}_i$. A monotone version can be computed if necessary as described in (4.80).

Another approach to adjusting the Kaplan-Meier estimator uses the inverse probability of censoring weighted principle (IPCW) due to Robins and coauthors (e.g., Robins and Finkelstein, 2000). This approach is based on analyzing the hazard of censoring by a Cox model in order to derive appropriate weights. However, it has been shown by Satten et al. (2001) that the additive model would be useful because it gives a much more flexible model for the censoring hazard than the Cox model.

The idea of the weighting procedure is simply to weight up individual counting processes according to their covariates in order to compensate for those who get censored. One defines an adjusted aggregated counting process, $N^*(t)$, and an adjusted number-at-risk processes, $Y^*(t)$, by

$$N^*(t) = \sum_{i=1}^{n} \int_0^t \frac{dN_i(u)}{\widehat{K}_i(u-)} \quad \text{and} \quad Y^*(t) = \sum_{i=1}^{n} \frac{Y_i(t)}{\widehat{K}_i(t-)},$$

respectively, where $\widehat{K}_i(t)$ is the estimated survival curve of individual i with respect to *the event of censoring*. This survival curve is estimated by formula (4.79) with the difference that the event is substituted by the occurrence of censoring and $\mathbf{x}_0$ is replaced by $\mathbf{x}_i$. So, one should note that formula (4.79) comes into both procedures defined here, but in different ways.

Having defined the adjusted counting and number-at-risk processes, we define an adjusted Nelson-Aalen estimator by analogy with the standard case:

$$\widehat{A}_{\text{adj}}(t) = \int_0^t \frac{J^*(s)}{Y^*(s)} dN^*(s),$$

where $J^*(t) = I(Y^*(t) > 0)$. The adjusted Kaplan-Meier estimate is then given by

$$\widehat{S}_{\text{adj},2}(t) = \prod_{T_j \leq t} \left(1 - \triangle \widehat{A}_{\text{adj}}(T_j) \right), \qquad (4.82)$$

where the T_j are the event times. Note that this will always estimate a proper monotone survival curve.

Example 4.11. Primary biliary cirrhosis. We consider for illustration the data on survival of 312 clinical trial patients with primary biliary cirrhosis (PBC) presented by Fleming and Harrington (1991). We perform an analysis of dependent censoring in the following way. We run an additive model with censoring as the "event" and death as "censoring," including the following covariates: age (in years), albumin, bilirubin, presence of edema, presence of hepatomegaly, and prothrombin time.

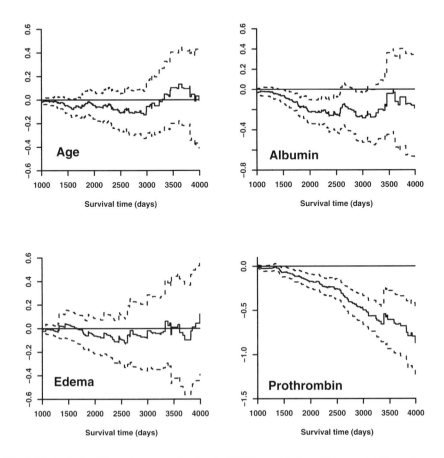

Fig. 4.12 *Analysis of dependent censoring for the PBC data with the additive model. Censoring is the "event" while death is considered "censoring."*

Some of the cumulative regression function estimates are shown in Figure 4.12 and give a clear indication of dependent censoring with respect to prothrombin time. This could be due to some individuals being censored when they had liver transplantation, and Fleming and Harrington (1991, page 154) warn that this may give "a small bias in a natural history model." However, when actually analyzing the data with respect to overall survival, the somewhat suprising result is that the adjusted Kaplan-Meier curves are very similar to the unadjusted one, and so we have not shown the curves here. In fact, one sometimes sees that weighting does not make much difference, but this is certainly not always the case.

To illustrate a case where some difference in the estimates is observed, we use the second part of the PBC data of Fleming and Harrington (1991). These data concern

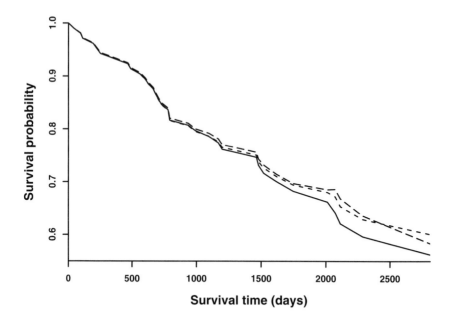

Fig. 4.13 *Second PBC data set. Survival curves adjusted for informative censoring by two methods. Unadjusted: drawn line; adjusted according to (4.81): dashed line with long dashes; adjusted according to (4.82): dashed line with short dashes.*

the survival of 106 additional patients who were not included in the clinical trial, but who still consented to have their basic measurements recorded and to be followed up for survival. Censoring can be analyzed by the additive model as earlier, giving a somewhat different picture (results not shown). We apply both approaches for adjusted estimation of the survival curve. The two adjusted curves are compared with the unadjusted Kaplan-Meier curve in Figure 4.13. One sees some difference; both adjusted curves are above the Kaplan-Meier curve, but the differences are still not very large. Note that only the corrected curves have a clear counterfactual interpretation, namely as estimates of the curve that would have been observed in the original population in the absence of censoring (for the concept of "counterfactual"; see Chapter 9). The uncorrected curve has no clear interpretation. Kaplan-Meier plots are usually not corrected in practice, and apparently the issue of selective censoring is often ignored. □

The principle of weighting by means of an additive model was introduced for non-Markov systems by Datta and Satten (2002) and further studied by Gunnes et al. (2007).

4.2.8 Excess mortality models and the relative survival function

In Section 3.2.5 we consider a model for the excess mortality, assuming that the mortality rate (hazard rate) of an individual i may be written as $\alpha_i(t) = \gamma(t) + \mu_i(t)$, where $\mu_i(t)$ is a *known* mortality rate for a person in the general population corresponding to subject i.

It may be of interest to study how the excess mortality depends on covariates. It has been pointed out that an additive model is especially useful in this context, since excess mortalities may easily become negative such that proportional hazards models may be useless (Zahl, 1995, 1996; Zahl and Tretli, 1997; Lambert et al., 2005). The additive model is then defined for the excess mortality $\gamma(t)$.

Example 4.12. Hip replacements. As an example, consider the hip replacement data (Example 1.5). In Example 3.11 we study the excess mortality of the patients, and Figures 3.14 and 3.15 show that the excess mortality is initially positive and then becomes negative. We now perform an additive hazards regression on the excess mortality, where the two covariates are indicators of whether the age at operation is less than or equal to 70 years or above 70 years. No constant term is included, and the estimated cumulative regression functions will then become the cumulative excess mortality estimator (3.44) computed separately in each of these groups. The cumulative rates have been smoothed by fractional polynomials in Stata, and the results are shown in Figure 4.14. For both groups the excess mortality is high just after the operation. Then it becomes close to one for those below 70 years at operation, while those above 70 years have an excess mortality clearly below one. Thus it seems as if the selection of patients who are allowed to have a hip replacement operation is stricter for ages above 70 years than for those below 70 years. □

Zahl (2003) points out how a combination of the additive and the proportional hazards model may be particularly useful for modeling excess mortality, and he explains how negative excess hazards may often be understood in terms of a frailty model.

Results from an excess mortality analysis are often presented in the form of a relative survival function; cf. Section 3.2.5. This is well known from cancer epidemiology (Ederer et al., 1961; Andersen and Væth, 1989). The relative survival function is traditionally interpreted as the survival probability in the population under study, divided by the expected survival probability had the population been free of the disease under study. This comparison with population mortality corresponds to performing an external adjustment. The same ideas may, however, be used for performing an internal adjustment in the additive model in general, as shall be explained now.

The additive model is a natural tool for extending the concept of relative survival function to standard survival data (without excess mortality); see Zahl and Aalen (1998). In fact, the exponential of minus the cumulative regression function may be interpreted as an adjusted relative survival function. It evaluates the covariate-specific contribution to the relative survival, comparing the survival of individuals that are similar except for differences in one covariate.

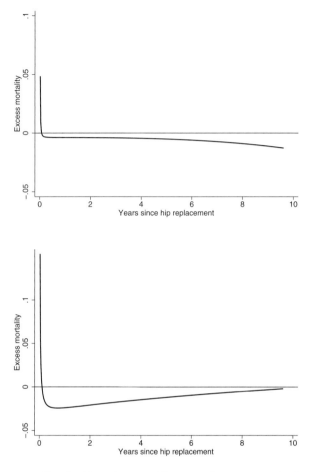

Fig. 4.14 *Excess mortality after hip replacement. Upper panel: age at operation less than or equal to 70 years; lower panel: age at operation above 70 years.*

Following Zahl and Aalen (1998) we define the adjusted relative survival function of covariate j as:

$$\widehat{R}_j(t) = \exp\{-\widehat{B}_j(t)\}.$$

The simplest interpretation of this concept arises when the covariate is dichotomous.

Example 4.13. Carcinoma of the oropharynx. We shall use the oropharynx data for illustration with all seven covariates of Example 4.6 included in the analysis. Let covariate j be the covariate condition dichotomized as to whether condition equals 1, or is greater than 1. The relative survival curve is given in Figure 4.15, showing that the relative survival of those with some disability compared to those with no disability declines to about 25% after about 400 days, and then stays at this level.

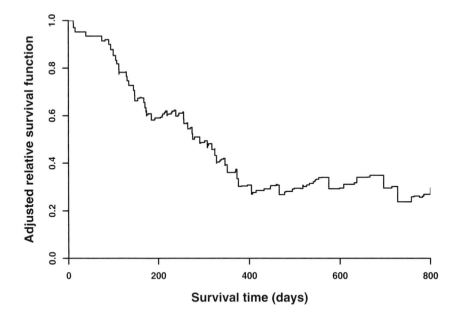

Fig. 4.15 *Adjusted relative survival function for oropharynx data, comparing the group with condition > 1 to the group with condition = 1.*

Running a Cox analysis on the same seven covariates gives a hazard ratio for the same comparison equal to 3.0. Clearly, the relative survival curve gives additional relevant information. The curve may not be monotone, but there is no obvious reason why it should be so either. ☐

4.2.9 Estimation of Markov transition probabilities

In Section 3.4 we discussed how one may estimate transition probabilities for the competing risks model and the three-state Markov illness-death model, as well as for general Markov chains, by the empirical transition matrix. When studying Markov chain models, one may be interested in considering the effects of covariates on the transition probabilities. It turns out that it is very convenient to assume an additive model for the transition intensities and combine this by means of the empirical transition matrix into an overall estimator of the transition probabilities as a function of the covariates (Aalen et al., 2001).

As in Section 3.4.3, we have a Markov chain with state space $\mathscr{I} = \{0, 1, \ldots, k\}$. However, we here assume that the transition intensity for an individual from state

$g \in \mathscr{I}$ to state $h \in \mathscr{I}$, $g \neq h$, depends on a vector of fixed or time-varying co-variates $\mathbf{x}_{gh}(t)$ that may be derived from the vector $\mathbf{x}(t)$ of basic covariates for the individual (typically by omitting some of its components). In order to emphasize the dependence on covariates, we will denote the $g \rightarrow h$ transition intensity by $\alpha_{gh}(t \mid \mathbf{x})$. Assuming that all covariates are external, we may then consider the Markov transition probabilities for given values of the fixed covariates and given paths of the time-varying ones.

We will assume that all the transition intensities may be given an additive specification. On the basis of data from a study we may then, as described in Section 4.2.1, estimate the vectors of cumulative regression functions for each of the relevant transitions in the Markov chain. We denote by $\widehat{\mathbf{B}}_{gh}(t)$ the vector of estimated cumulative regression functions for the $g \rightarrow h$ transition. Then, considering a specified covariate history $\{\mathbf{x}_0(u) : u \leq \tau\}$, we may compute the estimates of the corresponding cumulative transition intensities by

$$\widehat{A}_{gh}(t \mid \mathbf{x}_0) = \int_0^t \check{\mathbf{x}}_{0,gh}(u)^T d\widehat{\mathbf{B}}_{gh}(u) \qquad g, h \in \mathscr{I}, \quad g \neq h,$$

where $\check{\mathbf{x}}_{0,gh}(t)$ is the vector of covariates for the $g \rightarrow h$ transition with a leading 1 added [cf. (4.78)]. We define $\widehat{A}_{gg}(t \mid \mathbf{x}_0) = -\sum_{h \neq g} \widehat{A}_{gh}(t \mid \mathbf{x}_0)$.

We now consider the following question. If the matrix $\widehat{\mathbf{A}}(t)$ in the expression (3.80) for the empirical transition matrix is substituted by the matrix $\widehat{\mathbf{A}}(t \mid \mathbf{x}_0)$ consisting of the elements $\widehat{A}_{gh}(t \mid \mathbf{x}_0)$, what will be the properties of the corresponding estimator of the transition matrix? Because of the martingale property of $\widehat{\mathbf{A}}(t \mid \mathbf{x}_0)$, inherited from the estimates of the cumulative regression functions, there is reason to assume that this should perform well and have nice theoretical properties. Hence, we would get an estimator of the transition matrix that is dependent on covariate information.

The proposed estimator is given as

$$\widehat{\mathbf{P}}(s, t \mid \mathbf{x}_0) = \prod_{(s,t]} \left\{ \mathbf{I} + d\widehat{\mathbf{A}}(u \mid \mathbf{x}_0) \right\}. \tag{4.83}$$

Just as in (3.80) this product-integral is a finite product of matrices. Moreover, for the competing risks model and the three-state illness-death model, explicit expressions for the elements of (4.83) are given by formulas similar to (3.69), (3.70) and (3.74)–(3.76).

To study the statistical properties of the estimator (4.83), we introduce the cumulative intensity matrix $\mathbf{A}^*(t \mid \mathbf{x}_0)$ with elements

$$A_{gh}^*(t \mid \mathbf{x}_0) = \int_0^t J_{gh}(u) \, \alpha_{gh}(u \mid \mathbf{x}_0) \, du \qquad g \neq h,$$

and $A_{gg}^*(t \mid \mathbf{x}_0) = -\sum_{h \neq g} A_{gh}^*(t \mid \mathbf{x}_0)$. Here $J_{gh}(t)$ is an indicator showing whether there is a sufficient set of individuals at risk in state g at time t to ensure that the design matrix $\mathbf{X}_{gh}(t)$, used for estimating the cumulative regression functions for the $g \to h$ transition, has full rank (cf. Section 4.2.1). It then follows from (4.61) that $\widehat{\mathbf{A}}(t \,|\, \mathbf{x}_0) - \mathbf{A}^*(t \,|\, \mathbf{x}_0)$ is an exact matrix-valued martingale. Moreover, by (4.63), we may estimate the variance of $\widehat{A}_{hj}(u \,|\, \mathbf{x}_0) - A_{hj}^*(u \,|\, \mathbf{x}_0)$ by

$$\widehat{\sigma}_{hj}^2(u \,|\, \mathbf{x}_0) = \int_0^u \check{\mathbf{x}}_{0,hj}(v)^T \mathbf{X}_{hj}^-(v) \operatorname{diag}\{d\mathbf{N}_{hj}(v)\} \mathbf{X}_{hj}^-(v)^T \check{\mathbf{x}}_{0,hj}(v), \qquad (4.84)$$

where $\mathbf{N}_{hj}$ is the vector of counting processes for the $h \to j$ transition.

The situation is now completely analogous to that for the ordinary empirical transition matrix, and the arguments of Sections 3.4.4 and 3.4.5 carry through with only notational modifications. As a result, the elements of (4.83) are approximately normally distributed in large samples, and the variance estimator (3.85) applies provided we substitute $\widehat{\mathbf{P}}(s,t \,|\, \mathbf{x}_0)$ for $\widehat{\mathbf{P}}(s,t)$ and replace $\widehat{\sigma}_{hj}^2(u)$ [cf. (3.86)] by $\widehat{\sigma}_{hj}^2(u \,|\, \mathbf{x}_0)$ [cf. (4.84)] in the definition of $\widehat{\boldsymbol{\Psi}}(u)$ [cf. (3.87)]. Also the formulas (3.88)–(3.90) remain valid after similar modifications.

Example 4.14. Uranium miners. In Example 4.2 we studied lung cancer mortality in the Colorado Plateau uranium miners' cohort using the excess relative risk model (4.11). We will see how we may study the absolute risk of lung cancer death and how it depends on the exposure to radon and smoking using additive modeling.

As in Example 4.2 we consider the covariates $\mathbf{x}(t) = (x_{i1}(t), x_{i2}(t))^T$, where $x_{i1}(t)$ is cumulative radon exposure measured in working level months (WLM) up to two years prior to age t, and $x_{i2}(t)$ is cumulative smoking in number of packs smoked up to two years prior to t. By adopting a competing risks model with the three states 0: alive, 1: dead from lung cancer, and 2: dead from other causes, we will study how the absolute risk of lung cancer death (i.e., the probability of being in state 1) depends on the covariates.

We assume that the lung cancer mortality at age t takes the form

$$\alpha_{01}(t \,|\, \mathbf{x}_i) = \beta_0(t) + \beta_1(t)x_{i1}(t) + \beta_2(t)x_{i2}(t) + \beta_3(t)x_{i1}(t)x_{i2}(t), \qquad (4.85)$$

while the mortality $\alpha_{02}(t)$ from other causes, as an approximation, is assumed not to depend on the radon and smoking exposures. Note that our model for lung cancer mortality generalizes the semiparametric relative excess risk model (4.11) of Example 4.2.

In order to simplify the calculations, a nested case-control sample with 50 randomly selected controls per case was used to fit (4.85) as described in Section 4.3.5. Figure 4.16 shows the estimated cumulative regression functions (starting at age 40 years) per 100 WLM radon exposure and per 1000 packs of cigarettes smoked. The baseline is close to zero for ages below 70 years. The cumulative regression functions for radon, smoking, and the interaction are all fairly linear after age 50 years, corresponding to fairly constant excess risks: $\beta_1(t) \approx 0.0050$ per 100 WLM, $\beta_2(t) \approx 0.0025$ per 1000 packs, and $\beta_3(t) \approx 0.0030$. To estimate the cumulative mortality from other causes, a Nelson-Aalen estimator was used. Figure 4.17 shows the estimated cumulative mortality from other causes for the uranium miners together

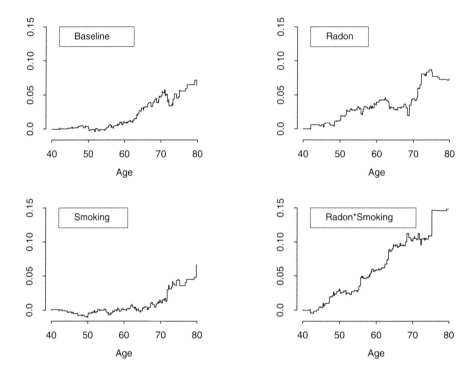

Fig. 4.16 *Estimated cumulative regression functions for the uranium miners.*

with the corresponding cumulative mortality for U.S. white males (Breslow and Day, 1987, appendix III).

We will assume that the cumulative radon and smoking exposures (lagged by two years) are not influenced by the lung cancer deaths, that is, that they are external time-dependent covariates. It then makes sense to consider the absolute risk of lung cancer death conditional on specific (i.e., deterministic) radon and smoking histories $\mathbf{x}_0(t) = (x_{01}(t), x_{02}(t))^T$. We will consider four such covariate histories. For all four situations considered, we assume a constant exposure intensity for radon between ages 20 and 50 years. Thus, the two-year lagged cumulative radon exposure $x_{01}(t)$ is zero for $t < 22$, then increases linearly up to the total dose of radon at $t = 52$, and is constant at the total dose thereafter. Smoking is described by the number of packs per day, and we assume that smoking begins at age 20 years and continues throughout life at the same level. The following four covariate histories are considered:

- Total radon dose 960 WLM, smoking 1 pack per day
- Total radon dose 960 WLM, smoking 1/2 pack per day
- Total radon dose 480 WLM, smoking 1 pack per day

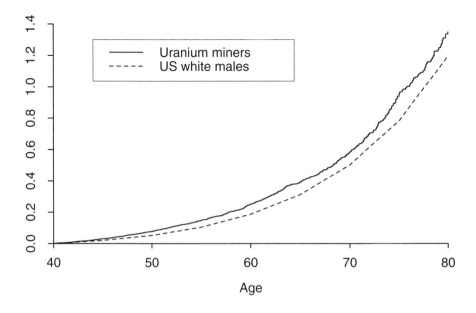

Fig. 4.17 *Estimated cumulative mortality from other causes for the uranium miners and the corresponding cumulative mortality for U.S. white males.*

- Total radon dose 480 WLM, smoking 1/2 pack per day

Figure 4.18 shows the cumulative hazards for these exposure histories (with lower integration limit set to 40). It is seen that for ages below 60 years, a doubling of the radon exposure has a larger effect on the lung cancer death rates than a doubling of the amount of smoking. After about 60 years of age, the situation is reversed. This is a consequence of our assumption that radon exposure stops at age 50 years, while smoking continues throughout life.

At least for ages below 70 years, the cumulative hazards in Figure 4.18 can roughly be interpreted as absolute risks of lung cancer deaths in the absence of deaths from other causes (e.g., Breslow and Day, 1987, section 2.2.b). Absolute risks of lung cancer deaths in the presence of deaths from other causes are shown in Figure 4.19 with 95% pointwise confidence intervals. The effect of taking deaths from other causes into account substantially changes the death risks, especially for the higher ages where the risk of death from other causes becomes high. □

Example 4.15. Bone marrow transplantation. In Example 3.16 we use the three-state illness-death model to study the course of disease for 137 patients with acute leukemia who had undergone a bone marrow transplantation (cf. Example 1.13). Of particular interest is the event of platelet recovery, that is, that the platelet count of a patient returns to a self-sustaining level.

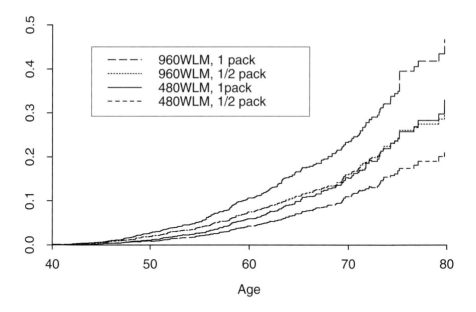

Fig. 4.18 *Estimated cumulative hazard for lung cancer death for four exposure histories.*

We describe the possible events for a patient by an illness-death model of the form shown in Figure 3.19, but with the states 0, 1, and 2 corresponding to "transplanted," "platelet recovered," and "relapsed or dead," respectively. A patient starts in state 0 at time $t = 0$ when he gets the bone marrow transplant. If his platelets recover, the patient moves to state 1, and if he then relapses or dies, he moves on to state 2. If the patient relapses or dies without the platelets returning to a normal level, he moves directly from state 0 to state 2.

Platelet recovery may be considered as a response to the bone marrow transplantation, and it is of interest to estimate the probability of being in response function, that is, the probability of being alive t weeks after transplantation with platelet count on a self-sustaining level (i.e., the probability of being in state 1). In Example 3.16 we estimated this function without considering any covariates. We here illustrate how we may find covariate-dependent estimates of the probability of being in response function.

To this end the patients are grouped into three risk groups based on their status at the time of transplantation: acute lymphoblastic leukemia (ALL), low-risk acute myeloctic leukemia (AML low-risk), and high-risk acute myeloctic leukemia (AML high-risk). These risk groups are represented by the indicators x_1 for AML low-risk and x_2 for AML high-risk patients. Other covariates we use are the indicator x_3 of FAB (French-American-British) classification M4 or M5 for AML patients, the indicator x_4 of whether a patient was given a graft-versus-host prophylactic

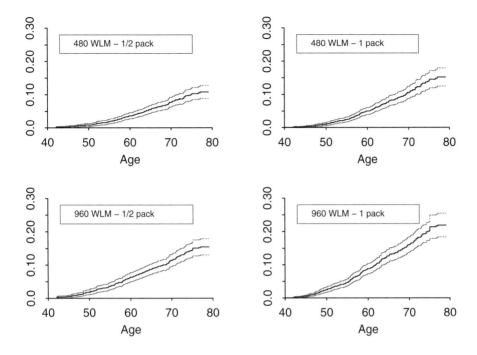

Fig. 4.19 *Estimated absolute risk of lung cancer death for four exposure histories.*

treatment combining methotrexate (MTX) with cyclosporine and possibly methyl-prednisolone, and three covariates for the patient's and donor's age: $x_5 =$ patient's age $- 28$, $x_6 =$ donor age $- 28$, and $x_7 = x_5 \times x_6$.

In order to estimate the probability of being in response for given values of the covariates, we first estimate the intensities for the three possible transitions $0 \to 1$, $1 \to 2$, and $0 \to 2$. (In this estimation, a number of tied $0 \to 1$ transitions as well as a few other tied events were broken at random.) The first two of these transition intensities is assumed to be of an additive form depending on the covariates:

- $0 \to 1$: MTX (x_4), patient's age (x_5), donor's age (x_6), and the interaction patient-donor age (x_7)
- $1 \to 2$: AML low-risk (x_1), AML high-risk (x_2), FAB classification (x_3), patient's age (x_5), and donor's age (x_6)

These covariates are the same as the ones used by Klein and Moeschberger (2003, section 9.5) in their analysis using proportional hazards models, with one exception. We do not include the interaction term x_7 in our model for the $1 \to 2$ transition, since this interaction is not important on the additive scale. The estimated cumulative

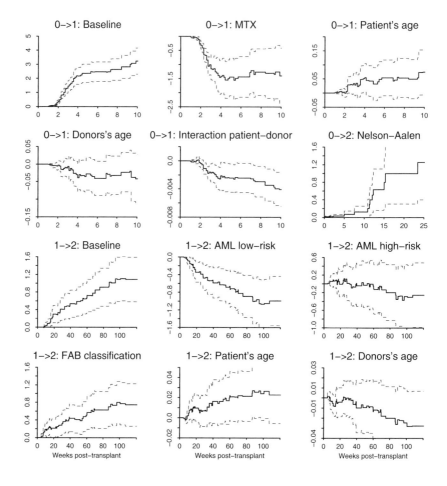

Fig. 4.20 *Cumulative regression functions for the three possible transitions for the bone marrow transplantation data.*

regression functions for the $0 \rightarrow 1$ and $1 \rightarrow 2$ transitions are shown in Figure 4.20 with pointwise 95% confidence intervals.

For the $0 \rightarrow 1$ transition, one should in particular note the estimates of the cumulative regression functions for MTX and the patient-donor age interaction. As seen from the slopes of these curves, both reduce the $0 \rightarrow 1$ transition intensity in the first 3–4 weeks after transplantation. Use of MTX seems to have no effect later, while the size of the age interaction is clearly reduced after about four weeks. For the $1 \rightarrow 2$ transition, all the estimated regression functions are fairly linear, corresponding to constant excess intensities of the covariates. Note in particular that AML

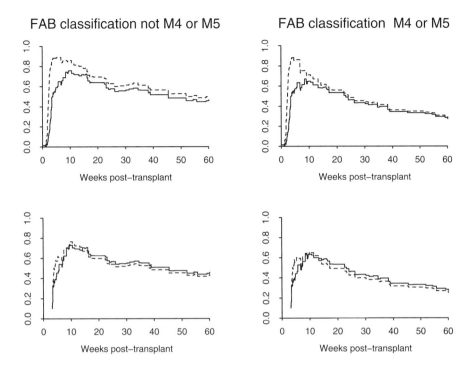

Fig. 4.21 *Estimated probability of being in response for AML high-risk patients without use of MTX (dashed lines) and with use of MTX (drawn lines). Left-hand panel: FAB classification not M4 or M5; right-hand panel: FAB classification M4 or M5. Upper panel:* $P_{01}(0,t \mid x_0)$; *lower panel:* $P_{01}(3,t \mid x_0)$.

low-risk patients have a reduced intensity for relapse or death, while this intensity is increased for patients with FAB classification M4 or M5.

As there are fewer than 20 patients who relapse or die without their platelets returning to a normal level, it is difficult in a meaningful way to estimate the effect of the covariates on the $0 \to 2$ transition using the additive model. We therefore adopt a model without covariates for this transition intensity, and estimate its integral using the Nelson-Aalen estimator. The estimate is shown in Figure 4.20. Note that the transition intensity is low and fairly constant the first 10 weeks after transplantation. Thereafter it increases to a higher level, but, as there are only 13 patients left in state 0 ten weeks after transplantation, the estimate becomes quite uncertain.

From the fitted model we may derive estimates of the probability of being in response for a patient with specific covariates. Figure 4.21 shows estimates for an AML high-risk patient assuming both patient and donor to be 28 years old ($x_{05} = x_{06} = x_{07} = 0$). The upper panel shows estimates of $P_{01}(0,t \mid x_0)$ for the first year after transplantation for the four combinations of MTX and FAB classification. In

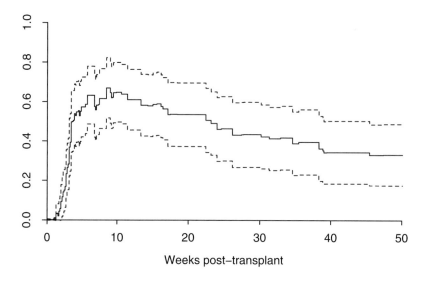

Fig. 4.22 *Estimated probability of being in response $P_{01}(0,t \mid \mathbf{x}_0)$ with pointwise 95% confidence intervals for an AML high-risk patient with FAB classification M4 or M5 using MTX.*

all cases, the estimated probability of being in response function increases steeply the first few weeks after transplantation as the platelets recover for most patients. Then the curves decrease due to relapse and death. Note that use of MTX reduces the probability of being in response, at least for the first 15–20 weeks, and that this probability is lower for patients with FAB classification M4 or M5 than for the other AML high-risk patients. The lower panel of Figure 4.21 shows the estimates of $P_{01}(3,t \mid \mathbf{x}_0)$. For all four combinations of the covariates, the probability of being in response is reduced for a patient whose platelets have not yet recovered three weeks after transplantation. However, use of MTX is no longer of any importance. This clearly illustrates the value of the present approach, which easily takes into account time-dependent effects of covariates.

Confidence intervals are not shown in Figure 4.21 in order not to overburden the figure. To give an indication of the estimation uncertainty, Figure 4.22 presents the estimate of $P_{01}(0,t \mid \mathbf{x}_0)$ with pointwise 95% confidence intervals for a patient using MTX with FAB classification M4 or M5. □

4.3 Nested case-control studies

Estimation for Cox regression and the other relative risk regression models of Section 4.1 is based on the partial likelihood (4.7), which at each observed event compares the covariate values of the individual experiencing the event to those of *all* individuals at risk at that time. Therefore, standard use of the relative risk regression models requires collection of covariate information on all individuals in a cohort even when only a small fraction of these actually experience the event of interest. This may be very expensive, or even logistically impossible, for large cohorts. Case-control sampling techniques, where covariate information is collected for all individuals experiencing the event of interest (cases), but only for a sample of the individuals not experiencing the event (controls) then offer useful alternatives that may drastically reduce the resources that need to be allocated to a study. Further, as most of the statistical information is contained in the cases, such studies may still be sufficient to give reliable answers to the questions of interest.

There are two important classes of case-control sampling designs: nested case-control sampling and case-cohort sampling. For both of these case-control designs, control sampling takes place within the framework of a well-defined cohort. For nested case-control sampling, a small number of controls is selected from those at risk at a case's event time, and a new sample of controls is selected for each case. For the case-cohort design, a subcohort is selected from the full cohort at the outset of the study, and the individuals in the subcohort are used as controls at all event times when they are at risk.

In this section we focus on the nested case-control designs. Throughout we assume, as described in the introduction to the chapter, that we have data from n individuals represented by the counting processes $N_1, N_2, \ldots, N_n$. Here $N_i(t)$ counts the number of occurrences of the event of interest for individual i in $[0,t]$, and it has an intensity process of the form [cf. (4.1)]

$$\lambda_i(t) = Y_i(t)\alpha(t \mid \mathbf{x}_i),   \tag{4.86}$$

where $Y_i(t)$ is an indicator taking the value 1 if individual i is at risk for the event of interest just before time t, and $\mathbf{x}_i(t) = (x_{i1}(t), \ldots, x_{ip}(t))^T$ is a covariate vector of fixed and/or time-varying covariates.

In the following, we first introduce the general framework for nested case-control sampling of Borgan et al. (1995) and describe how it specializes for simple random sampling and stratified (or counter-matched) sampling of the controls. Then, by using a counting process formulation, we show how inference methods for relative risk regression models paralleling those for cohort data apply for nested case-control data. We also indicate how the additive regression model of Section 4.2 may be fitted to nested case-control data.

We do not discuss case-cohort designs any further in this book. But the reader should be aware that these designs have many of the same virtues as the nested case-control designs, and that choosing between a nested case-control design and a case-cohort design for a specific study is often a matter of convenience; see, for example, the discussion in Langholz (2005).

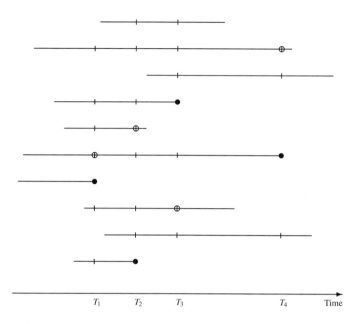

Fig. 4.23 *Illustration of nested case-control sampling, with one control per case, from a hypothetical cohort of ten individuals. Each individual is represented by a line starting at an entry time and ending at an exit time, corresponding to censoring or the occurrence of an event. Event times are indicated by dots (•), potential controls are indicated by bars (|), and the sampled controls are indicated by circles (○).*

4.3.1 A general framework for nested-case control sampling

For nested case-control sampling, one selects, whenever an event occurs, a (typically small) number of controls among those at risk. The set consisting of these controls together with the case is called a *sampled risk set*. Covariate information is collected on the individuals in the sampled risk sets but are not needed for the other individuals in the cohort.

Figure 4.23 illustrates the basic features of nested case-control sampling for a hypothetical cohort of 10 individuals when one control is selected per case (and assuming that at most one event may occur for an individual). Each individual in the cohort is represented by a horizontal line starting at some entry time and ending at some exit time. If the exit time corresponds to an event, this is represented by "•" in the figure. In the hypothetical cohort considered, four individuals are observed to experience an event. The potential controls for the four cases are indicated by "|" and are given as all individuals at risk at the times of the events, cases excluded. Among the potential controls one is selected as indicated by "○" in the figure. The four sampled risk sets are then represented by the four •, ○ pairs in Figure 4.23.

In order to describe in probabilistic terms how the sampling of controls is performed, we need to introduce the "cohort and sampling history" $\mathscr{H}_{t-}$, which contains information about events in the cohort (described by the "cohort history" $\mathscr{F}_{t-}$) as well as on the sampling of controls, up to, but not including, time t. Based on the parts of this history that are available to the researcher, one decides a sampling plan for the controls. Such a sampling plan may be specified as follows. Let $\mathscr{P}$ be the collection of all subsets of $\{1,2,\ldots,n\}$. Then, given $\mathscr{H}_{t-}$, if individual i experiences an event at time t, one selects the set $\mathbf{r} \in \mathscr{P}$ as the sampled risk set with (known) probability $\pi(\mathbf{r}|t,i)$. We assume that $\pi(\mathbf{r}|t,i) = 0$ if $i \notin \mathbf{r}$, so that only subsets containing i are selected. For notational simplicity we let $\pi(\mathbf{r}|t,i) = 0$ if $Y_i(t) = 0$. Note that

$$\sum_{\mathbf{r}\in\mathscr{P}} \pi(\mathbf{r}|t,i) = Y_i(t)$$

since $\pi(\mathbf{r}|t,i)$ is a probability distribution if $Y_i(t) = 1$, and 0 otherwise.

It is useful to have a factorization of the sampling probabilities $\pi(\mathbf{r}|t,i)$. To this end we write $Y(t) = \sum_{i=1}^{n} Y_i(t)$ for the number of individuals at risk and introduce

$$\pi(\mathbf{r}|t) = \frac{\sum_{l=1}^{n} \pi(\mathbf{r}|t,l)}{\sum_{l=1}^{n} Y_l(t)} = Y(t)^{-1} \sum_{l=1}^{n} \pi(\mathbf{r}|t,l). \tag{4.87}$$

That is, $\pi(\mathbf{r}|t)$ is the average of $\pi(\mathbf{r}|t,l)$ over all l. Note that

$$\sum_{\mathbf{r}\in\mathscr{P}} \pi(\mathbf{r}|t) = Y(t)^{-1} \sum_{l=1}^{n} \sum_{\mathbf{r}\in\mathscr{P}} \pi(\mathbf{r}|t,l) = Y(t)^{-1} \sum_{l=1}^{n} Y_l(t) = 1. \tag{4.88}$$

Thus $\pi(\mathbf{r}|t)$ is a probability distribution over sets $\mathbf{r} \in \mathscr{P}$. We then introduce the weights

$$w_i(t,\mathbf{r}) = \frac{\pi(\mathbf{r}|t,i)}{\pi(\mathbf{r}|t)} \tag{4.89}$$

and get the factorization

$$\pi(\mathbf{r}|t,i) = w_i(t,\mathbf{r})\,\pi(\mathbf{r}|t). \tag{4.90}$$

Note that the framework allows the sampling probabilities to depend in an arbitrary way on events in the past, that is, on events that are contained in $\mathscr{H}_{t-}$. The sampling probabilities may, however, not depend on events in the future. For example, one may not exclude as a potential control for a current case an individual that subsequently experiences an event. Also note that the selection of controls is done independently at the different event times, so that subjects may serve as controls for multiple cases, and cases may serve as controls for other cases that experienced an event when the case was at risk. For example, the case at time T_4 in Figure 4.23 had been selected as control at the earlier event time T_1. A basic assumption throughout is that not only the truncation and censoring, but also the sampling of controls, are independent in the sense that the additional knowledge of which individuals have entered the study, have been censored, or have been selected as controls before any time t do not carry information on the risks of events at t.

4.3.2 Two important nested case-control designs

The most common nested case-control design is *simple random sampling*, the classical nested case-control design of Thomas (1977). For this design, if individual i experiences an event at time t, one selects $m-1$ controls by simple random sampling from the $Y(t)-1$ potential controls in the risk set $\mathscr{R}(t) = \{l \mid Y_l(t) = 1\}$. In probabilistic terms the design is given by

$$\pi(\mathbf{r}|t,i) = \binom{Y(t)-1}{m-1}^{-1} I\{i \in \mathbf{r}, |\mathbf{r}| = m, \mathbf{r} \subset \mathscr{R}(t)\}$$

for any set $\mathbf{r} \in \mathscr{P}$. Here $|\mathbf{r}|$ is the number of elements in the set $\mathbf{r}$. The factorization (4.90) applies with

$$\pi(\mathbf{r}|t) = \binom{Y(t)}{m}^{-1} I\{|\mathbf{r}| = m, \mathbf{r} \subset \mathscr{R}(t)\}$$

and

$$w_i(t,\mathbf{r}) = \frac{Y(t)}{m} I\{i \in \mathbf{r}, |\mathbf{r}| = m, \mathbf{r} \subset \mathscr{R}(t)\}. \tag{4.91}$$

To select a simple random sample, the only piece of information needed from $\mathscr{H}_{t-}$ is the at risk status of the individuals. Often, however, some additional information is available for all cohort members; for example, a surrogate measure of the exposure of main interest may be available for everyone. Langholz and Borgan (1995) have developed a stratified nested case-control design, which makes it possible to incorporate such information into the sampling process in order to obtain a more informative sample of controls. For this design, which is called *counter-matching*, one applies the additional piece of information from $\mathscr{H}_{t-}$ to classify each individual at risk into one of, say, S, strata. We denote by $\mathscr{R}_s(t)$ the subset of the risk set $\mathscr{R}(t)$ that belongs to stratum s, and let $Y_s(t) = |\mathscr{R}_s(t)|$ be the number at risk in this stratum just before time t. If individual i experiences an event at t, we want to sample our controls such that the sampled risk set contains a prespecified number m_s of individuals from each stratum s; $s = 1, \ldots, S$. This is achieved as follows. Assume that the case i belongs to stratum $s(i)$. Then for $s \neq s(i)$ one samples randomly without replacement m_s controls from $\mathscr{R}_s(t)$. From the case's stratum $s(i)$ only $m_{s(i)} - 1$ controls are sampled. The case is, however, included in the sampled risk set so this contains a total of m_s from each stratum. Even though it is not made explicit in the notation, we note that the classification into strata may be time-dependent. A crucial assumption, however, is that the information on which the stratification is based has to be known "just before" time t.

In probabilistic terms, counter-matched sampling may be described as follows. For any set $\mathbf{r} \in \mathscr{P}$ that contains i, is a subset of $\mathscr{R}(t)$, and satisfies $|\mathbf{r} \cap \mathscr{R}_s(t)| = m_s$ for $s = 1, \ldots, S$, we have

$$\pi(\mathbf{r}|t,i) = \left\{ \binom{Y_{s(i)}(t) - 1}{m_{s(i)} - 1} \prod_{s \neq s(i)} \binom{Y_s(t)}{m_s} \right\}^{-1}.$$

Here the factorization (4.90) applies, with

$$\pi(\mathbf{r}|t) = \left\{ \prod_{s=1}^{S} \binom{Y_s(t)}{m_s} \right\}^{-1}$$

and

$$w_i(t,\mathbf{r}) = \frac{Y_{s(i)}(t)}{m_{s(i)}}.$$

Other sampling designs for the controls are discussed in Borgan et al. (1995), Langholz and Goldstein (1996), and Langholz (2007). Note also that a full cohort study is a special case of the general framework in which the full risk set is sampled with probability one, that is, $\pi(\mathbf{r}|t,i) = I\{\mathbf{r} = \mathscr{R}(t)\}$ for all $i \in \mathscr{R}(t)$, and $\pi(\mathbf{r}|t,i) = 0$ otherwise.

4.3.3 Counting process formulation of nested case-control sampling

It turns out to be very convenient to use counting processes to describe the data from a nested case-control study. To this end we denote by $T_1 < T_2 < \cdots$ the times when events are observed. Assuming that there are no tied event times, we let i_j be the individual who experiences an event at T_j and denote by $\widetilde{\mathscr{R}}_j$ the sampled risk set at that time. We then introduce the processes

$$N_{i,\mathbf{r}}(t) = \sum_{j \geq 1} I\{T_j \leq t, i_j = i, \widetilde{\mathscr{R}}_j = \mathbf{r}\} \qquad (4.92)$$

counting the observed events for individual i in $[0,t]$ with associated sampled risk set $\mathbf{r}$. Note that we may aggregate the processes $N_{i,\mathbf{r}}(t)$ over sets $\mathbf{r} \in \mathscr{P}$ to recover the counting process

$$N_i(t) = \sum_{\mathbf{r} \in \mathscr{P}} N_{i,\mathbf{r}}(t) = \sum_{j \geq 1} I\{T_j \leq t, i_j = i\}; \qquad i = 1, 2, \ldots, n. \qquad (4.93)$$

In a similar manner for a set $\mathbf{r} \in \mathscr{P}$ we may aggregate over individuals $l \in \mathbf{r}$ to obtain the process

$$N_{\mathbf{r}}(t) = \sum_{l \in \mathbf{r}} N_{l,\mathbf{r}}(t) = \sum_{j \geq 1} I\{T_j \leq t, \widetilde{\mathscr{R}}_j = \mathbf{r}\} \qquad (4.94)$$

counting the number of times in $[0,t]$ the sampled risk set equals the set $\mathbf{r}$.

The assumption that not only truncation and censoring, but also the sampling of controls, are independent ensures that the intensity processes of the counting processes N_i are given by (4.86), not only with respect to the "cohort history" $\mathscr{F}_{t-}$, but also with respect to the "cohort and sampling history" $\mathscr{H}_{t-}$. From this and (4.90) it follows that the intensity processes $\lambda_{i,\mathbf{r}}(t)$ of the counting process (4.92) takes the form

$$\lambda_{i,\mathbf{r}}(t) = \lambda_i(t)\,\pi(\mathbf{r}|t,i) = Y_i(t)\,w_i(t,\mathbf{r})\,\pi(\mathbf{r}|t)\,\alpha(t\,|\,\mathbf{x}_i), \qquad (4.95)$$

while

$$\lambda_{\mathbf{r}}(t) = \sum_{l \in \mathbf{r}} \lambda_{l,\mathbf{r}}(t) = \sum_{l \in \mathbf{r}} Y_l(t)\,w_l(t,\mathbf{r})\,\pi(\mathbf{r}|t)\,\alpha(t\,|\,\mathbf{x}_l) \qquad (4.96)$$

is the intensity processes of the counting process $N_{\mathbf{r}}(t)$; $\mathbf{r} \in \mathscr{P}$.

4.3.4 Relative risk regression for nested case-control data

We now assume that $\alpha(t\,|\,\mathbf{x}_i)$ in (4.86) is of the relative risk form

$$\alpha(t\,|\,\mathbf{x}_i) = \alpha_0(t)\,r(\boldsymbol{\beta},\mathbf{x}_i(t));$$

cf. (4.2). Estimation of $\boldsymbol{\beta}$ for nested case-control data is based on a partial likelihood similar to the one for the full cohort (cf. Section 4.1.1).

To derive this partial likelihood, we note that for the relative risk model the intensity processes (4.95) and (4.96) become

$$\lambda_{i,\mathbf{r}}(t) = Y_i(t)\,r(\boldsymbol{\beta},\mathbf{x}_i(t))\,w_i(t,\mathbf{r})\,\pi(\mathbf{r}|t)\,\alpha_0(t)$$

and

$$\lambda_{\mathbf{r}}(t) = \sum_{l \in \mathbf{r}} Y_l(t)\,r(\boldsymbol{\beta},\mathbf{x}_l(t))\,w_l(t,\mathbf{r})\,\pi(\mathbf{r}|t)\,\alpha_0(t).$$

The intensity process $\lambda_{i,\mathbf{r}}(t)$ may be factorized as

$$\lambda_{i,\mathbf{r}}(t) = \lambda_{\mathbf{r}}(t)\,\pi(i|t,\mathbf{r}),$$

where

$$\pi(i|t,\mathbf{r}) = \frac{\lambda_{i,\mathbf{r}}(t)}{\lambda_{\mathbf{r}}(t)} = \frac{Y_i(t)\,r(\boldsymbol{\beta},\mathbf{x}_i(t))\,w_i(t,\mathbf{r})}{\sum_{l \in \mathbf{r}} Y_l(t)\,r(\boldsymbol{\beta},\mathbf{x}_l(t))\,w_l(t,\mathbf{r})} \qquad (4.97)$$

is the conditional probability of observing an event for individual i at time t, given the past $\mathscr{H}_{t-}$ and that an event is observed for an individual in the set $\mathbf{r}$ at that time.

The partial likelihood for $\boldsymbol{\beta}$ is obtained by multiplying together the conditional probabilities (4.97) over all observed event times T_j, cases i_j, and sampled risk sets

$\widetilde{\mathscr{R}}_j$. Thus the partial likelihood for nested case-control data becomes

$$L(\boldsymbol{\beta}) = \prod_{T_j} \pi(i_j \,|\, T_j, \widetilde{\mathscr{R}}_j) = \prod_{T_j} \frac{r(\boldsymbol{\beta}, \mathbf{x}_{i_j}(T_j))\, w_{i_j}(T_j, \widetilde{\mathscr{R}}_j)}{\sum_{l \in \widetilde{\mathscr{R}}_j} r(\boldsymbol{\beta}, \mathbf{x}_l(T_j))\, w_l(T_j, \widetilde{\mathscr{R}}_j)}. \tag{4.98}$$

We note that (4.98) is similar to the full cohort partial likelihood (4.7). In fact, the full cohort partial likelihood is the special case of (4.98) in which the entire risk set is sampled with probability one and all weights are unity. Note that for simple random sampling of the controls, the weights (4.91) are the same for all individuals and hence cancel from (4.98), giving the partial likelihood of Oakes (1981). For computing, one may use standard software for Cox regression and other relative risk regression models, formally treating the label of the sampled risk sets as a stratification variable in the model (cf. Section 4.1.4) and using the weights $w_l(T_j, \widetilde{\mathscr{R}}_j)$.

The maximum partial likelihood estimator $\widehat{\boldsymbol{\beta}}$ is the value of $\boldsymbol{\beta}$ that maximizes (4.98). Using results for counting processes, martingales, and stochastic integrals, one may show by arguments paralleling the ones in Section 4.1.5 that $\widehat{\boldsymbol{\beta}}$ enjoys similar large sample properties as ordinary maximum likelihood estimators; for details, see Borgan et al. (1995) for Cox's regression model and Borgan and Langholz (2007) for general relative risk functions. In particular, in large samples, $\widehat{\boldsymbol{\beta}}$ is approximately multivariate normally distributed around the true value of $\boldsymbol{\beta}$ with a covariance matrix that may be estimated by $\mathscr{I}(\widehat{\boldsymbol{\beta}})^{-1}$, where $\mathscr{I}(\boldsymbol{\beta})$ is the expected information matrix

$$\mathscr{I}(\boldsymbol{\beta}) = \sum_{T_j} \mathbf{V}(\boldsymbol{\beta}, T_j, \widetilde{\mathscr{R}}_j)). \tag{4.99}$$

Here

$$\mathbf{V}(\boldsymbol{\beta}, T_j, \widetilde{\mathscr{R}}_j) = \frac{\mathbf{S}^{(2)}_{\widetilde{\mathscr{R}}_j}(\boldsymbol{\beta}, T_j)}{S^{(0)}_{\widetilde{\mathscr{R}}_j}(\boldsymbol{\beta}, T_j)} - \left(\frac{\mathbf{S}^{(1)}_{\widetilde{\mathscr{R}}_j}(\boldsymbol{\beta}, T_j)}{S^{(0)}_{\widetilde{\mathscr{R}}_j}(\boldsymbol{\beta}, T_j)} \right)^{\otimes 2},$$

where $\mathbf{v}^{\otimes 2}$ of a column vector $\mathbf{v}$ is the matrix $\mathbf{v}\mathbf{v}^\mathsf{T}$ and

$$S^{(0)}_{\widetilde{\mathscr{R}}_j}(\boldsymbol{\beta}, T_j) = \sum_{l \in \widetilde{\mathscr{R}}_j} r(\boldsymbol{\beta}, \mathbf{x}_l(T_j))\, w_l(T_j, \widetilde{\mathscr{R}}_j),$$

$$\mathbf{S}^{(1)}_{\widetilde{\mathscr{R}}_j}(\boldsymbol{\beta}, T_j) = \sum_{l \in \widetilde{\mathscr{R}}_j} \dot{\mathbf{r}}(\boldsymbol{\beta}, \mathbf{x}_l(T_j))\, w_l(T_j, \widetilde{\mathscr{R}}_j),$$

$$\mathbf{S}^{(2)}_{\widetilde{\mathscr{R}}_j}(\boldsymbol{\beta}, T_j) = \sum_{l \in \widetilde{\mathscr{R}}_j} \frac{\dot{\mathbf{r}}(\boldsymbol{\beta}, \mathbf{x}_l(T_j))^{\otimes 2}}{r(\boldsymbol{\beta}, \mathbf{x}_l(T_j))}\, w_l(T_j, \widetilde{\mathscr{R}}_j)$$

with $\dot{\mathbf{r}}(\boldsymbol{\beta}, \mathbf{x}_i(t)) = \partial r(\boldsymbol{\beta}, \mathbf{x}_i(t))/\partial \boldsymbol{\beta}$.

Alternatively, we may estimate the covariance matrix by $\mathbf{I}(\widehat{\boldsymbol{\beta}})^{-1}$, where

Table 4.6 Estimates (with standard errors) for five nested case-control samples of the uranium miners data. Estimated effects of radon is per 100 WLM and estimated effects of smoking is per 1000 packs of cigarettes.

Sampling design	Radon	Smoking
1:1 simple	0.830 (0.456)	0.405 (0.172)
1:3 simple	0.556 (0.215)	0.276 (0.093)
1:1 counter-matched	0.553 (0.215)	0.211 (0.089)
1:3 counter-matched	0.420 (0.137)	0.205 (0.068)
1:50 simple	0.404 (0.121)	0.185 (0.055)

$$\mathbf{I}(\boldsymbol{\beta}) = \left\{ -\frac{\partial^2}{\partial \beta_h \partial \beta_j} \log L(\boldsymbol{\beta}) \right\}$$

is the observed information martix. When the relative risk function is of the exponential form of Cox's model, the observed information matrix coincides with the expected information matrix, but this is not the case for other relative risk functions. As the expected information matrix depends only on quantities that are aggregates over the sampled risk sets, while the observed information matrix (formula not shown) depends specifically on the covariate values of the case, the expected information matrix tends to be the more stable of the two, and it is the one we recommend on a routine basis.

Example 4.16. Uranium miners. In Example 4.2 we studied the effect of radon exposure and smoking on the risk of lung cancer death for a cohort of uranium miners.

Although covariate information is available on all cohort subjects for this specific cohort, to illustrate the nested case-control methodology, we selected simple random and counter-matched samples with one and three controls per case. These data sets are denoted 1:1 and 1:3 simple random and counter-matched samples, respectively. The 23 tied failure times were broken randomly so that there was only one case per risk set. Counter-matching was based on radon exposure grouped into two or four strata according to the quartiles of the cumulative radon exposure for the cases (Langholz and Goldstein, 1996), and one control was sampled at random from each stratum except the one of the case.

As in Example 4.2 we fit the excess relative risk model

$$\alpha(t \mid \mathbf{x}_i) = \alpha_0(t)\{1 + \beta_1 x_{i1}(t)\}\{1 + \beta_2 x_{i2}(t)\},$$

where age is the basic time scale, $x_{i1}(t)$ is cumulative radon exposure measured in working level months (WLM), and $x_{i2}(t)$ is cumulative smoking in number of packs smoked, both lagged by two years. Table 4.6 shows the estimated regression coefficients with standard errors for the four case-control designs. For comparison, the table also shows the estimates when we sample 50 controls at random for each case, which gives almost the same estimates as for the full cohort. (In fact the cohort estimates of Example 4.2 were obtained in this way for computational convenience.)

We see that the case-control designs tend to give somewhat too large estimates for these data, but apart from the 1:1 simple design, all estimates are within the estimation uncertainty of the 1:50 design. Moreover the 1:1 counter-matched design provides more reliable estimates than the 1:3 simple design, while the 1:3 counter-matched design gives estimates that are almost comparable to the 1:50 design. This illustrates the efficiency gains one may obtain using counter-matched sampling of the controls. □

For Cox's regression model, the efficiency of the maximum partial likelihood estimator for nested case-control data (relative to the estimator based on the full cohort partial likelihood) has been studied by a number of authors. In particular Goldstein and Langholz (1992) showed that, when $\boldsymbol{\beta} = \mathbf{0}$, the asymptotic covariance matrix of the nested case-control estimator equals $m/(m-1)$ times the asymptotic covariance matrix of the full cohort estimator, independent of censoring and covariate distributions. Thus the efficiency of the simple nested case-control design relative to the full cohort is $(m-1)/m$ for testing associations between single exposures and disease. When $\boldsymbol{\beta}$ departs from zero, and when more than one regression coefficient has to be estimated, the efficiency of the nested case-control design may be much lower than given by the "$(m-1)/m$ efficiency rule" (e.g., Goldstein and Langholz, 1992; Borgan and Olsen, 1999).

As illustrated in Example 4.16, counter-matching may give an appreciable improvement in statistical efficiency for estimation of a regression coefficient of particular importance compared to simple nested case-control sampling. Intuitively, this is achieved by increasing the variation in the covariate of interest within each sampled risk set. The efficiency gain has been documented by both asymptotic relative efficiency calculations (e.g., Langholz and Borgan, 1995; Borgan and Olsen, 1999) and empirical studies. For example, in a study of a cohort of gold miners Steenland and Deddens (1997) found that a counter-matched design (with stratification based on duration of exposure) with three controls per case had the same statistical efficiency for estimating the effect of exposure to crystalline silica as a simple nested case-control study using ten controls.

The counting process formulation for nested case-control data is key to deriving the partial likelihood and the large sample properties of the maximum partial likelihood estimator. Moreover, by using a formulation similar to the one for cohort data, it paves the way for extending a number of statistical methods developed for the full cohort to nested case-control data.

In particular, the cumulative baseline hazard $A_0(t) = \int_0^t \alpha_0(u)du$ may be estimated by the Breslow-type estimator

$$\widehat{A}_0(t) = \sum_{T_j \leq t} \frac{1}{\sum_{l \in \widetilde{\mathscr{R}}_j} r(\widehat{\boldsymbol{\beta}}, \mathbf{x}_l(T_j)) w_l(T_j, \widetilde{\mathscr{R}}_j)} \tag{4.100}$$

(Borgan and Langholz, 1993; Borgan et al., 1995). Note that (4.100) is similar to the Breslow estimator (4.13) for the full cohort, except that we only sum over individuals in the sampled risk sets and we have to weight their relative risks. In fact, the full cohort Breslow estimator (4.13) is the special case of (4.100) in which the entire risk set is sampled with probability one and all weights are unity. Also the estimator (4.17) of the cumulative hazard for a given path of external time-dependent covariates and the estimator (4.18) of the corresponding survival probability may be extended in the obvious way to nested case-control data (Langholz and Borgan, 1997). All these estimators enjoy properties similar to the corresponding ones for the full cohort discussed in Section 4.1.6; see Borgan et al. (1995), Langholz and Borgan (1997), and Borgan (2002) for details.

In Section 4.1.3 we illustrated how martingale residual processes may be a convenient tool to check the fit of a relative risk regression model for cohort data. A similar procedure applies to nested case-control data, while a number of other methods available for checking models for cohort data may be more difficult to adopt to nested case-control data. Specifically, using the notation of Section 4.1.3, the group J martingale residual process for nested case-control data takes the form

$$\widehat{M}_J(t) = N_J(t) - \sum_{T_j \le t} \frac{\sum_{i \in \widetilde{\mathscr{R}}_j \cap J(T_j)} r(\widehat{\boldsymbol{\beta}}, \mathbf{x}_i(T_j)) \, w_i(T_j, \widetilde{\mathscr{R}}_j)}{\sum_{l \in \widetilde{\mathscr{R}}_j} r(\widehat{\boldsymbol{\beta}}, \mathbf{x}_l(T_j)) \, w_l(T_j, \widetilde{\mathscr{R}}_j)}. \tag{4.101}$$

This is similar to the martingale residual process (4.22) for the full cohort, which again is the special case in which the entire risk set is sampled with probability one and all weights are unity. All the results we derived in Section 4.1.3 carry over with only minor modifications to nested case-control data; details are provided in Borgan and Langholz (2007).

We have outlined an approach for analyzing nested case-control data based on the partial likelihood. For simple random sampling of the contols, alternative inference methods have been suggested (e.g. Samuelsen, 1997; Scheike and Juul, 2004). However, in our opinion, the modest efficiency gains obtained by these methods (for the simple nested case-control design) do not outweigh the practical complications in fitting the models and the loss of flexibility in tailoring the control sampling to the study needs (e.g., by using counter-matching).

4.3.5 Additive regression for nested case-control data: results

We now consider the situation where $\alpha(t \,|\, \mathbf{x}_i)$ in (4.86) has the additive form

$$\alpha(t \,|\, \mathbf{x}_i) = \beta_0(t) + \beta_1(t)x_{i1}(t) + \cdots + \beta_p(t)x_{ip}(t);$$

cf. (4.54). For nested case-control data we may estimate the cumulative regression functions $B_j(t) = \int_0^t \beta_j(u)du$ by an estimator that is similar to the one for the full cohort (cf. Section 4.2).

To define this estimator, we introduce the $|\widetilde{\mathscr{R}}_j| \times (p+1)$ matrix $\mathbf{X}(T_j, \widetilde{\mathscr{R}}_j)$ with rows $(1, x_{i1}(T_j), \ldots, x_{ip}(T_j)) Y_l(T_j) w_l(T_j, \widetilde{\mathscr{R}}_j)$; $l \in \widetilde{\mathscr{R}}_j$; with the first row, say, corresponding to the case and the remaining ones to the controls at time T_j. We also introduce $J(T_j,\widetilde{\mathscr{R}}_j)$ as the indicator of $\mathbf{X}(T_j,\widetilde{\mathscr{R}}_j)$ having full rank, and the least squares generalized inverse

$$\mathbf{X}^-(T_j,\widetilde{\mathscr{R}}_j) = \left(\mathbf{X}(T_j,\widetilde{\mathscr{R}}_j)^T \mathbf{X}(T_j,\widetilde{\mathscr{R}}_j)\right)^{-1} \mathbf{X}(T_j,\widetilde{\mathscr{R}}_j)^T.$$

We may then estimate the vector $\mathbf{B}(t) = (B_0(t),B_1(t),\ldots,B_p(t))^T$ of cumulative regression functions by

$$\widehat{\mathbf{B}}(t) = \sum_{T_j \le t} J(T_j,\widetilde{\mathscr{R}}_j)\mathbf{X}^-(T_j,\widetilde{\mathscr{R}}_j)\,\mathbf{e}_j, \qquad (4.102)$$

where $\mathbf{e}_j$ is the $|\widetilde{\mathscr{R}}_j|$-dimensional column vector consisting of zeros except for a leading 1 corresponding to the case. The covariance matrix of $\widehat{\mathbf{B}}(t)$ may be estimated by

$$\widehat{\boldsymbol{\Sigma}}(t) = \sum_{T_j \le t} J(T_j,\widetilde{\mathscr{R}}_j)\mathbf{X}^-(T_j,\widetilde{\mathscr{R}}_j)\operatorname{diag}(\mathbf{e}_j)\mathbf{X}^-(T_j,\widetilde{\mathscr{R}}_j)^T. \qquad (4.103)$$

Note that (4.102) and (4.103) are similar to the corresponding estimators (4.59) and (4.63) for the full cohort. In fact, the full cohort estimators are the special cases of (4.102) and (4.103) in which the entire risk set is sampled with probability one and all weights are unity.

In the next subsection we give a motivation for the estimators (4.102) and (4.103) using a counting process formulation and indicate how one may prove that the estimator (4.102) is approximately normal. Also the martingale test statistics and martingale residual processes considered for cohort data in Sections 4.2.3 and 4.2.4 may be adapted to nested case-control data using similar arguments (Borgan and Langholz, 1995, 1997).

The efficiency of the estimator (4.102) for nested case-control data relative to the full cohort estimator (4.59) is considerably lower than is the case for relative risk regression models (Zhang and Borgan, 1999). This is due to the nonparametric nature of the additive regression model, for which the increments of the regression functions have to be estimated at each event time using only the information provided by the sampled risk set. Thus the additive regression model requires more controls per case than the relative risk regression models to provide reliable estimates.

Even when information on the covariates is available for all individuals, it may be problematic to fit the additive model for large cohorts, due to difficulties in dealing with high-dimensional matrices. Then one option is to sample from the risk sets and apply the analysis methods described here using a fairly large number of controls per case. In fact, when fitting the additive regression model to the uranium miners data in Example 4.14 we used a simple random sample of 50 controls per case.

4.3.6 Additive regression for nested case-control data: theory

To motivate and study the properties of the estimator (4.102), it is convenient to reformulate it using counting process notation. To this end for $\mathbf{r} \in \mathscr{P}$ we introduce the $|\mathbf{r}|$-dimensional column vector $\mathbf{N_r}(t)$ with elements $N_{i,\mathbf{r}}(t)$, $i \in \mathbf{r}$. [Note that the vector-valued process $\mathbf{N_r}(t)$ is not the same as the scalar process $N_\mathbf{r}(t)$ given by (4.94).] We also introduce the $|\mathbf{r}| \times (p+1)$-dimensional matrix $\mathbf{X}(t, \mathbf{r})$ with rows $(1, x_{i1}(t), \ldots, x_{ip}(t)) Y_l(t) w_l(t, \mathbf{r})$; $l \in \mathbf{r}$; and the least squares generalized inverse $\mathbf{X}^-(t, \mathbf{r}) = (\mathbf{X}(t, \mathbf{r})^T \mathbf{X}(t, \mathbf{r}))^{-1} \mathbf{X}(t, \mathbf{r})^T$. With $J(t, \mathbf{r})$ the indicator of $\mathbf{X}(t, \mathbf{r})$ having full rank, the estimator (4.102) may be written

$$\widehat{\mathbf{B}}(t) = \sum_{\mathbf{r} \in \mathscr{P}} \int_0^t J(u, \mathbf{r}) \mathbf{X}^-(u, \mathbf{r}) \, d\mathbf{N_r}(u). \qquad (4.104)$$

Now

$$dN_{i,\mathbf{r}}(t) = \lambda_{i,\mathbf{r}}(t) dt + dM_{i,\mathbf{r}}(t),$$

where

$$\lambda_{i,\mathbf{r}}(t) = Y_i(t) w_i(t, \mathbf{r}) \pi(\mathbf{r} \,|\, t) \{\beta_0(t) + \beta_1(t) x_{i1}(t) + \cdots + \beta_p(t) x_{ip}(t)\}$$

[cf. (4.95)] and the $dM_{(i,\mathbf{r})}(t)$ are martingale increments. Using the $|\mathbf{r}|$-dimensional column vectors $\boldsymbol{\lambda_r}(t)$ and $\mathbf{M_r}(t)$ with elements $\lambda_{i,\mathbf{r}}(t)$ and $M_{i,\mathbf{r}}(t)$, respectively, for $i \in \mathbf{r}$, we have that

$$d\mathbf{N_r}(t) = \boldsymbol{\lambda_r}(t) \, dt + d\mathbf{M_r}(t) = \pi(\mathbf{r} \,|\, t) \mathbf{X}(t, \mathbf{r}) \, d\mathbf{B}(t) + d\mathbf{M_r}(t).$$

Then by (4.104), we may write

$$\widehat{\mathbf{B}}(t) - \mathbf{B}^*(t) = \mathbf{M}^*(t) \qquad (4.105)$$

where

$$\mathbf{B}^*(t) = \int_0^t \left\{ \sum_{\mathbf{r} \in \mathscr{P}} J(u, \mathbf{r}) \pi(\mathbf{r} \,|\, u) \right\} d\mathbf{B}(u) \qquad (4.106)$$

and

$$\mathbf{M}^*(t) = \sum_{\mathbf{r} \in \mathscr{P}} \int_0^t J(u, \mathbf{r}) \mathbf{X}^-(u, \mathbf{r}) \, d\mathbf{M_r}(u). \qquad (4.107)$$

We note that (4.105) is a decomposition similar to (4.61) for the full cohort.

Now if $\mathbf{X}(u, \mathbf{r})$ has full rank with high probability for all $\mathbf{r} \in \mathscr{P}$ and all $u \leq t$, then, by (4.88), $\sum_{\mathbf{r} \in \mathscr{P}} J(u, \mathbf{r}) \pi(\mathbf{r} \,|\, u) \approx 1$ and (4.106) is almost the same as $\mathbf{B}(t)$. This gives a justification for the estimator (4.102). Further, the optional variation process of the martingale (4.107) takes the form

$$[\mathbf{M}^*](t) = \sum_{\mathbf{r} \in \mathscr{P}} \int_0^t J(u, \mathbf{r}) \mathbf{X}^-(u, \mathbf{r}) \, \mathrm{diag}\{d\mathbf{N_r}(u)\} \mathbf{X}^-(u, \mathbf{r})^T,$$

which is seen to equal (4.103) thereby providing a justification for the proposed estimator for the covariance matrix. That $\widehat{\mathbf{B}}(t)$ is approximately normally distributed follows from the martingale central limit theorem using the decomposition (4.105); see Zhang and Borgan (1999) for details.

4.4 Exercises

4.1 Assume that the counting processes $N_i(t)$; $i = 1, 2, \ldots, n$; have intensity processes of the form $\lambda_i(t) = Y_i(t)\alpha_0(t)\exp(\boldsymbol{\beta}^T \mathbf{x}_i)$, where the $Y_i(t)$ are at risk indicators and the $\mathbf{x}_i = (x_{i1}, \ldots, x_{ip})^T$ are fixed covariates, and let $L(\boldsymbol{\beta})$ be the partial likelihood (4.7) with $r(\boldsymbol{\beta}, \mathbf{x}_i) = \exp(\boldsymbol{\beta}^T \mathbf{x}_i)$.

a) Derive the vector of score functions $\mathbf{U}(\boldsymbol{\beta}) = \log L(\boldsymbol{\beta})/\partial\boldsymbol{\beta}$.
b) Derive the observed information matrix $\mathbf{I}(\boldsymbol{\beta}) = -\mathbf{U}(\boldsymbol{\beta})/\partial\boldsymbol{\beta}^T$.

4.2 Assume that the counting processes $N_i(t)$; $i = 1, 2, \ldots, n$; have intensity processes of the form $\lambda_i(t) = Y_i(t)\alpha_0(t)\exp(\beta x_i)$, where the $Y_i(t)$ are at risk indicators and the x_i are binary covariates taking the values 0 or 1. Show that the score test for the null hypothesis $H_0 : \beta = 0$ is the log-rank test.

4.3 Assume that the counting processes $N_i(t)$; $i = 1, 2, \ldots, n$; have intensity processes of the form $\lambda_i(t) = Y_i(t)\beta_0(t)$, where the $Y_i(t)$ are at risk indicators. This is a special case of the additive regression model (4.55), and we may use (4.59) to estimate $B_0(t) = \int_0^t \beta_0(u)du$. Show that $\widehat{B}_0(t)$ is the Nelson-Aalen estimator.

4.4 Assume that the counting processes $N_i(t)$; $i = 1, 2, \ldots, n$; have intensity processes of the form $\lambda_i(t) = Y_i(t)\{\beta_0(t) + \beta_1(t)x_i\}$, where the $Y_i(t)$ are at risk indicators and the x_i are binary covariates taking the value 0 or 1. This is a special case of the additive regression model (4.55). Let $\mathbf{X}(t)$ be defined as in Section 4.2.1, and introduce $Y_.(t) = \sum_{i=1}^n Y_i(t)$, $Y^{(1)}(t) = \sum_{i=1}^n x_i Y_i(t)$, and $Y^{(0)}(t) = Y_.(t) - Y^{(1)}(t)$.

a) Show that
$$\mathbf{X}(t)^T\mathbf{X}(t) = \begin{pmatrix} Y_.(t) & Y^{(1)}(t) \\ Y^{(1)}(t) & Y^{(1)}(t) \end{pmatrix}.$$

b) Assume that $Y^{(0)}(t) > 0$ and $Y^{(1)}(t) > 0$. Show that
$$\left(\mathbf{X}(t)^T\mathbf{X}(t)\right)^{-1} = \begin{pmatrix} 1/Y^{(0)}(t) & -1/Y^{(0)}(t) \\ -1/Y^{(0)}(t) & 1/Y^{(0)}(t) + 1/Y^{(1)}(t) \end{pmatrix}.$$

c) Show that the estimator $\widehat{B}_0(t)$ for the cumulative baseline $B_0(t) = \int_0^t \beta_0(u)du$ is the Nelson-Aalen estimator for group 0 (i.e., for the individuals with $x_i = 0$). Also show that estimator $\widehat{B}_1(t)$ for the cumulative regression function $B_1(t) =$

$\int_0^t \beta_1(u)du$ is the difference between the Nelson-Aalen estimators for group 1 and group 0.

4.5   We consider the situation described in Exercise 4.4. In this exercise we consider the TST test for the null hypothesis $H_0 : \beta_1(t) = 0$ for $0 \leq t \leq t_0$. The TST test statistic is given as $Z_1(t_0)/\sqrt{V_{11}(t_0)}$. Here $Z_1(t_0)$ and $V_{11}(t_0)$ are given by (4.66) and (4.67), where the weight process $L_1(t)$ is given as described just after (4.68).

a) Show that

$$L_1(t) = \frac{Y^{(0)}(t)Y^{(1)}(t)}{Y_{\textbf{.}}(t)}$$

and that $Z_1(t_0)$ equals (3.51) with log-rank weight process. [*Hint:* Use (4.68) and the result in question (b) in Exercise 4.4.]

b) By (4.67) we have $V_{11}(t_0) = \int_0^{t_0} L_1^2(t) d\widehat{\sigma}_{11}(t)$, where we may either use (4.63) or (4.64) to compute $\widehat{\sigma}_{11}(t)$. Show that if we use (4.64), we have that $V_{11}(t_0)$ equals (3.55) with log-rank weight process.

4.6   Assume that we have right-censored survival data and that all covariates are time-fixed. Let $\mathbf{M}_{\mathrm{res}}(\tau)$ be the vector of martingale residuals for the additive regression model (Section 4.2.4). Prove that $\mathbf{X}(0)^T \mathbf{M}_{\mathrm{res}}(\tau) = 0$. [*Hint:* Show that $\mathbf{X}(0)^T d\mathbf{M}_{\mathrm{res}}(t) = 0$ for all t with $J(t) = 1$. To this end, note that since all covariates are time-fixed, we may write $\mathbf{X}(t) = \mathbf{Y}(t)\mathbf{X}$, where $\mathbf{Y}(t)$ is the $n \times n$ diagonal matrix with the at risk indicators $Y_1(t), Y_2(t), \ldots, Y_n(t)$ on the diagonal and $\mathbf{X}$ is the $n \times (p+1)$ matrix whose ith row has elements $1, x_{i1}, x_{i2}, \ldots, x_{ip}$. Further, for right-censored survival data $\mathbf{Y}(0)\mathbf{Y}(t) = \mathbf{Y}(t)$ for all t.]

4.7   Assume that the counting processes $N_i(t)$; $i = 1, 2, \ldots, n$; have intensity processes of the form $\lambda_i(t) = Y_i(t)\alpha_0(t)r(\boldsymbol{\beta}, \mathbf{x}_i)$, where the $Y_i(t)$ are at risk indicators and the $\mathbf{x}_i = (x_{i1}, \ldots, x_{ip})^T$ are fixed covariates. In Chapter 5 we show that the log-likelihood takes the form [cf. (5.5)]

$$l(\alpha_0, \boldsymbol{\beta}) = \sum_{i=1}^n \int_0^\tau \log \lambda_i(t)\, dN_i(t) - \int_0^\tau \lambda_{\textbf{.}}(t)dt,$$

where $[0, \tau]$ is the study time interval and $\lambda_{\textbf{.}}(t) = \sum_{i=1}^n \lambda_i(t)$.

a) Show that

$$l(\alpha_0, \boldsymbol{\beta}) = \int_0^\tau \log \alpha_0(t)\, dN_{\textbf{.}}(t) + \sum_{i=1}^n N_i(\tau)\log r(\boldsymbol{\beta}, \mathbf{x}_i) - \int_0^\tau S^{(0)}(\boldsymbol{\beta}, t)\alpha_0(t)dt,$$

where $N_{\textbf{.}}(t) = \sum_{i=1}^n N_i(t)$ and $S^{(0)}(\boldsymbol{\beta}, t) = \sum_{i=1}^n Y_i(t)r(\boldsymbol{\beta}, \mathbf{x}_i)$.

The log-likelihood in question (a) may be made arbitrarily large by letting $\alpha_0(t)$ be zero except from close to the observed event times where we let it peak higher and higher. However, if we consider an extended model, where the cumulative baseline hazard $A_0(t)$ may be any nonnegative, nondecreasing function, the log-likelihood achieves a maximum. For such an extended model, the log-likelihood is maximized if $A_0(t)$ is a step function with jumps at the observed event times $T_1 < T_2 < \cdots$. Assuming no tied event times, this gives the log-likelihood

$$l(A_0, \boldsymbol{\beta}) = \sum_{T_j} \log \triangle A_0(T_j) + \sum_{i=1}^{n} N_i(\tau) \log r(\boldsymbol{\beta}, \mathbf{x}_i) - \sum_{T_j} S^{(0)}(\boldsymbol{\beta}, T_j) \triangle A_0(T_j),$$

$$(4.108)$$

where $\triangle A_0(T_j)$ is the increment of the cumulative baseline hazard at T_j.

b) Show that for a given value of $\boldsymbol{\beta}$, the baseline hazard increments that maximize the log-likelihood (4.108) are given by

$$\triangle \widehat{A}_0(T_j, \boldsymbol{\beta}) = \frac{1}{S^{(0)}(\boldsymbol{\beta}, T_j)}. \qquad (4.109)$$

c) If we insert (4.109) in (4.108), we get a profile log-likelihood for $\boldsymbol{\beta}$. Show that this profile log-likelihood may be written

$$l(\widehat{A}_0, \boldsymbol{\beta}) = \sum_{i=1}^{n} N_i(\tau) \log r(\boldsymbol{\beta}, \mathbf{x}_i) - \sum_{T_j} \log S^{(0)}(\boldsymbol{\beta}, T_j) - N_.(\tau) = \log L(\boldsymbol{\beta}) - N_.(\tau),$$

where $L(\boldsymbol{\beta})$ is the partial likelihood (4.7). Thus an alternative interpretation of (4.7) is as a profile likelihood.

d) Explain that the results in questions (b) and (c) imply that the maximum partial likelihood estimator $\widehat{\boldsymbol{\beta}}$ and the Breslow estimator (4.13) are maximum likelihood estimators (in the extended model).

4.8 (Aalen, 1989) Write the additive model (4.54) in the following way

$$\alpha(t) = \beta_0(t) + \beta_1(t)X_1 + \cdots + \beta_p(t)X_p$$

where the X_j are random variables.

a) Assume that one of the covariates, say X_p, is dropped from the model, for example, because it is unmeasured. Show that, if the X_i are independent, you still have an additive model where only the baseline regression function $\beta_0(t)$ is changed.

b) Assume that the X_j are dependent and multivariate normally distributed. How would the model change in this case when the covariate X_p is dropped?

4.9 (Aalen, 1989) We consider the situation where there is just one covariate, X, that is measured with error, such that $X = U + e$ where e is a zero-mean error term independent of U. We assume that the additive model holds with respect to the true covariate, such that $\alpha(t) = \beta_0(t) + \beta_1(t)U$. Assume that U and e are both normally distributed with variances σ_u^2 and σ_e^2, respectively, and with the expectation of U equal to μ. Show that the model with respect to the observed covariate X is still an additive model and find the coefficients.

Chapter 5
Parametric counting process models

In biostatistics it has become a tradition to use non- and semiparametric methods, like those considered in the previous two chapters, to analyze censored survival data, while parametric methods are more common in reliability studies of failure times of technical equipment. In our opinion biostatistics would gain from the use of a wider range of statistical methods, including parametric methods, than is the current practice. In this chapter, we discuss the basic modeling and inferential issues for parametric counting process models. More advanced models and methods are discussed in Chapters 6, 7, 10, and 11.

As in Chapter 4, we consider counting processes $N_1, N_2, \ldots, N_n$, where $N_i(t)$ counts the number of occurrences of the event of interest for individual i in $[0,t]$; $i = 1, 2, \ldots, n$. For survival data the event can occur at most once for each individual, and $N_i(t)$ can only take the values 0 or 1. In this chapter we consider situations where the intensity processes of the $N_i(t)$ may be specified by a q-dimensional parameter $\boldsymbol{\theta} = (\theta_1, \ldots, \theta_q)^T$. We write the intensity process of $N_i(t)$ as $\lambda_i(t; \boldsymbol{\theta})$ to highlight this parametric specification.

Typically the individuals will not be under observation over the whole study time period, and to accommodate this, we assume that the intensity processes take the form

$$\lambda_i(t; \boldsymbol{\theta}) = Y_i(t)\alpha_i(t; \boldsymbol{\theta}). \tag{5.1}$$

Here $Y_i(t)$ is, as usual, an indicator taking the value 1 if individual i is under observation just before time t and the value 0 otherwise, while, assuming independent censoring, the intensity or hazard rate $\alpha_i(t; \boldsymbol{\theta})$ is the same as one would have in the situation without censoring (cf. Sections 1.4.2, 1.4.3, and 2.2.8).

For regression models $\alpha_i(t; \boldsymbol{\theta})$ will depend, in a parametric manner, on a covariate vector $\mathbf{x}_i(t) = (x_{i1}(t), \ldots, x_{ip}(t))^T$ whose components may be fixed or time-varying. Note that even though none of the specific examples in this chapter include internal time-dependent covariates, the general results presented are valid for both external and internal time-varying covariates.

We focus on likelihood inference in this chapter. In Section 5.1 we derive the likelihood for parametric counting process models, review the basic properties of the maximum likelihood estimator, and give some simple examples of parametric inference. Parametric regression models are considered in Section 5.2, with a focus on the so-called Poisson regression model. In Section 5.3, we give an outline of the derivations of the large sample properties of the maximum likelihood estimator.

5.1 Likelihood inference

As a motivation for the construction of likelihoods for general counting process models, we review in Section 5.1.1 some classic parametric models for survival times and consider in Section 5.1.2 data from such models subject to type I censoring (cf. Section 1.4.2). For this simple situation, the likelihood can be obtained as a product of the likelihood contributions for each individual. This approach does not apply to general counting process models, however, due to more complicated censoring schemes or the presence of time-dependent covariates. But in Section 5.1.3 we show how one may obtain a likelihood of the same form as a product of likelihood contributions over small intervals of time. As described in Section 5.1.4, this likelihood provides a basis for likelihood inference just as for the classical parametric models from introductory statistics courses. Some simple applications of parametric models are given in Section 5.1.5.

5.1.1 Parametric models for survival times

A number of parametric models for survival times have been proposed in the literature, in particular in the context of reliability studies of technical equipment. Here are a few of the more common ones. (Note that also other parameterizations of the distributions are used in the literature.)

- The *exponential distribution* depends on a single parameter v. The density and hazard rate are

$$f(t;v) = v\exp\{-vt\} \quad \text{and} \quad \alpha(t;v) = v \quad \text{for } t > 0.$$

- The *gamma distribution* depends on the scale parameter v and the shape parameter η. The density and hazard rate are

$$f(t;v,\eta) = \frac{v^\eta}{\Gamma(\eta)} t^{\eta-1} \exp\{-vt\}$$

$$\alpha(t;v,\eta) = \frac{v^\eta}{\Gamma(\eta,vt)} t^{\eta-1} \exp\{-vt\}$$

for $t > 0$, where $\Gamma(\eta, x) = \int_x^\infty u^{\eta-1} e^{-u} du$ is the (upper) incomplete gamma function.

- The *Weibull distribution* depends on the two parameters b and k. The density and hazard rate are

$$f(t;b,k) = bt^{k-1} \exp\{-bt^k/k\} \quad \text{and} \quad \alpha(t;b,k) = bt^{k-1} \quad \text{for } t > 0.$$

- The *Gompertz distribution* depends on the two parameters b and θ. The density and hazard rate are

$$f(t;b,\theta) = be^{\theta t} \exp\{b(1-e^{\theta t})/\theta\} \quad \text{and} \quad \alpha(t;b,\theta) = be^{\theta t} \quad \text{for } t > 0.$$

More examples of parametric models for survival times may be found in the textbooks by Kalbfleisch and Prentice (2002, section 2.2) and Klein and Moeschberger (2003, section 2.5).

All the parametric models mentioned here have hazard rates that are increasing, constant, or decreasing. Examples of parametric models, where the hazard rate may first increase and then decrease are given in Chapters 6, 10, and 11.

5.1.2 Likelihood for censored survival times

In this subsection we consider the simple situation with survival times subject to type I censoring, that is, censoring at fixed times. We start with the situation without censoring. Let $T_1, T_2, \ldots, T_n$ be independent and identically distributed nonnegative random variables with a parametric density function $f(t;\boldsymbol{\theta})$ and corresponding survival function $S(t;\boldsymbol{\theta}) = \int_t^\infty f(u;\boldsymbol{\theta})du$. The hazard rate is given by $\alpha(t;\boldsymbol{\theta}) = f(t;\boldsymbol{\theta})/S(t;\boldsymbol{\theta})$, while the survival function may be expressed by the hazard rate as $S(t;\boldsymbol{\theta}) = \exp\{-\int_0^t \alpha(u;\boldsymbol{\theta})du\}$ (cf. Section 1.1.2).

Let $c_1, c_2, \ldots, c_n$ be given (i.e., nonrandom) censoring times. Then, for each $i = 1, \ldots, n$, we do not (necessarily) observe the survival time T_i itself, but only the censored survival time $\widetilde{T}_i = \min(T_i, c_i)$ together with an indicator $D_i = I(\widetilde{T}_i = T_i)$ taking the value 1 if we observe the actual survival time, and the value 0 if only the censoring time is observed.

We will derive the likelihood for these data. If $D_i = 1$, the censored survival time $\widetilde{T}_i$ equals the actual survival time T_i, and the ith subject contributes $f(\widetilde{T}_i;\boldsymbol{\theta}) = \alpha(\widetilde{T}_i;\boldsymbol{\theta})\exp\left\{-\int_0^{\widetilde{T}_i} \alpha(t;\boldsymbol{\theta})dt\right\}$ to the likelihood. On the other hand, if $D_i = 0$, the censored survival time $\widetilde{T}_i$ equals the censoring time c_i. We then only know that the actual survival time of the ith subject exceeds $\widetilde{T}_i$, so its likelihood contribution becomes $S(\widetilde{T}_i;\boldsymbol{\theta}) = \exp\left\{-\int_0^{\widetilde{T}_i} \alpha(t;\boldsymbol{\theta})dt\right\}$. Thus the likelihood contribution of the ith subject may be written

$$L_i(\boldsymbol{\theta}) = f(\widetilde{T}_i;\boldsymbol{\theta})^{D_i} S(\widetilde{T}_i;\boldsymbol{\theta})^{1-D_i} = \alpha(\widetilde{T}_i;\boldsymbol{\theta})^{D_i} \exp\left\{-\int_0^{\widetilde{T}_i} \alpha(t;\boldsymbol{\theta})dt\right\}.$$

In the present situation, the likelihood contributions from the n individuals are independent. Thus the full likelihood $L(\boldsymbol{\theta}) = \prod_{i=1}^n L_i(\boldsymbol{\theta})$ becomes

$$L(\boldsymbol{\theta}) = \prod_{i=1}^n f(\widetilde{T}_i;\boldsymbol{\theta})^{D_i} S(\widetilde{T}_i;\boldsymbol{\theta})^{1-D_i}$$

$$= \prod_{i=1}^n \alpha(\widetilde{T}_i;\boldsymbol{\theta})^{D_i} \exp\left\{-\int_0^{\widetilde{T}_i} \alpha(t;\boldsymbol{\theta})dt\right\}.$$

As a preparation for the presentation of the likelihood for general counting process models in the next subsection, we rewrite the likelihood in terms of the counting processes $N_i(t) = I\{\widetilde{T}_i \le t; D_i = 1\}$ and their intensity processes $\lambda_i(t;\boldsymbol{\theta})$. The latter take the form $\lambda_i(t;\boldsymbol{\theta}) = Y_i(t)\alpha(t;\boldsymbol{\theta})$, where $Y_i(t) = I\{\widetilde{T}_i \ge t\}$ is an at risk indicator for subject i [cf. (5.1)]. We denote by τ the upper time limit for the study, so that $\widetilde{T}_i \le \tau$ for all i, and let $\triangle N_i(t) = N_i(t) - N_i(t-)$ be the increment of N_i at time t. Then $L_i(\boldsymbol{\theta})$ may be written as

$$L_i(\boldsymbol{\theta}) = \left\{\prod_{0<t\le\tau} \{Y_i(t)\alpha(t;\boldsymbol{\theta})\}^{\triangle N_i(t)}\right\} \exp\left\{-\int_0^\tau Y_i(t)\alpha(t;\boldsymbol{\theta})dt\right\}$$

$$= \left\{\prod_{0<t\le\tau} \lambda_i(t;\boldsymbol{\theta})^{\triangle N_i(t)}\right\} \exp\left\{-\int_0^\tau \lambda_i(t;\boldsymbol{\theta})dt\right\},$$

and the full likelihood takes the form

$$L(\boldsymbol{\theta}) = \left\{\prod_{i=1}^n \prod_{0<t\le\tau} \lambda_i(t;\boldsymbol{\theta})^{\triangle N_i(t)}\right\} \exp\left\{-\int_0^\tau \lambda_.(t;\boldsymbol{\theta})dt\right\}. \tag{5.2}$$

Here $\lambda_.(t;\boldsymbol{\theta}) = \sum_{i=1}^n \lambda_i(t;\boldsymbol{\theta})$ is the intensity process of the aggregated process $N_.(t) = \sum_{i=1}^n N_i(t)$ counting the occurrences of the event of interest among all the subjects.

5.1.3 Likelihood for counting process models

The arguments of the previous subsection are based on the assumptions that the survival times are independent and the censoring times are fixed. This ensures (i) that the likelihood contributions from the individuals are independent, and (ii) that no additional randomness is introduced by the censoring. For more complicated censoring schemes, one or both of these properties may fail to hold. If the censoring times are random, the full likelihood will also have contributions from the distribution of the censoring times, while censoring depending on events "in the past" may create dependence between the censored survival times for the n subjects even though their *uncensored* survival times are independent. The situation becomes even more complicated in the presence of time-dependent covariates. Then the full likelihood also depends on a model for the development of the time-dependent covariates. Nevertheless, as we will now demonstrate, we may still derive a likelihood that has the same form as (5.2) and that is not only valid for censored survival times but for counting process models in general. The remainder of this subsection may be skipped without loss of continuity by those readers not interested in these likelihood derivations.

Assume that we have counting processes $N_1(t), N_2(t), \ldots, N_n(t)$, with corresponding intensity processes $\lambda_1(t; \boldsymbol{\theta}), \lambda_2(t; \boldsymbol{\theta}), \ldots, \lambda_n(t; \boldsymbol{\theta})$. For later use we introduce the aggregated counting process $N_{\boldsymbol{\cdot}}(t) = \sum_{i=1}^{n} N_i(t)$ and note that this has intensity process $\lambda_{\boldsymbol{\cdot}}(t; \boldsymbol{\theta}) = \sum_{i=1}^{n} \lambda_i(t; \boldsymbol{\theta})$. As discussed earlier, we cannot in general derive the likelihood as a product of likelihood contributions for the individual counting processes $N_1, N_2, \ldots, N_n$. Rather we will heuristically derive the likelihood as a product of contributions, conditional on "the past," for each of a number of small time intervals.

To this end we partition the study time interval $[0, \tau]$ into a number of small time intervals $0 = t_0 < t_1 < t_2 < \cdots < t_K = \tau$, each of length dt. We denote by $\mathscr{F}_{t-}$ all data available to the researcher "just before" time t, that is, information on the events of interest (registered by the counting processes) as well as on censorings, delayed entries, and the values of fixed and time-varying covariates. Using the multiplication rule for conditional probabilities, we may then informally write the likelihood as

$$P(\text{data}) = \prod_{k=0}^{K-1} P(\text{data in } [t_k, t_k + dt) \mid \mathscr{F}_{t_k-})$$

$$= \prod_{k=0}^{K-1} \{P(\text{events of interest in } [t_k, t_k + dt) \mid \mathscr{F}_{t_k-}) \tag{5.3}$$

$$\times P(\text{other data in } [t_k, t_k + dt) \mid \text{events of interest in } [t_k, t_k + dt), \mathscr{F}_{t_k-})\}.$$

Note that to specify the full likelihood, we would have to specify the conditional distributions of the "other data" (censorings, covariates, etc.). To avoid this, we resort to the *partial* likelihood obtained by omitting these contributions to the full likelihood. This yields

$$\text{Partlik} = \prod_{k=0}^{K-1} P(\text{events of interest in } [t_k, t_k + dt) \mid \mathscr{F}_{t_k-}).$$

Informally we write $dN_i(t)$ for the increment of N_i over the small time interval $[t, t + dt)$ and $dN_{\boldsymbol{\cdot}}(t)$ for the corresponding increment of the aggregated process. Now, if dt is sufficiently small, at most one of the counting processes will have a jump in an interval of length dt. Thus, conditional on "the past" $\mathscr{F}_{t-}$, the occurrences of the event of interest in $[t, t + dt)$ may be considered as a multinomial trial with $n + 1$

possible outcomes: "$dN_i(t) = 1$"; $i = 1, 2, \ldots n$; and "$dN_{\bullet}(t) = 0$." Thus the partial likelihood may be written

$$\text{Partlik} = \prod_{k=0}^{K-1} \left(\prod_{i=1}^{n} P(dN_i(t) = 1 \mid \mathcal{F}_{t_k-})^{dN_i(t)} \right) P(dN_{\bullet}(t) = 0 \mid \mathcal{F}_{t_k-})^{1-dN_{\bullet}(t)}.$$

If we now let the number K of time intervals increase while dt becomes smaller, the finite product will approach a product-integral. Moreover, since

$$P(dN_i(t) = 1 \mid \mathcal{F}_{t-}) \approx \lambda_i(t; \boldsymbol{\theta}) dt \quad \text{and} \quad P(dN_{\bullet}(t) = 0 \mid \mathcal{F}_{t-}) \approx 1 - \lambda_{\bullet}(t; \boldsymbol{\theta}) dt,$$

this product-integral takes the form

$$\text{Partlik} = \prod_{0 < t \leq \tau} \left\{ \prod_{i=1}^{n} (\lambda_i(t; \boldsymbol{\theta}) dt)^{dN_i(t)} (1 - \lambda_{\bullet}(t; \boldsymbol{\theta}) dt)^{1-dN_{\bullet}(t)} \right\}.$$

This expression can be simplified further. As each counting process will have at most a finite number of jumps, the product-integral of the first part is just an ordinary finite product over the jump times of the counting processes. For the same reason the exponent $1 - dN_{\bullet}(t)$ in the last part of the product-integral equals 1 for all but a finite number of time points t, and it can therefore be omitted without altering the value of the product-integral. Finally, we may neglect the dt in the first part of the product-integral, for it will cancel anyway on forming likelihood ratios (which is what one does in a more formal derivation of likelihoods). Thus the partial likelihood may be given as

$$L(\boldsymbol{\theta}) = \left\{ \prod_{0 < t \leq \tau} \prod_{i=1}^{n} \lambda_i(t; \boldsymbol{\theta})^{\triangle N_i(t)} \right\} \prod_{0 < t \leq \tau} (1 - \lambda_{\bullet}(t; \boldsymbol{\theta}) dt)$$

$$= \left\{ \prod_{i=1}^{n} \prod_{0 < t \leq \tau} \lambda_i(t; \boldsymbol{\theta})^{\triangle N_i(t)} \right\} \exp \left\{ - \int_{0}^{\tau} \lambda_{\bullet}(t; \boldsymbol{\theta}) dt \right\}, \qquad (5.4)$$

where the second equality follows by the properties of the product-integral of an absolutely continuous function reviewed just after (A.10) in Appendix A.1.

In our derivation of the partial likelihood, we omitted the contributions to (5.3) corresponding to the distributions of censorings, late entries, and the values of fixed and time-varying covariates. If these distributions do *not* depend on $\boldsymbol{\theta}$, then (5.4) is actually the *full* likelihood for inference on $\boldsymbol{\theta}$. If the distributions *do* depend on $\boldsymbol{\theta}$, it is only a *partial* likelihood. In any case, as reviewed in the next subsection and proved in Section 5.3, estimators and test statistics based on (5.4) enjoy the usual large sample properties known from likelihood inference for the classical situation with independent and identically distributed random variables. For these reasons, we will not worry whether (5.4) is the full likelihood for $\boldsymbol{\theta}$ or only a partial likelihood, and we will in any case feel free to denote it "a likelihood."

A comment is in order on the difference between the partial likelihood (5.4) and Cox's partial likelihood (4.7) for relative risk regression models. We obtain the (partial) likelihood (5.4) by omitting contributions to the full likelihood (5.3) corresponding to censorings, late entries, and the values of fixed and time-varying covariates. To obtain Cox's partial likelihood (4.7) we further factorize (5.4) as

$$
L = \left\{ \prod_{i=1}^{n} \prod_{0 < t \le \tau} \lambda_i(t)^{\triangle N_i(t)} \right\} \exp \left\{ - \int_0^\tau \lambda_{\cdot}(t) dt \right\}
$$

$$
= \left\{ \prod_{i=1}^{n} \prod_{0 < t \le \tau} \left(\frac{\lambda_i(t)}{\lambda_{\cdot}(t)} \right)^{\triangle N_i(t)} \right\} \left\{ \prod_{0 < t \le \tau} \left(\lambda_{\cdot}(t;\boldsymbol{\theta})^{\triangle N_{\cdot}(t)} \right) \exp \left(- \int_0^\tau \lambda_{\cdot}(t) dt \right) \right\}.
$$

Here the leading factor is Cox's partial likelihood (cf. Section 4.1.1), while the second factor is the (partial) likelihood for the aggregated process. Thus Cox's partial likelihood is obtained from the (partial) likelihood (5.4) by disregarding the information about the total number of events and their times of occurrence.

5.1.4 The maximum likelihood estimator and related tests

In the previous subsection, we derived the likelihood (5.4) for the general class of counting process models described in the introduction to the chapter. Here we will use this likelihood to derive the maximum likelihood estimator for $\boldsymbol{\theta}$, review its statistical properties, and discuss test statistics related to this method of estimation. To this end, first note that by (5.4), the log-likelihood takes the form

$$
l(\boldsymbol{\theta}) = \log L(\boldsymbol{\theta}) = \sum_{i=1}^{n} \int_0^\tau \log \lambda_i(t;\boldsymbol{\theta}) dN_i(t) - \int_0^\tau \lambda_{\cdot}(t;\boldsymbol{\theta}) dt, \qquad (5.5)
$$

while the vector of score functions $\mathbf{U}(\boldsymbol{\theta}) = (U_1(\boldsymbol{\theta}), \dots, U_q(\boldsymbol{\theta}))^T$ has components

$$
U_j(\boldsymbol{\theta}) = \frac{\partial}{\partial \theta_j} l(\boldsymbol{\theta}) = \sum_{i=1}^{n} \int_0^\tau \frac{\partial}{\partial \theta_j} \log \lambda_i(t;\boldsymbol{\theta}) dN_i(t) - \int_0^\tau \frac{\partial}{\partial \theta_j} \lambda_{\cdot}(t;\boldsymbol{\theta}) dt. \quad (5.6)
$$

The log-likelihood function (5.5) may have a number of local maxima. Thus the likelihood equations $U_j(\boldsymbol{\theta}) = 0$, $j = 1, 2, \dots, n$, may have multiple solutions. However, we will not discuss conditions for a unique solution to these equations. Thus if more than one solution is found, one will have to check which gives the largest value of the log-likelihood (5.5) and hence corresponds to the maximum likelihood estimator $\widehat{\boldsymbol{\theta}}$.

In Section 5.3 we show that $\widehat{\boldsymbol{\theta}}$ enjoys "the usual properties" for maximum likelihood estimators, well known from the classical case of independent and identically distributed random variables. In particular, in large samples, $\widehat{\boldsymbol{\theta}}$ is approximately multivariate normally distributed around its true value with a covariance matrix that may be estimated by $\mathbf{I}(\widehat{\boldsymbol{\theta}})^{-1}$, where $\mathbf{I}(\boldsymbol{\theta}) = \left\{ -\frac{\partial^2}{\partial\theta_h\partial\theta_j} l(\boldsymbol{\theta}) \right\}$ is the observed information matrix. Alternatively, we may estimate the covariance matrix by $\mathscr{I}(\widehat{\boldsymbol{\theta}})^{-1}$, where the expected information matrix $\mathscr{I}(\boldsymbol{\theta}) = \{\mathscr{I}_{hj}(\boldsymbol{\theta})\}$ has elements of the form

$$\mathscr{I}_{hj}(\boldsymbol{\theta}) = \sum_{i=1}^{n} \int_0^\tau \left(\frac{\partial}{\partial\theta_h} \log\lambda_i(s;\boldsymbol{\theta}) \right) \left(\frac{\partial}{\partial\theta_j} \log\lambda_i(s;\boldsymbol{\theta}) \right) dN_i(s); \qquad (5.7)$$

cf. formula (5.23) in Section 5.3.

Also the usual likelihood-based statistical tests, that is, the likelihood ratio test, the score test, and the Wald test, apply as for the situation with independent and identically distributed random variables. For a review of these tests, see the summary in Section 4.1.1 on relative risk regression models.

5.1.5 Some applications

We now illustrate the results of the previous subsection with some simple examples of parametric models. More illustrations are given in Section 5.2 on parametric regression models and in Chapters 6, 7, 10 and 11.

Example 5.1. Exponential distribution and relapse of leukemia patients. We consider the situation with censored exponentially distributed survival times. Thus let $T_1, T_2, \ldots, T_n$ be independent and identically distributed with density function $f(t;v) = v\exp(-vt)$. We do not observe the survival times themselves, only the right-censored survival times $\widetilde{T}_i$ and the indicators $D_i = I(\widetilde{T}_i = T_i)$. Then the counting processes $N_i(t) = I(\widetilde{T}_i \leq t; D_i = 1)$, $i = 1, 2, \ldots, n$, have intensity processes $\lambda_i(t;v) = vY_i(t)$ with $Y_i(t) = I(\widetilde{T}_i \geq t)$.

We introduce the aggregated counting process $N_\cdot(t) = \sum_{i=1}^n N_i(t)$ and the process $Y_\cdot(t) = \sum_{i=1}^n Y_i(t)$ recording the number of individuals at risk. Then the score function (5.6) takes the form

$$U(v) = \frac{N_\cdot(\tau)}{v} - R(\tau),$$

where

$$R(\tau) = \int_0^\tau Y_\cdot(t)dt = \sum_{i=1}^n \widetilde{T}_i$$

is the total observation time for all the individuals counted together. The maximum likelihood estimator is the solution to the equation $U(v) = 0$, and it thus becomes

$$\widehat{v} = \frac{N_\cdot(\tau)}{R(\tau)}.$$

Table 5.1 Remission lengths in weeks for 42 children treated with 6-MP or placebo. Censored observations are indicated by a *. From Freireich et al. (1963).

Placebo	1	1	2	2	3	4	4	5	5	8	8
	8	8	11	11	12	12	15	17	22	23	
6-MP	6	6	6	6*	7	9*	10	10*	11*	13	16
	17*	19*	20*	22	23	25*	32*	32*	34*	35*	

Note that $\widehat{v}$ is the observed number of events ("occurrences") divided by the total observation time ("exposure"), and it thus has the form of an "occurrence/exposure rate."

The observed information takes the form

$$I(v) = -U'(v) = \frac{N_\cdot(\tau)}{v^2},$$

and it coincides with the expected information (5.7). By the general result of Section 5.1.4, it follows that in large samples $\widehat{v}$ is approximately normally distributed around the true value of v, with a variance that can be estimated by $\widehat{v}^2/N_\cdot(\tau) = \widehat{v}/R(\tau)$. Simulation studies have shown that about 100 events are needed to get a good approximation to the normal distribution. The approximation to the normal distribution is improved by using a log-transformation. In fact, only about 10 events are needed in order that $\log\widehat{v}$ is approximately normally distributed around $\log v$ with a variance that may be estimated by $1/N_\cdot(\tau)$ (Schou and Væth, 1980).

To illustrate these results we consider a classical data set on the length of remission of leukemia patients (Freireich et al., 1963). Forty-two children with acute leukemia responded to a primary treatment in the sense that most or all signs of the disease in the bone marrow disappeared, that is, the patients entered into partial or complete remission. The children were then randomized to remission maintenance therapy with the drug 6-mercaptopurine (6-MP) or placebo. The remission length, that is, the time from remission to relapse (when the disease resumed), is the survival time of interest in this study. The patients who were still in remission at the conclusion of the study provide right-censored survival times. The data, given in Table 5.1, have been used for illustration in a number of books (e.g., Cox and Oakes, 1984; McCullagh and Nelder, 1989; Andersen et al., 1993; Klein and Moeschberger, 2003).

Figure 5.1 shows Nelson-Aalen estimates of the cumulative relapse intensity (hazard rate for relapse) for the two groups of patients. It is seen that the relapse intensity is much higher for the placebo group than for the MP-6 group. Furthermore, the Nelson-Aalen plots are fairly linear, which shows that the time to relapse in each group may be adequately described by an exponential distribution.

The number of relapses are 21 and 9 in the placebo and MP-6 groups, respectively, while the total observation time in the two groups are 181 weeks and 359 weeks. Thus the estimated relapse intensities are

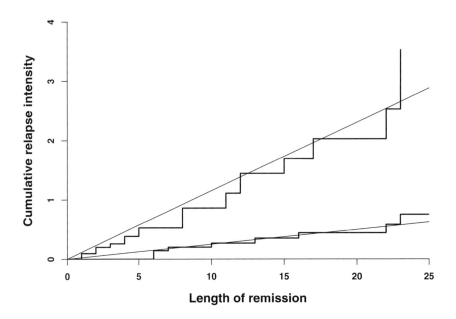

Fig. 5.1 *Nelson-Aalen estimates (step functions) of the cumulative relapse intensity for leukemia patients with the corresponding estimates under the exponential model (straight lines). Placebo: upper pair of curves; MP-6: lower pair of curves.*

$$\text{Placebo group: } \widehat{v}_{\text{plac}} = \frac{21}{182} = 0.115 \text{ per week}$$

$$\text{MP-6 group: } \widehat{v}_{\text{MP6}} = \frac{9}{359} = 0.025 \text{ per week}$$

In Figure 5.1 we have added straight lines with slopes $\widehat{v}_{\text{plac}}$ and $\widehat{v}_{\text{MP6}}$ corresponding to the cumulative hazards under the exponential model. The lines are in good agreement with the Nelson-Aalen plots.

The relapse intensity in the placebo group is almost five times higher than in the MP-6 group. To obtain a confidence interval for the intensity ratio $v_{\text{plac}}/v_{\text{MP6}}$, we may use the log-transform to get a 95% confidence interval for $\log v_{\text{plac}} - \log v_{\text{MP6}}$:

$$\log 0.115 - \log 0.025 \pm 1.96 \sqrt{\frac{1}{21} + \frac{1}{9}}.$$

Then by exponentiating these limits, we obtain a 95% confidence interval for the intensity ratio from 2.11 to 10.0 (cf. Exercise 5.1). $\square$

Example 5.2. Weibull distribution and cancer death in Norway. Here we consider the situation with censored lifetimes from a Weibull distribution. Thus the *uncen-sored* lifetimes $T_1, T_2, \ldots, T_n$ are independent and identically distributed with hazard rate $\alpha(t; b, k) = b t^{k-1}$. We observe $\widetilde{T}_i = \min(T_i, C_i)$ and $D_i = I(\widetilde{T}_i = T_i)$ for some independent right-censoring times C_i; $i = 1, 2, \ldots, n$.

We want to find the maximum likelihood estimators for the parameters of the Weibull model. To this end it may be convenient to reparameterize the Weibull haz-ard as $\alpha(t; \beta, k) = e^{\beta} (t/t_0)^{k-1}$ for suitable chosen t_0. Then the counting process $N_i(t) = I(\widetilde{T}_i \le t; D_i = 1)$ has intensity processes $\lambda_i(t; \beta, k) = e^{\beta} (t/t_0)^{k-1} Y_i(t)$, with $Y_i(t) = I(\widetilde{T}_i \ge t)$, while the aggregated counting process $N_{\boldsymbol{\cdot}}(t)$ has intensity process $\lambda_{\boldsymbol{\cdot}}(t; \beta, k) = e^{\beta} (t/t_0)^{k-1} Y_{\boldsymbol{\cdot}}(t)$, with $Y_{\boldsymbol{\cdot}}(t) = \sum_{i=1}^{n} Y_i(t)$.

The log-likelihood function (5.5) may be given as

$$l(\beta, k) = \beta N_{\boldsymbol{\cdot}}(\tau) + (k-1) \int_0^{\tau} \log(t/t_0) \, dN_{\boldsymbol{\cdot}}(t) - e^{\beta} \int_0^{\tau} (t/t_0)^{k-1} Y_{\boldsymbol{\cdot}}(t) \, dt. \quad (5.8)$$

In the Weibull case it is not possible to find analytic expressions for maximum like-lihood estimators, and one has to resort to numeric methods. The easiest is to max-imize the log-likelihood using software that handles censored Weibull data, or by using a program that may maximize a function of two or more variables.

By the general results of the previous subsection, the maximum likelihood esti-mator $(\widehat{\beta}, \widehat{k})$ is approximately bivariate normally distributed around the true para-meter values. Its covariance matrix may be estimated by $\mathscr{I}(\widehat{\beta}, \widehat{k})^{-1}$, where the ex-pected information matrix by (5.7) takes the form:

$$\mathscr{I}(\beta, k) = \begin{pmatrix} N_{\boldsymbol{\cdot}}(\tau) & \int_0^{\tau} \log(t/t_0) \, dN_{\boldsymbol{\cdot}}(t) \\ \int_0^{\tau} \log(t/t_0) \, dN_{\boldsymbol{\cdot}}(t) & \int_0^{\tau} (\log(t/t_0))^2 \, dN_{\boldsymbol{\cdot}}(t) \end{pmatrix}. \quad (5.9)$$

As an illustration, we use a Weibull hazard to model the cancer mortality among middle-aged Norwegian males and females using data from three Norwegian coun-ties (Example 1.12). As in Example 3.3, we left-truncate the survival times at 40 years and right-censor them at the age of 70 years. Note that the log-likelihood (5.8) and expected information (5.9) continue to hold for left-truncated and right-censored data, provided $Y_i(t)$ is the indicator that individual i is at risk "just before" time t, $l = 1, \ldots, n$.

We choose $t_0 = 55$ years when maximizing the log-likelihood, so that $b = e^{\beta}$ becomes the cancer hazard for a 55-year-old person. With this choice for males we get the estimates (with standard errors) $\widehat{\beta} = -6.09$ (0.108) and $\widehat{k} = 7.29$ (0.743), while for females the estimates become $\widehat{\beta} = -6.36$ (0.126) and $\widehat{k} = 6.36$ (0.912). Figure 5.2 shows the Nelson-Aalen estimates of the cumulative cancer hazard from Example 3.3 together with the cumulative Weibull hazards from the fitted models for males and females. It is seen that the Weibull model gives a nice fit for these data.

From the fitted Weibull models we see that the estimated cancer hazard at age 55 is $e^{-6.09} = 0.00227$ for males and $e^{-6.36} = 0.00173$ for females. Moreover, the

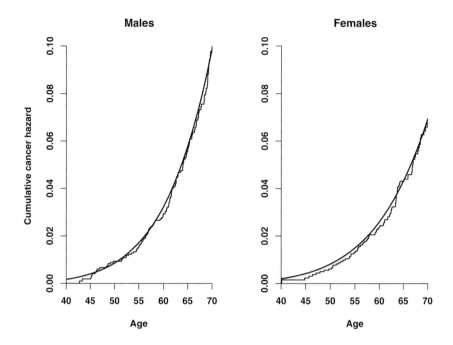

Fig. 5.2 *Nelson-Aalen estimates and estimates from a Weibull model of the cumulative cancer hazard for middle-aged Norwegian males (left) and females (right). Smooth lines: Weibull estimates; step functions: Nelson-Aalen estimates.*

cancer hazard increases with age as a power function, the estimated power (i.e., $k-1$) being about 6 for both genders (6.3 for males and 5.4 for females). □

Example 5.3. Standardized mortality ratio (SMR) and patients with hip replacement. In Example 3.4 we considered a nonparametric model for relative mortality. Here we consider the same situation, but now assuming the relative mortality to be constant over time. More specifically, we have left-truncated and right-censored survival times with death being the event of interest and assume that the hazard rate of the ith individual takes the form $\alpha_i(t;\theta) = \theta\mu_i(t)$. Here θ is a relative mortality common to all individuals, while $\mu_i(t)$ is the *known* mortality rate (hazard rate) at time t for a person from an external standard population corresponding to subject i (e.g., of the same sex and age as individual i). Then, for $i = 1,2,\ldots,n$, the process $N_i(t)$ registering death for individual i has intensity processes $\lambda_i(t;\theta) = \theta\mu_i(t)Y_i(t)$, where $Y_i(t)$ is an at risk indicator for the individual. The aggregated counting process $N_\cdot(t) = \sum_{i=1}^{n} N_i(t)$ has intensity process $\lambda_\cdot(t;\theta) = \theta\sum_{i=1}^{n}\mu_i(t)Y_i(t)$.

Here the score function (5.6) becomes

$$U(\theta) = \frac{N_\cdot(\tau)}{\theta} - R^\mu(\tau),$$

where

$$R^{\mu}(\tau) = \sum_{i=1}^{n} \int_{0}^{\tau} \mu_i(t) Y_i(t) dt.$$

Using the decomposition $N_{\cdot}(\tau) = \int_0^{\tau} \lambda_{\cdot}(t;\theta)dt + M_{\cdot}(\tau)$, we note that $EN_{\cdot}(\tau) = ER^{\mu}(\tau)$ when $\theta = 1$. Thus $R^{\mu}(\tau)$ can be interpreted as the "expected number of deaths" had the mortality in the group under study been the same as in the general population. Solving the equation $U(\theta) = 0$, we obtain the maximum likelihood estimator

$$\widehat{\theta} = \frac{N_{\cdot}(\tau)}{R^{\mu}(\tau)} \tag{5.10}$$

as the ratio between the observed and "expected" number of deaths. The estimator $\widehat{\theta}$ is known as the *standardized mortality ratio* (SMR).

The observed information takes the form

$$I(\theta) = -U'(\theta) = \frac{N_{\cdot}(\tau)}{\theta^2},$$

and it is the same as the expected information (5.7). Thus, in large samples, the standardized mortality ratio $\widehat{\theta}$ is approximately normally distributed around the true value of θ, with a variance that can be estimated by $\widehat{\theta}^2/N_{\cdot}(\tau) = \widehat{\theta}/R^{\mu}(\tau)$. As in Example 5.1, one gets a better approximation to a normal distribution by using a log-transform.

As an illustration, we consider the data from the Norwegian Arthroplasty Registry on 3503 females and 1497 males who had their first total hip replacement operation at a Norwegian hospital between September 1987 and February 1998; cf. Example 1.5. In Example 3.4 we consider a model for the relative mortality for these patients, allowing the relative mortality to depend on the time t elapsed since the operation. Our analysis in that example shows that, apart from a short period just after the operation, the relative mortality is fairly constant (cf. Figure 3.7). Thus, as a reasonable approximation, we may assume the relative mortality to be constant and estimate it using the standardized mortality ratio (5.10).

We observe a total of $N_{\cdot}(\tau) = 800$ deaths among the hip replacement patients (males and females combined), while $R^{\mu}(\tau) = 1109$ deaths had been expected if the mortality for these patients had been as in the general Norwegian population. The SMR for the hip replacement patients thus becomes

$$\widehat{\theta} = \frac{800}{1109} = 0.721,$$

with an approximate 95% log-transformed confidence interval from 0.652 to 0.791 (Exercise 5.1). This shows that the mortality of the hip replacement patients is about 70% of the mortality of the general population. □

Example 5.4. Piecewise constant hazards, occurrence/exposure rates and divorce in Norway. Let $N_1, N_2, \ldots, N_n$ count the occurrences of some event of interest for a sample of n individuals from a population, and assume that the intensity process of

$N_i(t)$ takes the form $\lambda_i(t) = Y_i(t)\alpha(t)$, where $Y_i(t)$ is an at risk indicator for the ith individual. Then the aggregated counting process $N_.(t)$ has intensity process $\lambda_.(t) = Y_.(t)\alpha(t)$, with $Y_.(t)$ being the number of individuals at risk "just before" time t.

In Sections 3.1.1 and 3.1.2 we discussed how, in the present situation, the Nelson-Aalen estimator can be used to estimate nonparametrically $A(t) = \int_0^t \alpha(s)ds$, while in Section 3.1.4 we saw how one, by smoothing the Nelson-Aalen estimator, can obtain a nonparametric estimate for $\alpha(t)$ itself. In this example, we will discuss an alternative approach for estimating $\alpha(t)$ without imposing strong parametric assumptions on its form; an approach closely related to the classical actuarial method for constructing lifetables.

To this end, consider a partition $0 = t_0 < t_1 < t_2 < \cdots t_K = \tau$ of the study time interval $[0, \tau]$ into K subintervals, and assume (as an approximation) that $\alpha(t)$ is constant over each of the subintervals. Thus, if θ_k is the constant value on the kth subinterval, we may write

$$\alpha(t) = \alpha(t; \boldsymbol{\theta}) = \sum_{k=1}^{K} \theta_k I_k(t), \qquad (5.11)$$

where $I_k(t) = 1$ for $t_{k-1} < t \leq t_k$, and $I_k(t) = 0$ otherwise. Formally speaking, (5.11) is a parametric model with parameter $\boldsymbol{\theta} = (\theta_1, \ldots, \theta_K)^T$, and we may use our general results to obtain the maximum likelihood estimators for the θ_k and derive their statistical properties.

Here the score functions (5.6) may be given as

$$U_j(\boldsymbol{\theta}) = \frac{O_j}{\theta_j} - R_j,$$

where

$$O_j = \int_0^\tau I_j(t) dN_.(t)$$

is the total number of events in the jth subinterval and

$$R_j = \int_0^\tau I_j(t) Y_.(t) dt$$

is the corresponding total observation time. Thus the maximum likelihood estimators are the occurrence/exposure rates

$$\widehat{\theta}_j = \frac{O_j}{R_j}.$$

The elements of the observed information matrix $\mathbf{I}(\boldsymbol{\theta}) = \{I_{hj}(\boldsymbol{\theta})\}$ takes the form $I_{hj}(\boldsymbol{\theta}) = \delta_{hj} O_j / \theta_j^2$, with δ_{hj} being a Kronecker delta. Thus the inverse of the observed information is the diagonal matrix

$$\mathbf{I}(\boldsymbol{\theta})^{-1} = \mathrm{diag}\left\{\frac{\theta_1^2}{O_1}, \ldots, \frac{\theta_K^2}{O_K}\right\}.$$

Table 5.2 Numbers of divorces and existing marriages by duration of marriages for marriages contracted in Norway in 1960, 1970, and 1980. The numbers of existing marriages are approximations obtained by disregarding mortality and emigration. (Source: Mamelund et al., 1997)

Duration of marriage	1960 cohort		1970 cohort		1980 cohort	
	Divorces	Number married	Divorces	Number married	Divorces	Number married
Initial		23651		29370		22230
0	10	23641	12	29358	4	22226
1	34	23607	27	29331	31	22195
2	97	23510	123	29208	164	22031
3	147	23363	317	28891	356	21675
4	191	23172	429	28462	467	21208
5	203	22969	454	28008	455	20753
6	181	22788	401	27607	460	20293
7	170	22618	401	27206	426	19867
8	162	22456	377	26829	415	19452
9	134	22322	313	26516	390	19062
10	149	22173	316	26200	423	18639
11	139	22034	323	25877	359	18280
12	149	21885	318	25559	400	17880
13	138	21747	291	25268	369	17511
14	152	21595	292	24976	342	17169
15	160	21435	323	24653	294	16875
16	167	21268	264	24389		
17	151	21117	272	24117		
18	153	20964	281	23835		
19	153	20811	282	23553		
20	155	20656	317	23236		
21	153	20503	284	22952		
22	160	20343	269	22683		
23	150	20193	269	22414		
24	144	20049	292	22122		
25	135	19914	231	21891		
26	109	19805				
27	132	19673				
28	122	19551				
29	88	19463				
30	74	19389				
31	64	19325				
32	66	19259				
33	61	19198				
34	59	19139				
35	38	19101				

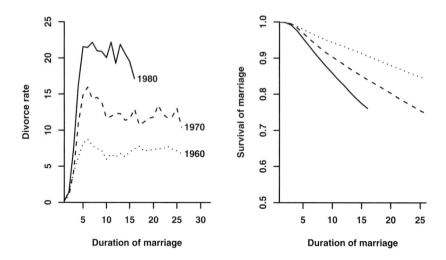

Fig. 5.3 *Rates of divorce per 1000 marriages per year (left panel) and empirical survival curves (right panel) for marriages contracted in 1960, 1970, and 1980.*

It follows that the occurrence/exposure rates $\widehat{\theta}_1, \widehat{\theta}_2, \ldots, \widehat{\theta}_K$ are approximately independent and normally distributed around their true values, and that the variance of $\widehat{\theta}_j$ can be estimated by $\widehat{\theta}_j^2/O_j = \widehat{\theta}_j/R_j$. As in Example 5.1, a better approximation to a normal distribution is obtained by using a log-transform.

To illustrate these results, we consider the rate of divorce (hazard rate for divorce) for the marriages contracted in Norway in 1960, 1970, and 1980; cf. Example 1.2. Table 5.2 gives the numbers of divorces for one-year marriage durations and the numbers still married by duration for the three marriage cohorts.

We consider the marriage rates to be constant over one-year marriage intervals and, as is often done for population data, we approximate the total observation time in a one-year interval by the average of the number married at the beginning and at the end of the interval. As an illustration, consider the duration 10 years for the marriage cohort of 1960. From Table 5.2 we see that the number of divorces were $O_{10,1960} = 149$, while we approximate the total observation time for this interval by $R_{10,1960} = (22322+22173)/2 = 22247.5$ person-years. Thus the estimated divorce rate after 10 years of marriage becomes

$$\widehat{\theta}_{10,1960} = \frac{O_{10,1960}}{R_{10,1960}} = \frac{149}{22247.5} = 6.7 \cdot 10^{-3},$$

that is, 6.7 divorces per 1000 marriages per year. The estimated divorce rates $\widehat{\theta}_{d,c}$ for other durations (d) and cohorts (c) are computed similarly.

The left-hand panel of Figure 5.3 shows divorce rates computed in this manner for the three marriage cohorts. The figure shows a clear increase in divorce risk with marriage cohort. Furthermore, for all marriage cohorts, the divorce rates increase with duration of marriage until about five years, where a slight decline and then leveling out occurs. The corresponding estimated survival curves, given in the right-hand panel of Figure 5.3 show how the proportions still married are decreasing to various degrees in the different marriage cohorts. Here the estimated t-year survival probability of a marriage in cohort c is given by $\widehat{S}_c(t) = \exp\left\{-\sum_{d \leq t} \widehat{\theta}_{d,c}\right\}$. $\square$

5.2 Parametric regression models

In this section we consider parametric regression models for counting processes. In order not to complicate the presentation, we will concentrate on the situation where all covariates are fixed over time. Thus assume that we have counting processes $N_1, N_2, \ldots, N_n$ registering some event of interest for n individuals. Moreover, assume that for each $i = 1, 2, \ldots, n$, the intensity process of $N_i(t)$ takes the form

$$\lambda_i(t) = Y_i(t)\alpha(t \mid \mathbf{x}_i), \tag{5.12}$$

with $Y_i(t)$ an at risk indicator, and $\mathbf{x}_i = (x_{i1}, \ldots, x_{ip})^T$ a vector of fixed covariates for the ith individual.

We will restrict our attention to relative risk regression models, where the individual hazard rates $\alpha(t \mid \mathbf{x}_i)$ may be specified as

$$\alpha(t \mid \mathbf{x}_i) = \alpha_0(t; \boldsymbol{\theta}) \, r(\boldsymbol{\beta}, \mathbf{x}_i), \tag{5.13}$$

where $\alpha_0(t; \boldsymbol{\theta})$ is a baseline hazard common to all the individuals, and $r(\boldsymbol{\beta}, \mathbf{x}_i)$ is a relative risk function describing the effect of the covariates. Model (5.13) is similar to model (4.2) of Section 4.1, with the exception that the baseline hazard takes a parametric form. We will focus on the situation where the baseline hazard is piecewise constant, giving rise to what is commonly denoted Poisson regression (even though no assumptions on Poisson distribution are needed).

5.2.1 Poisson regression

Assume that the counting processes have intensity processes given by (5.12) and (5.13), that is, by

$$\lambda_i(t; \boldsymbol{\theta}, \boldsymbol{\beta}) = Y_i(t)\alpha_0(t; \boldsymbol{\theta}) \, r(\boldsymbol{\beta}, \mathbf{x}_i). \tag{5.14}$$

For given parametric specifications of the baseline hazard $\alpha_0(t; \boldsymbol{\theta})$ and the relative risk function (cf. Section 4.1), one may obtain the maximum likelihood estimators $\widehat{\boldsymbol{\theta}}$

and $\widehat{\boldsymbol{\beta}}$ by maximizing the log-likelihood function (5.5) numerically. Useful specifications of the baseline hazard are, for example, $\alpha_0(t;v) = v$ and $\alpha_0(t;b,k) = bt^{k-1}$, giving rise to exponential regression and Weibull regression, respectively.

We will concentrate on the situation where the baseline hazard (as an approximation) may be assumed to be piecewise constant as a function of time. Thus in a similar manner as in Example 5.4, we let $0 = t_0 < t_1 < t_2 < \cdots < t_K = \tau$ be a partition of the study time interval $[0, \tau]$ and assume that the baseline hazard is constant over each of the K subintervals. Thus, if θ_k is the constant value of the baseline hazard on the kth subinterval, we may write

$$\alpha_0(t;\boldsymbol{\theta}) = \sum_{i=1}^{K} \theta_k I_k(t), \qquad (5.15)$$

where $I_k(t)$ is the indicator for the kth subinterval and $\boldsymbol{\theta} = (\theta_1, \ldots, \theta_K)^T$.

In order to write the likelihood in a convenient form, we introduce

$$O_{ik} = \int_0^\tau I_k(t) dN_i(t)$$

for the number of events for individual i in the kth subinterval, and

$$R_{ik} = \int_0^\tau I_k(t) Y_i(t) dt$$

for the total time individual i is at risk in this interval. Then the likelihood (5.4) may be written as

$$L(\boldsymbol{\theta}, \boldsymbol{\beta}) = \left\{ \prod_{i=1}^{n} \prod_{k=1}^{K} [\theta_k r(\boldsymbol{\beta}, \mathbf{x}_i)]^{O_{ik}} \right\} \exp\left\{ -\sum_{i=1}^{n} \sum_{k=1}^{K} \theta_k r(\boldsymbol{\beta}, \mathbf{x}_i) R_{ik} \right\}$$

$$= \prod_{k=1}^{K} \prod_{i=1}^{n} \left\{ [\theta_k r(\boldsymbol{\beta}, \mathbf{x}_i)]^{O_{ik}} \exp[-\theta_k r(\boldsymbol{\beta}, \mathbf{x}_i) R_{ik}] \right\}. \qquad (5.16)$$

We note that the likelihood is proportional to the likelihood one would get if the O_{ik} were assumed to be independent Poisson distributed random variables with means $\mu_{ik} = \theta_k r(\boldsymbol{\beta}, \mathbf{x}_i) R_{ik}$. For this reason piecewise constant hazard models are often denoted Poisson regression models, even though we do *not* assume the O_{ik} to be Poisson distributed or the R_{ik} to be fixed quantities.

The fact that (5.16) is proportional to a Poisson likelihood makes it possible to use software for Poisson regression [like `glm` in R (R Development Core Team, 2007)] to fit the model for some specifications of the relative risk function. In particular, for the exponential relative risk function, $r(\boldsymbol{\beta}, \mathbf{x}_i) = \exp(\boldsymbol{\beta}^T \mathbf{x}_i)$, we have

$$\mu_{ik} = \theta_k \exp(\boldsymbol{\beta}^T \mathbf{x}_i) R_{ik} = \exp\left\{ \psi_k + \boldsymbol{\beta}^T \mathbf{x}_i + \log R_{ik} \right\},$$

Table 5.3 Estimated regression coefficients with standard errors based on a Poisson regression analysis of the total mortality in three Norwegian counties.

Covariate x_j	$\widehat{\beta}_j$	$\mathrm{se}(\widehat{\beta}_j)$
Sex	−0.541	0.094
Former smoker	0.316	0.134
1–9 cigarettes per day	0.888	0.146
10–19 cigarettes per day	0.891	0.127
20 or more cigarettes per day	1.084	0.154

with $\psi_k = \log \theta_k$. Thus we may fit the model by Poisson regression software using logarithmic link function and treating the $\log R_{ik}$ as "offsets" in the model.

Poisson regression is especially useful for large cohort studies where all the covariates are categorical. If we assume that the covariate vectors can only take the L distinct values $\mathbf{x}^{(1)}, \mathbf{x}^{(2)}, \dots, \mathbf{x}^{(L)}$, the likelihood (5.16) simplifies to

$$L(\boldsymbol{\theta}, \boldsymbol{\beta}) = \prod_{k=1}^{K} \prod_{l=1}^{L} \left\{ \left[\theta_k \, r(\boldsymbol{\beta}, \mathbf{x}^{(l)}) \right]^{O_k^{(l)}} \exp\left[-\theta_k \, r(\boldsymbol{\beta}, \mathbf{x}^{(l)}) R_k^{(l)} \right] \right\}, \qquad (5.17)$$

where

$$O_k^{(l)} = \sum_{i:\mathbf{x}_i = \mathbf{x}^{(l)}} O_{ik}, \qquad \text{and} \qquad R_k^{(l)} = \sum_{i:\mathbf{x}_i = \mathbf{x}^{(l)}} R_{ik}.$$

This shows that the aggregated number of events, $O_k^{(l)}$, and the total observation time, $R_k^{(l)}$, per combination of covariate value and time interval are *sufficient* statistics. Thus without loss of information, inference can be based on these aggregated quantities, and this may give large computational savings when n is much larger than L.

Example 5.5. Mortality in three Norwegian counties. In Example 4.1 we used Cox regression to study the effect of sex and smoking habits on the total mortality in three Norwegian counties. We will here analyze the same data using Poisson regression.

To this end we assume that the mortality is constant over the five years age groups 40–44 years, 45–49 years, 50–54 years, 55–59 years, 60–64 years, and 65–69 years. We may then summarize the data into the number of deaths and person-years in each of the $6 \times 2 \times 5 = 60$ "cells" obtained by combining the six age groups, two genders, and five smoking groups. Thus we get an "effective data set size" of 60 records, which is a substantial reduction from the individual data with 4000 records used for Cox regression. (With modern computers, there are no problems handling the 4000 records, but the reduction in data size achieved by Poisson regression may be of importance for situations where the full data set contains millions of records.)

Figure 5.4 shows the estimated baseline mortality (i.e., the $\widehat{\theta}_k$s) obtained by Poisson regression. These estimate the mortality of a male nonsmoker. Table 5.3 gives the estimated regression coefficients for sex and smoking group with standard errors.

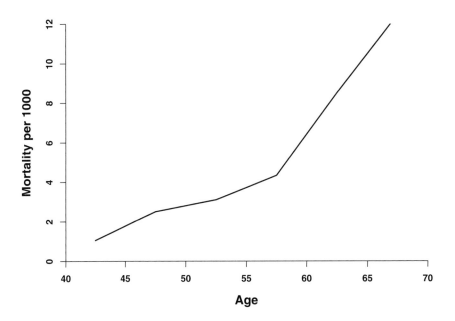

Fig. 5.4 *Estimates of baseline mortality based on a Poisson regression analysis of the total mortality in three Norwegian counties.*

All the regression estimates and their standard errors are almost indistinguishable from those of a Cox regression analysis (cf. Table 4.2). □

5.3 Proof of large sample properties

To avoid confusion, in this section we denote the true value of the vector of parameters by $\boldsymbol{\theta}_0$, while $\boldsymbol{\theta}$ will be used for the argument of the log-likelihood (5.5), the vector of score functions (5.6), and similar quantities. We will show that $\widehat{\boldsymbol{\theta}}$ is asymptotically multivariate normally distributed around $\boldsymbol{\theta}_0$ with a covariance matrix that may be estimated by the inverse of the observed or expected information matrix; cf. (5.19) and (5.23). For a detailed proof, including the necessary regularity conditions, see Andersen et al. (1993, section VI.1.2).

Note that by (5.6) the vector of score functions may be written

$$\mathbf{U}(\boldsymbol{\theta}) = \sum_{i=1}^{n} \int_0^{\tau} \frac{\partial}{\partial \boldsymbol{\theta}} \log \lambda_i(s;\boldsymbol{\theta}) dN_i(s) - \int_0^{\tau} \frac{\partial}{\partial \boldsymbol{\theta}} \lambda_{\bullet}(s;\boldsymbol{\theta}) ds, \qquad (5.18)$$

while the observed information matrix $\mathbf{I}(\boldsymbol{\theta}) = -\mathbf{U}(\boldsymbol{\theta})/\partial\boldsymbol{\theta}^T$ takes the form

$$\mathbf{I}(\boldsymbol{\theta}) = \int_0^\tau \frac{\partial^2}{\partial\boldsymbol{\theta}\,\partial\boldsymbol{\theta}^T}\lambda_\cdot(s;\boldsymbol{\theta})ds - \sum_{i=1}^n \int_0^\tau \frac{\partial^2}{\partial\boldsymbol{\theta}\,\partial\boldsymbol{\theta}^T}\log\lambda_i(s;\boldsymbol{\theta})dN_i(s). \quad (5.19)$$

We now use the decomposition $dN_i(s) = \lambda_i(s;\boldsymbol{\theta}_0)ds + dM_i(s)$ and find that the vector of score functions takes the form

$$\mathbf{U}(\boldsymbol{\theta}_0) = \sum_{i=1}^n \int_0^\tau \frac{\partial}{\partial\boldsymbol{\theta}}\log\lambda_i(s;\boldsymbol{\theta}_0)dM_i(s) \quad (5.20)$$

when evaluated at the true value of the parameter. The integrands in (5.20) are predictable processes. Thus the score is a (vector-valued) stochastic integral when evaluated at $\boldsymbol{\theta}_0$. In particular, $\mathbf{EU}(\boldsymbol{\theta}_0) = \mathbf{0}$. If on the right-hand side of (5.20) we replace the upper limit of integration by t, we get a stochastic process. This stochastic process is a martingale with a predictable variation process that, evaluated at τ, becomes

$$\langle\mathbf{U}(\boldsymbol{\theta}_0)\rangle(\tau) = \sum_{i=1}^n \int_0^\tau \left(\frac{\partial}{\partial\boldsymbol{\theta}}\log\lambda_i(s;\boldsymbol{\theta}_0)\right)^{\otimes 2}\lambda_i(s;\boldsymbol{\theta}_0)ds \quad (5.21)$$

[use (B.4) or (B.6) in Appendix B.2]. Note that, after some algebra, the observed information matrix (5.19), evaluated at $\boldsymbol{\theta}_0$, may be decomposed as

$$\mathbf{I}(\boldsymbol{\theta}_0) = \langle\mathbf{U}(\boldsymbol{\theta}_0)\rangle(\tau) - \sum_{i=1}^n \int_0^\tau \frac{\partial^2}{\partial\boldsymbol{\theta}\,\partial\boldsymbol{\theta}^T}\log\lambda_i(s;\boldsymbol{\theta}_0)\,dM_i(s). \quad (5.22)$$

Thus at the true parameter value, the observed information matrix equals the predictable variation process of the score function plus a stochastic integral. In particular, by taking expectations, it follows that the expected information matrix equals the covariance matrix of the score function, which is a key property in maximum likelihood estimation.

By the martingale central limit theorem (Section 2.3.3 and Appendix B.3), we may now show, under suitable regularity conditions, that $n^{-1/2}\mathbf{U}(\boldsymbol{\theta}_0)$ converges weakly to a multivariate normal distribution with mean zero and with a covariance matrix $\boldsymbol{\Sigma}$ that is the limit in probability of $n^{-1}\langle\mathbf{U}(\boldsymbol{\theta}_0)\rangle(\tau)$. Using (5.22) we may also show that both $n^{-1}\mathbf{I}(\boldsymbol{\theta}_0)$ and $n^{-1}\mathbf{I}(\widehat{\boldsymbol{\theta}})$ converge in probability to $\boldsymbol{\Sigma}$. From these results the large sample properties of $\widehat{\boldsymbol{\theta}}$ follow in the usual way. The main steps in the derivations are as follows. First note that $\widehat{\boldsymbol{\theta}}$ is the solution to the score equation $\mathbf{U}(\widehat{\boldsymbol{\theta}}) = \mathbf{0}$. Then Taylor expand the score equation around $\boldsymbol{\theta}_0$, to get

$$\mathbf{0} = \mathbf{U}(\widehat{\boldsymbol{\theta}}) \approx \mathbf{U}(\boldsymbol{\theta}_0) - \mathbf{I}(\boldsymbol{\theta}_0)(\widehat{\boldsymbol{\theta}} - \boldsymbol{\theta}_0).$$

From this we obtain

$$\sqrt{n}\left(\widehat{\boldsymbol{\theta}} - \boldsymbol{\theta}_0\right) \approx \left(n^{-1}\mathbf{I}(\boldsymbol{\theta}_0)\right)^{-1}n^{-1/2}\mathbf{U}(\boldsymbol{\theta}_0) \approx \boldsymbol{\Sigma}^{-1}n^{-1/2}\mathbf{U}(\boldsymbol{\theta}_0),$$

and it follows that $\sqrt{n}(\widehat{\boldsymbol{\theta}} - \boldsymbol{\theta}_0)$ converges weakly to a multivariate normal distribution with mean zero and covariance matrix $\boldsymbol{\Sigma}^{-1}\boldsymbol{\Sigma}\boldsymbol{\Sigma}^{-1} = \boldsymbol{\Sigma}^{-1}$. Thus $\widehat{\boldsymbol{\theta}}$ is approximately multivariate normally distributed around $\boldsymbol{\theta}_0$ in large samples. In order to estimate the covariance matrix of $\widehat{\boldsymbol{\theta}}$ we may use $\mathbf{I}(\widehat{\boldsymbol{\theta}})^{-1}$, where $\mathbf{I}(\boldsymbol{\theta})$ is the observed information matrix (5.19). Alternatively, we may estimate the covariance matrix by $\mathscr{I}(\widehat{\boldsymbol{\theta}})^{-1}$, where the (estimated) expected information matrix

$$\mathscr{I}(\boldsymbol{\theta}) = \sum_{i=1}^{n} \int_0^{\tau} \left(\frac{\partial}{\partial \boldsymbol{\theta}} \log \lambda_i(s; \boldsymbol{\theta}) \right)^{\otimes 2} dN_i(s) \tag{5.23}$$

is obtained by inserting $dN_i(s)$ for $\lambda_i(s; \boldsymbol{\theta}_0)ds$ in (5.21).

5.4 Exercises

5.1   Assume that the counting processes $N_i(t)$, $i = 1, 2, \ldots, n$, have intensity processes of the form $\lambda_i(t) = Y_i(t)e^{\beta}$, where the $Y_i(t)$ are at risk indicators. We introduce $R_i(t) = \int_0^t Y_i(u)du$, and let $N_.(t) = \sum_{i=1}^n N_i(t)$ and $R_.(t) = \sum_{i=1}^n R_i(t)$.

a) Show that the maximum likelihood estimator is $\widehat{\beta} = \log \widehat{v}$, with $\widehat{v} = N_.(\tau)/R_.(\tau)$.
b) Show that $\widehat{\beta}$ is approximately normally distributed around the true value of β with a variance that may be estimated by $1/N_.(\tau)$
c) Use the result in question (b) to derive an approximate 95% confidence interval for $v = \log \beta$. [*Hint:* Transform the standard 95% confidence interval for β.]

5.2   Assume that the counting processes $N_i(t)$, $i = 1, 2, \ldots, n$, have intensity processes of the form $\lambda_i(t) = Y_i(t)e^{\beta}(t/t_0)^{k-1}$, where the $Y_i(t)$ are at risk indicators.

a) Show that the log-likelihood function is given by (5.8), and derive the score functions for β and k.
b) Show that the expected information matrix [cf. (5.7)] takes the form given in Example 5.2.
c) Derive the observed information matrix. Do the observed and expected information matrices coincide?

5.3   Assume that the counting processes $N_i(t)$, $i = 1, 2, \ldots, n$, have intensity processes of the form $\lambda_i(t) = Y_i(t)be^{\theta t}$, where the $Y_i(t)$ are at risk indicators.

a) Find the log-likelihood function, and derive the score functions for b and θ.
b) Derive the observed and expected information matrices. Do they coincide?

5.4   Assume that the counting processes $N_i(t)$, $i = 1, 2, \ldots, n$, have intensity processes of the form $\lambda_i(t) = v Y_i(t)$, where the $Y_i(t)$ are at risk indicators and $v > 0$ is a parameter.

a) Show that the score function takes the form $U(v) = N_{\cdot}(\tau)/v - R_{\cdot}(\tau)$, where τ is the upper time limit of the study, $R_i(\tau) = \int_0^\tau Y_i(t)dt$, and a dot means summation over $i = 1, 2, \ldots, n$.
b) Show that, when evaluated at the true value of the parameter, the expected value of the score is zero and its variance equals $E(R_{\cdot}(\tau)/v)$. [*Hint:* $M_i(t) = N_i(t) - v\int_0^t Y_i(u)du$ is a martingale.]
c) Derive the observed information $i(v) = -U'(v)$ and show that its expected value, when evaluated at the true value of the parameter, equals $E(R_{\cdot}(\tau)/v)$.

Comment: The results in questions (b) and (c) show that, at the true parameter value, the expected score is zero and the variance of the score equals the expected information. General versions of these results are key in the proof of the large sample properties of the maximum likelihood estimator (cf. Section 5.3).

Chapter 6
Unobserved heterogeneity: The odd effects of frailty

Individuals differ. This is a basic observation of life and also of statistics. Some die old and some die young. Some are tall and some are small. In medicine one will find that a treatment that is useful for one person may be much less so for another person. And one person's risk factor may be less risky for another one. This variation is found everywhere in biology and other fields, like in reliability studies of technical devices. Even between genetically inbred laboratory animals one observes a considerable variation.

One aim of a statistical analysis may be precisely to understand the factors determining such variation. For instance one might perform a regression analysis with some covariates. However, it is a general observation that such analyses always leave an unexplained rest. There is some variation that cannot be explained by observable covariates, and sometimes this remaining variation may be large and important. Traditionally, this is considered an error variance, something that creates uncertainty but can otherwise be handled in the statistical analysis, and usually one would not worry too much about it. However, in some cases there are reasons to worry.

In survival analysis, one talks about unobserved heterogeneity to denote variation not explained by covariates. A more popular term is *frailty*, indicating that some individuals are more frail than others, that is, the event in question is more likely to happen for them. The term is not always appropriate, focusing too much on negative outcomes. For instance, in a study of fertility, it is not natural to characterize a woman who can easily conceive as "frail." However, the term frailty is so well established that we shall use it as a general term for describing unobserved heterogeneity.

More precisely, frailty shall mean a part of the unexplained variation. There will always be a basic randomness, even if individuals are identical in risk. The mathematical model defined in Section 6.2.1 makes this clear. In this chapter we consider the situation with at most a single event for each individual, and we return to the multivariate case in Chapter 7.

The concept of frailty has received great attention in survival analysis. Firstly, frailty may lead to artifacts and distort the observed results. Those individuals that are at high risk, the frail ones, will have the event in question occurring earlier than the others. Hence those remaining in the risk set after a while will have lower average frailty. This gives a selection effect that "drags down" the population hazard rate (Section 6.2.1). A population hazard rate that declines may give the superficial impression that the individual risk is declining. However, this may be entirely wrong. It is quite conceivable that the (unobservable) hazard for each individual is increasing, while the hazard observed in the population may decline.

Actually, this can be construed as a question of what the hazard rate really means. From the previous chapters it seen that the hazard rate is really important; it is the foundation of both the Cox model and, in a wider sense, of the counting process approach. Now we discover that the hazard rate is a complex, and even an elusive, concept. Clearly, its development over time is determined by individual development in risk. If the risk of the event occurring decreases for the individuals, then the hazard will decline. But it might equally well decline due to the mentioned selection effects. So, when observing the shape of the hazard rate it is not easy to know whether this is due to changes *within* individuals, or differences *between* individuals.

This has led to much confusion. For instance, it has been observed that divorce rates reach a maximum 5–7 years after the start of the marriage and then decline. This has been interpreted as meaning that most marriages go through a crisis after 5–7 years, with the risk declining if they get through this crisis (Bergman et al., 1977). However, what one observes might just as well be a selection effect due to frailty. There are several such examples.

Frailty may also lead to a number of more intricate artifacts, like declining relative risk, crossover phenomena, and false protectivity. We shall discuss these phenomena in Section 6.5.

But frailty is not only of interest due to its ability to cause confusion. There is also a more constructive side to this. When studying recurrent events over time or other kinds of multivariate survival data, one may find it useful to construct random effects models. In the event history setting, it is common to approach this by using frailty models. An important consideration in all of survival and event history analysis is that models shall easily handle right-censoring and left-truncation. The particular way one constructs frailty models fulfils this requirement. Hence, frailty theory constitutes an elegant way of constructing various kinds of random effects models, as done in Chapter 7.

Before going into the mathematics, we shall reflect a little over what frailty really means in practice, that is, what is the source of heterogeneity. Firstly, it could reflect biological differences present from the beginning. Some are born with a weaker heart, or a genetical disposition for cancer, for instance. Secondly, frailty could represent the induced weaknesses that result from the stresses of life. This points to frailty as a dynamic concept, something that is changing over time. Thirdly, it may be taken to include heterogeneity due to a disease being in a late or an early stage. A late diagnosis of cancer may give a high frailty of dying from it. This is different from frailty as an inherent biological property, since the late diagnosis may just be due to an unfortunate delay. It should be clear that frailty as used in survival analysis is a complex concept, covering many different kinds of heterogeneity among individuals.

The concept of frailty assumes the existence of essential differences between individuals. To what extent do such differences exist as regards for instance disease risk? Firstly, it is clear that the risk of many diseases depends on genetics. Many diseases, like testicular cancer, or schizophrenia, attack a small minority of people before the incidence starts declining strongly after a certain age, and strong familial association has been demonstrated. It is also shown that the metastatic potential of cancer, and hence the risk of not surviving the disease, may depend on genetic properties of the patient and of the primary tumor (e.g., Hunter, 2004).

The effect of heterogeneity, or frailty, has been recognized for a long time (e.g., Strehler and Mildvan, 1960). Also the social science literature has discussed this problem (e.g., Ginsberg, 1971, pages 251–255). In fact, a very early frailty model was the "mover-stayer" model of Blumen et al. (1955). The term "frailty" was introduced by Vaupel et al. (1979). Other early contributors are Heckman and Singer (1982), Vaupel and Yashin (1985), and Hougaard (1984). The subject has even hit the general scientific literature (Barinaga, 1992; Carey et al., 1992). An interesting overview of the frailty field with several new results is given in the doctoral dissertation of Wienke (2007).

Frailty models have some connection to Bayesian models, for example, the nonparametric Bayesian ones studied by Hjort (1990). However, the perspective here is not Bayesian but closer to an empirical Bayes point of view; see Section 7.2.3.

6.1 What is randomness in survival models?

Randomness is an elusive concept. In statistical models we operate with random noise, but we rarely go into what this really means. Rather, randomness is the big sack into which we put all the variation that cannot be explained in our model. In a sense, randomness is seen as a result of imperfection.

In event history and survival models the issue is more complex than in most statistical models. The event history is a stochastic process, and part of the randomness comes from the stochastic mechanism by which events unfold. In addition, one will have the ordinary randomness in the sense of measurement errors, variation between individuals, and so on.

As an example, consider the process of radioactive decay of some material. The radioactive pulses will come according to a Poisson process, and this randomness is not due to imperfection of our model but is determined by deep natural laws embedded in the theory of quantum mechanics. According to a widely accepted view in physics (Zeilinger, 2005), quantum randomness is of an essential nature, not a result of our own ignorance. In fact, this may be closely related to some issues of medical interest. Models of carcinogenesis assume that cancer develops through a number of transitions whereby certain cells become increasingly malignant. The transitions may be determined by mutations, which again might be effects of radiation. Hence, the underlying processes determining event histories may have essential random elements to them that we cannot get rid of by more precise observation. So even if the frailty of an individual was completely known, there would remain a basic amount of randomness.

The biological understanding of these issues is certainly very limited as yet. But one should not expect to be able to "explain" all variation, and it certainly is not a defeat that there remains a possibly large unexplained rest.

An interesting evaluation of the randomness issue is presented in the following quotation (Brennan, 2002):

> The inherent stochastic nature of cancer is supported by several observations in animal and twin studies. Genetically identical animals kept in as similar an environment as possible will not behave the same upon exposure to environmental carcinogens. While it is possible to estimate the proportion of animals that will develop a malignancy at a particular exposure level, there appears to be a random element determining which particular animals will develop tumors.

For further interesting comments along these lines, see Coggon and Martyn (2005). Randomness is also demonstrated in experiments reported in a *Nature* paper by Kirkwood and Finch (2002). It is shown that even genetically identical (i.e., isogenic) worms have great variation in their life times. The paper stresses the random and unpredictable nature of the cell damage that occurs with ageing. It is also stated in the paper that "one of the hallmarks of the ageing process is an increase in variability." A detailed study of frailty as a stochastic process is given in Chapter 11.

The stochastic nature of disease may seem to be in contradiction to the frailty effects discussed in this chapter. However, the truth is that you have both phenomena; sometimes the variation in risk between individuals is strong and important, in other cases what happens is mainly a matter of chance.

6.2 The proportional frailty model

The term "frailty theory" has been primarily associated with one particular mathematical formulation of frailty, namely the proportional frailty model. This is an extremely simple, but nevertheless very useful, formulation.

6.2.1 Basic properties

We assume that the hazard rate of an individual is given as the product of an individual specific quantity Z and a basic rate $\alpha(t)$:

$$\alpha(t|Z) = Z \cdot \alpha(t). \tag{6.1}$$

Here Z is considered as a random variable over the population of individuals, specifying the level of frailty. Note that both Z and $\alpha(t)$ are unobservable. What may be observed in a population is not the individual hazard rate, but the net result for a number of individuals with differing frailties, or what we term the population hazard rate. We shall explain how to calculate this.

Given Z, the probability of surviving up to time t is given by

$$S(t|Z) = e^{-ZA(t)}, \quad \text{where} \quad A(t) = \int_0^t \alpha(u)du.$$

The population survival function is found by integrating over the distribution of Z, that is,

$$S(t) = \mathrm{E}\{e^{-ZA(t)}\}. \tag{6.2}$$

It is very useful to introduce the Laplace transform of Z, defined by

$$\mathscr{L}(c) = \mathrm{E}\left(e^{-cZ}\right)$$

and to write:

$$S(t) = \mathscr{L}(A(t)). \tag{6.3}$$

The population hazard rate, denoted by $\mu(t)$, may be found by differentiating $-\log(S(t))$ (see Section 1.1.2):

$$\mu(t) = \alpha(t)\frac{-\mathscr{L}'(A(t))}{\mathscr{L}(A(t))}. \tag{6.4}$$

Hence, the difference between the individual hazard rate and the population hazard rate is determined by the second factor on the right-hand side of (6.4). This factor can be shown to always be a decreasing function.

The distribution of Z is an example of a mixing distribution; in fact, mixing models are used in a number of different contexts.

Clearly, the proportional frailty model (6.1) is mainly chosen for mathematical convenience. It represents a rather simplified view of how individual heterogeneity might act. There is no reason why heterogeneity should be determined at time zero, like it is in this model, or that it should act in a proportional manner. Certainly, reality is more complex. Nevertheless, the model is very effective in illustrating a number of issues concerning heterogeneity, as we shall see in the remainder of this chapter.

6.2.2 The Gamma frailty distribution

When applying the results of the previous subsection one would naturally seek to use frailty distributions with a tractable Laplace transform. One common choice is the gamma distribution, and in this subsection we give some results for this case.

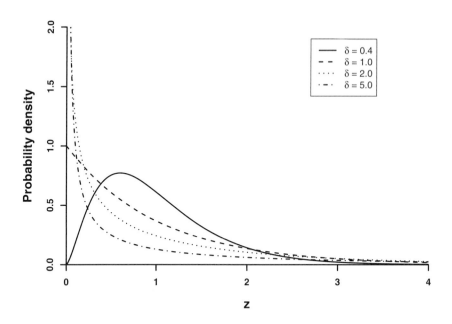

Fig. 6.1 *Probability densities of the gamma distribution with mean equal to one and different values of the variance δ.*

The density of the gamma distribution is given as

$$\frac{v^{\eta}}{\Gamma(\eta)} z^{\eta-1} \exp(-vz),\tag{6.5}$$

where v is a scale parameter and η a shape parameter. The Laplace transform is given as $\mathscr{L}(c) = (v/(v+c))^{\eta}$. This function has very nice properties, for instance, it can be explicitly differentiated any number of times, which is of importance when frailty is used for multivariate survival data. We shall return to this in the next chapter.

It is often natural to let the frailty distribution have mean equal to 1, that is, $v = \eta$. In this case it is useful to choose as a new parameter the variance, $\delta = 1/v$, of the frailty distribution. This parameter is a natural measure of the degree of heterogeneity of the population. The Laplace transform is then given as

$$\mathscr{L}(c) = \{1 + \delta c\}^{-1/\delta}.\tag{6.6}$$

The gamma distributions are very flexible, assuming many different shapes as shown in Figure 6.1.

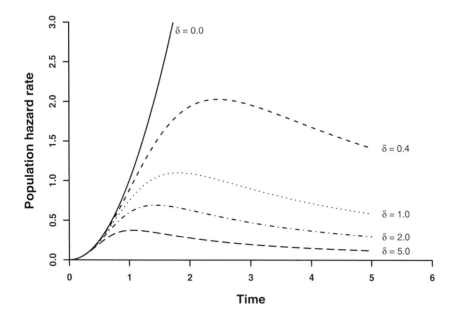

Fig. 6.2 *Population hazard rates with various values of the frailty variance δ. Basic hazard rate $\alpha(t) = t^2$.*

Combining equations (6.3) and (6.6) yields the following population survival function

$$S(t) = \{1 + \delta A(t)\}^{-1/\delta} \qquad (6.7)$$

and the population hazard rate

$$\mu(t) = \frac{\alpha(t)}{1 + \delta A(t)}. \qquad (6.8)$$

The basic effect of frailty is seen very clearly from (6.8). When $\delta = 0$ there is no frailty and $\mu(t)$ and $\alpha(t)$ are identical. As the variance δ increases, the denominator becomes larger, and it also increases with time, yielding the typical frailty shape of a hazard function that is "dragged down." One could easily have that $\mu(t)$ decreases after a while even if $\alpha(t)$ keeps increasing all the time. This is demonstrated in Figure 6.2, where we have chosen $\alpha(t) = t^2$. When $\delta = 0$ there is no frailty effect and the curve is identical to the basic rate. It is clearly seen how the presence of frailty pulls the hazard rate down, transforming an ever-increasing basic rate into a quite different shape.

Hazard rates that first increase and later decrease are seen very often in practice, and in many cases it is natural to imagine some frailty effect being present. Some examples of declining hazard are the following:

- Mortality rate of cancer patients, measured from the time of diagnosis typically first increases, then reaches a maximum, and then starts to decline. A likely reason is heterogeneity from the outset as regards the prospect of recovery.
- Mortality of patients with myocardial infarction starts to decline shortly after the infarction has taken place. Again, one would expect that survivors have had a less serious attack and so better chances from the beginning.
- The incidence rate of testicular cancer reaches a maximum at around 30 years of age and then starts to decline. A frailty analysis of such incidence data will be given in Section 6.9.

6.2.3 The PVF family of frailty distributions

Useful frailty distributions should have an explicit Laplace transform. A very general family of distributions of great use in frailty theory is the family of power variance function distributions, which we denote simply PVF distributions. This class of distributions was suggested by Hougaard; see the book by Hougaard (2000) for background and references. The PVF distributions are defined as those having the Laplace transform

$$\mathscr{L}(c;\rho,v,m) = \exp\left[-\rho\left\{1 - \left(\frac{v}{v+c}\right)^m\right\}\right] \tag{6.9}$$

with $v > 0$, $m > -1$, and $m\rho > 0$. The reason for suggesting the PVF frailty distributions is that they yield nice formulas for a number of interesting quantities. They also contain frailty distributions that have a positive probability at zero, something that is highly relevant when there is a group that is nonsusceptible to the risk in question. As will be shown, the class also contains a number of well-known distributions (sometimes as limiting cases). Practical applications of the PVF model may be found in Aalen (1988b, 1994).

The expectation and variance of a PVF distributed variable, Z, are as follows:

$$E(Z) = \frac{\rho m}{v}, \qquad Var(Z) = \frac{\rho m}{v}\frac{m+1}{v}. \tag{6.10}$$

The density is given by the infinite sum:

$$f(z;\rho,v,m) = \exp(-\rho - vz)\frac{1}{z}\sum_{n=1}^{\infty}\frac{\rho^n(vz)^{mn}}{\Gamma(mn)n!}, \tag{6.11}$$

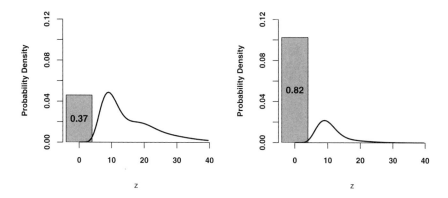

Fig. 6.3 *Size of atom at zero and plot of density for compound Poisson distributions with $\rho = 1$ and 0.2 in left and right panel, respectively ($m = 10$, $v = 1$). The sizes of the atoms are $\exp(-1) = 0.37$ and $\exp(-0.2) = 0.82$, respectively.*

which is, however, rarely used in practice. This density does not necessarily integrate up to 1, since there may be an atom at 0; see later for the compound Poisson distribution.

A number of important special cases are within the PVF family or arise as limits:

- *Gamma distribution.* If $\rho \to \infty$ and $m \to 0$ in such a way that $\rho m \to \eta$, then the Laplace transform approaches

$$\mathscr{L}_{\text{Gamma}}(c; v, \eta) = \left(\frac{v}{v+c}\right)^{\eta},$$

which is the Laplace transform of a gamma distribution with scale parameter v and shape parameter η. The gamma distribution is by far the most commonly used frailty distribution.

- *Compound Poisson distribution.* When $m > 0$ we have the compound Poisson distribution. This arises as the sum of independent gamma distributed variables, here with scale parameter v and shape parameter m, where the number of summands is Poisson distributed with expectation ρ. The compound Poisson distribution has the interesting property that there is a positive probability at zero, combined with a density on the positive real line. The atom at zero has probability $\exp(-\rho)$. Hence, it is a model for situations in which there is a group with zero frailty, often termed a nonsusceptible group. This is relevant in many situations, for instance, in cure models. Examples of compound Poisson distributions are given in Figure 6.3. The model may be seen as a cumulative damage model, where ρ determines the number of insults, the size of each insult being gamma distributed. Note the special case with $m = 1$, when the infinite series in (6.11)

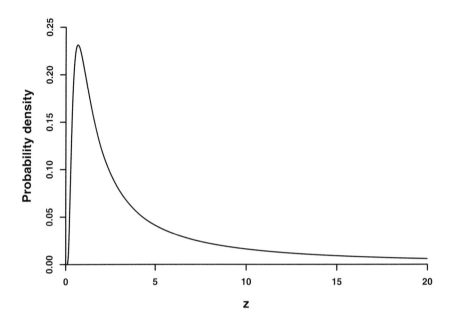

Fig. 6.4 *Probability density of the stable distribution (Lévy) with $a = 1$.*

sums up to

$$f(z;\rho,v,1) = \exp(-\rho - vz)\sqrt{\frac{v\rho}{z}}I_1(2\sqrt{v\rho z}),$$

where I_k denotes a modified Bessel function of the first kind (e.g., Weisstein, 2006a). Hence, this is the density of the continuous part of a compound Poisson distribution with $m = 1$.

- *Inverse Gaussian distribution.* When $m = -1/2$ and $\rho < 0$, we have an inverse Gaussian distribution. This arises as a first-passage time distribution in a Brownian motion with drift. It has an old history and was originally derived by the famous physicist Erwin Schrödinger (Schrödinger, 1915). The density is given as:

$$f(z;-\rho,v,-1/2) = \frac{\rho}{2\sqrt{\pi}\sqrt{v}}z^{-\frac{3}{2}}\exp\left(-zv+\rho - \frac{\rho^2}{4zv}\right).$$

- *Hougaard distribution.* This arises when $-1 < m < 0$ and $\rho < 0$. The distribution is continuous and unimodal. The inverse Gaussian distribution is a special case. The distribution was introduced to frailty theory by Hougaard; see Hougaard (2000) for more details on the distribution.
- *Stable distribution.* This arises when $-1 < m < 0$ and $v \to 0$, $\rho \to -\infty$ in such a way that $-v^m\rho$ approaches a positive constant a, yielding the following Laplace

transform

$$\mathcal{L}_{\text{stable}}(c;k,a) = \exp\left(-ac^k\right), \tag{6.12}$$

where $k = -m$ is between 0 and 1. The stable distributions are important in probability and statistics because the sum of identically distributed variables with this distribution again has a stable distribution. In general, the density is only given by an infinite sum, but an explicit formula exists in the special case when $m = -1/2$, yielding the Lévy distribution:

$$f_{\text{stable}}(z;1/2,a) = \frac{a}{\sqrt{\pi}}z^{-\frac{3}{2}}\exp(-\frac{a^2}{z}). \tag{6.13}$$

The stable distributions have very heavy tails, and the expectation does not exist. For an illustration, see Figure 6.4, which shows the density for a Lévy distribution with $a = 1$.

- *Poisson distribution.* If $m \to \infty$, $v \to \infty$ in such a way that $m/v \to 1$, then the PVF-distribution converges to a Poisson distribution with parameter ρ.
- *Normal distribution.* Assume that Z has the PVF Laplace transform (6.9). If $\rho \to \pm\infty$, $v \to \infty$ in such a way that ρ/v approaches a constant ξ, then

$$\frac{Z - \xi m}{\sqrt{\xi m(m+1)/v}}$$

has asymptotically a standard normal distribution (Jørgensen, 1987).

One sees that the simple formula (6.9) includes a large number of very different distributions as special cases or limits, from the normal distribution to extremely skewed stable distributions, including also distributions with atoms at 0 or the discrete Poisson distribution. Hence, this gives a rich set of possible distributions of heterogeneity.

A natural question is what kind of distributions one might meet in practice. What sort of shapes might one expect for actual frailty distributions? For known risk factors, like high blood pressure and cholesterol, a log-normal or gamma frailty distribution seems to give a reasonable fit. Superficially, one might expect unknown risk factors to have distributions similar to the known ones. However, this may not be the case. For instance, when it comes to genetic risks, the distributions may be very skewed. For some diseases one has found that a few individuals with single dominant genes might be at a much larger risk than the majority. Examples are the well-known breast cancer genes BRCA1 and BRCA2 (e.g., Antoniou and Easton, 2006).

When the frailty Z has the PVF Laplace transform (6.9), the following survival distribution may be derived from (6.3):

$$S(t) = \exp\left(-\rho\left\{1 - \left(\frac{1}{1+A(t)/v}\right)^m\right\}\right). \tag{6.14}$$

The population hazard rate is then given by

$$\mu(t) = \frac{\rho\, m}{v} \frac{\alpha(t)}{(1+A(t)/v)^{m+1}}.$$ (6.15)

One reason for the usefulness of the PVF distribution is that it results in this explicit formula for the population hazard rate. The formula is a simple generalization of the one for gamma distributed frailty given in (6.8). Notice that the frailty effect, that is, the extent to which the population hazard is "bent down," increases when m increases or v decreases. From (6.10) it is seen that the first fraction in the expression of $\mu(t)$ equals $E(Z)$, and one also sees that if the expectation is kept fixed while v decreases, then the variance of Z increases. Hence, an increasing variance leads to an increasing frailty effect, which should be expected.

It is also extremely useful that the parameter ρ appears as a *proportionality parameter* in the population hazard (6.15). This fits well with the elevated role of the proportionality assumption in survival analysis and turns out to be quite useful. We return to this in Section 6.5.4.

For statistical purposes it is apparent from (6.14) that the parameter v may be absorbed into $\alpha(t)$ or $A(t)$.

The functional form given in formula (6.15) may also be derived from viewpoints other than the frailty one assumed here. It is interesting to note that the same hazard function has been suggested by Bagdonavicius et al. (2004), corresponding to setting $-1 < m < 0$ (that is, the Hougaard frailty distribution in our framework). They use a particular parametrization in to analyse crossing survival curves.

6.2.4 Lévy-type frailty distributions

The Laplace transform for the PVF distribution given in (6.9) is a special case of the more general formula

$$\mathcal{L}(c) = \exp(-\rho\, \Phi(c)).$$ (6.16)

This is a valid Laplace transform for any nonnegative parameter ρ whenever $\Phi(c)$ is a Laplace exponent. Laplace exponents arise naturally in the theory of Lévy processes (see Appendix A.5), and we therefore denote a frailty distribution with Laplace transform (6.16) as a Lévy-type frailty distribution [to be distinguished from the Lévy distribution of (6.13)]. The PVF class of distributions is a special case, with

$$\Phi(c) = 1 - \left(\frac{v}{v+c}\right)^m.$$

The limiting cases of the PVF distributions also correspond to simple special cases of (6.16). Some examples are as follows:

Gamma distribution: $\Phi(c) = \eta(\log(v+c) - \log v)$;

Stable distribution: $\Phi(c) = \alpha c^k$.

A general compound Poisson distribution, extending that defined in Section 6.2.3, is given by

$$\Phi(c) = 1 - \mathcal{L}_0(c),$$

where $\mathcal{L}_0(c)$ is any Laplace transform of a nonnegative distribution.

The family defined by using (6.16) as a Laplace transform of a frailty distribution is very nice. The population hazard rate will simply be $\rho\alpha(t)\Phi'(A(t))$. The parameter ρ is a proportionality parameter, so there is a natural connection with the proportional hazards model; see Section 6.5.4.

A generalization of this to model time-dependent frailty by means of Lévy processes is studied in Chapter 11.

6.3 Hazard and frailty of survivors

When studying frailty, we imagine that we follow a population over time. Those who survive beyond a certain time will be more robust and have a different frailty distribution compared to the original frailty distribution. A useful model will allow us to have explicit expressions for the frailty distribution of survivors, and the best will be if the family of frailty distribution stay the same under frailty selection, with just a modification of the parameters. The PVF family satisfies this demand as will be shown here.

6.3.1 Results for the PVF distribution

Assume that the Laplace transform of the frailty Z is given by $\mathcal{L}(c)$. The individuals surviving beyond a certain time will have a frailty distribution that can be derived in the following way. The Laplace transform of the frailty distribution for survivors at time t is given by:

$$\mathcal{L}_t(c) = \mathrm{E}(e^{-cZ} \mid T > t) = \frac{\mathrm{E}(e^{-cZ}I\{T > t\})}{P(T > t)} \tag{6.17}$$

$$= \frac{\mathrm{E}(e^{-cZ - ZA(t)})}{\mathrm{E}(e^{-ZA(t)})} = \frac{\mathcal{L}(c + A(t))}{\mathcal{L}(A(t))}.$$

The Laplace transform of survivors $\mathcal{L}_t(c)$ belongs to the same parametric family as the original Laplace transform $\mathcal{L}(c)$ if the expression for this transform contains the term $c + \gamma$ for some parameter γ. This again corresponds to the density of the distribution of Z having a factor $\exp(-\gamma z)$. To explain what this means, consider the PVF Laplace transform (6.9). The formula for this transform contains the term $\nu + c$. The reason for this is that the PVF density given in (6.11) contains a factor

$\exp(-vz)$ that can be combined with the expression $\exp(-cz)$ when the Laplace transform is computed.

Applying formula (6.18) to the PVF Laplace transform (6.9) yields

$$\mathscr{L}_t(c;\rho,v,m) = \exp\left(-\rho\left(\frac{v}{v+A(t)}\right)^m\left\{1-\left(\frac{v+A(t)}{v+c+A(t)}\right)^m\right\}\right). \quad (6.18)$$

Hence, the frailty distribution of survivors is still a PVF distribution, with the following change in parameters:

$$\rho \rightarrow \rho\left(\frac{v}{v+A(t)}\right)^m, \quad v \rightarrow v+A(t), \quad m \rightarrow m.$$

For the particular case of the gamma distribution, the change of parameters is given as:

$$v \rightarrow v+A(t), \quad \eta \rightarrow \eta.$$

The preservation of the class of frailty distributions is a major advantage of the PVF family. Note that m is preserved, which is important since this parameter distinguishes major classes of the PVF family.

From (6.10), we may derive the expectation and variance of the frailty for survivors as follows:

$$E(Z) = \frac{\rho\,m v^m}{(v+A(t))^{m+1}}, \qquad Var(Z) = \frac{\rho\,m(m+1)v^m}{(v+A(t))^{m+2}}.$$

This yields

$$CV(Z) = (1+A(t)/v)^{m/2}\sqrt{\frac{m+1}{\rho m}},$$

where CV denotes the coefficient of variation, that is, the standard deviation divided by the expectation. As would be expected, both expectation and variance decrease with time. The coefficient of variation decreases with time t when $-1 < m < 0$, that is, for the Hougaard distributions, while it increases for $m > 0$, that is, the compound Poisson distributions. For the gamma distributions the coefficient of variation is constant. Hence, the population of survivors could be increasingly similar or dissimilar due to frailty selection, according to which specific frailty distribution is involved.

6.3.2 Cure models

For the compound Poisson frailty distribution, that is, when $m > 0$, there is a group with zero frailty (or a nonsusceptibe group). The probability of belonging to this group, which survives forever, is $\exp(-\rho)$. Models of this kind are often termed "cure" models, or models with "long-term survivors." There is a whole literature on this subject, where one issue is to estimate the proportion with zero frailty, which in the present model would amount to estimating ρ. An interesting book on cure models is Maller and Zhou (1996). One of their examples concerns recidivism of released prisoners. Many will return to prison later for new offenses, but a certain proportion of them seem to be immune, never committing a new crime.

Another frailty-type cure model would be to mix a probability for zero susceptibility with some frailty distribution for the susceptible ones. This was suggested, for instance, by Longini and Halloran (1996), generalizing a model by Farewell (1977). It might be thought that the shape of the frailty distribution of susceptibles should be independent of the probability of nonsusceptibility, like it is in the paper by Longini and Halloran (1996), while in the compound Poisson model the shape of the frailty density may depend on the size of the nonsusceptible group. However, it is likely that in practice there is a connection between the probability of susceptibility and the distribution of frailty (Aalen and Hjort, 2002). For instance, it is often the case that rare diseases have a high risk for a few individuals while more common diseases have less variation in risk between individuals.

Often one will find that estimating the proportion of nonsusceptibles, or long-term survivors, is not easy. Nevertheless, it is tempting to try this when, for instance, the Kaplan-Meier survival curves appear to be flattening out at a level above zero. An example will be given in the case study on testicular cancer in Section 6.9.

6.3.3 Asymptotic distribution of survivors

As seen earlier the PVF distributions are closed with respect to survival, such that the value of m is unchanged. For the Hougaard distributions, with $-1 < m < 0$ and $\rho < 0$, the ρ-value for survivors will go to minus infinity if $A(t)$ is unbounded. From the asymptotic result for PVF distributions it follows that the frailty distribution for survivors will approximate a normal distribution.

Aalen (1992, page 964) gives a result about the limiting frailty distribution among susceptible survivors when starting out with compound Poisson frailty ($m > 1$). In fact, when $t \rightarrow \infty$ and $A(t)$ is unbounded, the limiting distribution of $ZA(t)$ among the susceptible survivors at time t turns out to be a gamma distribution with expectation and variance both equal to m. This means that the frailty distribution of survivors can be approximated by a mixture of a probability at zero and a gamma density, similar to the model suggested by Longini and Halloran (1996).

An interesting general result on approximating frailty distributions for survivors with gamma distributions is given by Abbring and van den Berg (2007). These authors argue that this is a justification for the general validity of gamma frailty distribution.

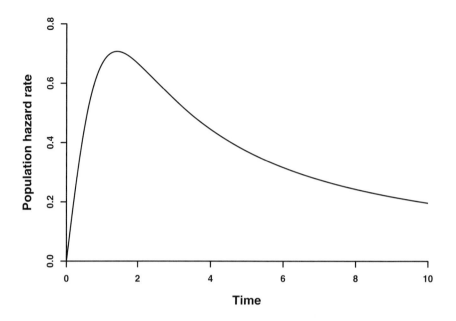

Fig. 6.5 *The hazard rate of a Burr distribution with $\delta = 1$, $b = 1$ and, $k = 2$.*

6.4 Parametric models derived from frailty distributions

Survival analysis, at least in biostatistics, is dominated by nonparametric and semi-parametric models. For many purposes, especially when using complex designs, parametric models are far more useful. Standard models, like the Weibull one, tend to have monotonic hazard rates. In practice, we often see hazard rates that are non-monotonic, for example, first increasing and then decreasing. Such hazard shapes can naturally be generated by frailty models; see Figure 6.5, which is based on the Burr distribution defined next.

6.4.1 A model based on Gamma frailty: the Burr distribution

The population hazard function in (6.8) defines a probability distribution of the time to event, which we shall denote by T. An important special case is when the basic rate is of Weibull form, that is, $\alpha(t) = bt^{k-1}$ for $k > 0$. First, note that the population survival, hazard, and density functions in this case are given by

$$S(t) = \left(1 + \frac{\delta b}{k} t^k\right)^{-1/\delta}, \quad \mu(t) = \frac{b t^{k-1}}{1 + \delta b t^k / k},$$

and

$$f(t) = b t^{k-1} \left(1 + \frac{\delta b}{k} t^k\right)^{-1-1/\delta}.$$

This is called a Burr distribution. It is also related to the F-distribution; in fact, apart from a scale transformation, it is a case of the generalized F-distribution defined in Johnson et al. (1995, chapter 27, section 8.1). For applications of this distribution in survival analysis, see Kalbfleisch and Prentice (2002, section 2.2.7), Aalen (1988b) and Aalen and Husebye (1991). In fact, Aalen (1988b) showed that this distribution gives a very good description of the survival of breast cancer patients in several different stages of disease.

The rth moment of the Burr distribution is given by

$$\begin{aligned}
\mathrm{E}(T^r) &= r \int_0^\infty t^{r-1} S(t)\, dt \\
&= r \int_0^\infty t^{r-1} \left(1 + \frac{\delta b}{k} t^k\right)^{-1/\delta} dt \\
&= \frac{1}{\delta} \left(\frac{k}{b\delta}\right)^{r/k} B\left(\frac{r}{k} + 1, \frac{1}{\delta} - \frac{r}{k}\right),
\end{aligned} \tag{6.19}$$

where B denotes the beta function (e.g., Weisstein, 2006b). Note that the argument is only valid when $\delta < k/r$, so that the moment does not exist beyond this value. Hence, when the heterogeneity δ is too large, even the expectation of the time to event does not exist.

When the variance of T exists, it may be of special interest to see how it is influenced by heterogeneity. The variance is given as:

$$\mathrm{Var}(T) = \frac{1}{\delta} \left(\frac{k}{b\delta}\right)^{2/k} \left\{ B\left(\frac{2}{k} + 1, \frac{1}{\delta} - \frac{2}{k}\right) - \frac{1}{\delta} B\left(\frac{1}{k} + 1, \frac{1}{\delta} - \frac{1}{k}\right)^2 \right\}.$$

Corresponding to more standard variance components models, one would expect that this variance increases with increasing heterogeneity δ. We have not proved this in general, but it may be demonstrated by plotting; see Figure 6.6.

6.4.2 A model based on PVF frailty

A general class of parametric distributions may be defined using PVF frailty distributions by letting $\alpha(t) = b t^{k-1}$ for $k > 0$, yielding:

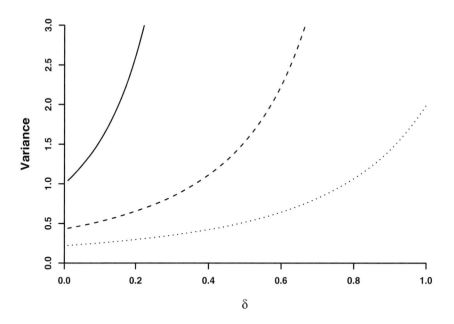

Fig. 6.6 *Gamma frailty distribution. The variance of time to event in the population as a function of heterogeneity δ. Basic rate t^2 (for lower curve), t (for intermediate curve) and 1 (for upper curve).*

$$S(t) = \mathcal{L}(bt;\rho,v,m) = \exp\left[-\rho\left\{1-\left(\frac{v}{v+bt^k/k}\right)^m\right\}\right].$$

The hazard rate is given as:

$$\mu(t) = \frac{\rho m}{v}bt^{k-1}\left(\frac{v}{v+bt^k/k}\right)^{m+1}.$$

6.4.3 The Weibull distribution derived from stable frailty

There is an interesting connection between the Weibull distribution and the stable frailty distribution. We start with a Weibull hazard $\alpha(t) = bt^{k-1}$, and using a stable frailty distribution we get the population survival function

$$S(t) = \mathcal{L}_{\text{Stable}}(bt^k/k; -m, a) = \exp(-a_1 t^{mk}),$$

where $a_1 = ab^m/k^m$ and $0 < m \leq 1$. This is again a Weibull distribution with hazard a_1mkt^{mk-1}; hence we have derived the interesting result of Hougaard (e.g., Hougaard, 2000) that *the Weibull distribution is preserved under stable frailty.*

Usually, the effects of frailty will only be felt after some time has passed, as illustrated in Figure 6.2. In the beginning, frailty will normally have very little effect. This is different for the stable frailty distribution, where the entire shape of the hazard rate is changed under frailty. For instance, let $k = 2$ and $m = 1/4$; then the original hazard is bt, and after frailty operating the hazard is proportional to $1/\sqrt{t}$. This phenomenon is due to the fact that the stable distribution has no expectation.

In fact, use of the stable distribution as a model for frailty goes back a long time; it was proposed by Lindsey and Patterson (1980) as a heterogeneity (in our terms, frailty) explanation of why one observes the so-called stretched exponential distribution. This distribution arises when $k = 1$, that is, the starting hazard rate is $\alpha(t) = b$, giving a population hazard rate a_1mt^{m-1}. The stretched exponential distribution, with this hazard rate, is very popular in physics and many other fields; for an excellent introduction, see Andersen et al. (2004).

6.4.4 Frailty and estimation

Statistical analysis of frailty models, where there is at most one event for each individual, will often be quite speculative due to lack of identifiability. There are exceptions from this, in particular when a parametric form can be postulated for the basic rate $\alpha(t)$. An application of this is given in the case study in Section 6.9. Usually, statistical analyses are more realistic when several events are observed for each individual, as discussed in Chapter 7.

Although frailty is usually seen as representing unobserved heterogeneity, there will in addition often be observed factors that create differences between individuals and that will naturally be considered in a regression analysis. One strategy that has been used is the combination of a Cox model and a frailty distribution, that is, a semiparametric regression model; see, for instance, Nielsen et al. (1992) and Zeng and Lin (2007). Although useful, this is technically complicated and it is in general far easier to use parametric models, for instance, with an underlying Weibull hazard rate.

One particular difficulty with the semiparametric model, as pointed out by Zeng and Lin (2007), may be to use counting process theory to prove asymptotic normality. However, as pointed out by Aalen in the discussion to Zeng and Lin (2007), one can start with a parametric model, for which there are no technical difficulties, and then likely use the sieve method (Bunea and McKeague, 2005) to handle a semiparametric situation.

6.5 The effect of frailty on hazard ratio

The comparison of groups is a basic element in epidemiology. It is, for instance, common to compare the mortality in different countries. One may look at, say, those who are 70 years old in two countries and compare their present-day mortality. An obvious, although often ignored, complication is that in one country the people of age 70 may be a much smaller proportion of their original birth cohort than in the other country. The chances of survival may have been very different and the groups that remain may not be comparable due to selection. This is the phenomenon one tries to understand in frailty theory. In fact, frailty gives rise to a number of somewhat surprising artifacts.

Consider a high-risk and a low risk group of individuals. The assumption of proportional hazards between two such groups is a common one in survival and event history analysis. The assumption has received an almost canonical status, despite the fact that it is mainly an assumption of convenience. If we assume proportionality to hold at the level of the individual hazard, what would then be observed at the population level? Typically, one will see that proportionality disappears. This well-known phenomenon has implications for the Cox model; it means that if one covariate is unobservable, or excluded from the model, then proportionality is destroyed. Hence the proportional hazards assumption is a vulnerable assumption. There are important exceptions to this, however, which we discuss in Section 6.5.4.

The discussion in this section is also relevant for the concept of relative risk in epidemiology. It is well known that various selection biases may hamper the interpretation of relative risk. Frailty theory can be seen as one way of making a mathematical theory for selection bias.

6.5.1 Decreasing relative risk and crossover

Assume that the basic hazard rates in two risk groups are $\alpha(t)$ and $r\alpha(t)$, respectively, where r is a number greater than 1; hence the latter group is the high risk group. With frailty variables Z_1 in group 1 and Z_2 in group 2, the simple proportional frailty model in (6.1) yields individual hazard rates equal to $Z_1 \alpha(t)$ and $Z_2 r\alpha(t)$, respectively.

Assuming that the frailty variables have the same gamma distribution with variance δ in both groups, the population hazard rates $\mu_1(t)$ and $\mu_2(t)$ may be derived from (6.8) to give the following ratio:

$$\frac{\mu_2(t)}{\mu_1(t)} = r\frac{1+\delta A(t)}{1+r\delta A(t)}. \tag{6.20}$$

This expression is decreasing in t, and if $A(t) \to \infty$ when t increases then the ratio approaches 1. Hence, the observed hazard ratio will be decreasing due to frailty.

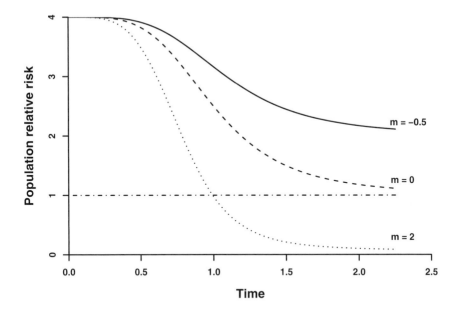

Fig. 6.7 *Hazard ratios for various PVF frailty distributions as given by (6.21). Parameters: $r = 4$, $v = 1$, $A(t) = t^4/4$.*

In this case the population hazard ratio will always remain above one, but with the more general PVF frailty, the ratio may cross below one.

If Z_1 and Z_2 both have the same PVF distribution, then the ratio of population hazards may be derived from (6.15) to yield:

$$\frac{\mu_2(t)}{\mu_1(t)} = r \left\{ \frac{v + A(t)}{v + rA(t)} \right\}^{m+1}. \tag{6.21}$$

In this case the hazard ratio will also decrease; eventually it reaches the limit r^{-m}.

For an illustration, see Figure 6.7. The limit is greater than 1 for the Hougaard distributions $(-1 < m < 0)$, it equals 1 in the gamma case $(m = 0)$, and it is less than 1 in the compound Poisson case $(m > 0)$. Hence, we have a *crossover*; that is the population relative risk falls below 1, even though, at the individual level, the relative risk is always above 1. Considering, for example, $m = 1$, the hazard ratio declines from 4 to 1/16. The reason for the crossover is the presence of a group with zero frailty in the compound Poisson case. In the high-risk group those with a positive frailty will "die out" faster than in the low-risk group, leaving the nonsusceptible ones to dominate. Hence, the crossover is quite reasonable.

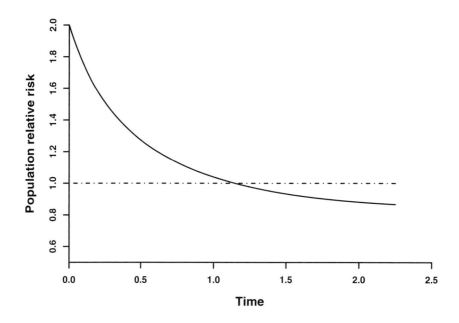

Fig. 6.8 *Hazard ratio with crossover as given by given by (6.22). Parameter: $\delta = 1$.*

Such crossovers may also occur when all individuals are susceptible if the relative risk at the individual level is declining. This could arise for biological reasons. Maybe the damage that a risk factor causes was most pronounced at young
ages and less pronounced at a more advanced age. To give an example, we shall
assume that the relative risk r is substituted by a function $r(t)$ that decreases but
is always above 1. This means that an individual in group 2 always has a higher
risk than a comparable individual in group 1. As a specific example, let $\alpha(t) = 1$
and $r(t) = 1 + e^{-t}$, that is, $r(t)$ declines from 2 towards 1. With gamma distributed
frailty this yields the hazard ratio

$$\frac{\mu_2(t)}{\mu_1(t)} = \frac{(1+e^{-t})(1+\delta t)}{1+\delta t + \delta(1-e^{-t})}. \tag{6.22}$$

This function is plotted in Figure 6.8 and is seen to cross below 1.

Connected to the previous example, one might ask for which $r(t)$ you would
actually observe proportional hazards on the population level. If, for instance, the
individual relative risk increases according to $r(t) = 2(1 + \delta t)$, then the population
relative risk is exactly equal to 2 for all t.

The crossover phenomenon has a close relationship to Simpson's paradox; for an
introduction to this paradox, see, e.g., Pearl (2000). In fact, the selection that occurs

over time creates a skew distribution with respect to the frailty variable, and this skewness is responsible for the paradoxical crossing effect, just like in Simpson's paradox.

6.5.2 The effect of discontinuing treatment

As a more practical example, consider a clinical trial comparing survival between two different treatment groups. There is often an issue as to how long the treatment should be continued. To evaluate this, one may look at the cumulative hazard rates and decide to discontinue treatment when they have become parallel for the two groups. How wise is this approach? Is it possible to conclude anything about the present treatment effect at time t by comparing $\mu_1(t)$ and $\mu_2(t)$? Here we assume that group 1 is treated, while group 2 is the comparison or control group, presumably with a higher risk if treatment is efficient. Assume the same model as in the previous subsection, with basic hazard rates $\alpha(t)$ in the treated and $r\alpha(t)$ $(r > 1)$ in the control group, with frailty variables having the same variance in both groups.

Assuming there is an immediate effect of stopping treatment, the hazard rate of the low-risk (or treatment) group would then jump up to that of the high-risk group, that is, group 1 would have hazard $\alpha(t)$ up to some time, t_1, and $r\alpha(t)$ after this time. Up to t_1 the ratio $\mu_2(t)/\mu_1(t)$ is found from (6.20). From (6.8), one may deduce the following ratio at a time $t > t_1$:

$$\frac{\mu_2(t)}{\mu_1(t)} = \frac{1 + \delta A(t_1) + r\delta(A(t) - A(t_1))}{1 + r\delta A(t)} = 1 - (r-1)\frac{\delta A(t_1)}{1 + r\delta A(t)}. \tag{6.23}$$

Note that this ratio is smaller than 1, and so the original treatment group will suddenly have a higher population hazard than the nontreatment group when treatment is discontinued. Hence, the changing effect of treatment cannot be directly discerned from the observed hazard rates.

To get a clearer picture, we have plotted the inverse ratio $\mu_1(t)/\mu_2(t)$ in Figure 6.9. One sees how the treatment effect starts with a halving of the hazard ratio corresponding to $1/r = 0.5$, with the treatment effect gradually diminishing due to frailty. When treatment is discontinued at $t = 1$, the treatment group suddenly gets a higher hazard than the control group, with the hazard ratio jumping to 1.5 before gradually declining towards 1. Hence, the effect of discontinuing treatment can be surprising and cannot be predicted without understanding the frailty effects.

Of course, the assumption made here that stopping treatment has an immediate effect is a bit unrealistic. However, one could easily modify this to handle the situation where the treatment effect fades out gradually after treatment is stopped. Anyway, the analysis presented here gives a good idea of qualitative effects that might arise.

So far we have assumed a constant effect of active treatment at the level of the basic hazard rate. A more realistic situation might be that the treatment effect is

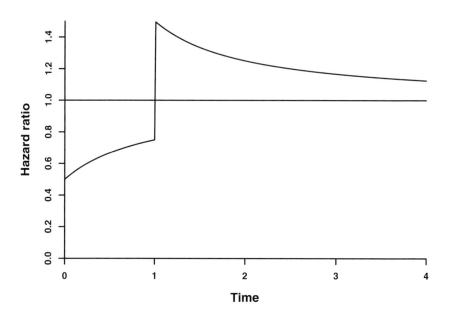

Fig. 6.9 *The hazard ratio $\mu_1(t)/\mu_2(t)$ between a treatment and a control group, when treatment is discontinued at time 1. Parameters: $\alpha(t) = 1$, $\delta = 1$, $r = 2$.*

gradually declining over time. In cancer, for instance, it is well known that many treatments may have an initially promising effect, which then declines. As an example, we shall assume that the basic hazard rate of the treated group is $\alpha(t)$ and in the control group $r(t)\alpha(t)$, with $r(t) = 1 + r(1 - t/5)$, for some $r > 0$ and $t \leq 5$. This risk function declines from $r + 1$ to 1 as t goes from 0 to 5. We can again compute the ratio between the population hazards, $\mu_2(t)/\mu_1(t)$. We do not give the formula here but just present a couple of figures. In Figure 6.10, the hazard ratio between the control and treatment group is shown at both the basic level and the observable population level, and the second one is seen to decline much faster and to cross the unit line. Hence, we have a crossover phenomenon, as discussed in the previous section.

Imagine now that one might be tempted to discontinue the treatment when the population hazard rate was equal in the two groups, that is, when the lower curve in Figure 6.10 crossed 1; this happens at $t = 2.32$. If the effect of treatment disappears abruptly at discontinuation, the observed population hazard ratio between the treated and control groups, will be as seen in Figure 6.11, that is, the hazard in the treatment group increases immediately to a higher level. Hence, equal population hazards in the treatment and control groups do not mean that the treatment does not work any

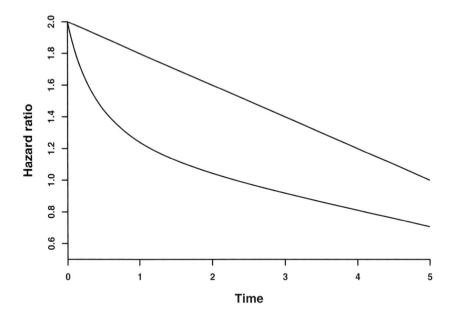

Fig. 6.10 *The hazard ratio between a control and a treatment group at the level of the basic hazard rate (upper curve) and the level of the population rate (lower curve), when* $r(t) = 1 + r(1 - t/5)$. *Parameters:* $\alpha(t) = 1, \delta = 1, r = 1$.

more. Again, one might be surprised, and clearly one cannot determine whether to continue the treatment by just looking at the observable hazards.

6.5.3 *Practical implications of artifacts*

Hazard ratios and other versions of relative risks play a fundamental role in clinical medicine and epidemiology. We now see that the interpretation of changes over time in these quantities is indeed difficult due to frailty selection and shows several artifacts. These are not just academic questions. It is often found that observed relative risk decreases over time. It is, for instance, well established that in young and middle-aged people high blood pressure or high cholesterol level increases coronary heart disease mortality. With increasing age, however, much of this effect seems to disappear; the risk factors are not strong predictors for old people (Rastas et al., 2006). This may be due to physiological change in individuals of advancing age. However, as pointed out by Jacobsen et al. (1992) and Rastas et al. (2006), it can also be a selection effect of the frailty type. Those individuals who are most vulnerable

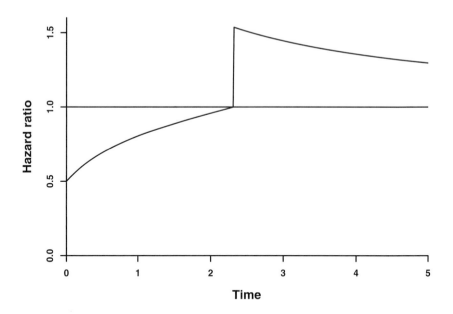

Fig. 6.11 *The hazard ratio $\mu_1(t)/\mu_2(t)$ between a treatment and a control group when $r(t) = 1 + r(1 - t/5)$. Treatment is discontinued when the hazard ratio reaches 1. Parameters: $\alpha(t) = 1$, $\delta = 1, r = 1$.*

to the effect of the risk factors may succumb relatively early, and the remaining ones may be less vulnerable. There are also several examples of crossover phenomena, for instance the "black/white mortality crossover" discussed by Manton et al. (1981), who suggest a frailty explanation. In medicine, there are "paradoxical" phenomena that may be frailty effects. For instance, Langer et al. (1989, 1991) found that for very old men survival improved with increasing diastolic blood pressure, although there might be several possible explanations of this; see Boshuizen et al. (1998) and Rastas et al. (2006).

An interesting application to the study of excess mortality among cancer patients has been made by Zahl (1997). Based on Norwegian data for malignant melanoma, he finds that from 15 years after cancer diagnosis, the patients actually have a lower mortality than the general population. This is a crossover phenomenon since initially the mortality rate is, of course, much increased, and Zahl fits frailty models to the data.

As pointed out by Vaupel et al. (1979) and Vaupel (2005), frailty considerations are of paramount importance when comparing mortality data for different periods. As an example, in Figure 6.12, we compare male population mortality in Norway for the year 1991 with the period 1901–5, that is, we have divided the mortality of

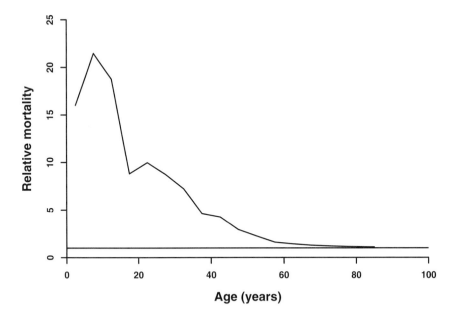

Fig. 6.12 *Relative mortality for Norwegian men in the period 1901–5 compared to 1991. The horizontal line indicates relative risk 1, corresponding to equal mortality. (Source: Statistics Norway.)*

the first period with the mortality of the second period, thereby computing a relative mortality. As expected, at younger ages the mortality at the beginning of the century was much higher than the mortality toward the end of the century, for example, a relative mortality up to 20 for young children. However, the difference seems to disappear entirely at the older ages. Superficially, this is strange since there has been a tremendous progress in social conditions and medical treatment throughout the century, also for elderly people. Therefore, a part of the decline in relative risk over age must be due to frailty. By the end of the century, one would expect that far more people had survived to old age than 90 years earlier, and so the elderly population in our days contain more frail people than it did in earlier times. This also makes contemporary comparisons between different countries in the world difficult if they have had a very different history regarding likelihood of survival.

6.5.4 Frailty models yielding proportional hazards

When comparing two groups in survival analysis, it is often assumed that the hazards are proportional; this is, for instance, the basic assumption in the Cox model. It is rather obvious that one would rarely think that all individuals in one group have the same fixed hazard, which is proportionally, say, higher than for all individuals in the other group. Rather, it is natural to think that in both groups there is a variation in hazard between individuals and that the proportionality assumption represents some kind of average comparison between the groups. The individual variation within each group may be described through frailty models. It is an interesting question which kind of frailty models might actually yield proportional hazards. Some results from Aalen and Hjort (2002) will be presented here.

Typically, frailty models for a high-risk and a low-risk group will not yield proportional hazards, as, for instance, shown in Section 6.5.1, where frailty implies converging population hazards. However, the overall picture is not as simple as this.

Firstly, as discovered by Hougaard (e.g., Hougaard, 2000), there is one family of frailty distributions, namely the stable one, which preserves proportionality. If the frailty has a stable distribution with Laplace transform given in (6.12), then the population hazard can be derived from (6.4) as $\alpha(t)aA(t)^{k-1}$. Following the setup of Section 6.5.1, we assume that frailty in the two groups has the same stable distribution, while the basic hazard rates are $\alpha(t)$ and $r\alpha(t)$, respectively, where $r > 1$. It follows that the ratio of population hazards is given as

$$\frac{\mu_2(t)}{\mu_1(t)} = r^k.$$

Hence the original proportionality constant r is transformed to another constant r^k. Since k is a number between 0 and 1, it follows that r^k is always less than r. For instance, if the frailty has a Lévy distribution as given in (6.13), then the population hazard ratio will equal $\sqrt{r}$.

One limitation of this result is that the nonnegative stable distributions are rather special, with very heavy tails and no expectation or variance existing. This does not preclude them from being useful in many cases. It would be nice if an extension of this result existed that could be applied to the Cox model, so that the proportional regression structure of this model was preserved if a covariate was unobserved or observed with measurement error. Such an extension of Hougaard's result does not appear to exist, however.

Interestingly, there is a second approach that yields proportionality in a natural way. In the model considered in Section 6.5.1, it has been presumed that the frailty distribution of the high-risk and low-risk groups coincide and that only the proportionality constant r distinguishes the groups. When considering the situation more closely, one may realize that this may not be the most natural setup. Rather, one would assume in many cases that the frailty structure of low-risk and high-risk groups are different. For instance, considering the risk of lung cancer for smokers and nonsmokers, one would not assume that the risk distribution of nonsmokers is simply a scaled-down version of the risk distribution for smokers. Biologically, the reason risk varies among individuals would be different in nonsmokers (genetical, environmental) from smokers (with amount of smoking playing an important role).

Therefore, we shall assume different frailty distributions in the two risk groups but common basic hazard rate $\alpha(t)$. From formula (6.4), it follows that proportional population hazards between two groups with frailty Laplace transform $\mathcal{L}_1(c)$ and $\mathcal{L}_2(c)$ will imply the following:

$$\alpha(t)\frac{-\mathcal{L}_2'(A(t))}{\mathcal{L}_2(A(t))} = k\,\alpha(t)\frac{-\mathcal{L}_1'(A(t))}{\mathcal{L}_1(A(t))}$$

for some constant k. Integrating on both sides, changing the variable, and using the side conditions $\mathcal{L}_1(0) = 1$ and $\mathcal{L}_2(0) = 1$ yields:

$$\mathcal{L}_2(c) = \mathcal{L}_1(c)^k. \tag{6.24}$$

In particular, this relation holds for a PVF family with $k = \rho$. More generally, it is valid for a Lévy-type distribution with Laplace transform defined in (6.16).

If Z_1 and Z_2 have Laplace transforms $\mathcal{L}_1(c)$ and $\mathcal{L}_2(c)$, which satisfy (6.24), then the following relationships hold:

$$E(Z_2) = k\,E(Z_1), \qquad \mathrm{Var}(Z_2) = k\,\mathrm{Var}(Z_1), \qquad \mathrm{CV}^2(Z_2) = \frac{1}{k}\,\mathrm{CV}^2(Z_1), \quad (6.25)$$

where CV^2 denotes the squared coefficient of variation. Hence, in the present frailty framework, proportional hazards implies proportional expectations and variances, while CV^2 is inversely proportional. This means that a high-risk group has a larger expected frailty, with a smaller coefficient of variation than a low-risk group. One way of thinking about the model defined by (6.24) with varying values of k, is that several more or less independent sources contribute to the frailty of an individual. The contribution may be from genetics, environment, lifestyle, and other influences. A high-risk individual can be perceived as having more sources contributing to the risk than a low-risk individual, that is, a higher value of the parameter k.

One natural choice is to let the low-risk group have a PVF frailty distribution with parameter ρ_1 and the high-risk group with parameter ρ_2 where $\rho_2 > \rho_1$, while the remaining parameters, ν and m, as well as the basic hazard rate $\alpha(t)$, are the same in both groups. As before, let the population hazard rates be $\mu_1(t)$ and $\mu_2(t)$ and let CV_1^2 and CV_2^2 be the squared coefficients of variation of the low-risk and high-risk groups respectively. One can then write

$$\frac{\mu_2(t)}{\mu_1(t)} = \frac{\mathrm{CV}_1^2}{\mathrm{CV}_2^2} = \frac{\rho_2}{\rho_1}. \tag{6.26}$$

This gives an interesting way of viewing the proportionality factor in a Cox model. As mentioned earlier, it is unlikely that there is actual proportionality at the individual level between two groups. Rather, one would think that there is a variation in risk in each group and that proportionality, if it happens to exist in a given case, emerges as some kind of average property. Equation (6.26) shows that in the present setting the proportionality factor is simply the fraction between the squared coefficients of variation of the risk distributions in the two groups.

Equation (6.26) may explain why in a Cox analysis one will usually not observe very large hazard ratios. Assume that the high-risk group has $CV_2^2 = 1$. As an example, let the hazard ratio be equal to 5; then the low-risk group has $CV_1^2 = 5$. Such large coefficients of variation imply very skewed distributions; see, e.g., Moger et al. (2004). In particular, looking at the compound Poisson case of the PVF distributions and letting ρ approach zero, the corresponding CV^2 increases, and the distribution will essentially become concentrated in two points (corresponding to 0 or 1 event in the compound Poisson process). This is what one would expect if a major gene was operating. In the absence of such strong genetic effects one would not expect to see very large hazard ratios; see also Aalen (1991) for related considerations.

In particular, one may consider a PVF family with $k = \rho$. Part of the individual variation may be explained through covariates, so it may be natural to put a regression structure on the parameter ρ. Let a vector covariate, $\mathbf{x}$, be given. A Cox model will then arise if ρ satisfies

$$\rho = \exp(\boldsymbol{\beta}^T \mathbf{x})$$

for a coefficient vector $\boldsymbol{\beta}$. We then assume that the covariates only influence the model through the parameter ρ.

6.6 Competing risks and false protectivity

An interesting phenomenon related to those discussed earlier, is the occurrence of *false protectivity* (Di Serio, 1997). This may arise when frailty effects are combined with competing risks. For simplicity, we consider only two competing risks, corresponding to the two events B and C, with hazard rates at the individual level equal to $Z_B \alpha_B(t)$ and $Z_C \alpha_C(t)$, and cumulative hazard rates $Z_B A_B(t)$ and $Z_C A_C(t)$; see the illustration in Figure 6.13. We will concentrate on event B. In the competing risks setting, individuals at risk for experiencing this event at time t are only those

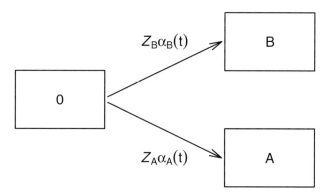

Fig. 6.13 *Competing risks model with frailty effects.*

for which neither event B nor C has happened before t. False protectivity may arise when the frailty variables Z_B and Z_C are correlated. A simple model for correlation has been considered by Yashin and Iachine (1995) and Zahl (1997). They postulate that the frailty variables are sums of suitable gamma distributed variables:

$$Z_B = Y_1 + Y_3, \qquad Z_C = Y_2 + Y_3.$$

Let us assume that Y_1, Y_2, and Y_3 are independent and gamma distributed with Laplace transforms $\mathscr{L}_1(c)$, $\mathscr{L}_2(c)$, and $\mathscr{L}_3(c)$, respectively. The joint Laplace transform is then given as

$$\mathscr{L}(c_1, c_2) = E\exp(-c_1 Z_B - c_2 Z_C) = \mathscr{L}_1(c_1)\mathscr{L}_2(c_2)\mathscr{L}_3(c_1 + c_2).$$

The probability of surviving both risks equals

$$\begin{aligned}
S(t) &= E\exp(-Z_B A_B(t) - Z_C A_C(t)) \\
&= \mathscr{L}(A_B(t), A_C(t)) \\
&= \mathscr{L}_1(A_B(t))\mathscr{L}_2(A_C(t))\mathscr{L}_3(A_B(t) + A_C(t)).
\end{aligned}$$

The probability of surviving to time t and then having event B occur in the small time interval $[t, t + dt)$ equals

$$E\{Z_B \alpha_B(t)\exp(-Z_B A_B(t) - Z_C A_C(t))\}dt = -\alpha_B(t)\mathscr{L}^{(1)}(A_B(t), A_C(t))dt,$$

where $\mathscr{L}^{(1)}(c_1, c_2)$ is the partial derivative of $\mathscr{L}$ with respect to the first argument. The rate, $\mu_B(t)$, at which event B occurs for survivors at time t is then found by dividing by $S(t)$, giving

$$\begin{aligned}
\mu_B(t) &= \alpha_B(t)\frac{-\mathscr{L}^{(1)}(A_B(t), A_C(t))}{\mathscr{L}(A_B(t), A_C(t))} \\
&= \alpha_B(t)\frac{-\mathscr{L}_1'(A_B(t))}{\mathscr{L}_1(A_B(t))} + \alpha_B(t)\frac{-\mathscr{L}_3'(A_B(t) + A_C(t))}{\mathscr{L}_3(A_B(t) + A_C(t))}. \quad (6.27)
\end{aligned}$$

The first part of this equation is the same as in (6.4), while the second part comes from the joint frailty component Y_3 of the two competing risks. The interesting thing to note here is that the second part also depends on $\alpha_C(t)$ (through its integral). Hence, any factor that influences the risk of C will, on the population level, also be seen to influence the risk of B even when $\alpha_B(t)$ is independent of this risk factor. This creates a false association, which can be quite confusing.

To get a clearer picture, let the Ys be gamma distributed, each with expectation equal to 1/2, and with variances δ_1, δ_2, and δ_3 respectively. (This ensures that the expected frailty is 1.) This yields

$$\mathscr{L}_1(c) = (1 + 2\delta_1 c)^{-1/(4\delta_1)}$$

and similarly for the other Laplace transforms. From (6.27), it follows that

$$\mu_B(t) = \alpha_B(t) \left\{ \frac{1}{2 + 4\delta_1 A_B(t)} + \frac{1}{2 + 4\delta_3 (A_B(t) + A_C(t))} \right\}. \tag{6.28}$$

Assume now that there is a covariate that has a causal influence on event C, but not on B. More precisely, say that $\alpha_C(t)$ increases positively with the covariate, but that $\alpha_B(t)$ is not influenced by it. From (6.28) it is apparent that the population hazard rate of B is then negatively dependent on the covariate. Hence, there is a false impression of a protective effect of the covariate on event B; this is just what Di Serio (1997) calls false protectivity.

In the medical literature, the kind of bias studied here is termed "differential survival bias." There is medical evidence showing that cigarette smoking has a protective effect against Parkinson's disease and Alzheimer's disease. It has been suggested that this is an example of false protectivity, but the issue is not at all clear. For discussions of this in the medical literature, see Morens et al. (1996) and Almeida et al. (2002).

6.7 A frailty model for the speed of a process

Until now, we have looked at frailty as modifying the hazard rates of events. More generally, we might study processes of a more complex nature and ask whether the speed varies between individuals. We use a simple Markov model, as shown in Figure 6.14 where the constant transition intensities are multiplied by a frailty variable Z. By the results of Section A.2.5, we have the following transition probabilities conditional on the frailty variable:

$$P_{00}(t \mid Z) = e^{-Z\alpha_1 t}, \quad P_{01}(t \mid Z) = \frac{\alpha_1}{\alpha_1 - \alpha_2}(e^{-Z\alpha_2 t} - e^{-Z\alpha_1 t}), \quad P_{11}(t \mid Z) = e^{-Z\alpha_2 t}.$$

Integrating out the frailty variable as previously, we get:

$$P_{00}(t) = \mathscr{L}(\alpha_1 t), \quad P_{01}(t) = \frac{\alpha_1}{\alpha_1 - \alpha_2}(\mathscr{L}(\alpha_2 t) - \mathscr{L}(\alpha_1 t)), \quad P_{11}(t) = \mathscr{L}(\alpha_2 t).$$

The conditional hazard rate of entering state 2 at time t, given that the individual was present in state 1 at time $s < t$, is:

Fig. 6.14 *A series Markov model with two transitions and transition intensities with a common frailty.*

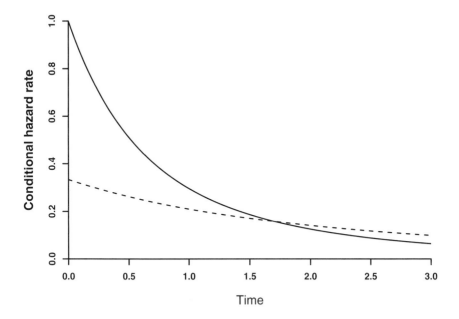

Fig. 6.15 *The function $\mu_2(t - s \mid s)$ with parameters $\alpha_1 = 1$, $\alpha_2 = 1$, and $\delta = 1$, and for two choices of the first transition time: $s = 1$ (curve starting at level 1) and $s = 5$.*

$$\mu_1(t - s \mid s) = \frac{E(P_{01}(s \mid Z)P_{11}(t - s \mid Z)\alpha_2 Z)}{P_{01}(s)} = \alpha_2 \frac{\mathscr{L}'(\alpha_1 s + \alpha_2(t - s)) - \mathscr{L}'(\alpha_2 t)}{\mathscr{L}(\alpha_2 s) - \mathscr{L}(\alpha_1 s)}.$$

One could also be interested in the conditional hazard rate of entering state 2 at time t, given that the individual entered state 1 at time $s < t$, which is:

$$\mu_2(t - s \mid s) = \frac{E(P_{00}(s \mid Z)\alpha_1 Z P_{11}(t - s \mid Z)\alpha_2 Z)}{E(P_{00}(s \mid Z)\alpha_1 Z)} = \alpha_2 \frac{-\mathscr{L}''(\alpha_1 s + \alpha_2(t - s))}{\mathscr{L}'(\alpha_1 s)}.$$

Assuming gamma frailty and using the Laplace transform in (6.6), we get

$$\mu_2(t - s \mid s) = \alpha_2 \frac{1 + \delta}{1 + \delta \alpha_1 s} \left(\frac{1 + \delta \alpha_1 s}{1 + \delta(\alpha_1 s + \alpha_2(t - s))} \right)^{2 + 1/\delta}.$$

Note that the limit is α_2 when the frailty variance δ goes to zero, just as it should be.

The function $\mu_2(t - s \mid s)$ is illustrated in Figure 6.15. One sees a strong frailty effect on the rate for further transition according to when the first transition took place. When the first transition took place early, the hazard rate of the second transition is much higher than when the first transition took place late.

A general study of a Markov chain with random speed is given in Aalen (1988a).

Results of this type may be of considerable medical interest; e.g., Kopans et al. (2003) published a paper in the journal *Cancer* in which they attempted to explain the fact that young women who get breast cancer tend to have more aggressive disease with faster-growing tumors than older women. They suggested that some tumors grow more rapidly than others, and that these tumors become clinically detectable at a younger age. Hence, younger women would be expected to have faster-growing tumors. A related result has been demonstrated in the aftermath of the Chernobyl nuclear accident in 1986. Williams and Baverstock (2006) show how thyroid tumors occurring early have been more clinically aggressive than those occurring later. Considerable individual variation in the rate of cancer growth has also been demonstrated in the context of breast cancer screening; see Weedon-Fekjær et al. (2007).

6.8 Frailty and association between individuals

When there is a variation between individuals as expressed by a frailty distribution, one should think that there is also an association between related individuals. They would be expected to have similar frailty, and hence if one individual has an event then the likelihood of this occurring to a related one should be increased. This idea is carried further in Chapter 7, especially in the shared frailty model, but some preliminary considerations shall be made here.

The probability of an event happening over a time interval (t_1, t_2) in the frailty model (6.1) for given frailty Z is expressed as:

$$p(Z) = 1 - \exp(-Z(A(t_2) - A(t_1))).$$

Consider two individuals with, possibly correlated, frailties Z_1 and Z_2, and such that conditional on Z_1 and Z_2 the probability of occurrence of the event is independent between individuals. Let the events B_1 and B_2 indicate that the event happens for the respective individuals. One may compute

$$R = \frac{P(B_2 \mid B_1)}{P(B_2)} = \frac{P(B_1 \text{ and } B_2)}{P(B_1)P(B_2)} = \frac{\mathrm{E}(p(Z_1)p(Z_2))}{\mathrm{E}(p(Z_1))\mathrm{E}(p(Z_2))}$$

$$= 1 + \mathrm{cor}(p(Z_1), p(Z_2))\,\mathrm{CV}(p(Z_1))\,\mathrm{CV}(p(Z_2)),$$

where cor denotes correlation and CV the coefficient of variation. The left-hand side, $R = P(B_2|B_1)/P(B_2)$, is a kind of relative risk, albeit the conditional probability is divided by the marginal one, which is most useful here. This formula brings out the fact that a high-risk association, expressed by R, between two individuals depends on two conditions. On the one hand, there should be a high correlation between the risk functions $p(Z_1)$ and $p(Z_2)$. On the other hand, the coefficients of variation of the risk functions should be rather large, that is, the risk levels should vary a good deal in the population as a function of the factor considered. Even a perfect correlation of 1 is not sufficient to ensure a high value of R if the variation in risk in the population is small. In fact, even a coefficient of variation of 1 is quite high corresponding to a considerably skewed distribution of $p(Z)$. Hence, two individuals with identical frailty, for example, identical twins, if the risk is determined solely by genes, might easily have an R less than 2. That is, if the event occurs for one individual, the risk of the second may have just a modest increase. In cancer, one often sees relative risk within families around 2, which would be interpreted as possibly indicating a modest genetic effect. However, it might easily be consistent with a strong genetic effect.

Again we see that frailty considerations produce surprising results, and that one should understand the variation in risk between individuals and bring that into consideration. Similar results to the ones presented here are discussed in Vaupel and Yashin (1999); see also Aalen (1991).

6.9 Case study: A frailty model for testicular cancer

We shall give an application of the theory to the analysis of incidence of testicular cancer (Example 1.8). When attempting to fit a frailty model to univariate survival data, one will usually be confronted with an identifiability problem. It is obvious that many different frailty models can result in a given population hazard rate. The simplest case would be to assume no frailty and let the individual and population hazard rates coincide. In general this identifiability problem can only be avoided in the multivariate case where several events are observed for an individual or a group of related individuals (cf. Chapter 7).

However, in some situations one can make assumptions that allow a model to be estimated. A key question is to make reasonable assumptions about the basic rate $\alpha(t)$. When studying the incidence rate of cancer, one is helped by the fact that there exists a mathematical theory of carcinogenesis. Starting with the Armitage-Doll multistage model of the early 1950s (Armitage and Doll, 1954), a number of more sophisticated recent models have been made (Kopp-Schneider, 1997; Portier et al., 2000). These models seem to imply that the basic rate can, at least approximately, be assumed to be of Weibull type, that is, $\alpha(t) = bt^{k-1}$. In particular, Armitage & Doll's model implies that k is the number of stages in the multistage model. It is important to notice that the carcinogenesis models operate on the individual level. If they shall be applied to populations, a frailty effect has to be included. We shall here look at the case of testicular cancer.

In fact, the frailty effect is often ignored when making deductions from cancer incidence. Here, as elsewhere, it is important to realize that the hazard rate does not say anything directly about the development at the individual level, but it is mixed up with selection effects.

Testicular cancer has a particular age distribution compared to other types of cancer. The disease is rare before 15 years of age, but after puberty there is a strong increase in incidence, until it reaches the highest level in the age group 30–34 years. The incidence then falls sharply with age. In addition, there has been a strong increase in testicular cancer over time; in Norway the incidence rate has increased at least threefold over the last 35 years (Aalen and Tretli, 1999). The importance of understanding these features has been emphasized in the epidemiological literature. For example, Klotz (1999) formulates a number of "outstanding epidemiological issues" of testicular cancer epidemiology. One of those is: "Why is the incidence of testicular cancer highest among young adults, with little risk among older men, even though spermatogenesis is a lifelong process?" We shall present an analysis given by Aalen and Tretli (1999).

For testicular cancer it has been suggested that variation between individuals in risk, or frailty, may be created early in life, and that the risk of testicular cancer is determined by events in fetal life. As suggested by Klotz (1999) the carcinogenic effect may be due to "estrogen-mimicking chemicals" acting as "hormonal disrupters transplacentally to interfere with gonadal development in utero." A number of pollutants may have this "estrogen-mimicking" effect, including DDT, PCBs, nonylphenol, bisphenola, and vinclozolin.

Hence, some individuals may be damaged in fetal life, and this damage may be required to develop testicular cancer. This is the basis for our frailty model. The damaged individuals are the frail ones, with the amount of frailty varying between individuals, while those who are not damaged are not prone to develop testicular cancer. The frail proportion is allowed to change with calendar time, to account for the observation of an increased incidence, which may be due to increasing pollution with the suspected chemicals.

We shall use a compound Poisson frailty distribution where parameters m and v of the underlying gamma distribution are assumed constant over birth cohorts, while the parameter of the underlying Poisson distribution, ρ, is dependent on the birth cohort. This is, of course, a simplified model, but the data do not allow too many degrees of freedom in the specification of the model.

Incidence data have been collected by the Cancer Registry of Norway. In the statistical analysis, all testicular cancer reported between 1953 and 1993 were included and divided in two groups according to whether they are seminomas or nonseminomas (which are two different types of tumors). The data are divided into 19 five year birth cohort intervals, starting with 1886–90 and going up to 1976–80, and 13 age intervals, 0–14 years, 15–19 years, 20–24 years, …, 65-69 years, and 70 years and above. Let O_{ij} be the number of observed testicular cancer cases in birth cohort i and age group j, and let R_{ij} be the number of person-years at risk. The partition points of the age intervals, starting with 15 years and going up to 70 years, are denoted by $t_1, \ldots, t_{12}$. We exclude the age interval 0–14 years, and assume that time $t = 0$ in the frailty model is the 13th birthday of the individual.

The likelihood function, based on a Poisson regression model (see Section 5.2), is given as

$$L = \prod_{i=1}^{i=19} \prod_{j=2}^{j=13} \mu_{ij}^{O_{ij}} \exp(-\mu_{ij}). \qquad (6.29)$$

Here, μ_{ij} is the expected number of cases, defined as the average hazard rate per year for birth cohort i and age interval j multiplied by the number of person-years:

$$\mu_{ij} = R_{ij}\{\log(S(t_{j-1} - 13) - \log(S(t_j - 13)\}/5.$$

The logic behind this is that the natural logarithm of the survival function is the cumulative hazard rate, so that the expression following R_{ij} becomes the average hazard rate. The survival function $S(t)$ is defined by (6.14), which shows the dependence on the parameters. In our case, $A(t) = t^k/k$, and the parameter b in the frailty model is subsumed in the parameter $1/v$.

Parameters to be estimated are the Weibull shape parameter k, and the scale and shape parameters v and η of the underlying gamma distributions, as well as the Poisson parameters ρ_i, which depend on the birth cohort. A penalized likelihood term was used to account for changes in the latter parameter over calendar time. The idea is that each cohort has its own specific value, that is ρ_i for cohort i. In order to smooth the variation in the ρ_is and prevent them from varying too much, one introduces a term into the likelihood that becomes large if neighboring ρ_is are too dissimilar. More precisely, the following penalization term is subtracted from the log-likelihood (we define $\rho_0 = 0$):

$$\frac{1}{2}\sigma \sum_{i=1}^{19} (\rho_i - \rho_{i-1})^2.$$

The parameter σ determines the extent of smoothing to be applied. In our analysis we put $\sigma = 4$.

Two types of testicular cancer, seminomas and nonseminomas, are distinguished, and parameter estimates are shown in Table 6.1. The estimated parameter k is very close to 4 for the seminomas and very close to 3 for the nonseminomas, indicating according to the multistage model that, respectively, four and three changes are required before cancer develops. Multistage interpretations of epidemiological findings are with necessity somewhat speculative, but the difference between the two histological groups is of interest.

Proportions of susceptibles, $1 - \exp(-\rho_i) \approx \rho_i$ as a function of birth cohort, are shown in Figure 6.16 for seminomas and nonseminomas separately. The strong increase in incidence of testicular cancer is clearly documented. Figures 6.17 and 6.18 show observed and expected incidence rates. Clearly, the fit is very good, although this does not in itself prove the validity of the model.

Table 6.1 *Maximum lilkelihood estimates of parameters in the frailty model.*

Parameter	Seminomas			nonseminomas		
	Gamma		Weibull	Gamma		Weibull
	v	η	k	v	η	k
Estimate	9.1×10^4	0.37	4.06	1.0×10^3	0.43	2.95
Standard error	3.6×10^4	0.12	0.21	0.24×10^3	0.12	0.18

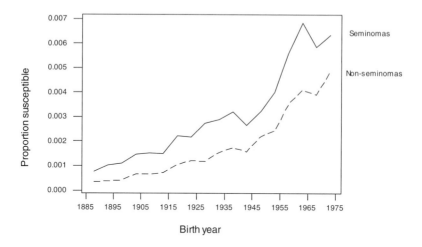

Fig. 6.16 *Plot of ρ as a function of birth cohort for seminomas and nonseminomas. Figure from* Aalen and Tretli (1999).

In this case study, the frailty point of view is founded in biological theory and so is not just a theoretical construct. Still, it is always better to estimate a frailty model when multivariate survival data are available. For instance, it is quite clear that association within families would be very interesting for saying something about individual variation in risk, as one might expect that the risk is to some extent mediated through genetic properties. It turns out that there is a familial clustering in testicular cancer, and this will be studied by a multivariate frailty model in Section 7.5.

6.10 Exercises

6.1 Prove the moment formulas (6.10) of the PVF distribution.

6.2 Prove the formulas in (6.25).

6.3 Calculate the hazard function in (6.4) for the various special cases in Section 6.2.3.

6.4 Show that $-\mathscr{L}'(c)/\mathscr{L}(c)$ is always a decreasing function for a general Laplace transform of a nonnegative distribution. This function, with a special value of c, is a

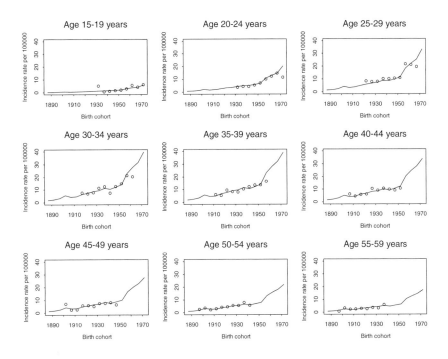

Fig. 6.17 *Testicular cancer: Expected (solid curve) and observed incidence over birth cohorts by age group.*

factor in (6.4). [*Hint:* Prove it first for special cases of $\mathscr{L}(c)$, and try then to prove it for a general Laplace transform.]

6.5 Prove formula (6.19). [*Hint:* Compute the integral by using the transformation $v = bt^{k+1}/\{(k+1)/\delta + bt^{k+1}\}$.]

6.6 Derive formula (6.23).

6.7 In order to get to know the various frailty distributions in the PVF family it is a good idea to plot them for various parameters. Plot the compound Poisson distribution when $m = 1$, the inverse Gaussian distribution, and the Lévy distribution (stable distribution with $m = -1/2$), and compare them. [The modified Bessel function of the first kind may be computed by several programs, e.g., Mathematica, Matlab, or the fOptions package of R.]

6.8 The Laplace transform for survivors in formula (6.18) is also valid for the gamma distribution by letting $\rho \to \infty$ and $m \to 0$. Derive this limit. By considering the density of the gamma distribution, explain how this density changes for survivors when time increases.

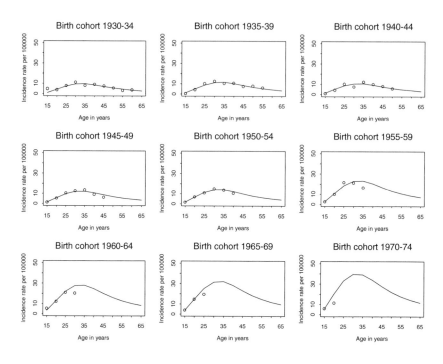

Fig. 6.18 *Testicular cancer: Expected (solid curve) and observed incidence over age groups by birth cohort.*

Chapter 7
Multivariate frailty models

Data in survival analysis are usually assumed to be univariate, with one, possibly censored, lifetime for each individual. All the standard methodology, including Kaplan-Meier plots and Cox analysis, is geared toward handling this situation. However, multivariate survival data also arise naturally in many contexts. Such data pose a problem for ordinary multivariate methods, which will have difficulty handling censored data.

There are two typical ways multivariate survival data can arise. One, which may be termed the *recurrent events* situation, is when several successive events of the same type are registered for each individual, for instance, repeated occurrences of ear infections. The other, which may be termed the *clustered survival data* situation, is when several units that may fail are collected in a cluster. Examples of the clustered survival data situation may be the possible failure of several dental fillings for an individual or the lifetimes of twins. The cluster structure may in fact be rather complex, including, for instance, related individuals in family groups. Sometimes one would assume common distributions for the individual units in a cluster; in other cases, like when considering the lifetimes of father and son, the distributions may be different.

For recurrent events one must make assumptions about the underlying stochastic process. We shall consider two main situations: first, the one where, given the frailty, the process is of renewal type. By this we mean that the times between consecutive events for a particular individual are independent random variables with identical distributions. This case is very similar to the clustered data situation, since the interevent times are identically distributed conditional on having the same frailty. Second, we consider the case where the underlying process given the frailty is Poisson. Unconditionally, this is a doubly stochastic Poisson process (or a Cox process).

An example of recurrent events of the renewal type is the data of Example 1.11 on the cyclic pattern of motility (spontaneous movements) of the small bowel in humans. We will analyze these data in Example 7.1. Recurrent events of the Poisson type are given by the bladder tumor data of Example 1.10, which are analyzed in Example 7.3. The clustered data situation may be exemplified by the data in Example 1.9 on the duration of amalgam fillings in teeth. Each patient has a number of fillings that may fail, and one can imagine that some patients have a larger risk of failure than others due to varying dental hygiene and other factors. We are going to analyze this data set in Example 7.2.

There are four basic approaches to analyzing multivariate survival data:

- Marginal models
- Frailty models
- Random effects models for transformed times
- Dynamic models

The marginal models are very pragmatic and do not attempt to yield any insight into the deeper structure of the processes. In marginal models the focus is on just parts of the available data, but they may still be quite useful for evaluating, say, treatment effects. On the other hand, the frailty, random effects and dynamic models are supposed to give more realistic descriptions of the full data set. In this chapter we consider the frailty and random effects models, while marginal and dynamic models are treated in Chapter 8. We shall consider mainly parametric models, in fact we think parametric modeling is often the best approach for these complex data.

We cannot cover all aspects of multivariate frailty modeling here. A more comprehensive introduction to multivariate survival data from a frailty point of view is given in the book by Hougaard (2000). Furthermore, one should note the close relationship to Bayesian models where prior distributions on hazard rates constitute a close analogy to the frailty distributions (in spite of the former being viewed as subjective and the latter as objective). Such Bayesian models have been used extensively for analyzing geographical variation. Osnes and Aalen (1999) analyze how cancer survival varies over municipalities in Norway where geographically close municipalities are assumed to have similar hazard rates, but this will not be treated here.

7.1 Censoring in the multivariate case

In the multivariate case, censoring takes on a new dimension because it may depend on previous events for the same individual, and also possibly on previous events for other individuals. The nature of the censoring process is important for writing up likelihood functions, or for any other statistical analysis. One must note that censoring usually refers to the actual time scale (the unique time of the real world), while in the statistical analysis, time scales are often reshuffled such that, for instance, the starting point for each individual is considered as time zero. If censoring is influenced by decisions being taken or by occurrences in other processes, then these influences must follow the actual real time. This is a complication that we return to briefly in Section 7.1.2.

Previously, censoring was discussed in Sections 1.4 and 2.2.8 and the basic assumption of independent censoring is still in charge. We shall continue here with more discussion of the multivariate case.

7.1.1 Censoring for recurrent event data

The small bowel motility data (Example 1.11) are well suited for discussing censoring. In the particular study of Aalen and Husebye (1991), censoring occurred at
a fixed time, so the situation is relatively simple. Other studies in the same field,
however, use different censoring schemes. In one study, observation continues until
a fixed number of complete periods have been observed (Ouyang et al., 1989); see
also (iii) below. In another study observation of an individual is stopped at the first
event after a fixed time (Kellow et al., 1986), see also (v) below. A problem in some
studies of small bowel motility, and certainly in other fields as well, is that the rules
for stopping observation are not clearly described, indicating a lack of awareness
of the importance of this aspect of the design. We shall discuss this relative to the
natural time scale on which the events unfold.

Imagine a number of individual processes, censored at possibly different times.
Some possible examples of censoring schemes are as follows:

  (i) The censoring times are fixed in advance, for instance, by the end of a prede-
      termined observation period.
 (ii) The censoring times are random variables independent of the event processes.
      Each individual may have a separate censoring time, or there may be one in
      common for all.
(iii) Each process is censored when a fixed number of events have occurred.
(iv) Each process is censored either after a fixed number of events or when the
      observation time exceeds a certain limit, whichever comes first.
 (v) Each process is censored at the first event occurring after a fixed time.

There is an important distinction between the first two and the last three examples. In the latter ones, the censoring time depends on the development within the
individual processes themselves, thus creating a seemingly more complicated situation. In practice, however, the statistical analysis can be performed in basically
the same manner for all these situations. However, this is only because all these
censoring times are stopping times (see Section 2.2.8).

In practice, censoring is not always performed at stopping times. Assume, for
instance, that the study ends at a fixed time, resulting in incomplete observation
intervals at the end. It might be tempting to simply ignore these in order to achieve
a simple analysis. This would amount to stopping observation at the end of the last
complete renewal interval for each individual. However, this is not a stopping time in
the martingale sense, because one looks into the future before one decides to stop.
It is also intuitively clear that such a procedure would bias the estimated average
length of the inter-event times downward, since a long interval is more likely to
contain the censoring time than a short one (this phenomenon is called *length bias*).
Nevertheless, ignoring the incomplete intervals has been a common practice in small
bowel motility studies; see the discussion in Aalen and Husebye (1991), although
sometimes, of course, the numerical effect of this may not be very great.

The start of observation should also take place at a stopping time. In the small
bowel motility example, for instance, often the registration for an individual starts in

the middle of a cycle. Commonly, one will then wait for the first event and ignore the incomplete period before this event. This is quite satisfactory since the first observed event is a stopping time. Note the asymmetry of the situation: *an incomplete period may safely be ignored at the start but not at the end of observation.* (Although it gives no bias, ignoring the incomplete interval at the start still entails a slight loss of information. It can be included in the analysis, of course, but that is usually a slight complication.)

When censoring occurs at stopping times, it may be shown that the likelihood is the same as if censoring had occurred at some fixed time for each individual. Precise mathematical statements of this in the context of event history analysis may be found in Andersen et al. (1993). A similar result is also well known from general theory concerning likelihood-based statistical methods; see for instance Lee (1989, chapter 7).

7.1.2 Censoring for clustered survival data

For clustered data, many of the same considerations are valid. The main difference is that not only the past of a single unit may influence the censoring, but the past of several units belonging to a cluster. One difficulty may arise here, namely that the events are in fact occurring on one time scale but being analyzed on another. Consider, for instance, the amalgam fillings data. The fillings are inserted at different times, but when the data are being analyzed the insertion time of each filling is set to zero. If censoring is dependent on previous events on the original (true) time scale, then one may discover that censoring depends on future events in the time scale of the analysis. This issue of several possible time scales may also arise in the recurrent event situation, for example, when considering a renewal process. Therefore, a stopping time at the original scale may not be a stopping time at the scale of analysis, and so independent censoring on one time scale may not be so on another scale. Probably, this does not invalidate the estimates or tests, but the martingale assumption that plays such an important role in our derivations may not hold any more.

In fact, for parametric models there will be no difficulty since the likelihood can always be given in the original time scale where censoring is independent. This is discussed by Andersen et al. (1993, example X.1.9) and denoted *Arjas' real-time approach.* Since in parametric models one will simply maximize the likelihood, the statistical analysis is still valid.

The situation is more difficult for the non- and semiparametric analyses, for example, the Kaplan-Meier and Nelson-Aalen estimators or Cox regression, which are not computed on the original (true) time scale, but on another time scale where censoring may not be independent. For these estimators, martingale arguments play an important role and time scales cannot simply be changed. However, one can always approximate through parametric models and thereby justify the estimators; thus exploiting the previously mentioned fact that for parametric models there are no particular difficulties. One then divides the time scale into a number of intervals and replaces the continuous hazard with a step function that is constant on each interval, so that the estimation problem consists in estimating successive constant hazard rates, which is an easy task. The Nelson-Aalen estimator is then the limit of a cumulative hazard estimator when the interval lengths go to 0, and similarly for the Kaplan-Meier estimator. The same argument holds for nonparametric tests, and so the validity of the estimators and tests should be no problem as long as censoring is independent on the original time scale. The martingale property of the estimators may still not hold, but asymptotic results may in principle be derived via the parametric models. Of course, we recognize that some theoretical justification would be needed for these tentative suggestions. Essentially, we are proposing a sieve argument; see, for instance, Bunea and McKeague (2005) for an introduction to sieve theory.

7.2 Shared frailty models

As indicated in the introduction to this chapter, from a modeling point of view we may consider two different situations for multivariate survival data. In this section we consider the situation where several survival times are related to each other as either multiple units in a cluster or inter-event times in a renewal process. In Section 7.3 we consider models for recurrent event data of the Poisson process type.

As shown in Chapter 6, frailty models illuminate a number of important aspects of univariate survival data. When turning to multivariate data, the proportional frailty model assumes even greater importance as a very useful tool for handling such data in the presence of censoring. Basically, frailty models in this context are random effects models, analogous to those well known from linear normal model theory. However, the frailty models are better adapted to handle censored data than the normal models, although these may also be of use, as we shall see in Section 7.6. The dependence may be modeled through a frailty variable, such that all survival times that are related to each other (as either multiple units in a cluster or inter-event times in a renewal process) have the same level of frailty attached to them.

Such an approach to modeling dependence accommodates censored data very well and yields quite simple analyses. While for univariate data one will typically have an identifiability problem when attempting to estimate frailty models, there will generally be no such problem in the multivariate case.

It is assumed that to all survival times that are related to each other (as either multiple units in a cluster or inter-event times in a renewal process) there corresponds one fixed value of the frailty variable Z. Hence frailty measures the specific risk level for a cluster or an individual renewal process, and given Z the survival times are independent. In addition, we assume the proportional frailty model where the hazard rate for a survival time is given as $Z\alpha(t)$, where $\alpha(t)$ is the basic rate; see (6.1). This model is often called the shared frailty model, meaning that the same frailty is shared by all the survival times pertaining to an individual renewal process or a cluster.

We shall first derive the joint distribution for a shared frailty model. Then we derive the likelihood function for a set of multivariate survival data.

7.2.1 Joint distribution

We shall first develop the joint distribution of a set of survival times relating to a given renewal process or cluster. We consider a slightly more general model than in the previous subsection, allowing the basic rate to differ between the survival times. Consider n survival times $T_1, \ldots, T_n$ related to a single renewal process or cluster, with basic rates $\alpha_1(t), \ldots, \alpha_n(t)$ and frailty variable Z, so that the survival times are independent given Z. By formula (6.3), the marginal survival function of T_i is given by $S_i(t) = \mathscr{L}(A_i(t))$, where $A_i(t) = \int_0^t \alpha_i(u)du$ and $\mathscr{L}(c)$ is the Laplace transform of Z. By the same argument as used for that formula, it follows that the joint unconditional survival function of $T_1, \ldots, T_n$ is:

$$S_{\text{tot}}(t_1, \ldots, t_n) = P(T_1 > t_1, \ldots, T_n > t_n) = \mathscr{L}(A_1(t_1) + \cdots + A_n(t_n)).$$

Note that we can write

$$\mathscr{L}^{-1}(S_{\text{tot}}(t_1, \ldots, t_n)) = \mathscr{L}^{-1}(S_1(t_1)) + \mathscr{L}^{-1}(S_2(t_2)) + \cdots + \mathscr{L}^{-1}(S_n(t_n)),$$

which is the definition of an *Archimedean copula* (e.g., Hougaard, 2000, section 13.5). Such copulas enjoy nice properties and are popular in statistical analysis. Using, for instance, the gamma Laplace transform (6.6), we have:

$$\mathscr{L}(c) = \{1 + \delta c\}^{-1/\delta}, \quad \mathscr{L}^{-1}(y) = (y^{-\delta} - 1)/\delta.$$

Here, the joint survival function is simply:

$$S_{\text{tot}}(t_1, \ldots, t_n) = \{1 + \delta(A_1(t) + \cdots + A_n(t))\}^{-1/\delta}.$$

The shared frailty model also has nice properties with respect to censoring, as shown in the next subsection.

7.2.2 Likelihood

The likelihood function is quite easily derived for shared frailty models. If one makes parametric assumptions, a standard maximum likelihood analysis can be performed. While in the previous subsection we considered complete survival times, we now allow them to be censored, using the notation T for the actual survival time and $\widetilde{T}$ for the censored time.

Consider m independent renewal processes or clusters, and for renewal process or cluster i, let $\widetilde{T}_{ij}$, $j = 1,\ldots,n_i$, denote the observation times for the n_i units, or the inter-event times of the n_i recurrent events. These observation times may be the actual survival times or they can be censored survival times. The basic rate $\alpha(t)$ is a fixed function and is supposed to be the same for all i, but each renewal process or cluster has a separate independent frailty variable Z_i. Let $\mathbf{Z}$ denote the vector of the Z_is. All frailty variables are identically distributed; let the generic variable be denoted Z.

Furthermore, let D_{ij}, $j = 1,\ldots,n_i$, be binary variables that are equal to 1 if survival time (i, j) is uncensored and equal to 0 if it is censored. Let $D_{i\bullet} = \sum_j D_{ij}$ denote the number of uncensored survival times for renewal process or cluster i, and let $V_i = \sum_j A(\widetilde{T}_{ij})$, with $A(t) = \int_0^t \alpha(s)ds$, be the corresponding total cumulative hazard.

Following Andersen et al. (1993, section IX.3) we make the following assumption on the censoring: conditional on $\mathbf{Z} = \mathbf{z}$ censoring is independent and noninformative on $\mathbf{z}$. The independent censoring assumption is the well-known one, but in addition it is clear that censoring should not contain any information on the frailty variable; otherwise the nature of the frailty as something completely unobserved would be destroyed.

Define the history of renewal process or cluster i as $H_i = (\widetilde{T}_{ij}, D_{ij}, j = 1,\ldots,n_i)$, and let $P(H_i)$ denote the joint distribution of these variables. The contribution to the likelihood from renewal process or cluster i may be derived as follows. Conditional with respect to Z_i, the survival times are independent with intensity $Z_i\alpha(t)$ and their conditional likelihood can be deduced from formula (5.4):

$$P(H_i \mid Z_i) = \prod_{j=1}^{n_i} \left[(Z_i\alpha(\widetilde{T}_{ij}))^{D_{ij}} \exp(-Z_iA(\widetilde{T}_{ij})) \right]. \qquad (7.1)$$

The unconditional likelihood follows by taking the expectation with respect to the frailty variable Z:

$$P(H_i) = \prod_{j=1}^{n_i} \left[\alpha(\widetilde{T}_{ij})^{D_{ij}} \right] E_{Z_i} \left\{ Z_i^{D_{i\bullet}} \exp(-V_iZ_i) \right\}. \qquad (7.2)$$

As in Chapter 6, a suitable reformulation of (7.2) can be given by introducing the Laplace transform of Z, defined as $\mathscr{L}(c) = E\{\exp(-cZ)\}$. Applying the fact that the rth derivative $\mathscr{L}^{(r)}(c)$ equals $(-1)^r E\{Z^r \exp(-cZ)\}$, formula (7.2) can be written as:

$$P(H_i) = \prod_{j=1}^{n_i} \left[\alpha(\widetilde{T}_{ij})^{D_{ij}} \right] (-1)^{D_{i\bullet}} \mathscr{L}^{(D_{i\bullet})}(V_i).$$

This is the likelihood contribution of renewal process or cluster i. The total log-likelihood of m independent renewal processes or clusters is then given as

$$\log L = \sum_{i=1}^{m} \left[\sum_{j=1}^{n_i} D_{ij} \log(\alpha(\tilde{T}_{ij})) + \log\{(-1)^{D_{i\bullet}} \mathscr{L}^{(D_{i\bullet})}(V_i)\} \right], \qquad (7.3)$$

which may be maximized over a set of parameters to derive estimates.

7.2.3 Empirical Bayes estimate of individual frailty

Like in other variance component models, one may be interested in estimating the levels of the random factor, which is here the frailty Z, for each renewal process or cluster. This is done as follows by an empirical Bayes approach. Let $f(z)$ denote the density of Z. By Bayes formula and applying (7.1) and (7.2), we have:

$$f(z \mid H_i) = \frac{P(H_i \mid Z_i = z) \cdot f(z)}{P(H_i)} = \frac{z^{D_{i\bullet}} \exp(-V_i z)}{E_{Z_i} \left\{ Z_i^{D_{i\bullet}} \exp(-V_i Z_i) \right\}} f(z).$$

The conditional Laplace transform of Z_i given the history H_i is given by

$$\mathscr{L}_{H_i}(c) = E(e^{-cZ_i} \mid H_i) = \frac{E_{Z_i} \left\{ \exp(-cZ_i) Z_i^{D_{i\bullet}} \exp(-V_i Z_i) \right\}}{E_{Z_i} \{ Z^{D_{i\bullet}} \exp(-V_i Z) \}}$$

$$= \frac{E_{Z_i} \left\{ Z_i^{D_{i\bullet}} \exp(-(c + V_i) Z_i) \right\}}{E_{Z_i} \left\{ Z_i^{D_{i\bullet}} \exp(-V_i) \right\}} = \frac{\mathscr{L}^{(D_{i\bullet})}(c + V_i)}{\mathscr{L}^{(D_{i\bullet})}(V_i)}.$$

In brief, we have found:

$$\mathscr{L}_{H_i}(c) = \frac{\mathscr{L}^{(D_{i\bullet})}(c + V_i)}{\mathscr{L}^{(D_{i\bullet})}(V_i)}. \qquad (7.4)$$

By differentiating this Laplace transform with respect to c and evaluating the negative of the derivative at $c = 0$, one derives the conditional expectation of Z for renewal process or cluster i given its observed history:

$$\hat{Z}_i = E(Z_i \mid H_i) = -\frac{\mathscr{L}^{(D_{i\bullet}+1)}(V_i)}{\mathscr{L}^{(D_{i\bullet})}(V_i)}. \qquad (7.5)$$

If the relevant parameters in the model can be estimated from the data, this formula yields an empirical Bayes estimate of the frailty for renewal process or cluster i.

As for univariate frailty models, the Laplace transform of the frailty distribution is seen to play a role in these expressions. The likelihood becomes especially simple if the Laplace transform can be differentiated explicitly any number of times. This can be done, for instance, for the gamma distribution and for some other distributions, as will be seen in Section 7.2.5. For the PVF family (see Section 6.2.3) no closed form of the higher-order derivatives exists, but an iterative formula that can be easily programmed is given by Hougaard (2000).

7.2.4 Gamma distributed frailty

We have the following Laplace transform for a gamma frailty distribution with scale parameter v and shape parameter η:

$$\mathcal{L}(c;v,\eta) = \left(\frac{v}{v+c}\right)^{\eta}.$$

The sth derivative of this function is found by direct evaluation as follows:

$$\mathcal{L}^{(s)}(c;v,\eta) = v^{\eta}(-1)^{s}\{v+c\}^{-\eta-s}\prod_{q=1}^{s}(\eta+q-1). \tag{7.6}$$

In frailty modeling it is usually suitable to let the frailty variable have expectation 1, and we follow the presentation in Section 6.2.2 where δ denotes the variance and the Laplace transform equals:

$$\mathcal{L}(c) = \mathcal{L}(c;1/\delta,1/\delta) = \{1+\delta c\}^{-1/\delta}.$$

The following formula follows as a special case of (7.6):

$$\mathcal{L}^{(s)}(c) = \delta^{s}(-1)^{s}\{1+\delta c\}^{-1/\delta-s}\prod_{q=1}^{s}\left(\frac{1}{\delta}+q-1\right).$$

In order to apply the likelihood function, the simplest approach is a complete parametric specification. In addition to a parametric class of frailty distributions, like the gamma one given here, one must give a specification of the basic rate. Here we choose a Weibull specification, writing $\alpha(t) = bt^{k-1}$ for $k > 0$ and a positive parameter b. The log-likelihood function is a special case of (7.3) and is given by:

$$\log L(b,k,\delta) = \sum_{i=1}^{m}\left[\sum_{j=1}^{n_i} D_{ij}\{\log(b) + (k-1)\log(\widetilde{T}_{ij}) + \log(1+\delta(\sum_{r=1}^{j} K_{ir}-1))\}\right]$$

$$-\sum_{i=1}^{m}\left[(\delta^{-1}+\sum_{j=1}^{n_i} D_{ij})\log\{1+\frac{\delta b}{k}(\sum_{j=1}^{n_i}\widetilde{T}_{ij}^{k})\}\right].$$

If covariates, denoted by the vector $\mathbf{x}$, are to be included, this can be done in a proportional fashion with b replaced by $b\exp(\boldsymbol{\beta}^T\mathbf{x})$.

Explicit empirical Bayes results are easily derived. From (7.4) it follows that the conditional Laplace transform of Z given the history is

$$\mathscr{L}_{H_i}(c) = \left(1 + \frac{c}{\delta^{-1} + V_i}\right)^{-\delta^{-1} - D_{i\bullet}}. \tag{7.7}$$

This is again the Laplace transform of a gamma distribution, but this time with shape parameter $\delta^{-1} + D_{i\bullet}$ and scale parameter $\delta^{-1} + V_i$. This gives the following empirical Bayes estimate [see also formula (7.5)]:

$$\widehat{Z}_i = \mathrm{E}(Z_i | H_i) = \frac{\delta^{-1} + D_{i\bullet}}{\delta^{-1} + V_i}, \tag{7.8}$$

when the relevant parameters are estimated. The corresponding conditional variance follows from the variance formula of the gamma distribution:

$$\mathrm{Var}(Z_i | H_i) = \frac{\delta^{-1} + D_{i\bullet}}{(\delta^{-1} + V_i)^2}. \tag{7.9}$$

Finally, consider a future time T of an observation for renewal process or cluster i. One may argue that its conditional distribution given the history, H_i, has intensity $Z_{H,i}\alpha(t)$, with $Z_{H,i}$ having the modified gamma distribution with Laplace transform in (7.7). Assuming a Weibull form of $\alpha(t)$, the conditional moments of T given the history may essentially be derived just as (6.19), yielding the formula:

$$\mathrm{E}(T^r | H_i) = (\delta^{-1} + D_{i\bullet}) \left\{\frac{k}{b}(\delta^{-1} + V_i)\right\}^{r/k} B\left(\frac{r}{k} + 1, \delta^{-1} + D_{i\bullet} - \frac{r}{k}\right). \tag{7.10}$$

The special case $r = 1$ may be used to estimate the expected value of a future time of a particular renewal process or cluster given its history.

Example 7.1. Movements of the small bowel. As an example of data with recurrent events, we shall here analyze the data of Aalen and Husebye (1991) concerning the cyclic pattern of motility (spontaneous movements) of the small bowel in humans; see Example 1.11. This phenomenon is very important for digestion. In the mentioned paper, we studied the migrating motor complex (MMC), which travels down the small bowel at irregular intervals (these intervals last from minutes to several hours). Several such intervals in a row are registered for each person, with a censored one at the end, signifying the end of observation. The task is to estimate the average duration of the intervals and the variation within and between individuals.

A total of 19 healthy individuals (15 males and 4 females) of median age 26 years were examined. Intraluminal pressures were recorded continuously from 5.45 P.M. to 7.25 A.M. the following day (13 hours and 40 minutes), by two transducers mounted on a thin catheter located in the proximal small bowel. At 6 P.M., a standardized mixed meal was given to each individual. This induces what is termed fed state activity, characterized by irregular contraction, lasting from 4 to 7 hours. The fed state is followed by a fasting state, during which a cyclic motility pattern occurs. Three phases of this may be defined; however, only the activity front (phase III) is easy to distinguish. Phase III therefore defines the fasting cycle, also called MMC, and the time interval between two phase IIIs is termed an MMC period. The start of

Table 7.1 *Small bowel motility: duration of MMC periods (in minutes) for 19 individuals.*

Individual	Completely observed periods	Censored period
1	112 145 39 52 21 34 33 51	54
2	206 147	30
3	284 59 186	4
4	94 98 84	87
5	67	131
6	124 34 87 75 43 38 58 142 75	23
7	116 71 83 68 125	111
8	111 59 47 95	110
9	98 161 154 55	44
10	166 56	122
11	63 90 63 103 51	85
12	47 86 68 144	72
13	120 106 176	6
14	112 25 57 166	85
15	132 267 89	86
16	120 47 165 64 113	12
17	162 141 107 69	39
18	106 56 158 41 41 168	13
19	147 134 78 66 100	4

fasting motility was defined by the first phase III. Several MMC periods occurred in each individual (mean number 4.2) with a censored MMC period terminating the records. Censoring was due to the termination of measurement at 7.25 A.M. The data are shown in Table 7.1.

We assume that the recurrent events in this case form a renewal process. This implies, for instance, that there is no trend in the process. Instead, it is a stable cyclic phenomenon. Many clinical studies with recurrent events will probably not be of this kind; however, sometimes renewal processes occur naturally, as in the case of biorhythms.

Hence, our data fit the shared frailty framework. The censoring variables D_{ij} for individual i are all equal to 1 except for the final one, which is zero. We use a parametric model with gamma frailty and Weibull basic hazard and find maximum likelihood estimates.

In order to get detailed information on the variability of the estimates, a preliminary bootstrap analysis was carried out. The estimates of the parameter b turned out to have a distribution that was skewed to the right, while the estimates of k and δ had rather symmetric distributions. A logarithmic transformation removed the skewness for the first parameter. Hence, we present the maximum likelihood estimates as: $\log(\widehat{b}) = -10.0$ (1.0), $\widehat{k} = 2.28$ (0.22), and $\widehat{\delta} = 0.146$ (0.12). The standard errors of the estimates of $\log(b)$, k, and δ can be interpreted, approximately, in accordance with a normal distribution, meaning, for instance, that ± 2 standard errors should give about a 95% confidence interval.

The estimated shape parameter δ^{-1} of the gamma distribution equals 6.85, which is quite large, that is, the heterogeneity is small. This implies that the frailty

Table 7.2 *Small bowel motility: Empirical Bayes estimate for each individual from formula (7.8) with standard deviation calculated from formula (7.9). Expected conditional inter-event time and standard deviation calculated from formula (7.10).*

Individual	Empirical Bayes estimate with standard deviation		Expected inter-event time with standard deviation	
1	1.46	(0.38)	89	(43)
2	0.72	(0.24)	123	(61)
3	0.55	(0.18)	137	(68)
4	1.08	(0.34)	103	(51)
5	0.93	(0.33)	111	(55)
6	1.40	(0.35)	90	(43)
7	1.07	(0.31)	102	(50)
8	1.13	(0.34)	100	(49)
9	0.91	(0.28)	110	(54)
10	0.85	(0.29)	114	(57)
11	1.27	(0.37)	95	(46)
12	1.11	(0.34)	101	(49)
13	0.87	(0.28)	113	(55)
14	1.01	(0.31)	105	(52)
15	0.62	(0.20)	131	(64)
16	1.03	(0.30)	104	(51)
17	0.92	(0.28)	110	(54)
18	1.03	(0.29)	104	(50)
19	1.04	(0.30)	104	(50)

distribution is close to a normal one, with expectation 1 and standard deviation $\sqrt{\delta} = 0.38$. Hence, we have a simple description of the variation between individuals in the level of the intensity, saying that the proportionality factor could vary at least from 0.2 to 1.8 between individuals. In order to judge whether the variation between individuals is significant, the null hypothesis $\delta = 0$ has been tested by a likelihood ratio test, yielding a chi-squared value of 2.58 with one degree of freedom, that is, a P-value equal to 10.8%. This confirms that the variation between individuals is not statistically significant.

There is a technical issue concerning the validity of the chi-square test when the null hypothesis is on the boundary of the natural interval of variation of δ. Actually, this interval can be extended somewhat to incorporate negative values for δ, see Aalen and Husebye (1991), so that the asymptotic theory would still be valid since 0 can be viewed as an internal point. Alternatively, one may use results of Dominicus et al. (2006), showing that one may simply halve the P-value from the chi-squared test to get a one-sided P-value. In our case, this would be 5.4%, which is above the 5% level, but still more borderline.

Finally, for each individual we may give an estimate of its frailty based on the history of that individual. Empirical Bayes estimates are given in Table 7.2. The empirical Bayes estimates for each individual are calculated from formula (7.8) with standard deviation from formula (7.9). In addition, the expected conditional inter-event time and standard deviation given the past history are calculated from formula

(7.10). It is clearly seen how interesting individual information on the basis of past history can easily be computed.

This example is also analyzed in an interesting study by Lindqvist (2006). □

Example 7.2. Amalgam fillings. As an example of clustered survival data, we will analyze the lifetime of amalgam fillings; see Example 1.9. Note the high dimensionality of the data (up to 38 units, i.e., fillings, for an individual), and the lack of balance (the smallest number of fillings for an individual being 4). This, together with a large number of censored observations, makes for a situation where one has to model the dependence in just the right way in order to manage the analysis. The multivariate frailty approach handles the situation elegantly.

The basic hazard rate $\alpha(t)$ is assumed to be of Weibull shape with t measured in years from insertion of the specific filling. Note that the time scale relates to individual fillings, so that time 0 for different fillings of a certain patient may be at widely differing ages for that patient. As earlier, we use a gamma frailty distribution. The maximum likelihood estimates of the parameters, with s.e. in parenthesis, are: $\log(\widehat{b}) = -4.21$ (0.25), $\widehat{k} = 0.43$ (0.10), and $\widehat{\delta} = 0.85$ (0.31). There was a clearly significant frailty effect, the Wald chi-square test for H_0: $\delta = 0$ being 7.23 with one degree of freedom, giving a P-value of 0.7%. As mentioned in Section 7.4, the P-value may be halved according to one possible way of handling null hypotheses on the boundary.

In order to demonstrate the results of the model, survival curves for the fillings of four selected patients are shown in Figure 7.1. The step functions in the plots are Kaplan-Meier survival curves computed from the fillings for each individual patient. These are compared to smooth survival functions computed as follows: For each patient an empirical Bayes estimate of the frailty, denoted $\widehat{Z}_i$, is computed. The individual survival curve for this patient is then estimated by the Weibull survival function $\exp(-\widehat{Z}_i \widehat{b} t^{\widehat{k}} / \widehat{k})$.

The two left-hand plots in Figure 7.1 show the results for patients with large numbers of fillings inserted (37 and 26, respectively, for upper and lower plot). Apparently, the fit is good. The two right-hand plots of Figure 7.1 show results for two patients with few fillings inserted (10 and 4, respectively, for upper and lower plot) and few observed failures (0 and 3, respectively). One sees here that the survival curves computed from the parametric model deviate somewhat from the Kaplan-Meier plot, indicating the well-known empirical Bayes shrinkage toward the mean.

The general impression from considering all data is that the simple parametric frailty model give a well-fitting and compact description. □

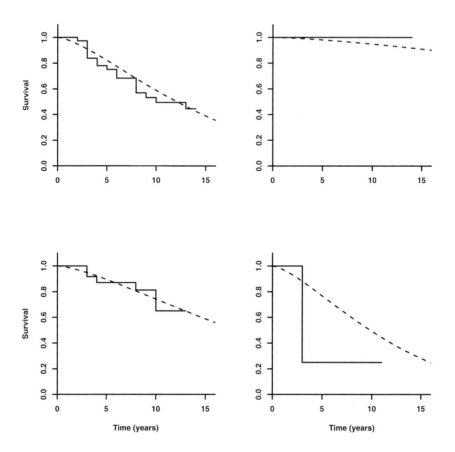

Fig. 7.1 *Survival of amalgam fillings for four selected patients. The step functions are individual Kaplan-Meier survival curves. These are compared to empirical Bayes survival functions computed from the frailty model.*

7.2.5 Other frailty distributions suitable for the shared frailty model

The gamma distribution is the prototype used in many frailty models because of the nice mathematical properties of the gamma family. Two of these properties can be formulated generally as follows:

1. If $\mathscr{L}(c)$ belongs to a family of Laplace transforms, then $\mathscr{L}(c+b)/\mathscr{L}(b)$ belongs to the same family for any positive constant b.
2. If $\mathscr{L}(c)$ belongs to a family of Laplace transforms, then normalized derivatives, $\mathscr{L}^{(r)}(c)/\mathscr{L}^{(r)}(0)$, of any order also belong to the same family.

These two requirements guarantee that expressions like (7.4) yield a nice explicit formula and that the resulting Laplace transform belongs to the same class as the original one.

Usually, we think of $\mathscr{L}(c)$ as the Laplace transform for a parametric family of distributions, and property 1 then corresponds to the density having an exponential factor $\exp(-vz)$, with v being a parameter in the family. Property 2 corresponds to there being a multiplicative factor z^η with η being a parameter. This is because repeated differentiation of the Laplace transform yields terms $z^{\eta+1}$, $z^{\eta+2}$, etc. The mentioned properties are obviously fulfilled by the gamma distributions where the density is precisely a normalized product of $\exp(-vz)$ and z^η type terms. However, there exist extensions of the gamma distributions that are easily handled, and we shall mention two of those.

Generalizing the gamma distributions these classes allow greater flexibility, for example, a choice of more heavy tails. Although these classes would traditionally be regarded as somewhat intractable since they use Bessel and hypergeometric functions, one should note that such functions are standard in many computer programs. In fact, the traditional view in statistics of what are tractable distributions should definitely change since many previously intractable functions are today easily available in many computer programs.

The first is the *generalized inverse Gaussian distribution*. The density is constructed by multiplying a factor $\exp(-\psi/z)$ onto the gamma density (6.5):

$$\frac{(v/\psi)^{\eta/2}}{2K_\eta(2\sqrt{\psi v})} z^{\eta-1} \exp(-vz - \frac{\psi}{z}), \qquad z \geq 0, \tag{7.11}$$

where $v, \psi > 0$ while η can be any real number. The density is normalized to integrate to 1 by using $K_\eta(x)$, which is a modified Bessel function of order η (e.g., Wolfram, 1999). Introductions to generalized inverse Gaussian distributions are given by Jørgensen (1982) and Chhikara and Folks (1989); the latter reference also explains its use in mixing of a Poisson distribution. The Laplace transform is

$$\mathscr{L}(c;\eta,v,\psi) = (\frac{v}{c+v})^{\eta/2} \frac{K_\eta(2\sqrt{\psi(c+v)})}{K_\eta(2\sqrt{\psi v})}. \tag{7.12}$$

Differentiation with respect to c yields the normalized derivative:

$$\mathscr{L}^{(k)}(c;\eta,v,\psi)/\mathscr{L}^{(k)}(0;\eta,v,\psi) = \mathscr{L}(c;\eta+k,v,\psi), \tag{7.13}$$

illustrating closure property 2. Hence, the conditional Laplace transform of the frailty given the past, as given in (7.4), is as follows:

$$\mathscr{L}_{H_i}(c) = \frac{\mathscr{L}^{(D_{i\bullet})}(c+V_i)}{\mathscr{L}^{(D_{i\bullet})}(V_i)} = \frac{\mathscr{L}(c+V_i;\eta+D_{i\bullet},v,\psi)}{\mathscr{L}(V_i;\eta+D_{i\bullet},v,\psi)} = \mathscr{L}(c;\eta+D_{i\bullet},v+V_i,\psi).$$

Hence, just like for the gamma distribution, the conditional Laplace transform of the frailty distribution given the past history is found by adding V_i to a scale parameter and $D_{i\bullet}$ to a shape parameter.

The second distribution to be considered here is the *Kummer distribution*. The density is constructed by multiplying a factor $(\delta + z)^{-\beta}$ onto the gamma density:

$$\frac{\delta^{\beta-\eta}}{\Gamma(\eta)U(\eta,\eta-\beta+1,v\delta)}\, z^{\eta-1}\exp(-vz)(\delta+z)^{-\beta}, \qquad z \geq 0, \qquad (7.14)$$

where $\eta, v, \delta > 0$ and β is any real number. The Kummer distribution was suggested by Armero and Bayarri (1997). The density is normalized by means of U, which is a confluent hypergeometric function of the second kind, often called the Kummer function (e.g., Wolfram, 1999).

The Laplace transform of the Kummer distribution is

$$\mathscr{L}(c;\eta,v,\beta,\delta) = \frac{U(\eta,\eta-\beta+1,(c+v)\delta)}{U(\eta,\eta-\beta+1,v\delta)}. \qquad (7.15)$$

Differentiation with respect to c yields the normalized derivative [from the properties of the confluent hypergeometric function (e.g., Lebedev, 1972)]:

$$\mathscr{L}^{(k)}(c;\eta,v,\beta,\delta)/\mathscr{L}^{(k)}(0;\eta,v,\beta,\delta) = \mathscr{L}(c;\eta+k,v,\beta,\delta),$$

again illustrating closure property 2. The conditional Laplace transform given the past is given by:

$$\mathscr{L}_{H_i}(c) = \mathscr{L}(c;\eta+D_{i\bullet},v+V_i,\beta,\delta).$$

We may also allow $v = 0$ if $\beta > \eta$, which with $\delta = 1$ yields a beta distribution of the second kind (Kendall and Stuart, 1977). Another interesting special case of the Kummer distribution is the Pareto distribution, which arises when $\eta = 1$, $v = 0$, and $\beta > \eta$ and is very heavy tailed.

7.3 Frailty and counting processes

So far we have concentrated on individuals or clusters, with several survival times related to each either as multiple units or as inter-event times in a renewal process. We shall now assume a more general viewpoint and consider counting processes with intensity processes dependent on a frailty variable.

Assume that a simple frailty structure is valid for each individual counting process, $N_i(t)$, that is, the individual intensity given the frailty variable Z_i can be written

$$\alpha_i(t) = Z_i Y_i(t)\alpha(t).$$

Here $\alpha(t)$ is a common intensity and a fixed function independent of the past, and the $Y_i(t)$ are observable predictable processes. The variables Z_i, $i = 1,...,n$, are independent identically distributed frailty variables giving the multiplicative factor that determines the risk of an individual. Let $\mathbf{Z}$ denote the vector of the Z_is. Censoring may occur and we assume the following: conditional on $\mathbf{Z} = \mathbf{z}$ censoring is independent and noninformative on $\mathbf{z}$.

An important application of this structure will be recurrent events modeled by several parallel Poisson processes with intensities $Z_i \alpha(t)$. The Poisson processes may be right-censored, and the processes $Y_i(t)$ will then assume the value 1 as long as the process is observed and 0 afterwards.

The clustered survival data situation with a number of independent individuals having multiple units at risk (e.g., the amalgam fillings data) may also be included in the present setup. In this case, $Y_i(t)$ will represent the number of units still at risk for cluster i.

What is the observable intensity process of an individual counting process if the Z_i are unknown (which one would usually assume)? By the innovation theorem (Section 2.2.7), the answer is given by the formula

$$\lambda_i(t) = Y_i(t)\alpha(t)\mathrm{E}(Z_i \,|\, \mathscr{F}_t).$$

By slightly modifying the derivation of formula (7.5), one gets

$$\lambda_i(t) = Y_i(t)\alpha(t)\frac{-\mathscr{L}^{(N_i(t-)+1)}(A_i^*(t))}{\mathscr{L}^{(N_i(t-))}(A_i^*(t))}, \qquad (7.16)$$

where $A_i^*(t) = \int_0^t Y_i(t)\alpha(s)\,ds$ and $\mathscr{L}(c)$ is the Laplace transform of the frailty variable Z_i. The Laplace transforms in Section 7.2.5 will be particularly suitable for finding explicit versions of the formula.

For the case where the Z_i are gamma distributed with scale parameter ν and shape parameter η, we get:

$$\lambda_i(t) = Y_i(t)\,\alpha(t)\frac{\eta + N_i(t-)}{\nu + A_i^*(t)}. \qquad (7.17)$$

This formula can also be found in Andersen et al. (1993, section IX.4). The formula is related to (7.8) and shows how the intensity is updated according to occurrences in the observed process. A large number of occurrences, for instance, yields a large $N_i(t-)$ and thus a large $\lambda_i(t)$, implying that the rate of new occurrences in the future is also higher.

Nielsen et al. (1992) [see also Andersen et al. (1993, chapter IX)] considered the semiparametric gamma frailty model, where $\alpha(t)$ may be any nonnegative function, and discussed the extension where covariates are included in a Cox-type regression model $\alpha(t) = \alpha_0(t)\exp\{\boldsymbol{\beta}^T x_i(t)\}$. They proposed estimating the parameters of the model (including the increments of the cumulative baseline hazard) using maximum likelihood estimation and the EM algorithm, and they conjectured that the estimates should enjoy the usual "likelihood properties." This was later proved by Murphy (1995) and Parner (1998). The semiparametric approach for models with frailties is discussed in great generality by Zeng and Lin (2007).

Table 7.3 *Parameter estimates from a semiparametric gamma frailty model for the bladder data.*

Covariate	Estimate	Standard error	Hazard ratio	Wald statistic	P-value
Treatment	−0.634	0.334	0.530	1.90	0.058
Number of initial tumors	0.238	0.093	1.269	2.55	0.011
Size of largest tumor	−0.030	0.114	0.971	0.26	0.790

In the much simpler parametric case, that is, when $\alpha(t)$ follows a parametric model, the full likelihood function may be derived by insertion of the intensity process (7.16) or (7.17) into formula (5.4). Estimates may then be found by maximizing the likelihood directly, and standard counting process tools can be used for proving the properties of estimators.

Example 7.3. Recurrent superficial bladder tumors. We shall analyze a data set from a randomized clinical trial on recurrent superficial bladder tumors (Example 1.10) using the semiparametric gamma frailty model. We use the data set as presented in table 9.2 of Kalbfleisch and Prentice (2002), but with only the first four recurrence times for each patient. In the trial, 38 patients received treatment (thiotepa) and 48 received placebo. The outcome of interest was occurrence of superficial tumors in the bladder, and all tumors were removed prior to randomization. Each individual could experience a sequence of tumors during the follow-up time, and all times of these recurrent events were recorded. The data are analyzed as Poisson processes with gamma frailty and the following covariates: treatment (thiotepa vs. placebo), number of initial tumors, and size of the largest initial tumor. We use the `coxph` program in R (R Development Core Team, 2007) with gamma frailty, and we get the estimated regression coefficients in Table 7.3. There is a clear effect of the number of initial tumors and an indication for an effect of treatment (even though it is not significant at conventional levels). There is also a significant frailty effect, and the variance of the gamma frailty is 1.07. The data will also be analyzed by a marginal model and a dynamic model in Examples 8.3 and 8.6. □

7.4 Hierarchical multivariate frailty models

The multivariate structure considered so far is of a very simple kind; basically, there are two levels: the unit and the repeated measures carried out on the unit. In general, one may have hierarchical structures with several levels, for instance, individuals, siblings, families, and neighborhoods. Several models have been proposed for handling this type of data, as discussed, for instance in Hougaard (2000). However, the suggestions appear a bit ad hoc, and in this section we propose a general model with some advantages.

7.4.1 A multivariate model based on Lévy-type distributions

Assume that at the bottom (first) level there is variation determined by the frailty variable X_1; for example, this could be the variation between individuals belonging to one family. At the next (second) level, the variation is determined by X_2; for instance this is the variation between families. Above this comes another (third) level, with variation X_3, etc. We assume that all the frailty variables have expectation 1.

We shall use the Lévy-type frailty distributions defined in Section 6.2.4. Let $X_l(\rho)$ be a Lévy process with time parameter ρ; the Laplace transform of X_l is then $\exp(-\rho \Phi_l(c))$, where $\Phi_l(c)$ is the Laplace exponent, satisfying $\Phi_l(0) = 0$. In our application, we furthermore assume that the expectation is one when $\rho = 1$, that is, we have $\Phi_l'(0) = 1$ for all l. Note that $\mathrm{Var} X_l(\rho) = -\Phi_l''(0)\rho$. In the following, whenever the time parameter is one, we will for simplicity write $X_l(1) = X_l$

The hierarchical structure is constructed in the following way. Consider only the variation at the top (third) level. The frailty at this level is given by the variable $Z_3 = X_3(1) = X_3$, whose Laplace transform is

$$\mathscr{L}_3(c) = \mathrm{E}\exp(-cX_3) = \exp(-\Phi_3(c)).$$

Assume, for the moment, that the value of X_3 is given. Then, at the next level down (the second level), the frailty is $Z_2 = X_2(X_3)$, that is, the second level frailty is $X_2(\rho)$, where ρ is randomized by X_3. At this level, the Laplace transform of the frailty Z_2 is thus

$$\mathscr{L}_2(c) = \mathrm{E}\exp(-cX_2(X_3)) = \mathrm{E}[\mathrm{E}\exp(-cX_2(X_3)) \mid X_3]$$
$$= \mathrm{E}\exp(-X_3\Phi_2(c)) = \exp(-\Phi_3(\Phi_2(c))).$$

This is a so-called power mixture (Abate and Whitt, 1996). Similarly, for the first level, we randomize the time parameter of $X_1(\rho)$ by Z_2, that is, we define the frailty as $Z_1 = X_1(X_2(X_3))$. This yields the the Laplace transform:

$$\mathscr{L}_1(c) = \mathrm{E}\exp(-Z_2\Phi_1(c)) = \exp(-\Phi_3(\Phi_2(\Phi_1(c)))). \qquad (7.18)$$

This approach is easily extended to more levels. Hence, the structure is generated by applying function iteration to the Laplace exponent.

Simple results are valid for the expectation and variance (assuming the latter exists for the model in question). Consider the frailty variable Z_2, with Laplace transform $\mathscr{L}_2(c)$. We have

$$\mathrm{E} Z_2 = \Phi_3'(\Phi_2(0))\Phi_2'(0) = 1$$

and

$$\mathrm{Var} Z_2 = -\Phi_3''(\Phi_2(0))(\Phi_2'(0))^2 - \Phi_3'(\Phi_2(0))\Phi_2''(0) = -\Phi_3''(0) - \Phi_2''(0),$$

that is,

$$\mathrm{Var}Z_2 = \mathrm{Var}X_2 + \mathrm{Var}X_3.$$

It similarly follows that for the random variable Z_1 with Laplace transform $\mathscr{L}_1(c)$, we have:

$$\mathrm{E}Z_1 = 1, \quad \mathrm{Var}Z_1 = \mathrm{Var}X_1 + \mathrm{Var}X_2 + \mathrm{Var}X_3.$$

Hence, we have a hierarchical model of nonnegative random variables that all have expectation 1, but where the variance can be decomposed into a sum coming from different sources. This is different from other models where a decomposition would typically affect both expectation and variance. But it is very useful in a frailty context where the expectation should be kept constant and just the variance decomposed.

7.4.2 A multivariate stable model

Assume we have a stable frailty distribution (Section 6.2.3), that is, $\Phi_l(c) = a_l c^{\beta_l}$. (Note that in this case the expectation and variance do not exist.) In this case the function composition involved in formulas like (7.18) is very simple, for example, we have

$$\Phi_3(\Phi_2(\Phi_1(c))) = a_1^{\beta_2\beta_3} a_2^{\beta_3} a_3 c^{\beta_1\beta_2\beta_3},$$

which again is the Laplace exponent of a stable distribution. This underscores the nice properties of the stable distribution, as presented by Hougaard (2000).

7.4.3 The PVF distribution with $m = 1$

This is a compound Poisson distribution (Section 6.2.3) with Laplace exponent $\Phi(c; \rho, v) = \rho c / (v + c)$. It is easily shown that

$$\Phi(\Phi(c; \rho_1, v_1); \rho_2, v_2) = \Phi(c; \frac{\rho_1\rho_2}{\rho_1 + v_2}, \frac{v_1 v_2}{\rho_1 + v_2}). \tag{7.19}$$

Hence, the family is closed with respect to composition of the Laplace exponent.

7.4.4 A trivariate model

The model presented here can be used to construct multivariate survival distributions. This is best illustrated by a simple example. Hougaard (2000, page 355) suggests a trivariate model for the lifetimes (T_1, T_2, T_3) of a sibling group where individuals 2 and 3 are monozygotic twins and individual 1 is an ordinary sibling (single birth). As in the general model, let X_1 correspond to the variation between ordinary siblings, and X_2 to the variation between different sets of siblings. The frailties are given as $X_1^{(1)}(X_2)$ for individual 1, and as $X_1^{(2)}(X_2)$ for individuals 2 and 3, where different superscripts indicate independent processes. The joint Laplace transform of the frailties is given by:

$$\mathcal{L}(c_1, c_2, c_3)$$
$$= E[E\exp(-c_1 X_1^{(1)}(X_2) - c_2 X_1^{(2)}(X_2) - c_3 X_1^{(2)}(X_2)) \mid X_2]$$
$$= E\exp(-X_2(1)(\Phi_1(c_1) + \Phi_1(c_2 + c_3))) = \exp(-\Phi_2(\Phi_1(c_1) + \Phi_1(c_2 + c_3)))$$

Hence the joint survival distribution is:

$$S(t_1, t_2, t_3) = \exp(-\Phi_2(\Phi_1(A_1(t_1)) + \Phi_1(A_2(t_2) + A_3(t_3)))).$$

This generalizes formula (10.3) in Hougaard (2000), which is what we would get for the stable distribution. In fact, Hougaard's construction in his formula (10.1) is different from ours; the one proposed here seems more natural. Note the simple structure elucidated here: at each level the independent contributions are added up, and then a Φ-function is applied to reach the next level. This is a general structure that can be used to write up more complex models as well.

7.4.5 A simple genetic model

Consider once more three lifetimes (T_1, T_2, T_3), but this time for a mother, a father, and a child, respectively. We briefly illustrate how a hierarchical model can be combined with an additive frailty model to produce a trivariate model for such nuclear families. Such a model can be used to separate the effects of genes and environment on the frailty of an individual, similar to standard biometrical models that decompose a continuous biological trait directly (Bulmer, 1985). In a nuclear family, there are essentially only the correlations between mother and child and between father and child, that can be used for estimation (Lunde et al., 2007; Magnus et al., 2001). This means that more complex patterns of heritability cannot be resolved (Gjessing and Lic, 2008), but a standard measure of heritability can still be computed.

Assuming the mother and father are independent, we can write their frailties as $Z_1 = X_1^{(1)}(X_2^{(1)}(1))$ and $Z_2 = X_1^{(2)}(X_2^{(2)}(1))$, where again different superscripts indicate independence. The child has her own level-1 frailty $X_1^{(3)}$, independent of the other level-1 frailties, but her value of $X_2^{(3)}$ is correlated to both $X_2^{(1)}$ and $X_2^{(2)}$. As before we assume that $EX_j^{(i)} = 1$ for all i, j, and write $\sigma_1^2 = \text{Var}\{X_1^{(i)}\}$, $\sigma_2^2 = \text{Var}\{X_2^{(i)}\}$ for the variances at levels 1 and 2, respectively.

The level-1 frailties may be seen as individual environmental influences, factors that contribute to the frailty level but are not shared among family members. We will now assume that the level-2 frailties correspond to "standard" autosomal genes with an additive effect on frailty, that is, we rule out effects of sex-linked genes, gene-gene interactions, etc. (Bulmer, 1985).

The most natural way to describe the level-2 correlations is to use an additive genetic model (Petersen, 1998), with

$$X_2^{(1)} = MO_1 + MO_2,$$
$$X_2^{(2)} = FA_1 + FA_2,$$
$$X_2^{(3)} = MO_1 + FA_1.$$

Each of $MO_1 = MO_1(1)$, MO_2, FA_1, and FA_2 are independent Lévy frailty variables with a common Laplace exponent Φ_G, chosen so that $EMO_i = EFA_i = 1/2$ and $VarMO_i = VarFA_i = 1/2\sigma_2^2$. This construction intuitively describes MO_1 as those maternal genes that are passed on to the child, and MO_2 as those that are not, respectively; FA_1 and FA_2 are the corresponding components for the father.

From this, it follows that

$$\text{cor}(X_2^{(1)}, X_2^{(2)}) = 0,$$
$$\text{cor}(X_2^{(1)}, X_2^{(3)}) = 1/2,$$
$$\text{cor}(X_2^{(2)}, X_2^{(3)}) = 1/2,$$

which is consistent with the standard models for quantitative genetic traits (Bulmer, 1985). Furthermore, by conditioning on the highest level,

$$\begin{aligned}
\text{Cov}\{Z_1, Z_3\} &= \text{Cov}\left\{X_1^{(1)}(X_2^{(1)}), X_1^{(3)}(X_2^{(3)})\right\} \\
&= \text{E}\left\{\text{Cov}\left\{X_1^{(1)}(X_2^{(1)}), X_1^{(3)}(X_2^{(3)})|X_2^{(1)}, X_2^{(3)}\right\}\right\} \\
&\quad + \text{Cov}\left\{\text{E}\left\{X_1^{(1)}(X_2^{(1)})|X_2^{(1)}, X_2^{(3)}\right\}, \text{E}\left\{X_1^{(3)}(X_2^{(3)})|X_2^{(1)}, X_2^{(3)}\right\}\right\} \\
&= 0 + \text{Cov}\left\{X_2^{(1)}, X_2^{(3)}\right\} = \text{Var}(MO_1) = \frac{1}{2}\sigma_2^2.
\end{aligned}$$

That is, the covariance between the total frailty of the mother and the child is equal to the genetic covariance at level 2. Similarly, $\text{Cov}\{Z_2, Z_3\} = (1/2)\sigma_2^2$. Since $\text{Var}(Z_i) = \text{Var}(X_1^{(i)}) + \text{Var}(X_2^{(i)}) = \sigma_1^2 + \sigma_2^2$,

$$\text{cor}(Z_1, Z_3) = \frac{1}{2}\frac{\sigma_2^2}{\sigma_1^2 + \sigma_2^2}.$$

This makes it natural to define

$$h^2 = \frac{\sigma_2^2}{\sigma_1^2 + \sigma_2^2}$$

as the proportion of the frailty variance explained by the additive genetic effects. This is usually termed the *heritability* of a trait. Similarly, $e^2 = \sigma_1^2/(\sigma_1^2 + \sigma_2^2)$ is the proportion explained by individual environmental effects, so that $h^2 + e^2 = 1$. For further details and an application, see Moger et al. (2008).

When decomposing trait variability (here, frailty variability), it should always be kept in mind that heritability is not an absolute concept. In particular, it depends on how the effects are incorporated into the model. In our example, the heritability measures the proportion of *frailty* variance explained by additive genetic effects. If one were to measure the proportion of, for instance, *event time* variability explained by genes, the result is likely to differ. In addition, heritability is also measured relative to the population under study. In a population with high environmental heterogeneity, genes will by necessity explain less of the variability. By the same token, if the environment itself is very homogeneous, the heritability will always be close to one. For related comments, see Section 6.8.

7.5 Case study: A hierarchical frailty model for testicular cancer

A frailty model for the incidence of testicular cancer was studied in Section 6.9. Fitting a model to univariate data, as was done there, can be viewed as somewhat speculative, although the general conclusion that emerges is consistent with biological knowledge. In this section we include family data for testicular cancer; this gives an example of multivariate survival data where identifiability is much less of a problem than in the univariate case. The basic observation is that brothers of men with testicular cancer have a much higher risk of getting this disease themselves than in the general population of men; in fact, the risk turns out to be more than seven-fold. The presentation is based on Moger et al. (2004). We shall construct a hierarchical model corresponding to the theory of Section 7.4.1.

In Section 6.9 we fitted a compound Poisson frailty distribution, that is, a distribution with Laplace transform

$$\mathcal{L}(c; \rho, \nu, m) = \exp\left[-\rho\{1 - \left(\frac{\nu}{\nu + c}\right)^m\}\right], \tag{7.20}$$

where $m > 0$; see Section 6.2.3. The idea is now that this distribution varies between families and that $\mathcal{L}(c; \rho, \nu, m)$ corresponds to the first-level Laplace transform $\mathcal{L}_1(c)$ of Section 7.4.1. We assume that the family variation is due only to the parameter ρ. Hence ρ is now considered a random variable varying between the families, and we denote its Laplace transform by $\mathcal{H}_\rho(c)$. Combining this with (7.20) yields the Laplace transform:

$$\mathcal{L}_2(c) = \mathcal{H}_\rho\left(\rho\{1 - \left(\frac{\nu}{\nu + c}\right)^m\}\right),$$

which is the second-level Laplace transform of Section 7.4.1.

The compound Poisson frailty implies that only a subset of individuals are susceptible, that is, they have a positive frailty. The probability of being susceptible is $p = 1 - \exp(-\rho)$. The strength of the genetic association within a family may be evaluated by computing the conditional probability of a brother of a susceptible man also being susceptible:

$$P_{\text{fam}} = P(\text{brother susceptible} \mid \text{man susceptible})$$

$$= \frac{P(\text{brother and man susceptible})}{P(\text{man susceptible})}$$

$$= \frac{\int (1 - \exp(-\rho))^2 f(\rho) d\rho}{\int (1 - \exp(-\rho)) f(\rho) d\rho} = \frac{1 - 2\mathcal{H}_\rho(1) + \mathcal{H}_\rho(2)}{1 - \mathcal{H}_\rho(1)}, \qquad (7.21)$$

where $f(\rho)$ is the density of ρ. The conditional risk of susceptibility given a susceptible brother can be compared to the overall risk as follows:

$$R = \frac{P_{\text{fam}}}{P(\text{man susceptible})} = \frac{1 - 2\mathcal{H}_\rho(1) + \mathcal{H}_\rho(2)}{(1 - \mathcal{H}_\rho(1))^2}.$$

This is similar to, but not exactly the same as, the relative risk, or risk ratio, in which the ratio is considered between two exclusive groups. In this case, that would be brothers of a susceptible and a nonsusceptible man. However, the measures will be quite similar at the levels of association considered for testicular cancer, since the disease is rare. We will therefore denote R as the relative risk, keeping in mind that it deviates somewhat from the common definition.

Since ρ is a random variable, this is also the case for p. The squared coefficient of variation of p is given as:

$$CV^2 = \frac{\text{Var}(p)}{\{\text{E}(p)\}^2} = \frac{\text{E}(p^2)}{\{\text{E}(p)\}^2} - 1 = \frac{1 - 2\mathcal{H}_\rho(1) + \mathcal{H}_\rho(2)}{(1 - \mathcal{H}_\rho(1))^2} - 1 = R - 1.$$

We then get the useful relation

$$R = CV^2 + 1,$$

which is analogous to a result in Aalen (1991). A large value of R will imply a large CV^2 for the distribution of p. In Moger et al. (2004), it is shown that such distributions will necessarily be very skewed, with some families having very high risk compared to the moderate or small risk of the others. This again points in the direction of genetic effects, or to very strong and specific environmental effects.

The model has been applied to analyze a data set on testicular cancer among brothers. All testicular cancer patients treated at the Norwegian Radium Hospital were registered over a period of more than ten years. The patient cohort consists of 895 consecutive patients referred for treatment for testicular germ cell tumor at the hospital in the period of 10.5 years between 1st January 1981 and 30th June 1991 and 27 patients treated at the University Hospital in Bergen, Norway, during the same period. Information on surviving testicular cancer patients and their families with distant relatives was obtained by means of a questionnaire. Included in the information is year of birth, age at cancer or censoring, and whether the cancer was seminoma or nonseminoma. There were 797 questionnaires (86.4 %) with family information returned.

Since brothers of testicular cancer patients are more prone to get the disease, this example will focus on them. The total number of individuals in the cohort of brothers is 1694. There are 277 probands who have no brothers, and 520 probands with at least one brother. (By "proband" we mean those in the original patient cohort.) The number of affected brothers is 14. There are 3 doubly ascertained sibships of brothers, meaning that two related cases were identified independently, leading back to the same family. This yields a total of 808 separate cancer cases. The median age at diagnosis is 36.7 years for seminomas and 28.7 years for nonseminomas.

The model introduced earlier assumes a common value of ρ for all brothers in a family, with the value varying between families. In Section 6.9, based on Aalen and Tretli (1999), the family variation was not included and the ρ in formula (7.20) was considered as a parameter. This parameter was estimated separately for seminoma and nonseminoma cancer for each birth cohort. To use this information in the present analysis, let μ_{iS} and μ_{iN}, with S indicating seminoma and N nonseminoma, be the estimates for the parameters ρ_i from Section 6.9 corresponding to the birth cohort of brother i. Note that the values of μ_{iS} and μ_{iN} would be specific to each brother in a family. Consider $\mu_{iS}Z$ and $\mu_{iN}Z$, where Z is gamma distributed with both shape and scale parameter $1/\delta$, like in Section 7.2.4, such that the expectation is 1. The quantities $\mu_{iS}Z$ and $\mu_{iN}Z$ will have expectations corresponding to the respective estimates from Aalen and Tretli (1999), but all brothers in a family have a common value of Z, regardless of when they were born. One then remains within the model introduced earlier and the quantities μ_{iS} and μ_{iN} can be seen as covariates.

Since only families with at least one case of testicular cancer are included in the data, one has to account for a possible ascertainment bias, by which we mean that families with more cases are more likely to be included in a study than those with a single or few cases. This requires a very careful analysis, which is given in Moger et al. (2004). The maximum-likelihood estimate for the shape parameter $1/\delta$, adjusted for ascertainment bias, is 0.151. The estimate corresponds to a relative risk of about 7.4 in brothers of testicular cancer patients compared to the general population (Moger et al., 2004). The 95% confidence interval for $1/\delta$ is (0.08, 0.29). Since the value of the shape parameter is small and must be positive, the interval is calculated for $\ln(1/\delta)$ and transformed to get the interval for $1/\delta$. The confidence interval for $1/\delta$ gives an approximate confidence interval for the relative risk from 4.3 to 13.

The small value of the shape parameter implies a very skewed variation in susceptibility between families. Most families have no or very little susceptibility, while a few families have a great susceptibility. Hence the total probability of being susceptible to testicular cancer will be very unevenly distributed over families. This may point toward a genetic explanation; possibly single genes are operating in some families to produce the high risk observed. Other models of familial association, for instance, based on environmental or polygenic factors, are not likely to lead to such high relative risks; see also Aalen (1991) and Khoury et al. (1988). In fact, a segregation analysis of family data of testicular cancer has been carried out, suggesting that a major recessive gene may be involved (Heimdal et al., 1997). However, definitive conclusions about this cannot be drawn since a factor like the mother's uterine climate, for instance, specific maternal hormone levels, could have a similar effect to a genetic factor in the models.

7.6 Random effects models for transformed times

In Section 7.2 we consider frailty models for the situation were several survival times are related to each other either as multiple units in a cluster, or as inter-event times in a renewal process. An alternative to the frailty formulation is to transform the survival times (or inter-event times) in such a way that they become approximately normally distributed and then apply classical linear random effects models (also called variance components models). That is, we apply a transformation of the type

$$g(T_{ij}) = \mu + \boldsymbol{\beta}^T \mathbf{x}_{ij} + U_i + E_{ij}, \tag{7.22}$$

where μ is a constant, $\mathbf{x}_{ij}$ is a vector of covariates with coefficients $\boldsymbol{\beta}$, and the Us and Es are all stochastically independent and normally distributed with zero mean and standard deviations σ_u and σ_e, respectively. The random variable U_i then models the variation between individual renewal processes or clusters, while the E_{ij} model the variation within individual renewal processes or clusters. The statistical aim is to evaluate these sources of variation by estimating σ_u and σ_e and to estimate μ and $\boldsymbol{\beta}$.

An important special case is the *accelerated failure time model*, which arises when the link function g is logarithmic:

$$\log(T_{ij}) = \mu + \boldsymbol{\beta}^T \mathbf{x}_{ij} + U_i + E_{ij}.$$

In the accelerated failure time model the random terms may be nonnormal, but here we shall just assume normality.

7.6.1 Likelihood function

We first consider only the case with renewal processes with a possibly censored inter-event time at the end of the observation period, as illustrated with the small bowel motility data (Example 1.11). We compute the likelihood iteratively for each individual renewal process. For this purpose, one needs the conditional distribution of an inter-event time T_{ij} given all previous inter-event times T_{il}, $l = 1, \ldots, j-1$. From standard normal distribution theory [for instance chapter 4 of Johnson and

Wichern (1982)] one may deduce that the conditional distribution of $g(T_{ij})$ is normal with expectation and standard deviation

$$\xi_{ij} = v_{j-1}\mu + (1 - v_{j-1})\frac{1}{j-1}\sum_{l=1}^{j-1} g(T_{il}) \quad \text{and} \quad \tau_j = \sqrt{v_{j-1}\sigma_u^2 + \sigma_e^2},$$

respectively, where

$$v_j = \frac{1}{\sigma_u^2} \Big/ \Big(\frac{1}{\sigma_u^2} + \frac{j}{\sigma_e^2}\Big).$$

Multiplying over successive inter-event times, yields the following likelihood contribution from renewal process i:

$$(2\pi)^{(-n_i/2)} \prod_{j=1}^{n_i-1} \Big(\frac{1}{\tau_j}\Big) \exp\Big\{ -\sum_{j=1}^{n_i-1} \frac{1}{2\tau_j^2}(g(T_{ij}) - \xi_{ij})^2 \Big\}$$

$$\times \{1 - \Phi[(g(T_{in_i}) - \xi_{n_i})/\tau_{n_i}]\},$$

where the final term is for the censored interval at the end of the observation period, and Φ denotes the standard cumulative normal distribution function. Hence, the log-likelihood function contribution from renewal process i is equal to (ignoring additive parts not depending on the parameters):

$$\log L_i(\mu, \boldsymbol{\beta}, \sigma_u, \sigma_e) = \sum_{j=1}^{n_i-1} \Big\{ -\frac{1}{2\tau_j^2}(g(T_{ij}) - \xi_{ij})^2 - \log \tau_j \Big\}$$

$$+ \log\{1 - \Phi[(g(T_{in_i}) - \xi_{n_i})/\tau_{n_i}]\}.$$

Estimation of the parameters may be performed by summing this function over the individual renewal processes and maximizing the resulting expression. Statistical analysis is based on ordinary asymptotic theory for maximum likelihood estimation, where the number of individual renewal processes is assumed to be reasonably large.

Example 7.4. Movements of the small bowel. Preliminary analysis indicated that there was no need for transforming the data, hence the link function g equals the identity. The log-likelihood was maximized to yield the estimates (with standard errors): $\hat{\mu} = 107$ (6.9), $\hat{\sigma}_u = 16.2$ (8.6), and $\hat{\sigma}_e = 49.3$ (4.3). The estimate of μ equals 107 minutes and answers the question how to assess the average length of MMC periods over individuals. If the calculation is performed without the incomplete periods, which has sometimes been done in similar studies, the result is 102 minutes. This is an underestimate, although not a dramatic one. In other situations, however, the difference may be much greater. In another study (Husebye, personal communication) based on 15 healthy individuals, ignoring the incomplete periods resulted in an underestimate of 14 minutes (12%) for the MMC period.

One further observes that the variation within individuals appears to be much larger than the variation between individuals. To test for the latter, a likelihood ratio test has been applied to the null hypothesis $\sigma_u = 0$, yielding the chi-square value 1.41 with one degree of freedom which is clearly nonsignificant (*P*-value 0.24). One problem in this analysis is that the null hypothesis is on the boundary of the parameter space. But following Dominicus et al. (2006), we could just halve the *P*-value to account for this, which gives a *P*-value of 0.12. The domination of intra-individual variation is biologically important when the length of the MMC period is used as a medical parameter. The lack of significance of the inter-individual variation is also confirmed in the shared frailty analysis of Section 7.1.

From a medical point of view, the large variation found within individuals is an important result. This has a major impact on the interpretation of the MMC period, implying, for instance, that differences between individuals and changes due to disease will be hard to prove if only one MMC period is observed. The numerical effect of censored periods will obviously be greater if the recording period is shorter or the duration of MMC periods longer than in the present study. □

7.6.2 General case

We now consider the transformation model in (7.22) with general censoring indicators D_{ij}, $j = 1, \ldots, n_i$, see Section 7.2.2. A closed-form likelihood cannot be given, but one can condition with respect to U_i, and get the following likelihood contribution for individual renewal process or cluster i:

$$L_i(\mu, \boldsymbol{\beta}, \sigma_u, \sigma_e) = \int_{-\infty}^{+\infty} \prod_{j=1}^{n_i} \left[\left\{ \frac{1}{\sigma_e} \phi(\gamma_{ij}(w)) \right\}^{D_{ij}} \right.$$
$$\left. \times \left\{ 1 - \Phi(\gamma_{ij}(w)) \right\}^{1-D_{ij}} \right] \frac{1}{\sigma_u} \phi\left(\frac{w}{\sigma_u}\right) dw,$$

where $\phi(x)$ is the density of the standard normal distribution and

$$\gamma_{ij}(w) = \frac{g(\widetilde{T}_{ij}) - \mu - \boldsymbol{\beta}^T \mathbf{x}_{ij} - w}{\sigma_e}.$$

The full likelihood is derived by multiplying this expression over the index i, that is,

$$L = \prod_{i=1}^{m} L_i(\mu, \boldsymbol{\beta}, \sigma_u, \sigma_e).$$

When maximizing the likelihood, the integral can be computed by numerical techniques (Skrondal and Rabe-Hesketh, 2004).

Empirical Bayes estimates of individual Us may be derived as follows. For an individual i with history H_i, we get

$$g(u \mid H_i) = \frac{P(H_i \mid U = u) \cdot g(u)}{P(H_i)},$$

implying that

$$E(U \mid H_i) = \frac{\int_{-\infty}^{+\infty} \prod_{j=1}^{n_i} \left[\{\phi(\gamma_{ij}(w))\}^{D_{ij}} \{1 - \Phi(\gamma_{ij}(w))\}^{1-D_{ij}} \right] w \phi\left(\frac{w}{\sigma_u}\right) dw}{\int_{-\infty}^{+\infty} \prod_{j=1}^{n_i} \left[\{\phi(\gamma_{ij}(w))\}^{D_{ij}} \{1 - \Phi(\gamma_{ij}(w))\}^{1-D_{ij}} \right] \phi\left(\frac{w}{\sigma_u}\right) dw}.$$

7.6.3 Comparing frailty and random effects models

In the tradition of statistics, it is common to handle multivariate data by some type of random effects model. The random effects for transformed times may be a natural choice to use in a medical research paper, because it belongs to the old and well-established tradition of normality-based statistical methods. There is also the advantage that one gets direct estimates of the components of variance. On the other hand, there is no doubt that approaches based on frailty are conceptually more natural when considering occurrence of events in time. The great success of survival and event history analysis shows the fruitfulness of intensity-based methods. Hence the frailty analysis has much to recommend it. One important point to its credit is the fact that individual empirical Bayes estimates can be explicitly given; no such explicit formulas can be given for the variance component model when censoring is present. Also, much more general censoring can be easily handled with frailty models than with accelerated failure time models.

An attractive property of the classical random effects models is that the random effect does not alter the structure of the error term, which stays normal in any case. This does not hold for the frailty models, as demonstrated in Chapter 6.

7.7 Exercises

7.1 Prove formula (7.10).

7.2 Plot the density of the Kummer distribution given in formula (7.14) for some sets of the parameters. The confluent hypergeometric function of the second kind goes into the normalization factor; this function may be computed by several programs, such as Mathematica, Matlab, or the fOptions package of R. Compare to the gamma distribution and show that the Kummer distribution may be more skewed.

7.3 Plot the density of the generalized inverse Gaussian distribution given in formula (7.11). Compare with plots of the gamma distribution and the Kummer distribution in the previous exercise. What kind of differences do you find?

7.4 Consider the Pareto frailty distribution as a special case of the Kummer distribution by inserting $\eta = 1$ and $\nu = 0$ into formula (7.14). Show that the Pareto density is $(\beta - 1)\delta^{\beta-1}(\delta + z)^{-\beta}$. If individual i has had $D_{i\bullet} = 5$ events with a total cumulative hazard $V_i = 10$, then what is the conditional distribution of the frailty? Draw plots of the unconditional and conditional frailty distributions in the same diagram.

7.5 Prove formula (7.19). From Chapter 6, write the probability density of the PVF distribution with $m = 1$. Compute the density and make plots for various parameter values.

Chapter 8
Marginal and dynamic models for recurrent events and clustered survival data

We shall consider observation of clustered survival data or processes with recurrent events as defined in the introduction to Chapter 7. In the case of recurrent events we focus on the concept of dynamic models, which represent an attempt to understand in explicit terms how the past influences the present and the future. We may think of this as causal influences, but statistical dependence on the past may also be a reflection of heterogeneity. Instead of setting up a random effects, or frailty, model, one may alternatively condition with respect to past events to get a counting process model with a suitable intensity process. Frailty will induce dependence, such that, for example, the rate of a new event is increased if many events have been observed previously for this individual, since this would indicate a high frailty. The formulas (7.16) and (7.17) can be seen as expressions transforming the frailty structure into conditional representations given the past.

The existence of dynamic models follows from a general theorem for semimartingales, namely the Doob-Meyer decomposition, which states, essentially, that any semimartingale can be decomposed into a martingale and a compensator (Section 2.2.3). The martingale represents the "noise" or unpredictable changes, while the compensator represents the influence of the past history. A counting process is a submartingale, and hence a semimartingale, and the compensator is just the cumulative intensity process.

As mentioned, it is often the case that the rate of new events in a process increases when there have been many events in the past. However, this is not always the case; there may be situations where many events in the past decrease the rate of new events. Such a protective effect will be discussed in a case study concerning diarrhea in children; see Section 8.7.

As an alternative to dynamic models, we have the marginal models that do not give a complete description of the past. In fact, by a marginal model one would usually mean that one ignores the dependence on the past in a process. This may be reasonable if the focus is really on a treatment effect, say, or on the effect of specific risk factors. But ignoring the past entails some technical complications, and one will miss the opportunity to understand the details of the underlying process.

The concepts of dynamic and marginal models shall be related to the counting process setup. An intensity process for a counting process with recurrent events will often depend on the past; if it does not, this means that the process is Poisson. If we have insufficient information of the past to model the intensity process, we may still model the rate of events occurring in the process in a more incomplete sense. In that case we speak about the rate function as the analogy of the intensity process, but the martingale property of the residual process, in this case the difference between the counting process and the cumulative rate function, will no longer hold. Typically, marginal models will employ rate functions. A dynamic model, using information from the past, may be an actual modeling of the intensity process, or it may be a model of the rate function if the past information is incomplete.

For additive regression models, there is a close connection between marginal (rate) models and the more complete intensity models. In fact, the former may be embedded in the intensity models, and by appropriate orthogonalization results for the marginal model may be derived from the intensity model.

A very comprehensive introduction to analysis of recurrent event data is given in the book by Cook and Lawless (2007), who discuss a number of different approaches in this field.

8.1 Intensity models and rate models

To introduce the rate models and describe how they differ from intensity models for counting processes, we briefly recapitulate some concepts. For individual i, let $N_i(t)$ be the process counting the number of occurrences of a recurrent event up to time t. According to the Doob-Meyer decomposition, the process $N_i(t)$ may be decomposed into the sum of a martingale $M_i(t)$ and a cumulative intensity process $\Lambda_i(t)$ (Section 2.2.5). Thus, provided that the cumulative intensity process is absolutely continuous, the existence of an intensity process $\lambda_i(t)$ is guaranteed. The interpretation of the intensity process is

$$E(dN_i(t)|\mathscr{F}_{t-}) = \lambda_i(t)dt, \qquad (8.1)$$

where the history $\mathscr{F}_{t-}$ contains information on (fixed and time-dependent) covariates as well as censorings and observed events in all counting processes prior to time t. Normally, an intensity process will depend on past occurrences in the counting processes, giving dynamic models as we understand it here. The exception to this are the Poisson processes. A model based on the intensity processes in this sense will be called an *intensity model*.

However, we can also ignore the dependence on the past and model a *rate function* $r_i(t)$ rather than the intensity process $\lambda_i(t)$, getting a *rate model* or a *marginal model*. To define the rate function, we first look at the problem without censoring. To this end, let $\widetilde{N}_i(t)$ denote the fully observed counting process for individual i. Then the rate function is given by

$$r_i(t)dt = E(d\widetilde{N}_i(t)|\mathscr{E}_{t-}),\qquad(8.2)$$

rate function where $\mathscr{E}_{t-}$ contains information on (all or some of) the *external* co-variates prior to time t but no information on internal covariates or events prior to this time. Note that for the rate model $\widetilde{N}_i(t) - \int_0^t r_i(s)ds$ will in general not be a martingale.

We do not (necessarily) observe all events for individual i; rather, the events of individual i are only observed when a left-continuous at risk indicator process $Y_i(t)$ takes the value one. In particular, if right-censoring at time C_i is the only reason for incomplete observation, $Y_i(t) = I(C_i \geq t)$. The process counting the observed events for individual i may now be written $N_i(t) = \int_0^t Y_i(s)d\widetilde{N}_i(s)$. Here we assume that the at risk process $Y_i(t)$ does not depend on internal covariates or previously observed events for individual i. Then by (8.2)

$$E(dN_i(t)|Y_i(t),\mathscr{E}_{t-}) = E(Y_i(t)d\widetilde{N}_i(t)|Y_i(t),\mathscr{E}_{t-})$$
$$= Y_i(t)E(d\widetilde{N}_i(t)|Y_i(t),\mathscr{E}_{t-}) = Y_i(t)r_i(t)dt.\qquad(8.3)$$

Note that the assumption we make on the censoring is stronger than the common assumption of independent censoring (Sections 1.4 and 2.2.8), where censoring is also allowed to depend on events in the past. Further, comparing (8.1) and (8.3), we note that the difference between the rate function and the intensity process lies in the conditioning. For the intensity process, we condition on both internal and external covariates *and* on the observed events prior to time t, while for the rate function we *only* condition on the at risk status and the external covariates.

One should realize that the intensity model and the rate model as defined earlier are, in a sense, extreme cases. One could have models using some of the dependence on the past, which are also rate models. As for the intensity models, they are usually built in a rather pragmatic fashion, using sufficient covariates, including functions of the past, so that the residual processes do not show any significant "overdispersion" compared with what should have been expected from martingale theory. In Section 8.3.2 we show how this is done.

It is important to note that the standard counting process results do not work for rate functions due to the lack of martingale property of $N_i(t) - \int_0^t Y_i(s)r_i(s)ds$. Often the estimation procedures are still valid, but the variance estimates from martingale theory are not correct (they will typically underestimate the true variance) and have to be substituted with sandwich type estimators. Important work on statistical analysis for rate functions has been done by Lawless and Nadeau (1995), Cook and Lawless (2007), Lin et al. (2000), Scheike (2002), Chiang et al. (2005), and Martinussen and Scheike (2006). Empirical process theory turns out to be useful for deriving asymptotic results, but we shall not go into the details of this here.

However, there is an important exception to the limitation of martingale theory for rate models: we shall show how additive models connect rate functions and intensity processes so that martingale methods may be valid also for rate models; see Sections 8.3.2 and 8.4.

8.1.1 *Dynamic covariates*

Dynamic covariates (Aalen et al., 2004) sum up important aspects of the previous development of the processes that may contain prognostic information. These could be repeated occurrences of disease, repeated awakenings during a night in a study of sleeplessness, etc. Examples of dynamic covariates for recurrent events data can be:

- *time since last event.* This could be used as a check of the Markov property. If the process is Markovian there should be no dependence on time since the last event.
- *number of previous events in the process.* This could also be seen as a check of frailty. In case of frailty effects one should expect that an excessive number of previous events would predict a greater intensity of events also in the future.

For clustered event data, dynamic covariates may also be useful:

- *estimated cumulative hazard.* If the counting process $N_i(t)$ counts the number of events in cluster i (like in the amalgam filling data; see Example 1.9), then the number of previous events is not appropriate. One has to consider also the number of units at risk at any time in a given cluster. A reasonable dynamic covariate in this case would be the estimated cumulative hazard of the event, estimated by a Nelson-Aalen estimator for each cluster. For instance, in the amalgam filling data, for each patient one could compute the Nelson-Aalen estimate of failure of amalgam fillings and use this as a dynamic covariate.

The examples refer to previous events in the same individual counting process. One could, however, also imagine cross-effects between the various counting processes, so that the intensity in a process depends on what has happened in other parallel processes. For instance, there are several events that might spread between individuals, for example, an infectious disease. Many events may be influenced by what happens in a social environment. A high rate of divorce may lower the barriers to divorce in married couples. A high tolerance to alcohol among peers may increase the risk of binge drinking. This kind of cross-dependence can also be included in our analysis.

Of course, there are numerous ways to construct dynamic covariates, and in any given case it must be evaluated what are the most reasonable choices. One problem with dynamic covariates is that they are sometimes of no use before some occurrences have taken place, hence they cannot be used from the very beginning. Practically, the problem may be handled by either ridge regression, see Section 8.3.2, or not starting estimation until a few events have occurred. There is nothing incorrect in this as long as one starts at an optional stopping time.

A very attractive feature of the martingale structure underlying the counting process theory is that dynamic covariates in the Cox and the additive model do not have to be treated any differently from other covariates with regard to estimation procedures; see also Chapter 4. In fact, the covariate functions can be arbitrary predictable processes (apart from regularity conditions, of course).

Dynamic covariates have been used by previous authors, for instance, Kalbfleisch and Prentice (2002), Aalen et al. (2004), Peña (2006), Peña et al. (2007), Gandy and Jensen (2004), and Miloslavsky et al. (2004). Dynamic covariates are, of course, internal covariates and special cases of time-dependent covariates; but the terminology "dynamic" focuses on their role as explicitly picking up past developments in the counting processes. It is well known that one has to be careful handling such internal covariates jointly with fixed covariates; see Kalbfleisch and Prentice (2002, page 199). Dynamic covariates are "responsive" in the terminology of Kalbfleisch and Prentice. Their values may, for instance, be influenced by treatment assignment, and in the statistical analysis dynamic covariates may "steal" from the effect of treatment and other fixed covariates and weaken their effect. Kalbfleisch and Prentice (2002, chapter 9) still use the number of previous events as covariate, but without sufficient clarification, in our opinion, of their validity and interpretation. One solution for handling this is treated in detail in Fosen, Borgan et al. (2006) and will be discussed in Section 8.4. We believe the use of dynamic covariates represents a useful and important procedure, but one has to be careful to avoid bias.

It is also important to realize that when dynamic covariates are inserted into a regression model for an intensity process, this amounts to defining a stochastic process and one has to take care that this process is in fact well defined. This is a nontrivial issue, especially with dynamic covariates in the Cox model where one could easily end up defining a process that may explode in finite time; see Section 8.6.3. Several authors use dynamic covariates in a Cox model, including Miloslavsky et al. (2004) and Kalbfleisch and Prentice (2002), so there is certainly a need to discuss this issue.

Dynamic covariates may also be used in a different sense, namely as indicating the structure in a series of time-dependent covariate measurements. An interesting example is given in de Bruijne et al. (2001), where time since last covariate measurement is used.

8.1.2 Connecting intensity and rate models in the additive case

For a Cox model it is well known that the proportional hazards structure is not preserved under marginalization, that is, when a covariate is regarded as a random variable and integrated out, This is demonstrated in Section 6.5.1. The additive structure is nicer in this respect, as demonstrated here.

First we add the general remark that covariates are typically considered as fixed quantities in regression models. However, in practice they are almost always random quantities, not determined or fixed in advance. This issue is ignored in most regression models but could be quite important for some interpretations.

In this subsection we assume for simplicity that no censoring is present.

Leaving out a fixed covariate

Assume the covariates X_1 and X_2 are independent random variables and that we have a counting process $N(t)$ with the following true model for the intensity process:

$$\lambda(t) = \beta_0(t) + \beta_1(t)X_1 + \beta_2(t)X_2. \tag{8.4}$$

Here the history $\mathscr{F}_{t-}$ is generated by the past jumps of the counting process as well as the covariates X_1 and X_2. Note that the statement that this is an intensity process is equivalent to saying that the counting process is a Poisson process, conditional on X_1 and X_2, since there is no dependence on the past in the definition of the intensity. Unconditionally, the process will be a doubly stochastic Poisson process.

Now assume that X_2 is unknown, that is, a frailty variable. The rate function $r(t)$ with X_1 as the only covariate may be found by the following informal argument (which is similar to the innovation theorem):

$$
\begin{aligned}
r(t) &= \frac{1}{dt}\mathrm{E}(dN(t) \mid X_1) = \frac{1}{dt}\mathrm{E}(\mathrm{E}(dN(t) \mid X_1, X_2) \mid X_1) \\
&= \mathrm{E}\{\beta_0(t) + \beta_1(t)X_1 + \beta_2(t)X_2 \mid X_1\} \\
&= \beta_0(t) + \beta_1(t)X_1 + \beta_2(t)\mathrm{E}(X_2).
\end{aligned}
$$

Hence, the marginal model is still linear in X_1 with the same regression function $\beta_1(t)$, and we can write the rate function as:

$$r(t) = \beta_0^*(t) + \beta_1(t)X_1.$$

Why is this not an intensity process but merely a rate function? The reason is that in calculating $r(t)$ we condition only with respect to X_1 and not with respect to the past of the counting process. To find the intensity process with a history generated by the past jumps of the counting process as well as just the covariate X_1 would be very complicated, and that is a justification for using the simpler rate function.

Notice that we would also get an additive model for $r(t)$ if X_1 and X_2 are correlated and normally distributed (due to the linear conditional mean), but then the regression function of X_1 would be changed (Aalen, 1989).

Leaving out the dynamic covariate $N(t-)$

We now start with an additive model where the dynamic covariate is the number of previous events, $N(t-)$, and find that additivity is preserved under marginalization. Assume the following intensity process:

$$\lambda(t) = \beta_0(t) + \beta_1(t)X + \beta_2(t)N(t-),$$

where the history $\mathscr{F}_{t-}$ is now generated by the past jumps of the counting process as well as the covariate X. Notice that $\lambda(t)$ defines a birth process with immigration (Chiang, 1968).

Define the cumulative rate function $R(t) = E(N(t)|X)$. The rate function $r(t) = R'(t)$ of the marginal model given only X is found as:

$$
\begin{aligned}
r(t) &= \frac{1}{dt}E(dN(t)\mid X) = \frac{1}{dt}E\left\{E(dN(t)\mid X,N(t-))\mid X\right\} \\
&= E(\beta_0(t) + \beta_1(t)X + \beta_3(t)N(t-)\mid X) \\
&= \beta_0(t) + \beta_1(t)X + \beta_3(t)E(N(t)\mid X).
\end{aligned}
$$

This yields the differential equation

$$
R'(t) = \beta_0(t) + \beta_1(t)X + \beta_2(t)R(t),
$$

with the initial condition $R(0) = 0$ corresponding to $N(0) = 0$. The solution of this linear differential equation is straightforward; we skip the details. Noting that the rate function equals $r(t) = R'(t)$, the following expression is found for it:

$$
\begin{aligned}
r(t) = \beta_0(t) + \beta_2(t)\int_0^t \beta_0(v)\exp(\int_v^t \beta_2(u)du)\,dv \\
+ X\left\{\beta_1(t) + \beta_2(t)\int_0^t \beta_1(v)\exp(\int_v^t \beta_2(u)du)\,dv\right\}.
\end{aligned}
$$

Hence, the rate function is still additive in X although the coefficient has changed. If, for example, $\beta_1(t)$ and $\beta_2(t)$ are both positive, then the coefficient of X is larger in the marginal than in the dynamic model.

Note that the linearity in X depends on the particular dynamic covariate used here and would not hold in general.

The difference in the coefficients found here between the marginal and dynamic models is very important. As an example, let $\beta_0(t) = 0$, $\beta_1(t) = 1$, and $\beta_2(t) = 1$. Then

$$
r(t) = X\exp(t).
$$

Hence the regression function of X equals $\beta_1(t) = 1$ in the dynamic model and $\exp(t)$ in the marginal model, which clearly demonstrates the large difference in the estimated influence of X, which there may be between a full dynamic and a marginal model.

Hence we have two possibly very different coefficients for the effect of X, and one might ask which is more "correct". Usually, the answer would be that the coefficient in the marginal model is the correct one. The dynamic covariate, so to speak, carries the influence of the fixed covariate, X, and one might say that the dynamic covariate is "stealing" the effect of the fixed one. This is a well-recognized problem when analyzing the effect of treatments in survival analysis, where treatment effects may be underestimated in models with internal time-dependent covariates (Kalbfleisch and Prentice, 2002). The problem is that one might be interested in a dynamic model to get a full picture of the process, but the effects of some of the covariates may be wrongly estimated; typically, they are underestimated.

However, the question of the correct effect of treatment might also be more complex. In Section 8.6 we show that the dynamic model throws a more detailed light on the treatment mechanism than does the marginal model.

One should also realize that this is really a causal problem. The point is that some covariates may be in the causal pathways of other covariates, and if they are all included in the same analysis, the correct effects are not being estimated for all of them. A simple solution is available within the additive model context and shall be presented in Section 8.4.

8.2 Nonparametric statistical analysis

Before going into the regression models, we shall have a look at a simple plotting procedure for marginal models and for intensity models with dynamic covariates. We shall focus on the Nelson-Aalen estimator, which is the more generalizable quantity since the Kaplan-Meier estimator is limited to the survival situation.

Recurrent event data may be analyzed in several different ways. Peña et al. (2001) studied Kaplan-Meier and Nelson-Aalen plots for inter-event times. Lawless and Nadeau (1995), on the other hand, looked at the cumulative mean function, which is simply the average rate of events occurring in the individual processes, and studied a Nelson-Aalen estimator for this. We focus on the latter situation here, generalizing it to a dynamic setting. However, we first present a simple version for clustered data.

8.2.1 A marginal Nelson-Aalen estimator for clustered survival data

Consider m independent clusters, each having a number of units, and let $N_i(t)$ be the process counting the failures for cluster i, while $Y_i(t)$ are the number of units at risk within the cluster. The hazard rate in cluster i is $\alpha_i(t)$, and we want to estimate the average cumulative hazard rate over the clusters. Let $\mathscr{R}(t)$ be the set of clusters having at least one unit at risk at time t and let $m(t)$ be the number of such clusters. More precisely, we want to estimate the integral over the following average hazard rate:

$$\bar{\alpha}(t) = \frac{1}{m(t)} \sum_{i \in \mathscr{R}(t)} \alpha_i(t).$$

Note that this is the mean over those clusters that are actually observable at time t and that $\mathscr{R}(t)$ may change over time.

We want to estimate the cumulative average hazard rate $\int_0^t \bar{\alpha}(s)ds$. The estimation procedure is simply to compute a Nelson-Aalen estimator for each cluster and then average as follows:

$$L(t) = \int_0^t \frac{1}{m(s)} \sum_{i \in \mathcal{R}(s)} \frac{1}{Y_i(s)} dN_i(s).$$

The corresponding marginal survival curve may be estimated as $\exp(-L(t))$ (or a corresponding product-integral). By standard counting process arguments the following estimate of the standard error of $L(t)$ may be derived:

$$s(t) = \sqrt{\int_0^t \frac{1}{m(s)^2} \sum_{i \in \mathcal{R}(s)} \frac{1}{Y_i^2(s)} dN_i(s)}. \tag{8.5}$$

Note that this is based on a fixed effect model, where the hazard rates $\alpha_i(t)$ are supposed to be given quantities.

One caveat should be noted concerning $L(t)$; the estimate only makes sense as long as the great majority of clusters still have units at risk, that is, $m(t)$ should not be much less than m. Otherwise, the mean would only be representative of a selected group, probably with low failure intensities since these would be most likely to have a long survival.

Marginal nonparametric analyses are useful to some extent. However, the fitting of smooth parametric models, possibly with random effects to signify individual variation, is a natural and valuable supplement. One important approach is based on the shared frailty models treated in Section 7.2.

Example 8.1. Duration of amalgam fillings. The data have been described in Example 1.9 and analyzed in Example 7.2. An important issue is to get an overall estimate of the lifetimes of amalgam fillings. Here an individual patient is a "cluster", while his fillings are the "units".

A plot of $L(t)$ for the present data set is shown in Figure 8.1 with pointwise 95% confidence intervals. The plot is drawn up to 16 years and could be considered reliable up to about 14 years when 25 of 32 patients still have fillings at risk. After this time the risk set dwindles quickly. □

8.2.2 A dynamic Nelson-Aalen estimator for recurrent event data

The marginal Nelson-Aalen plot was constructed as an average plot for each individual process, and is naturally connected to the clustered event situation exemplified by the amalgam data. When we have recurrent event data, we can construct dynamic Nelson-Aalen plots dependent on the individual processes. As an example, we shall use the sleep data (Example 1.14). In the definition of the Nelson-Aalen estimator:

$$\widehat{A}(t) = \int_0^t Y(u)^{-1} dN(u),$$

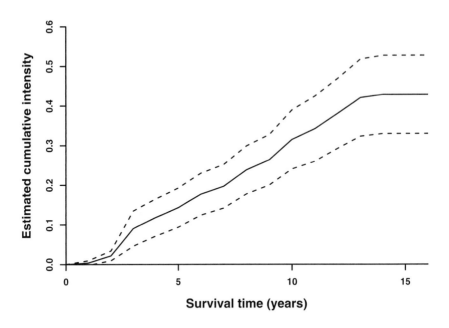

Fig. 8.1 *Mean Nelson-Aalen plot for the life time of amalgam fillings.*

$N(t)$ is typically the sum of the individual counting processes, and $Y(t)$ is then the number of processes at risk. Notice that the risk set can be defined in a dynamic fashion, for example, as those individuals that have a certain history up to time t. This gives a very flexible way of judging the connections between the past and the future.

A special case will be when the risk set consists of all observable processes at a given time. Then the Nelson-Aalen estimator will estimate the cumulative average rate of events in the processes; it is also called the cumulative mean function (Lawless and Nadeau, 1995).

The Nelson-Aalen estimates described here will not necessarily satisfy the martingale property. That depends on whether the intensity process has the multiplicative intensity form as described in Section 3.1.2. For processes with recurrent events there may be dependence on the past, and hence the Nelson-Aalen estimate must in general be defined within a rate model; that is, we may have a multiplicative rate model in the terminology of Martinussen and Scheike (2006). Estimation of variances, and asymptotic theory, may be handled outside of the martingale theory. For instance, one can use the theory of empirical processes; see Ghosh and Lin (2000).

However, it is important to be aware that rate models and marginal models may in many cases also be handled within the framework of counting processes and martingales, so that even Nelson-Aalen estimators for rate models may be studied with martingale theory. This can be done when the intensity process can be formulated (possibly approximately) by an additive model, see Sections 8.3.2 and 8.4.

Example 8.2. Sleep data. Yassouridis et al. (1999) describe an experiment where a number of people have been observed during one night (Example 1.14). Every 30 seconds their sleep status was registered. In addition, the cortisol level was measured every 20 minutes. We have analyzed a set of 27 individuals. Following Yassouridis et al., we define the time for each individual as time since first time asleep.

We wish to study the number of awakenings during the night. We define those at risk in three different ways, where "average" denotes the statistical mean:

1. $Y(t)$ is the number asleep just before time t
2. $Y_1(t)$ is the number asleep at time t who have had a larger than (or equal to) average number of awakenings before time t
3. $Y_2(t)$ is the number asleep at time t who have had a smaller than average number of awakenings before time t

Note that $Y(t) = Y_1(t) + Y_2(t)$.

The three Nelson-Aalen plots are shown in Figure 8.2 and show clearly that those who have been awake several times in the past are more likely to wake up again soon. Based on smoothing the increments of the Nelson-Aalen plot by the lowess procedure (Stata), we also estimate the hazard functions; see Figure 8.3. Clearly, the difference between the curves is large to begin with, indicating that previous awakenings are especially influential early in the night, but not so much later. Estimation of hazard rates based on the Nelson-Aalen plot is discussed in Section 3.1.4; see also Chiang et al. (2005).

The present estimation can be seen as a special case of the additive hazards model (which for recurrent events is treated in more detail in Section 8.3.2). Define for the individual process i the indicator function $I_i(t)$ equal to 1 if the number of awakenings at time t is at least equal to the mean for all processes, and $I_i(t)$ equal to zero otherwise. Then the rate function can be written as:

$$r_i(t) = \beta_1(t)I_i(t) + \beta_2(t)(1 - I_i(t)),$$

and the estimated cumulative regression functions corresponding to $\beta_1(t)$ and $\beta_2(t)$ are precisely the Nelson-Aalen plots for cases 2 and 3. □

8.3 Regression analysis of recurrent events and clustered survival data

We shall consider the situation in which either clustered survival data or a number of processes with recurrent events are studied. What we primarily want to do

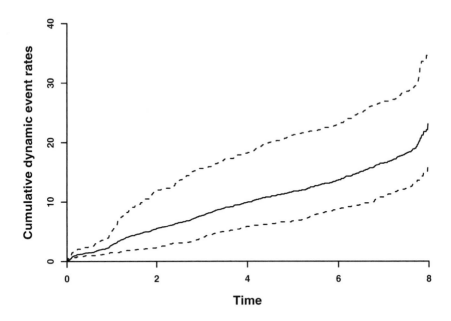

Fig. 8.2 *Dynamic Nelson-Aalen plots. Middle curve: cumulative event rate of waking up for all people asleep; upper curve: cumulative event rate of waking up for people asleep who have had an above average number of awakenings previously; lower curve: cumulative event rate of waking up for people asleep who have had a below-average number of awakenings previously.*

is to make a joint analysis, getting a correct picture of both the effect of fixed co-variates, including possibly treatment, and the dynamic properties of the underlying process when recurrent events are observed. An alternative to a dynamic analysis is a marginal analysis, which shall also be discussed.

There is a considerable amount of literature on regression analysis of multivariate survival data. However, there is still a need of clarification in this area, in particular as regards the relationship between marginal and dynamic models. Metcalfe and Thompson (2007) discuss and compare two of the major approaches for analyzing recurrent survival data, one being marginal and the other dynamic in character. Although their discussion is useful, it also illustrates a lack of clarification in this field, in particular as regards the fact that you get different regression estimates with the two different models. Our ambition is to give a clear representation of this issue; see Section 8.4.

In the field of recurrent event data, one often refers to the Andersen-Gill model, which is a general counting process formulation of the Cox model. However, when introducing dynamic covariates in the Andersen-Gill model with the purpose of making a representation of the dependence structure in the processes, it is not clear

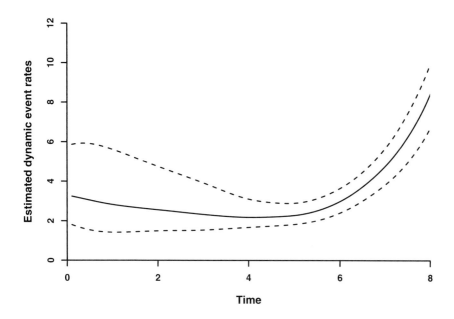

Fig. 8.3 *Estimated dynamic event rates. Middle curve: rate of waking up for all people asleep; upper curve: rate of waking up for people asleep who have had an above-average number of awakenings previously; lower curve: rate of waking up for people asleep who have had a below-average number of awakenings previously.*

that the resulting model is meaningful. In fact, one must beware of several pitfalls: the risk of models that are not well defined and may explode in finite time, and the risk of underestimation of treatment effects; see Sections 8.4 and 8.6.3.

An illuminating discussion of dynamic modeling is given in the papers by Peña (2006) and Peña et al. (2007).

8.3.1 Relative risk models

A marginal Cox model

We may adopt a relative risk type model for the rate function, that is, assume that (8.2) takes the form

$$r_i(t) = Y_i(t)\alpha_0(t)r(\boldsymbol{\beta}, \mathbf{x}_i(t)). \tag{8.6}$$

Table 8.1 *Parameter estimates from a marginal Cox regression model for the bladder data.*

Covariate	Estimate	Standard Error	Wald Statistic	P-value
Treatment	−0.465	0.268	−1.73	0.083
Number of initial tumors	0.175	0.063	2.77	0.006
Size of largest tumor	−0.044	0.079	−0.554	0.580

The most common choice would be $r(\boldsymbol{\beta}, \mathbf{x}_i(t)) = \exp\{\boldsymbol{\beta}^T \mathbf{x}_i(t)\}$, but other alternatives could be used as well. Most commonly, we would have fixed covariates; if they are time-dependent they will have to be external.

For the model specification in (8.6) we may still estimate the regression coefficients by maximizing (4.7). However, for the marginal model, (4.7) is no longer a partial likelihood, but only a pseudo-likelihood. A detailed theory for the marginal Cox model has been developed by Lin et al. (2000). Counting process theory cannot be used here since the martingale property is no longer fulfilled. Instead, empirical process theory is applied. We shall not present the details here, but instead refer to the paper by Lin et al. (2000). Sandwich type estimates may be used for standard errors (also called robust standard errors), and these are given in computer programs like Stata and R/S-Plus. Of course, standard errors of the regression coefficients may also easily be estimated by the bootstrap.

Example 8.3. Recurrent superficial bladder tumors. In Example 7.3 we analyze the bladder tumor data of Example 1.10 by a semiparametric gamma frailty model. We now use a marginal Cox model, so we drop the frailty component and only include the covariates treatment (thiotepa vs. placebo), number of initial tumors, and size of the largest initial tumor. We use the `coxph` program in R (R Development Core Team, 2007) with robust standard errors, and we get the estimated regression coefficients in Table 8.1. Also in the marginal model, we get a clear effect of the number of initial tumors and an indication of an effect of treatment (although not significant). The estimated effects are smaller than those reported in Table 7.3 for the semiparametric gamma frailty model. In this connection, one should note, however, that the interpretation of the parameter estimates are not the same for the two models. The marginal estimates are "population average" effects, whereas the covariate effects from the frailty model can be seen as effects on the individual level. The data shall also be analyzed by a dynamic additive model in Example 8.6. □

A dynamic Cox model

For recurrent event data, where the event in question may occur more than once for each individual, the intensity process $\lambda_i(t)$ will typically depend on the number of prior events for individual i and the times of their occurrences. One option is then to model this dependence using internal time-dependent, or dynamic covariates. If one has included sufficient information about the past to actually model the intensity process, martingale theory can be used. If there is only partial information about the relevant aspects of past development, we would still have a marginal model similar to the one described in the previous section, but with an extended rate function where past developments also play a role.

Let us assume that we can actually model the intensity process. As shown in Section 8.6.3, the process defined by a Cox model with $\lambda_i(t) = \alpha_0(t)\exp(\beta N_i(t-))$ will for $\beta > 0$ have a positive probability of explosion in finite intervals. Modifying the dynamic covariate by using $N_i(t-)/t$ or $\log(N_i(t-))$ instead brings us into a situation where explosion may be avoided, although this is not guaranteed. Cox models with the number of previous events $N_i(t-)$ as a covariate have been suggested by several authors, including Miloslavsky et al. (2004) and Kalbfleisch and Prentice (2002), without the issue of proper existence of the model being discussed.

Anyway, in spite of these theoretical difficulties, a Cox model can still be applied with some caution. One may hope that the data will force the estimates into regions where things work reasonably well. In the analysis, we can apply an ordinary Cox model, but with some of the covariates being functions of the individual's past history, that is, special cases of internal time-dependent covariates. The estimation can be carried out in the usual way, since the counting process approach allows covariates dependent on past history; see Section 4.1. Analysis of a set of data is given in Example 8.4.

8.3.2 Additive models

The additive regression model was defined and studied in Section 4.2, and the theory there also covers internal time-dependent covariates, including dynamic covariates. Here we consider the application of this model to the clustered survival data and the recurrent event data situation. The additive model can be defined for both intensity processes and rate functions, and the estimation will be carried out in the same manner. Just like for the Cox model, the difference between the two approaches arises when standard deviations of estimates and test statistics have to be estimated; see Martinussen and Scheike (2006, section 5.6). However, in the additive case, a rate model may often be embedded in an intensity model by appropriate orthogonalization, so that variance results and asymptotic theory may be valid for the rate model as well.

The material here is taken from Aalen et al. (2004). Extensions of additive models to longitudinal data can be found in Borgan et al. (2007) and Diggle et al. (2007).

In the estimation we have to distinguish again between the recurrent events situation and the clustered survival data.

(i) *Recurrent events; no clustering.* Let $\lambda_i(t)$ be the intensity process (or the rate function) for individual i, and consider the model (4.55), which we write anew here:

$$\lambda_i(t) = Y_i(t)\left\{\beta_0(t) + \beta_1(t)x_{i1}(t) + \cdots + \beta_p(t)x_{ip}(t)\right\}.$$

The new element is that the covariate processes may be functions of what happened previously in the counting processes. The estimation is carried out as in Section 4.2, the estimator $\mathbf{B}(t) = (B_0(t), B_1(t), \ldots, B_p(t))^T$ of the cumulative regression functions being defined by

$$\widehat{\mathbf{B}}(t) = \int_0^t J(u) \mathbf{X}^-(u) \, d\mathbf{N}(u).$$

The design matrix $\mathbf{X}(t)$ has as its ith row the elements $Y_i(t), Y_i(t)x_{i1}(t), \ldots, Y_i(t)x_{ip}(t)$ and the generalized inverse is given by

$$\mathbf{X}^-(t) = \left(\mathbf{X}(t)^T \mathbf{X}(t)\right)^{-1} \mathbf{X}(t)^T.$$

In some cases $\mathbf{X}(t)^T \mathbf{X}(t)$ may be singular, and the generalized inverse is not well defined. This may, for instance, occur early in processes with dynamic covariates, since the process has to run a while for the covariates to be meaningful (if, for example, the covariate is the number of previous events, then initially this is zero for all processes and no estimation can be performed). One solution is to understand $\mathbf{X}^-(t)$ to be identically equal to zero when $\mathbf{X}(t)^T \mathbf{X}(t)$ is singular. Hence, estimation takes a pause, returning zero values, in the case of singularity. However, in singular or near-singular situations, one may alternatively use ridge regression to avoid stopping the estimation. The generalized inverse is then modified in the following way:

$$\mathbf{X}^-(t) = \left(\mathbf{X}(t)^T \mathbf{X}(t) + k\mathbf{I}\right)^{-1} \mathbf{X}(t)^T,$$

where $\mathbf{I}$ is the identity matrix and k is a suitable constant. Ridge regression is a standard tool in ordinary linear regression with well-known properties in that context. We shall find it useful here as well.

Often one may want to center the covariates. That is, for all columns in $\mathbf{X}(t)$, except the first, one subtracts the mean of those individuals at risk at any given time. Then column one of $\mathbf{X}(t)$ is orthogonal with respect to the remaining columns, implying that the first element of $\widehat{\mathbf{B}}(t)$ is the same estimate as one would get if the covariates were not included in the analysis. This implies that the first element of $\widehat{\mathbf{B}}(t)$, that is, the estimated cumulative baseline intensity, is the Nelson-Aalen estimator. We shall center all covariates, thus ensuring the validity of this interpretation.

This orthogonalization can be continued; one may pick out, say, a fixed covariate and then orthogonalize the remaining covariates with respect to this. Then the estimated regression function of the chosen covariate will coincide with what had been estimated in a marginal additive model with only this covariate present. This is a way of bringing marginal models into the structure of counting processes and martingales; see also Section 8.4.

(ii) *Clustered survival data.* The case with several units at risk for each individual is also important. The amalgam fillings example is of this kind, with the units for a certain patient being his fillings at risk. One alternative is to aggregate over units, such that the counting process N_i for individual i, that is, the ith element of $\mathbf{N}$, is the sum of all counting processes for the individual's units at risk. The intensity process (or the rate function) may be written:

$$\lambda_i(t) = K_i(t)Y_i(t)\left\{\beta_0(t) + \beta_1(t)x_{i1}(t) + \cdots + \beta_p(t)x_{ip}(t)\right\},$$

where $K_i(t)$ is the number of units at risk at time t for individual i. As before, let $\mathbf{X}(t)$ be the $n \times (p+1)$ matrix, where $Y_i(t), Y_i(t)x_{i1}(t), \ldots, Y_i(t)x_{ip}(t)$ are the elements of the ith row, and let $\mathbf{K}(t)$ be the diagonal matrix with the ith diagonal element equal to $K_i(t)$. In vector form we may write

$$\lambda(t) = \mathbf{K}(t)\mathbf{X}(t)\boldsymbol{\beta}(t).$$

A reasonable estimator of $\mathbf{B}(t)$ in this case is

$$\widetilde{\mathbf{B}}(t) = \int_0^t J(u)\,\mathbf{X}^-(u)\mathbf{K}(u)^{-1}\,d\mathbf{N}(u).$$

Simple modifications of known results for $\widehat{\mathbf{B}}(t)$ from Section 4.2 may yield corresponding results for $\widetilde{\mathbf{B}}(t)$.

It is not always natural to aggregate. Considering the amalgam fillings, it might be the case that individual fillings for a patient had tooth-specific covariates and that the counting processes could therefore not be aggregated without loss of information. In that case, there would be a counting process for each filling. The dynamic covariates would then have to contain information on the previous fates of all fillings belonging to a specific person. This creates a dependence within clusters of counting processes.

Residual processes

The residual processes as defined and studied in Section 4.2.4 are valid also for this more general situation, with several events in each process and dynamic covariates. However, we can now study the residual processes for individual counting processes, which was not possible when there was at most one event in each process. We have to look more carefully into the estimation of the variance of the residual processes.

In Section 4.2.4 the vector of residual processes was defined by

$$\mathbf{M}_{\text{res}}(t) = \int_0^t J(u)\,d\mathbf{N}(u) - \int_0^t J(u)\mathbf{X}(u)d\widehat{\mathbf{B}}(u),$$

and in Equation (4.75) it was shown how this is a martingale when the additive model in question is actually a model of the intensity process. If we only model a rate function, then the residual process still makes sense, but it is no longer a martingale. In fact, in this section we demonstrate how to check whether our model may reasonably be viewed as a model of the intensity process or if some essential element is lacking so that it is only a rate function that we estimate.

Assume that we have an additive model of the intensity process. In that case, $\mathbf{M}_{\text{res}}(t)$ is a martingale, and the predictable variation process of this martingale may be derived from the theory of stochastic integrals, to yield the matrix

$$\langle \mathbf{M}_{\text{res}} \rangle(t) = \int_0^t J(u)(\mathbf{I} - \mathbf{X}(u)\mathbf{X}^-(u)) \operatorname{diag}(\boldsymbol{\lambda}(u)du)(\mathbf{I} - \mathbf{X}(u)\mathbf{X}^-(u))^T.$$

In order to estimate this process, and hence the variance of the residual processes, one must substitute the intensity vector $\boldsymbol{\lambda}(s)\,ds$ by the estimate derived from the regression model: $\mathbf{X}(u)\mathbf{X}^-(u)\,d\mathbf{N}(u)$. Hence the following estimated covariance matrix of the martingale residual processes is suggested:

$$\mathbf{V}(t) = \int_0^t J(u)(\mathbf{I} - \mathbf{X}(u)\mathbf{X}^-(u)) \operatorname{diag}(\mathbf{X}(u)\mathbf{X}^-(u)\,d\mathbf{N}(u))(\mathbf{I} - \mathbf{X}(u)\mathbf{X}^-(u))^T.$$
$$(8.7)$$

Standardized residual processes are now defined by dividing the residual processes by their estimated standard deviation at any time t. Furthermore, by applying kernel or lowess estimation to $\mathbf{M}_{\text{res}}(t)$ and $\mathbf{V}(t)$, one may estimate standardized residuals with a local interpretation.

By asymptotic theory, normal distributions will appear when a reasonable number of events occur, that is, when t is not too small. Hence, one would expect most standardized residuals to have values between -2 and $+2$ if the model is true; therefore, plotting the standardized residual processes will give information on model fit. We can also plot the standard deviation of the standardized residual process, which should then be close to 1 if the model holds. Note that this is a way of checking whether the estimated $\lambda_i(t)$ really estimates an intensity process. In case of deviation, we must conclude that we just estimate a rate function. For an illustration of this approach, see Section 8.6.

In practice, the question of whether the intensity model holds must be approached in a pragmatic way. We cannot know what the true Doob-Meyer decomposition really is, and there will probably be many different estimated intensity processes that could give a good fit. So the issue is to find one that can serve the purpose. If the fit is not good, it may mean that some important effect on the intensity process has not been revealed, and one must be content with having a rate function.

Testing the martingale property

As a continuation of the preceding theory, we can also perform statistical testing to see whether we can reject a null hypothesis that the residuals satisfy the martingale property, which would be more or less the same as testing that we really have an intensity process as opposed to a rate function. Over a given time interval $[0, t_0]$ we define the test as follows:

$$\mathbf{V} = \int_0^{t_0} \mathbf{U}(s)d\mathbf{M}_{\text{res}}(s),$$

where $\mathbf{U}(t)$ is a $q \times n$ matrix of predictable processes. Using formula (8.7) we get the following variance estimate:

$$\widehat{\text{cov}}\,\mathbf{V} = \int_0^{t_0} J(s)\mathbf{U}(s)(\mathbf{I} - \mathbf{X}(s)\mathbf{X}^-(s)) \qquad (8.8)$$
$$\times \text{diag}(\mathbf{X}(u)\mathbf{X}^-(u)\,d\mathbf{N}(u))(\mathbf{I} - \mathbf{X}(s)\mathbf{X}^-(s))^T\mathbf{U}(s)^T.$$

The test statistic

$$\chi^2(\mathbf{V}) = \mathbf{V}^T\widehat{\text{cov}}(\mathbf{V})^{-1}\mathbf{V}$$

will be chi-squared distributed with q degrees of freedom under the null hypothesis that the residual processes are the appropriate martingales that follow from counting process theory.

Variance estimators for rate functions

When the rate function uses less information about the past than would be needed for an intensity model, the variance can be estimated without using martingale theory. The residuals presented here are related to a robust variance estimator for the additive model suggested by Scheike (2002). When the martingale property does not hold, Scheike proposes the following estimator of the variance of the estimated regression function:

$$\widehat{\text{cov}}\,\widehat{\mathbf{B}}(t) = \sum_{i=1}^{n} \widehat{\mathbf{Q}}_i(t)\widehat{\mathbf{Q}}_i(t)^T,$$

with

$$\widehat{\mathbf{Q}}_i(t) = \int_0^t J(u)\left(\mathbf{X}(u)^T\mathbf{X}(u)\right)^{-1}X_i(u)^T\left(dN_i(u) - X_i(u)d\widehat{\mathbf{B}}(u)\right),$$

where $X_i(u)$ is the ith row of $\mathbf{X}(t)$. This is a sandwich type estimator.

Embedding a rate model in an intensity model

An alternative method in some cases would be to embed the rate model in an intensity model. Assume that the true intensity model is

$$\lambda_i(t) = Y_i(t)\{\beta_0(t) + \beta_1(t)x_{i1}(t) + \cdots + \beta_p(t)x_{ip}(t)\},$$

but that one is interested in estimating just the rate model

$$r_i(t) = Y_i(t)\{\beta_0^*(t) + \beta_1^*(t)x_{i1}(t)\}.$$

The correspondence between the two situations is taken care of by orthogonalization, which has to be carried out at each event time. Subtracting first the mean of all covariates at every time t yields covariate functions that are orthogonal to the constant term; assume this is already done in the formulations of $\lambda_i(t)$ and $r_i(t)$. The next step is to orthogonalize covariate 2 and onward with respect to the first covariate, yielding an intensity model

$$\lambda_i(t) = Y_i(t) \left\{ \beta_0^*(t) + \beta_1^*(t)x_{i1}(t) + \beta_2(t)x_{i2}^*(t) + \cdots + \beta_p(t)x_{ip}^*(t) \right\}. \tag{8.9}$$

The orthogonalization of the $p-1$ last covariates changes the first two β-functions, but not the rest, and also ensures that the first two β-functions are the same as in the rate model when a least squares estimation approach is used. If all the covariates are observed, one can estimate model (8.9). The estimate of the cumulative regression corresponding to $\beta_1^*(t)$ would be the same as for the rate model and its standard error would follow from martingale estimation of the variance based on model (8.9). The orthogonalization principle is further developed in Section 8.4 and Chapter 9.

Testing for influence: The hat matrix

The hat matrix is an important quantity in ordinary linear regression. It is defined as $\mathbf{X}(t)\mathbf{X}^-(t)$, and the diagonal is of interest as a measure of influence. It may be quite informative to plot the diagonal elements, $h_{jj,\text{cum}}(t)$, of the cumulative hat matrix:

$$H_{\text{cum}}(t) = \sum_{T_i \leq t} \mathbf{X}(T_i)\mathbf{X}^-(T_i), \tag{8.10}$$

where T_i are the successive times when jumps occur in any process. The idea is to look for processes with particularly high values. From ordinary linear regression (see, e.g., Wetherill (1986)), it is well known that the average value of the diagonal elements of a hat matrix is k/n, where n is the number of observations and k is the rank of the hat matrix. In fact, k coincides with the sum of the diagonal elements of the hat matrix. Elements above k/n are said to have high leverage, and it is common to select points for investigation if a diagonal element of the hat matrix is greater than $2k/n$. In our case, we can apply this theory at every jump time, and so the criterion for an outlying process will be when

$$h_{jj,\text{cum}}(t) > 2 \sum_{T_i < t} \frac{\text{rank}(\mathbf{X}(T_i)\mathbf{X}^-(T_i))}{n_i},$$

where n_i is the number of processes at risk at time T_i.

An application of the hat matrix is presented later in Example 8.5 under the analysis of sleep patterns.

Example 8.4. Movements of the small bowel. The small bowel motility data are described in Examples 1.11 and 7.1. For each individual there is a counting process running in clock time starting with the first phase III and counting the later phase IIIs. The intensity of the occurrence of a phase III shall be analyzed, with covariates as follows:

Table 8.2 *Small bowel motility: Cox analysis with dynamic covariates.*

Covariate	Estimate	Standard Error	Hazard Ratio	Wald Statistic	P-value
Number of previous events	−0.00	0.11	1.00	−0.01	0.99
Time since last event	1.77	0.29	5.88	6.05	1.4×10^{-9}

- Covariate 1: This is a time-dependent covariate counting the number of previous phase IIIs for the individual. The intention is to decide whether there is dependence between the inter-event times for an individual.
- Covariate 2: This is another time-dependent covariate measuring time since the last occurrence of a phase III. The object is to check whether the process is Markovian. The covariate is dichotomized to be smaller (or equal to) or larger than 50 minutes.

To estimate the influence of the covariates in a meaningful way, it is clear that some events must already have occurred since the covariates are defined relative to previous events. Here we decided to start estimation at midnight when a few events had already occurred.

First we use a dynamic Cox model with the two covariates. The results of the analysis are shown in Table 8.2. The influence of the number of previous events is seen to be virtually nil. This fits with the frailty analysis in Example 7.1, which indicates that there is very little variation between individuals as regards the occurrence of phase IIIs. In practice, this means that the intraindividual variation dominates. Furthermore, one sees from the table that the time since the previous event for the individual in question has a strong effect on the intensity of a new event; the longer the time, the more likely a new event. This shows that the process is non-Markovian, a conclusion that again fits well with the frailty analysis in Example 7.1, where the inter-event time for each individual is estimated to have an increasing Weibull hazard.

We also perform an additive analysis, applying the same covariates as for the Cox model. The results of the analysis are shown in Figure 8.4. The cumulative baseline intensity appears to be approximately a straight line, indicating a constant intensity of new phase IIIs. As in the Cox analysis there is no influence of the number of previous events (using the test for the additive model (Section 4.2.3), suitably normalized, yields the test statistic −0.33 based on observations from midnight to 6.30). One also sees from the figure that time since the last previous event for the individual in question has a strong effect on the intensity of a new event; the longer the time, the more likely a new event (normalized test statistic is 6.15, $p < 0.001$, from midnight to 6:30). There do not seem to be significant time-dependent changes. Hence, the results in this case are very similar to those from the Cox model, but a more detailed picture is given. □

Example 8.5. Analysis of sleep patterns. We analyze the sleep data; see Examples 1.14 and 8.2. In these data from Yassouridis et al. (1999), 27 individuals had

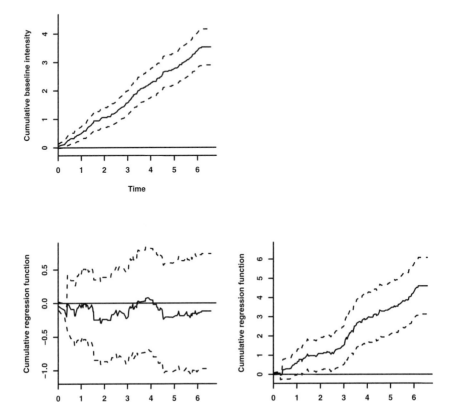

Fig. 8.4 *Occurrence of phase III events in small bowel motility: cumulative baseline intensity (upper panel), cumulative regression function of covariate measuring previous number of phase III events (lower left panel) and cumulative regression function of covariate measuring time since last phase III event (lower right panel). Pointwise 95% confidence limits. The time axis goes from midnight until 6:30.*

their sleep status registered every 30 seconds over one night. In addition, the cortisol level was measured every 20 minutes. We present some statistical analyses from Aalen et al. (2004).

An analysis of the data using a multiplicative hazards model is described by Yassouridis et al. (1999) and by Fahrmeir and Klinger (1998). Here, we do not perform an extensive analysis of the data, but use them for illustrative purposes to indicate the potential of our approach.

We have confined ourselves to the transitions from the state "asleep" to the state "awake", corresponding to the process that counts the cumulative number of

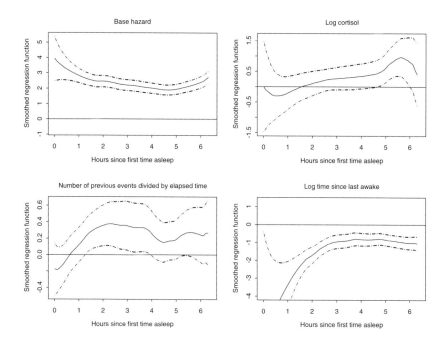

Fig. 8.5 *Sleep data: smoothed baseline hazard function and smoothed regression functions to-gether with pointwise 95% confidence limits. Smoothing parameter (bandwidth) is 1.67 hour. The figure is republished from Aalen et al. (2004).*

transitions of this kind after the first time the individual falls asleep. The num-ber at risk at each time point is the number of persons who are asleep just prior to this time. We here estimate the regression functions by kernel smoothing tech-niques. For this purpose we have used the methods suggested in Aalen (1993) and Keiding and Andersen (1989). We have used ridge regression with parameter 0.001 and smoothing bandwidth of 1.67 hours.

The following time-varying covariates have been used:

- covariate 1: logarithm of cortisol level
- covariate 2: cumulative number of times awake, divided by elapsed time
- covariate 3: logarithm of time since last awake

The two latter covariates are dynamic covariates. The smoothed regression func-tions (not cumulative functions) are shown in Figure 8.5. The figure shows that cortisol has a positive effect (i.e., increasing the likelihood of waking up) during the later part of the night. The number of previous times awake also has a positive effect on the hazard of waking up during most of the night. The length of the current sleeping period has a negative effect; the longer it has lasted, the less likely is it

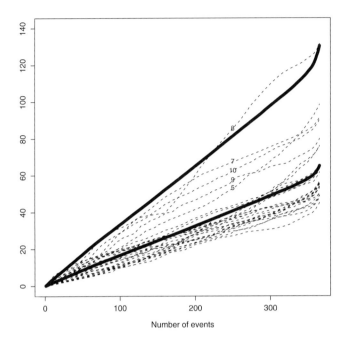

Number of events

Fig. 8.6 *Sleep data: The 27 individual cumulative hat processes (diagonal hat matrices) as a function of the time intervals where the events happen, together with the expected cumulative hat matrix (the lower thick solid line) and the outlying process criterion (the upper thick solid line). The figure is republished from Aalen et al. (2004).*

for the individual to wake up. This effect seems to be most pronounced early in the night.

To see whether any individual has a large influence on the results, we plot the diagonal elements of the cumulative hat matrix in Equation (8.10). In Figure 8.6, we see that one of the 27 individual cumulative hat processes exceeds the outlying process criterion, meaning that this observation is the one having the highest influence on the analysis. When looking more closely at the individual data, one sees that the person has unusually many awakenings early in the night. □

8.4 Dynamic path analysis of recurrent event data

Path analysis is a classical method extending traditional linear regression analysis. The basic idea is to assume a causal structure between the variables and then to carry out a series of regression analyses in order to estimate direct and indirect effects of

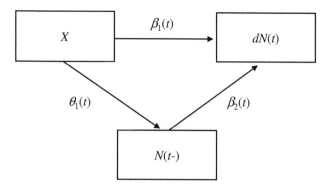

Fig. 8.7 *Full path diagram of the dynamic model.*

variables along the causal paths. We extend this method to stochastic processes and call it dynamic path analysis. An introduction to the special case of recurrent events is given here; a more general introduction is found in Sections 9.2 and 9.3.

8.4.1 General considerations

It was pointed out earlier that when using dynamic covariates, as well as other time-dependent covariates, one meets the problem that such covariates may be responsive (Kalbfleisch and Prentice, 2002). For instance, they may "steal" parts of the effects of a treatment, so that the estimated effect of treatment on the occurrence of events is reduced. This was illustrated in Section 8.1.2, where it was shown that the effect of a covariate in a dynamic and a marginal model could be very different. In order to understand this phenomenon, one has to think in a causal fashion and bring in the concepts of direct and indirect effects.

We shall consider the case with recurrent events and study the effect of a fixed covariate X on the event process, together with the dynamic covariate $N(t-)$, which is the number of previous events in an individual process. Since the fixed covariate X influences the occurrence of an event $dN(t)$, it will obviously also influence the cumulative number of events $N(t-)$. This dependence relationship is shown by the arrows in Figure 8.7. The full effect of X on $dN(t)$ will in a sense be the "sum" of the direct effect and the indirect effect through $N(t-)$. The question is whether such a "path diagram" can be formalized in a useful way. For nonlinear models, like Cox and logistic regression, this is difficult due to nonlinearity, and indeed no useful formalization seems to be known. For additive, or linear, models, one can get a simple structure that is similar to classical path analysis; for an introduction to the latter, see Loehlin (2004). The new method is called dynamic path analysis and was first presented by Fosen, Borgan et al. (2006) and Fosen, Ferkingstad et al. (2006).

The different paths indicated in Figure 8.7 may be given a natural interpretation. For instance, it is possible that a treatment X has an initial effect on the rate of events, but after a while the effect disappears. Then this initial effect has led to a buildup of events expressed in the dynamic covariate $N(t-)$, which again may speed up the occurrence of new events. Hence, X still has an indirect effect after the direct effect is gone. It is this type of interpretation that the path analysis may help with, as will be demonstrated more clearly in the following data example.

We shall consider n independent individual processes. Assume that the full model with both the fixed and the dynamic covariate is given by

$$dN(t) = \{\mathbf{Y}(t)\beta_0(t) + \mathbf{X}^c(t)\beta_1(t) + \mathbf{D}(t)\beta_2(t)\} dt + d\mathbf{M}(t), \tag{8.11}$$

where this is understood as a proper counting process model specifying the intensity process, with $\mathbf{M}(t)$ being a martingale. Here $\mathbf{Y}(t)$ is a vector of indicator variables, indicating whether the individual is still in the risk set. The vector $\mathbf{X}^c(t)$ contains zeros for those not in the risk set and is otherwise centered by subtracting the mean X-value of those still in the risk set from the original covariate values. Centering is useful because $\mathbf{Y}(t)$ is then orthogonal to $\mathbf{X}^c(t)$. Note that in spite of the fact that X is a fixed covariate for each individual, the vector of covariates $\mathbf{X}^c(t)$ will still depend on t because the risk set changes over time. Finally, the vector $\mathbf{D}(t)$ contains the centered $N(t-)$ values of the processes at risk at time t and has zeros for those not at risk.

By projecting the vector $\mathbf{D}(t)$ on $\mathbf{X}^c(t)$, we get

$$\mathbf{D}(t) = \theta_1(t)\mathbf{X}^c(t) + \theta_2(t)\mathbf{D}^x(t),$$

where $\mathbf{D}^x(t)$ is the orthogonal component. Inserting this into formula (8.11) yields

$$\begin{aligned} dN(t) = &\{\mathbf{Y}(t)\beta_0(t) + \mathbf{X}^c(t)(\beta_1(t) + \theta_1(t)\beta_2(t)) \\ &+ \theta_2(t)\mathbf{D}^x(t)\beta_2(t)\} dt + d\mathbf{M}(t). \end{aligned} \tag{8.12}$$

Next, we consider projecting $dN(t)$ on $\mathbf{X}(t)$. The result of this is also given in (8.12), with the orthogonal component being

$$d\mathbf{M}^*(t) = \theta_2(t)\mathbf{D}^x(t)\beta_2(t)dt + d\mathbf{M}(t).$$

Hence, we can write:

$$dN(t) = \{\mathbf{Y}(t)\beta_0(t) + \mathbf{X}^c(t)(\beta_1(t) + \theta_1(t)\beta_2(t))\} dt + d\mathbf{M}^*(t) \tag{8.13}$$

In such a marginal model, the remainder term $\mathbf{M}^*(t)$ cannot be expected to be a martingale because the model is incomplete; therefore a "star" is attached. The process would be a martingale if there were no dynamic effects present, but here we assume the presence of such effects. The model in Equation (8.13) must be understood as a rate model in the sense defined by Scheike (2002); see also Martinussen and Scheike (2006).

In fact, the connection between the full dynamic model and the marginal model can be illustrated in a nice graphical way. From Figure 8.7 we see that the coefficient of $\mathbf{X}^c(t)$ in Equation (8.13) is the sum of the direct effect from X on $dN(t)$ and the product of the coefficients along the indirect path going through $N(t-)$, that is:

$$total\ effect = direct\ effect + indirect\ effect, \tag{8.14}$$

whereby the total effect is the same as the marginal effect. This can also be written in integrated form as:

$$total\ effect\ over\ time\ interval\ (0,t) = \int_0^t (\beta_1(s) + \theta_1(s)\beta_2(s))ds$$

$$= B_1(t) + \int_0^t \theta_1(s)\beta_2(s)ds.$$

For a more general network, we can write (8.14) as:

$$total\ effect = integrated\ sum\ of\ products\ of\ effects\ along\ all\ paths.$$

These are just the relationships that hold in standard path analysis, because in both cases the analysis is based on linear projections. The difference here is that this is valid at each event time and that estimation is being performed at each such time and then combined in an additive model. A more detailed study of these ideas is given in Chapter 9.

It is important to note that the models considered here are statistical models, set up for the pragmatic purpose of estimating the effects. They are not models that pretend to give a complete description of how the processes are generated. The input to statistical models are the bits and pieces of the processes that are actually observed, while the input to a generating model would be the complete processes. For instance, the covariates will rarely be observed completely but may be measured just at certain irregular times, for example, at patients' visits. The additive model studied here has the advantage that one may put into it the data that are actually available. There is no detailed modeling; what one needs to know is essentially the time ordering of observations.

Example 8.6. Recurrent superficial bladder tumors. Following Fosen, Borgan et al. (2006), we shall analyze the bladder tumor data used in Examples 7.3 and 8.3. The data were studied with a dynamic covariate, but by a different procedure from here, in Kalbfleisch and Prentice (2002, section 9.4.3). This time we use the complete data, as presented in table 9.2 of Kalbfleisch and Prentice (2002). In the trial, 38 patients received treatment (thiotepa) and 48 received placebo. Each individual could experience a sequence of tumors during the follow-up time, which was 31 months on average, and all times of these *recurrent events* were recorded. In total, there were 45 and 87 recurrent events in the treatment group and the placebo group, respectively. Included in the data set are covariates regarding treatment and information on initial tumors prior to randomization.

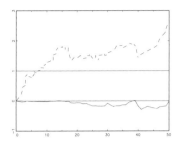

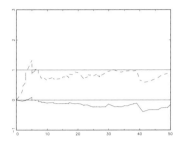

(a) *The marginal model: covariates (fixed): size of the largest initial tumor, initial number of tumors, and treatment.*

(b) *The dynamic model: covariates: size of the largest initial tumor, initial number of tumors, treatment, and the previous number of tumors.*

Fig. 8.8 *Mean (lower curve) and standard deviation (upper curve) of the standardized martingale residuals of the recurrent bladder tumor data, using the additive hazard regression model. Figure reproduced with permission from Fosen, Borgan et al. (2006), Copyright Wiley-VCH Verlag GmbH & Co. KGaA.*

Let the process $N(t)$ count the number of recurrent tumors for an individual up to time t, where time is set to zero at randomization. We shall use the four covariates *treatment*, *number of initial tumors*, *size of the largest initial tumor*, and *number of previous tumors* until time t. The three first covariates are fixed at time zero, while the last one equals $N(t-)$, which is a dynamic covariate. The aim was to find both the effect of the treatment and other baseline covariates as well as to study the effect of the number of previous tumors. We will also address another question: to what extent is the effect of treatment working directly and to what extent is it working indirectly through $N(t-)$?

We fit a marginal additive model with the three fixed covariates, as well as a dynamic model also including the dynamic covariate $N(t-)$ (previous number of tumors). The time-varying mean and standard deviation of the individual standardized residual processes, as defined in Section 8.3.2, are shown in Figures 8.8(a) and 8.8(b). The standard deviation should be approximately equal to 1 at all times if the model is adequately capturing the pattern of the data. We see that the standard deviation is increasing with time to above 2 in the marginal model (left panel), clearly revealing that there are patterns in the data that the marginal model fails to catch. However, in the dynamic model (right panel) the standard deviation is almost constant in time and close to 1, indicating that the dynamic model fits well. Hence, from the point of view of solely judging the treatment effect, the marginal model may appear the most appropriate, but when considering whether a correct description of the data set is given, the dynamic model appears most correct. The question is how to reconcile these differences.

First we fit a dynamic regression model, that is, including all covariates. The cumulative regression functions are shown in Figure 8.9. We see that there is a clear effect of $N(t-)$ on survival, and after time 30 there is also a clear effect of

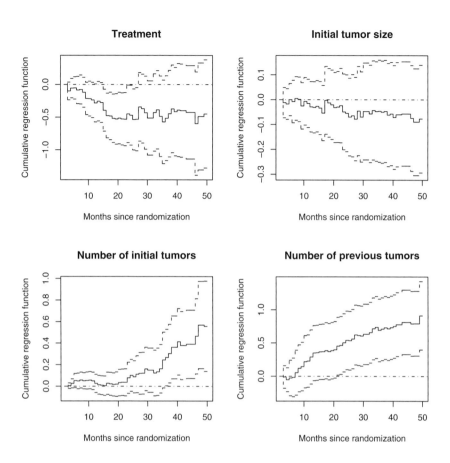

Fig. 8.9 *The cumulative regression functions of the additive hazard regression analysis of the recurrent bladder tumor data where the size of the largest initial tumor, the initial number of tumors, treatment, and the previous number of tumors are included as covariates. Figure reproduced with permission from Fosen, Borgan et al. (2006), Copyright Wiley-VCH Verlag GmbH & Co. KGaA.*

the number of initial tumors. There seems to be no effect of the size of the largest initial tumor and some effect of treatment. However, since $N(t-)$ is an intermediate covariate, the figure only shows the direct effect of treatment. The total treatment effect, obtained from the marginal model, is shown in Figure 8.10 and is seen to have a larger magnitude than the direct effect.

The connections between the models are clarified in the path diagram of Figure 8.11. The details may be found in Fosen, Borgan et al. (2006). The effect of treatment is partly an indirect effect going through the number of previous tumors. It is found that treatment is working directly only for the first part of the follow-up, whereas after this time, all the treatment effect has been absorbed by $N(t-)$. Thus,

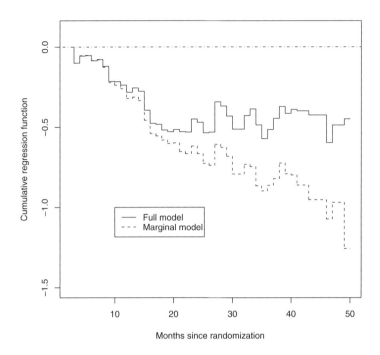

Fig. 8.10 *The cumulative regression functions of the covariate treatment in the additive hazard regression analysis of the recurrent bladder tumor data. For the model with and without the time-dependent covariate "initital number of tumors" and where the size of the largest initial tumor, the initial number of tumors, and treatment are included as the other covariates. Figure reproduced with permission from Fosen, Borgan et al. (2006), Copyright Wiley-VCH Verlag GmbH & Co. KGaA.*

after 20 months all future tumors can be explained by $N(t-)$ and covariates other than treatment.

The bladder cancer data have also been analyzed by Kalbfleisch and Prentice (2002, sections 9.4.3 and 9.5.2). In their tables 9.3 and 9.5, they presented different Cox models and showed that the inclusion of the dynamic covariate $N(t-)$ decreased the estimated effect of treatment with thiotepa. It was pointed out that the treatment effect in the presence of the time-dependent covariate could be interpreted as the effect for individuals having identical value of the time-dependent covariate at a certain time. However, this interpretation does not answer the question of which treatment effect is the correct one. With the dynamic path analysis tool we recognize $N(t-)$ as intermediate, telling us that the model without $N(t-)$ gives the total treatment effect, while the model with $N(t-)$ gives the direct effect of treatment.

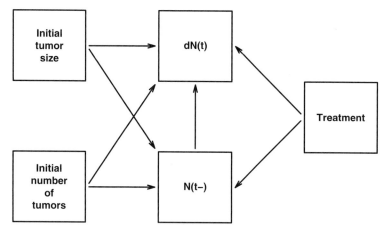

Fig. 8.11 *Path diagram for the recurrent bladder tumor data.*

We can also see how these effects change over time, which would not be visible in a frailty model.

Essentially, the same data were also used by Metcalfe and Thompson (2007), using various versions of the Cox model. Therneau and Grambsch (2000) give a thorough discussion of the recurrent event data issue, also using the bladder tumor data. The general impression from the latter two references, as well as Kalbfleisch and Prentice (2002), is that the appropriate analysis of recurrent event data within the Cox model framework is somewhat unsettled. No clear recommendations are forthcoming. We believe the additive model gives a clearer guide to the appropriate analysis. □

8.5 Contrasting dynamic and frailty models

The frailty and random effect models from Chapter 7 are popular and useful models. However, the usual frailty models make very specific assumptions about the effect of time, assuming that there is a fixed frailty for each individual. This may not be true, of course; it is quite possible that the frailty may be a stochastic process that changes over time. A more fundamental issue is whether there is any frailty at all. Maybe the dependence on the past that is interpreted as a frailty selection-type effect is really a causal effect of the past. Or maybe we have a mixture of both types of effect.

This dilemma has been recognized for a long time. For instance, Feller (1971, pages 57-58) discusses the phenomenon of spurious contagion, pointing out that the Polya process, which is a specific case of a process with intensity (7.17), may be defined as either a process of contagion, that is, previous cases increasing the likelihood of new cases, or a mixture of noncontagious Poisson processes. He states: "We have thus the curious fact that a good fit of the same distribution may be interpreted in two ways diametrically opposite in their nature as well as in their practical implications."

It is of interest to note that in economics one has considered a similar problem. For instance, the 2000 Nobel Prize winner James J. Heckman, in his Nobel lecture (Heckman, 2000, page 287), discusses the problem of distinguishing "heterogeneity and state dependence", which is similar to what we discussed above. Interestingly, Heckman asserts that it is possible in certain situations to make such a distinction statistically. However, this needs certain assumptions; otherwise one cannot know whether what one observes is due to individual heterogeneity or causal dynamic effects. One should also note that in some situations, one may have natural replications, which could help the situation considerably. In the sleep example, one could make observations over several nights, and this would clearly give a better possibility of estimating the natural sleep pattern of each individual, and hence of distinguishing heterogeneity from state-dependent effects.

An interesting study of the distinction between frailty effects and causal effects has been made in the field of psychiatry. One typically observes that the risk of recurrence increases with the number of previous episodes in diseases like depression and bipolar disorder. To a large extent, this is likely to be a frailty effect, but there may also be some causal effects of previous episodes. Attempts have been made to distinguish between these two possibilities; see, for instance, Kessing and Andersen (2005), who indicate some effects of previous episodes. However, one cannot draw definitive conclusions about this merely from data on occurrences.

In fact, any kind of dependence observed in a process can, in a mathematical sense, always be interpreted as effects of the past; that is, it can be given a *dynamic* interpretation. What the Doob-Meyer decomposition tells us is that there is essentially always an intensity process, which, of course, is a function of the past, however the counting processes comes about. For instance, there might be an underlying random effect, or frailty model, of a possibly complex and general nature; nevertheless the whole setup can be reformulated by intensity processes depending on the past.

On the other hand, any kind of dependence can also be given a frailty-type interpretation, with no effects of the past involved. When a certain number of individual processes are observed, there is the theoretical possibility that they came from a population of Poisson processes with a number of fixed intensity functions. There might be no dependence on the past, even if there appears to be.

Generally, any dynamic effects may simply reflect unobserved underlying variables, or they may represent real causal effects of past events. Without information besides the processes themselves, we don't know.

In spite of these difficulties in interpretation, a dynamic statistical analysis as given in this chapter may yield considerable insight into the nature of the data. We have focused mainly on an additive analysis, which is simple to carry out in practice. Since no likelihood is needed, the analysis does not require a particularly structured setting. Whenever a number of events are observed over time, one may introduce information about the past. The method is a pragmatic way of using the information available at any given time. It is easier to carry out than alternatives, like frailty models, which may be difficult to fit.

8.6 Dynamic models – theoretical considerations

When defining dynamic models we are in fact studying stochastic processes, and we have to take care that these are well defined, which is not a trivial issue. Furthermore, one should understand how various models are connected to one another.

8.6.1 A dynamic view of the frailty model for Poisson processes

We return to the situation of Section 7.3 where the individual intensity given the frailty variable Z_i is $\alpha_i(t) = Z_i \, \alpha(t)$, where $\alpha(t)$ is a common baseline intensity and a fixed function, that is, independent of the past, while the Z_i, $i = 1,...,n$, are independent identically distributed variables giving the multiplicative factor that determines the risk of an individual. When the Z_i are gamma distributed with scale parameter v and shape parameter η, we know from formula (7.17) that

$$\lambda_i(t) = \alpha(t)\frac{\eta + N_i(t-)}{v + A(t)},$$

where $A(t) = \int_0^t \alpha(s)\,ds$. Note that the previous number of events, $N_i(t-)$, comes into the intensity in an *additive*, or linear, fashion, and we have shown that it is useful to apply additive models.

An interesting question is to what extent the linear dependence on $N_i(t-)$ is due to the gamma distribution, or whether it has, at least approximately, a more general validity. Using formula (7.16), we may insert the Laplace transforms of (7.12) and (7.15) to get the following intensity processes. For the generalized inverse Gaussian distribution we have

$$\lambda_i(t) = \alpha(t)\frac{K_{N_i(t-)+\eta+1}\left(2\sqrt{\psi(A(t)+v)}\right)}{K_{N_i(t-)+\eta}\left(2\sqrt{\psi(A(t)+v)}\right)}. \tag{8.15}$$

In Jørgensen (1982, page 24), it is shown that the generalized inverse Gaussian distribution (7.11) can be approximated with a gamma distribution for large η. Since large values of $N_i(t-)$ would mean a large η-parameter in the conditional frailty distribution given the past, the above intensity process can be approximated with the one corresponding to a gamma distribution. This again implies an approximate linear dependence on $N_i(t-)$, which justifies an additive model.

For the Kummer distribution we consider only the special case of the beta distribution of the second kind ($\beta > \eta$, $\nu = 0$, $\delta = 1$). Then we get

$$\lambda_i(t) = \alpha(t)\frac{U(\eta + N_i(t-) + 1, \eta + N_i(t-) - \beta + k + 2, A(t))}{U(\eta + N_i(t-), \eta + N_i(t-) - \beta + k + 1, A(t))}. \tag{8.16}$$

Numerical investigations indicate an approximate linear dependence on $N_i(t-)$ in this case as well.

What we have shown here is that frailty models may alternatively be viewed as dynamic models. Clearly, the set of possible dynamic models is much richer than the classical frailty models since the dependence on the past can be described in a large number of different ways. In order to make statistical analyses on data, one must specify certain structured statistical models, and we here concentrate on Cox models and additive models.

8.6.2 General view on the connection between dynamic and frailty models

When applying dynamic models, one must be aware of the stochastic properties of the process one is defining. It may be tempting to just put up any kind of dynamic model that comes to mind and apply it to statistical data, but such an approach may not be feasible. The intensity process in formula (7.17) has similarities to a birth process. This is an important kind of stochastic process, and we shall elaborate this connection somewhat. We now look at one specific process and suppress the subscript i.

When formula (7.17) was derived, we started with a frailty model and deduced the intensity process. We shall now go in the opposite direction, starting with the intensity process and deducing the equivalence to an underlying frailty model.

Chiang (1968) defines a Polya process by letting the intensity process be

$$\lambda(t) = \lambda\frac{1 + aN(t-)}{1 + \lambda at}. \tag{8.17}$$

Assume that $N(0) = j$. Chiang shows that a Polya process $N(t)$ has a negative binomial distribution with Laplace transform

$$\mathscr{L}_P(c) = \exp(c/a)\{1 + (\exp(c) - 1)(1 + \lambda at)\}^{-(j+1/a)},$$

which, apart from an additive constant j, gives a negative binomial distribution.

One can derive the distribution of $\lambda(t)$ by a simple transformation, using formula (8.17), and let time t go to infinity. This yields the following limiting Laplace transform for the intensity process:

$$(1 + c\lambda)^{-(j+1/a)}.$$

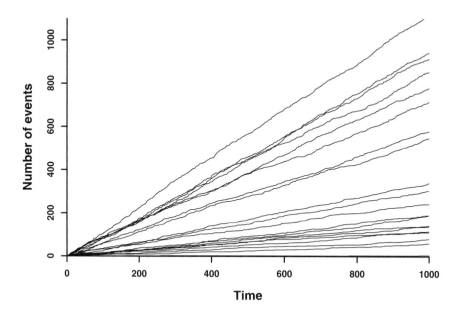

Fig. 8.12 *Simulating 20 copies of a process with intensity $\lambda_i(t) = (1 + N_i(t-))/(1+t)$.*

This is the Laplace transform of a gamma distribution with shape parameter $j + 1/a$ and scale parameter $1/\lambda$. The results we have found are similar to formula (7.17) with constant baseline intensity. What we see here is thus the same relation as in the previous section with a gamma distributed frailty, but in the opposite direction. It is evident that the frailty distribution is recapitulated in the limit in the dynamic model.

This is related to almost sure convergence results for Polya processes. It can be derived, for instance, from results in Pfeifer and Heller (1987) that $\lambda(t)$ converges *almost surely* to a distribution when t goes to infinity, which in this case will just be the frailty distribution. This means that the individual processes $\lambda_i(t)$ will settle into different paths, converging for each i to a specific frailty. Simulations show that the specific path taken by an individual process is typically determined by the random evolution at a very early stage. Once this early stage is passed, there is much less room for variation. A demonstration of this is given in Figure 8.12, using the intensity process formula (8.17) with $a = 1$ and $\lambda = 1$ and assuming $N(0) = 0$. The simulation is based on a discrete approximation to the counting process, with increments denoted $\{N_j^D, \ j = 0, \ldots, 1000\}$, where N_j^D is Poisson distributed with expectation $(N_{j-1}^D + 1)/(1 + j)$, which is similar to (8.17) with $a = 1$ and $\lambda = 1$. The starting value used is $N_0^D = 0$.

8.6.3 Are dynamic models well defined?

When defining dynamic models, one should beware of the phenomenon of "dishonest" processes, that is, processes that may explode in a finite time interval. In fact, Cox and Miller (1965, page 163) point out that by defining the intensity process as $\lambda_i(t) = N_i(t-)^2$, one gets a dishonest process. Clearly, this also creates potential problems for applying a Cox model with $\lambda_i(t) = \alpha(t)\exp(\beta N_i(t-))$ with $\beta > 0$, where the growth in the intensity process due to increasing number of events is stronger than for the squared function. Nevertheless, we shall consider such models in a practical statistical setting and see that they may be useful. However, there is a caveat here: we don't really know how our processes may behave. When the dependence on the number of events is linear, on the other hand, there appears to be no problem. The following topics are also discussed in Aalen and Gjessing (2007).

A general criterion for models to be nonexplosive is the Feller condition, which can be used on processes where the intensity is only dependent on the number of previous occurrences, that is, $\lambda_i(t) = g(N_i(t-))$ for a nonnegative function g. Then nonexplosiveness on finite intervals is guaranteed if and only if

$$\sum_{i=1}^{\infty} \frac{1}{g(i)} \quad \text{diverges.}$$

This condition immediately holds for a linear function $g(\cdot)$, so the linear or additive model will be a safe choice. For a quadratic form $g(i) = i^2$, on the other hand, it is clear that the sum converges, and there will be explosion with a certain probability. For an exponential form $\lambda_i(t) = \exp(\beta N_i(t-))$, it is clear that $\sum_{i=1}^{\infty}\exp(-\beta i)$ converges for all positive β, implying once more a positive probability for explosion on finite time intervals.

For the Cox model $\lambda_i(t) = \alpha(t)\exp(\beta N_i(t-))$, one may attempt to adjust $\alpha(t)$ to compensate for the increase in the exponential part. However, it is immediately clear that if $\alpha(t)$ is bounded below by a positive number on some finite interval, then there will always be a positive probability of explosion.

The situation is more difficult when $\lambda_i(t)$ has a more complex form. For instance, one may have a Cox model of the type $\lambda_i(t) = \alpha(t)\exp(\beta N_i(t-)/t)$. More generally, we shall consider a form $\lambda_i(t) = h(N_i(t-),t)$. The question is what kind of functions h will lead to a well-defined process N_i and which functions will cause explosions. Mathematically, this is related to the Lipschitz condition used in differential equations theory (Birkhoff and Rota, 1989). Typically, a local Lipschitz condition in the first argument of h guarantees a unique solution up to a certain point in time but does not exclude the possibility of explosion at a later time. A global Lipschitz condition, however, sets a growth restriction on the behavior of h, and guarantees a unique nonexplosive solution. Similar conditions exist for stochastic differential equations driven by Wiener or Poisson processes, that is, including the counting processes considered here (Protter, 1990). Additive processes usually satisfy global Lipschitz conditions, whereas exponentially based processes only satisfy local conditions. In general, one has to be careful when defining dynamic models to ensure that they are actually well defined, and this may be a nontrivial issue.

Assume that h is a convex function in the first argument. From counting process theory and with the use of Jensen's inequality, we have:

$$EN_i(t) = E\int_0^t h(N_i(s-),s)ds = \int_0^t Eh(N_i(s-),s)ds \geq \int_0^t h(EN_i(s-),s)ds.$$

Consider a function $f(t)$ satisfying

$$f(t) = \int_0^t h(f(s),s)ds. \tag{8.18}$$

We can now use a general comparison theorem for differential equations (Birkhoff and Rota, 1989). From the theorem it follows that $EN_i(t) \geq f(t)$. Hence, we may solve (8.18) and study whether the solution is explosive. The equation may be rewritten as:

$$f'(t) = h(f(t),t). \tag{8.19}$$

If the solution to this differential equation explodes, then the process will be expected to explode with positive probability. (The opposite direction is more complex; the process may explode even though the preceding equation has a non-explosive solution.)

Note that the solution to (8.19) is just what is often termed the deterministic solution, as opposed to the stochastic solution determined by a counting process with intensity process $\lambda_i(t) = h(N_i(t-),t)$. There is a complex and interesting relationship between deterministic and stochastic solutions. This subject is of great interest in areas like population dynamics and the mathematical theory of epidemics; see, e.g., Allen (2003). One sometimes finds that stochastic and deterministic solutions is close, but this is not always the case.

Let us first consider the Cox type model $\lambda_i(t) = \alpha(t)\exp(\beta N_i(t-))$ mentioned earlier. The special case of (8.19) relevant here is:

$$f'(t) = \alpha(t)\exp(\beta f(t)).$$

A solution to this equation with initial condition $f(0) = c$ yields

$$f(t) = -\frac{1}{\beta}\log(e^{-\beta c} - \beta\int_0^t \alpha(s)ds). \tag{8.20}$$

If $\beta \leq 0$ there is no explosion. For $\beta > 0$ we see that the deterministic solution explodes if $\int_0^t \alpha(s)ds$ reaches $e^{-\beta c}/\beta$ at some finite time. In particular, for $c = 0$ this means that if $\beta\int_0^t \alpha(s)\,ds = 1$ has a solution with t finite, $f(t)$ explodes.

In order to give a concrete illustration we shall consider the special case of (8.19) corresponding to the Cox type model

$$\lambda_i(t) = \alpha\exp(\beta N_i(t-)/t). \tag{8.21}$$

This is a sensible model since it implies that it is the average number of events per time unit that is of importance. We will assume that the process starts at time 1 (just to avoid the singularity at 0) and that $EN(1) = c \geq 0$ is the initial value. The relevant equation is

$$f'(t) = \alpha \exp(\beta f(t)/t), \quad t \geq 1, \tag{8.22}$$

with $f(1) = c$. When $\beta \leq 0$ no explosion occurs, so we will focus on $\beta > 0$. Aided by Mathematica (Wolfram, 1999) or by substituting $a(t) \overset{\text{def}}{=} f(t)/t$ and separating the equation, we find the following implicit solution:

$$\int_c^{\frac{f(t)}{t}} \frac{1}{g(u)} \, du = \log(t), \tag{8.23}$$

where $g(u) \overset{\text{def}}{=} \alpha e^{\beta u} - u$, $u \geq 0$. The solution fulfills the initial condition $f(1) = c$. From (8.23) one may decide whether $f(t)$ explodes or not by observing that we cannot necessarily solve this equation for $f(t)$ for all values of α, β, and t. For some combinations of α and β the left-hand side may remain bounded as $t \to \infty$.

To analyze Equation (8.23) in detail, consider the auxiliary function $g(u)$, the denominator of the integrand. We have $g'(u) = \alpha \beta e^{\beta u} - 1$ and $g''(u) = \alpha \beta^2 e^{\beta u}$. Note that $g(0) = \alpha > 0$. Since g'' is strictly positive, g is convex and has a unique minimum, which we denote u_0. By setting $g'(u_0) = 0$ we find that $u_0 = -\log(\alpha \beta)/\beta$ and $g(u_0) = 1/\beta - u_0$. There are now three possibilities:

1. $g(u) > 0$ for all $u \geq 0$. Then the integrand of (8.23) is nonsingular and we have $\int_c^\infty 1/g(u) \, du < \infty$. For large enough t, (8.23) cannot have a solution for f, and explosion occurs.
2. $g(u_0) = 0$, that is, g is tangential to the x-axis. Then $1/g(u)$ has a nonintegrable singularity at $u = u_0$. If $0 \leq c < u_0$, there will be no explosion, if $c > u_0$ an explosion will occur. In the very special case $c = u_0$ the solution (8.23) is not valid but is replaced by the simple solution, $f(t) = ct$, and no explosion.
3. $g(u_0) < 0$. Then g has two zeros u_1 and u_2, $u_1 < u_2$, and the integrand has nonintegrable singularities at these two values. Accordingly, if $0 \leq c < u_1$ or $u_1 < c < u_2$ there is no explosion. If $c > u_2$ the solution explodes. If $c = u_1$ or $c = u_2$ there is no explosion, as earlier.

In conclusion, there is an explosion if g has no zero or if c is larger than the largest zero of g. This translates into saying that there is an explosion in finite time if either $g(u_0) > 0$ or $(g(c) > 0$ and $g'(c) > 0)$, that is, $\alpha \beta > e^{-1}$ or $(\alpha e^{\beta c} > c$ and $\alpha \beta e^{\beta c} > 1)$. Note, in particular, that when the starting level c is large enough, the second condition is necessarily fulfilled. As a numerical illustration, put $c = 1$ and $\alpha = 1$. Then, explosion in finite time occurs if $\beta > e^{-1} = 0.368$.

The preceding discussion concerns lower bounds for the expected values. Note that in both (8.20) and the more complicated case discussed here, the starting level c plays a critical role. If c is large, then explosion will typically occur. When considering the random counting processes, there will be some probability that several events occur in a short time interval. Hence the level, corresponding to c, may become quite

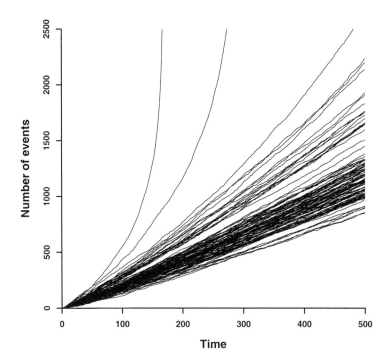

Fig. 8.13 *Simulating 100 copies of a discrete process with Cox type intensity; see details in text.*

large by accident, yielding later explosion. Simulation experiments indicate that the early stages are particularly vulnerable; here, random events play an important role and decide whether a process will explode or become stable. A similar phenomenon has been observed in the theory of epidemics.

It is also noted from simulation experiences that the counting process either explodes or tends to stabilize as a Poisson process with the rate κ defined as the trivial solution to Equation (8.22).

An illustrating simulation of the dynamic Cox model with intensity process (8.21) is shown in Figure 8.13. The simulation is based on a discrete approximation to the counting process, with increments denoted $\{N_j^D, \ j = 0, \ldots, 500\}$, where N_j^D is Poisson distributed with expectation $\exp(N_{j-1}/(2.5j))$, which is similar to (8.21) with $\alpha = 1$ and $\beta = 0.4$. The starting value used is $N_0^D = 0$. One clearly sees the explosion in a few cases, although the parameters are fulfilling the requirement $\alpha\beta > e^{-1}$ indicated earlier.

8.7 Case study: Protection from natural infections with enterotoxigenic *Escherichia coli*

Diarrhea is one of the leading causes of child morbidity and mortality; it is responsible for approximately 2.5 million deaths annually among children younger than 5 years, mostly in developing countries (World Health Organization, 1998; Kosek et al., 2003). Although the number of deaths has fallen from approximately 4.5 million in 1980, probably in part due to improved treatment with oral dehydration therapy, the incidence of diarrhea has been stable at an average of about 3.2 episodes per child-year in developing countries (Kosek et al., 2003). Enterotoxigenic *E. coli* (ETEC) is among the most commonly identified enteropathogens, and certain subgroups of ETEC strains contribute substantially to the burden of diarrheal disease in children in developing countries (Steinsland et al., 2002; Porat et al., 1998; Levine et al., 1993).

One of the complicating factors in the ongoing efforts to develop a vaccine against ETEC diarrhea is the relatively large number of different ETEC strains. Two main characteristics of ETEC are the protein enterotoxins and the colonization factors (CFs). A single strain of ETEC may produce one or more of the three toxins LT, STp and STh. In addition, it may have one or more of several different colonization factors, which are surface proteins, 21 of which have so far been described (Gaastra and Svennerholm, 1996). Most CFs are hair-like structures that help the bacteria attach to the wall of the small intestine. Once attached and colonizing the gut wall, the ETEC will release the toxins, possibly causing diarrhea. The CFs are main components of vaccine constructs that are currently being developed.

The protection against new infections mediated by natural infections with the different ETEC strains may form a basis against which vaccine efficacy may be compared. It is thus of great value to study the natural ETEC infection patterns in children to assess the extent of this protection, even though it cannot be designed as a randomized study. This was done in a large birth cohort study between January 25, 1996, and April 28, 1998 in suburbs of Bissau, Guinea-Bissau, where ETEC infections are endemic. Two hundred children younger than three weeks were enrolled in the study, at an even rate during all of 1996. They were followed until the age of two years, with weekly collection of stool specimens, irrespective of whether they had diarrhea. The study was prematurely ended in April 1998 due to the outbreak of a civil war, resulting in a median length of follow-up of 18.4 months. The stool specimens were analyzed for the presence of ETEC, classified with regard to their toxin-CF profile. For a full description of the study and the protection analyses, see Steinsland (2003) and the references therein. In particular, more information about the study population and area can be found in Mølbak (2000) and about infection rates and pathogenicity in Steinsland et al. (2002). The results presented here are primarily extracted from Steinsland et al. (2003).

Although symptomatic infections (i.e., infections with diarrhea) are of even greater interest for developing an ETEC vaccine, in the following we will only consider what strain(s) of ETEC were present in a given sample, disregarding clinical presentation. The main issue is to establish whether earlier infections induce a natural protection against new infections. Basically, this can be done by including a time-dependent covariate, indicating whether an earlier infection has taken place (Moulton et al., 1998). The idea is that if such an infection induces protection, children with an earlier infection should have a lower rate of (re)infection compared to children without previous infections.

The analysis is challenging for several reasons. First, it is important to control for age. Age-dependent changes of the gastrointestinal tract (Steinsland et al., 2002) as well as change in general resistance to infections will result in a changing propensity to infections. The children also become more mobile as they grow older, thus increasing exposure to various infections. Since older children typically have had several earlier infections, age effects could conceivably confound protection estimates. Closely related to age is breast-feeding status. Although important for the child's resistance to infections, we will not consider breast feeding here since almost all of the study children were breast-fed throughout the study period.

A second issue is that ETEC infections tend to cluster in time, related to local epidemics. We account for this by choosing calendar time as our main time axis, so that, for instance, increased infection pressure in the warmer and more humid season will directly influence the baseline hazard common to all children (Moulton and Dibley, 1997). Since an epidemic typically involves a single strain of ETEC (Steinsland et al., 2002), it may also be important to allow different baseline hazards for the different strains.

A third issue is that some children generally are more exposed to infections due to their environment or they are genetically more predisposed to becoming colonized. This corresponds to an individual frailty, influencing the baseline hazard for each child separately. This frailty effect may bias protection estimates toward zero, or may even cause a first infection to apparently *increase* the risk of a new one, only because an early first infection serves as an indication that the child is particularly frail.

Fourth, an ETEC strain may elicit protection against new infections with the same strain but no cross-protection against other ETEC strains. For our purposes, an ETEC strain was defined simply by its toxin-CF combination. The primary question was whether a particular CF induced protection against new ETEC infections with strains carrying the *same* CFs but with a different toxin-CF profile, since this would indicate an anticolonizing effect of a vaccine based on CFs. Because there are far fewer CFs than ETEC strains, it is hoped that vaccines based on CFs may protect against several different strains. In addition, the possible extra protective effect of an ETEC infection against a new infection with precisely the same toxin-CF profile should be measured.

Fifth, there were periods during the follow-up when the children were unavailable for sampling, mostly due to factors unrelated to the study issues. If absence of the child were informative, for instance, if the child was hospitalized for treatment of prolonged diarrhea or died from dehydration, this might complicate matters even further. For a description of the drop-out pattern, see Steinsland et al. (2003). Figure 8.14 shows the follow-up and infection pattern for every 20th child in the study.

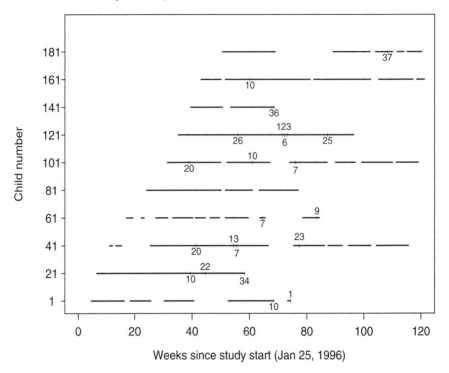

Fig. 8.14 *Follow-up and infection pattern for every 20th child included in the Guinea-Bissau study. Horizontal lines indicate period of follow-up for each child. Numbers above and below lines indicate infecting strain. Horizontal positioning of numbers indicates time at which the infection was observed. Numbers alternate above and below line to enhance readability.*

To disentangle these effects, we construct a dynamic model with time-dependent covariates as described in this chapter. Our definition of ETEC strains as a toxin-CF combination resulted in 45 unique strains. Let $\alpha_{ij}(t)$ be the baseline intensity for child i, $i = 1, 2, \ldots, 200$ and for strain j, $j = 1, 2, \ldots, 45$ at calendar time t, and let $N_{ij}(t)$ be the process counting the number of infections with strain j for child i before time t. The absence pattern for child i is described by the process $Y_i(t)$, which is zero when the child is unavailable for sampling, and one otherwise. Note that Y may alternate between zero and one when, for instance, the parents do sporadic work outside town and bring their child with them. The individual susceptibility and infection pressure of the child is described by a random gamma frailty variable Z_i. Alternatively, this could be accomplished by including a time-dependent covariate counting, for instance, the total number of earlier infections (see Section 8.6.1), regardless of type. For the final results, our choice of frailty description has a negligible effect on the protection estimates. To measure the protective effects, two time-dependent covariates were defined. Let $x_{ij}^{CF}(t)$ be zero until the first infection with a strain carrying the same CF as strain j, and one thereafter. That is, $x_{ij}^{CF}(t) = \mathbf{1}\{\sum_{k \in CF(j)} N_{ik}(t) \geq 1\}$, where $CF(j)$ denotes the set of strains having the same CF as j. Thus, x_{ij}^{CF} is a CF-specific indicator for individual i. Similarly, let $x_{ij}^{S}(t)$ be an indicator for strain-specific infections, being zero until the first infection of strain j for individual i, and one thereafter, that is, $x_{ij}^{S}(t) = \mathbf{1}\{N_{ij}(t) \geq 1\}$. In addition, age should be included as a time-dependent covariate. It could be included as a linear function $x_i^{AGE}(t) = t - t_{0i}$, where t_{0i} is the (calendar) time of birth. A more flexible approach is to include it as a smooth function $f(t - t_{0i})$ where the function f can be estimated, for instance, as a smoothing spline or by a group of indicator variables $x_i^{AGE_k}(t) = \mathbf{1}\{t - t_{0i} \in I_k\}$, where $I_1, I_2, \ldots$ is a series of intervals spanning the range from 0 to 2 years, for instance, in three-month categories (leaving out the covariate for the reference interval).

The resulting full model for the intensity is then

$$\lambda_{ij}(t) = \alpha_j(t) Y_i(t) Z_i \exp\left(\beta_{CF} x_{ij}^{CF}(t) + \beta_S x_{ij}^S(t) + \beta_{AGE} x_i^{AGE}(t)\right).$$

The separate baseline hazards for each strain are accounted for by stratifying the analysis on strain (j). It is important to realize, however, that the model is not stratified in the ordinary sense, that is, that children are subdivided into separate groups. Rather, each child is under risk in all strata at the same time. This avoids the reduced power potentially caused by many thin strata. In fact, it is only when there are tied events, that is, that two different children get the same or different infections at the same time, that the partial likelihood of the stratified analysis differs from the unstratified, since an infected individual is considered a case only in the relevant stratum.

The estimation was done in S-PLUS version 6.2 for Windows (S-PLUS®, 2002), using the `coxph` function. The `coxph` function has no implementation of time-dependent covariates as specified functions of study time. Rather, this must be dealt with by artificially censoring all individuals at all times where somebody has an event, and computing the updated time-dependent covariates at that time. The artificially censored individuals are re-included immediately after censoring. The censoring is done in each stratum (i.e. for each strain) separately. This produces a large number of lines in the data file for each child, resulting in an excessively large, but still manageable, data file. The data records are in the (start, stop] format (Therneau and Grambsch, 2000, page 187).

The hazard ratio estimates were $\exp(\beta_{CF}) = 1.12$ (95% CI 0.82 to 1.53) for the colonization factors and $\exp(\beta_S) - 0.44$ (95% CI 0.26 to 0.73) for the strain-specific protection. Age was adjusted for by including a smoothing spline with 4 degrees of freedom (the default in S-PLUS). The age effect estimate is shown in Figure 8.15. We notice a marked increase in hazard of infection with increasing age, at least for the first year. The effect seems to level off in the second year, although the estimates become more unreliable as age increases. The decrease in precision for the higher age range is partly due to the decreasing number of children available in that age range (Figure 8.14). One possible interpretation of the increasing hazard with age is that as the children grow older they also become more mobile and thus more easily exposed to infections.

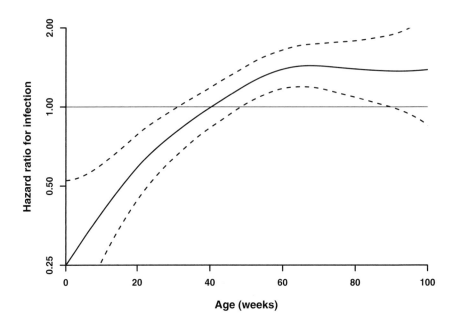

Fig. 8.15 *The effect of age on overall risk of infection. The middle (solid) line is the hazard ratio estimated from a smoothing spline with four degrees of freedom. Dashed lines are approximate 95% pointwise confidence limits. The horizontal line at 1.00 indicates reference level.*

With the present parameterization, the hazard rate ratio estimate of 0.44 for the strain-specific protection should be interpreted as saying that a child has only 44% of the hazard of being infected with a specific strain, say strain 20, if they have previously been infected with that strain, relative to the hazard of individuals with an earlier infection with a different strain containing the same CF as strain 20. This protection is estimated as an "average" across all strains.

The estimate of 1.12 for the CF-specific protection is then the "protection" induced by an earlier infection with the same CF (but different strain), compared to children with no earlier infection with the same CF (but possibly with different strains with different CFs). In fact, this number suggests a slightly increased risk of a new infection rather than a protection, although this is far from statistically significant (p-value 0.48). Without overinterpreting this increase, it might be the effect of a less-than-optimal adjustment for frailty, which would result in an apparent increase in risk following an infection due to the selection effect of frail individuals mentioned earlier.

Somewhat surprisingly, even though the age effect in itself is very strong, the actual implementation of the age covariates matters little to the protection estimates; age does not seem to have a marked confounding effect on protection.

To demonstrate that neither age adjustments nor frailty corrections have much to say for the final result, we present the estimates from a selection of models in Table 8.3. It is seen that all models produce similar results.

Table 8.3 *Estimates of hazard rate ratio of new infections for children with an earlier infection with the same CF but different strain and infections with the same strain. Model adjustments/corrections are: 1: standard gamma frailty; 2: frailty replaced by a categorical variable counting 0, 1, 2, 3 or 4+ earlier infections of any type; 3: no frailty, but robust variance estimation accounting for within-child dependencies; 4: no frailty and standard estimates of variance; 5: frailty correction but no age adjustment. All models are stratified on strain. Age (6 categories) is adjusted for in all but the last model, and all frailty corrections assume a gamma frailty.*

	CF Alone			Same Strain		
Model	$\exp(\beta)$	95% CI	P-Value	$\exp(\beta)$	95% CI	P-Value
1	1.12	(0.82, 1.53)	0.48	0.44	(0.26, 0.73)	0.0014
2	1.13	(0.82, 1.55)	0.47	0.45	(0.27, 0.74)	0.0018
3	1.21	(0.88, 1.66)	0.25	0.45	(0.27, 0.76)	0.0030
4	1.21	(0.89, 1.64)	0.23	0.45	(0.27, 0.75)	0.0019
5	1.18	(0.86, 1.61)	0.30	0.43	(0.26, 0.71)	0.0011

As a further investigation of the protective effects, we looked to see if there was any change in the effects over time, that is, whether time since infection influenced the protection in any particular way. This was done by replacing the time-dependent covariates $x_{ij}^{CF}(t)$ and $x_{ij}^{S}(t)$ with variables measuring time since infection. The covariate effects were included using smoothing splines. Omitting the details, Figure 8.16 shows how the protective effects vary over time. Although pointwise confidence intervals are fairly wide, it seems clear that a child will develop maximum protection at least within a few months after an infection, but only against the same strain. It should be kept in mind, however, that the stiffness of the spline curves will probably prevent the "same strain" curve from descending rapidly from the start. In addition, the problem of distinguishing between two closely spaced but separate infections with the same strain and a single infection with that strain will come into play when evaluating how fast strain-specific protection sets in. Thus, it may well be the case that the protection is present even sooner after an infection than the curves suggest. Another interesting feature is that the protective effect seems to remain stable over time. Although the wide confidence intervals should be kept in mind, this solidifies the evidence for a same-strain protection.

The conclusion to be drawn from our analysis is that a vaccine based solely on CFs may not yield adequate protection against infection with the many different ETEC strains circulating in child populations like the one in the currently described Bissau cohort. It appears that there may be other, strain-specific, antigens that may provide such protection. This finding, combined with the far-from-promising results of the first large efficacy trial of a CF-based ETEC vaccine in young children (Savarino et al., 2003) indicates that the present strategy for developing an ETEC vaccine for children in developing countries may need reevaluation.

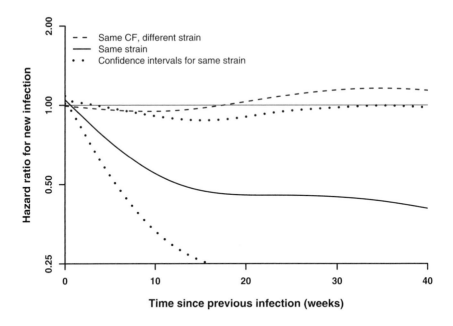

Fig. 8.16 *Protective effect as a function of time since last infection. Solid line shows the effect estimate for same strain infections, with 95% pointwise confidence interval (dotted). Dashed line shows the effect of a previous infection by a different strain with the same colonization factor.*

8.8 Exercises

8.1 Demonstrate the validity of the standard error formula (8.5).

8.2 In the additive model $\lambda_i(t) = Y_i(t)(\beta_0(t) + \beta_1(t)x_{i1}(t) + \cdots + \beta_p(t)x_{ip}(t))$ assume that all covariates are centered (i.e., the mean values of all individuals at risk have been subtracted) at all event times. Show that $\widehat{B}_0(t)$ is identical to the Nelson-Aalen estimator.

8.3 Assume that $p = 2$ and that $x_{i1}(t)$ and $x_{i2}(t)$ are indicator variables such that $x_{i1}(t) = 1 - x_{i2}(t)$. Show that $\widehat{B}_1(t)$ and $\widehat{B}_2(t)$ are the Nelson-Aalen estimators of two complementary groups.

8.4 Assume that X_1 and X_2 in (8.4) are correlated and normally distributed. Calculate the rate function with $r(t)$ with X_1 as the only covariate and show that the rate function is linear in X_1 (Aalen, 1989).

Chapter 9
Causality

One of the exciting new developments in the field of statistics is a renewal of interest in the causality concept. This has led to several new approaches to defining and studying causality in statistical terms. Causality is based on a notion of the past influencing the present and the future. This has very natural links to the type of stochastic processes considered in this book, and it is therefore appropriate to incorporate material on causality.

We shall start with a discussion of causality concepts seen from a statistical point of view. We then continue with various models where causal thinking is applied, ranging from Granger-Schweder causality to counterfactual causality.

9.1 Statistics and causality

We tend to think that everything that happens has a cause. This may not be true, however. Even in science, causality does not rule everywhere, as has been demonstrated in quantum mechanics. In fact, in an essay in *Nature* (Zeilinger, 2005), it is stated:

> The discovery that individual events are irreducibly random is probably one of the most significant findings of the twentieth century. ... But for the individual event in quantum physics, not only do we not know the cause, there is no cause. The instant when a radioactive atom decays, or the path taken by a photon behind a half-silvered beamsplitter are objectively random. There is nothing in the Universe that determines the way an individual event will happen. Since individual events may very well have macroscopic consequences, including a specific mutation in our genetic code, the Universe is fundamentally unpredictable and open, not causally closed.

It is interesting to note that even events with potential medical consequences, such as a mutation that gives cancer, may happen in a fundamentally noncausal way. Nevertheless, one would feel that most events are caused in some sense, and that one may influence some of these causes with a beneficial effect, for instance, by applying a medical treatment.

Traditionally, the statistical community has shown a very cautious attitude toward talking about causal connections; see Pearl (2000) for a discussion of this. This is based on an awareness of the fundamental difference between a statistical association and a real causal connection. However, there is a contradiction between this attitude and the actual practice of statistics, including the way our field is viewed by other scientists. For instance, any statistician working in medical research, which is arguably the largest single area of application of statistics, will have noticed that the interest in statistics in this field is to a large extent due to the belief that statistics *can* help in proving causality. In the first issue of the *New England Journal of Medicine* in 2000, there was an editorial about the *eleven* most important developments in the last millennium (Editors, 2000). One of these was "Application of statistics to medicine." The examples given were typically of a nature where statistics had actually contributed to proving causality; witness, for instance, the following quotation from the editorial:

> One of the earliest clinical trials took place in 1747, when James Lind treated 12 scorbutic ship passengers with cider, an elixir of vitriol, vinegar, sea water, oranges and lemons, or an electuary recommended by the ship's surgeon. The success of the citrus-containing treatment eventually led the British Admiralty to mandate the provision of lime juice to all sailors, thereby eliminating scurvy from the navy. The origin of modern epidemiology is often traced to 1854, when John Snow demonstrated the transmission of cholera from contaminated water by analyzing disease rates among citizens served by the Broad Street Pump in London's Golden Square. He arrested the further spread of the disease by removing the pump handle from the polluted well.

Pearl (2000) has strong objections to the extreme caution toward discussing causality that has been common in statistics, stating, for instance: "This position of caution and avoidance has paralyzed many fields that look to statistics for guidance, especially economics and social science." One major danger of avoiding the subject of causality in statistical education and statistical literature, is that one never gets any insight into this fascinating concept, which has such an old history in philosophy and science. The fact is that statistics plays a major role in looking for causal connections in many fields and statisticians who know next to nothing about causality as a larger concept will be far less useful than they could have been.

Often, causality is understood in the sense that there is some underlying mechanism one might try to uncover. Much work in science is directed towards understanding such mechanisms, elucidating them in various experimental studies. However, in biology and medicine, the level of mechanistic understanding will necessarily be limited. When proposing a new medication, for instance, there may be some mechanism that one thinks one understands and that suggests that the medication should work. Nevertheless, one is entirely dependent on clinical trials to see whether it actually works in practice. This is because the multitude of causal connections in the body that result from taking the medication are complex, and also partly unknown, so that one cannot reason one's way through all the possibilities. Therefore, statistical studies are a necessary supplement to a mechanistic understanding. This holds for the qualitative effects to be observed, and even more for the size of the effects.

Furthermore, in many fields, there is almost no proven or universally accepted mechanistic understanding. This is especially true of much of psychiatry and the social sciences, and this implies that causal deductions may be even more dependent on statistical studies.

9.1.1 Schools of statistical causality

Considerations of the preceding type have led to a movement to give the causality concept a more prominent place in statistics and to consider much more actively what the possibilities and limitations of making deductions about causality are. Some prominent authors in this field are J. Pearl, J. M. Robins, and D. B. Rubin. These authors have many relevant publications, including Pearl (2000), Robins et al. (2000), and Rubin (1974). The idea is to build causal considerations, often from external sources, into the model, and to judge in a much more precise way what can really be said. The concept of a "causal effect" is prominent, and although such an effect may not be exactly estimated, one can often give bounds for its assessment or do sensitivity analyses.

There are three major "schools" regarding statistical causality:

- Graphical models
- Predictive causality
- Counterfactual causality

The so-called graphical models represent an old tradition, going back at least to Wright (1921) under the name path analysis. The idea is to make diagrams showing how variables influence one another. For example, the measured intelligence of a child depends on its inherited abilities but also on the parents' socioeconomic status and education, and this may be expressed in a diagram. This way of thinking has been further developed in recent decades, and now a very extensive theory exists for graphical models (Pearl, 2000). A limitation of this approach is that the aspect of time is not explicitly drawn in. Clearly, causality unfolds in time, and this should be explicitly incorporated. In fact, we shall present a new approach, called *dynamic path analysis*, that combines the graphical models with a continuous development in time (Fosen, Borgan et al., 2006; Fosen, Ferkingstad et al., 2006). Classical path analysis usually presumes advance knowledge of graphical structure. In dynamic path analysis, this is somewhat helped with the time ordering that occurs in the processes. There is a strong relationship to so-called local dependence graphs as developed by Didelez (2007).

A fundamental idea in causal thinking, namely the distinction between direct, indirect, and total effects that was introduced by Wright (1921) with his path analysis, will be incorporated and extended in dynamic path analysis. The concepts of direct and indirect effects also have a general validity beyond path analysis. The major focus on nonlinear models in event history analysis have pushed these important concepts to the background.

Attempts at incorporating time when discovering causal effects are further found in the theories of predictive causality. This is related to probabilistic causality as defined by Suppes (1970). Several developments can be collected under the heading "predictive causality." One direction is the work of Arjas and coauthors; see, for instance, Arjas and Eerola (1993). Another development is the concept of Granger causality (Granger, 1969), which is well known in economics, and the closely related idea of local dependence (Schweder, 1970). Here, we focus on an integrated version of the two latter ideas, which we call Granger-Schweder causality; see Section 9.4. The predictive causal ideas are based on stochastic processes and therefore are very well suited for this book. In this chapter, we introduce a different point of view from the classical one in the causality field. We show how graphical models may be intimately connected to predictive causality, thus removing the traditional distinction between the first two schools mentioned earlier.

The third major area is counterfactual causality (Robins, 1986; Rubin, 1974; Pearl, 2000), which focuses primarily on analyzing the effect of medical treatments and preventive health measures. The idea is to hopefully get an unbiased picture of the effects of particular actions, like taking a medication. Even in a properly conducted clinical trial, there will often be difficulties like noncompliance, and one aim of the counterfactual theories is to handle this. In epidemiological studies, counterfactual approaches may, for instance, help to clarify and analyze time-dependent confounding. We introduce the counterfactual point of view in Section 9.5.

It is important to note that even though one may have a modest attitude regarding the possibility of deducing causality from statistical data, one may still let the analysis be led by causal thinking or reasoning. By "causal thinking" we mean that one is actively using presumed causal connections between the variables involved. For instance, it is very common to run a logistic regression or Cox regression, where the independent variables are lumped together and their internal connections are ignored. Quite often, there is an obvious causal pattern since some independent variables are measured or known earlier than others. For example, the conditions during childhood cannot be influenced by events occurring later in life; it will have to be the other way round. The reason the structure among the independent variables is often ignored may be partly due to technical difficulties with making path models for logistic and Cox regression because of nonlinearity. More generally, there is a distinction between regression models and structural models, where the latter incorporate causal connections. It is a weakness that so much of biostatistics is dominated by regression models instead of the more insightful structural models. We shall demonstrate in Section 9.2 that the additive model lends itself very naturally to a structural interpretation.

As a general comment, we note that while some of the variables measured in clinical or epidemiological trials may be expected to have a causal effect, this is certainly not true for many other variables. For instance, in clinical trials it is typically the case that most important covariates are measures of the extent of disease, like measures of tumor size or metastasis in cancer. Survival analysis should distinguish between causally active measures, like a treatment or some risk factor, and the more passive measures of disease extent. This may even have implications for how data are analyzed; see Section 10.3.8, where covariates can be modeled through either a causal drift parameter or a stage (or distance) parameter.

There is a relationship to the concepts of internal and external covariates as discussed by Kalbfleisch and Prentice (2002). These concepts were also discussed in the introduction to Chapter 4. The internal covariates are generated by the process itself, while the external covariates come from the outside, unaffected by the process, or they are effects fixed from the beginning. Measure of tumor size would naturally be perceived as an internal covariate if it is measured repeatedly, while treatment might be perceived as external if fixed at the start of the study or as internal if the treatment is modified according to the development of the illness. A similar set of concepts plays an important role in economics, namely endogenous (internal) and exogenous (external).

It is natural to think that causal input comes from the outside, bringing something new into the process, and so is associated with external covariates. This is especially so if causality is associated with intervention, which is often the case. However, a causal viewpoint may also mean that one tries to understand the internal workings of a system.

9.1.2 Some philosophical aspects

Considering causality, one should not ignore the very interesting philosophical aspects of this concept. The empirical view of causality dominating in statistics is closely related to Hume's classical causality concept from 1740, where he stresses how a causal understanding depends on observation of a constant conjunction between phenomena (Hume, 1965). As pointed out by Pötter and Blossfeld (2001), the constant conjunction assumption is not realistic in medicine and social science, where connections do not appear to be deterministic, but rather stochastic, and they suggest that a modern version of the "constant conjunction" idea could be the probabilistic causality of Suppes (1970). There also seems to be a relationship to predictive causality in a more general sense. Hume also briefly suggested a counterfactual causality concept.

An alternative to the empiricist view championed by Hume is a more mechanistic view of causality, focusing on understanding the underlying mechanism. This is more in tune with modern natural science, where the aim is to understand the workings of nature. By "understanding" one means deriving models to lay bare the mechanisms governing nature. A philosophical discussion of mechanisms in causality may be found in Machamer et al. (2000); see also Aalen and Frigessi (2007).

Very readable discussions of causality, including philosophical aspects can be found in the social science literature; see Gorard (2002) and Pötter and Blossfeld (2001). A nice statistical paper is the one by Cox and Wermuth (2004), who discuss various causality concepts, distinguishing between different levels of causality. Their level 0 represents the simple statistical association. Level 1 concerns the effect of an intervention, which is closely related to the counterfactual causality concept. The understanding of causal mechanisms corresponds to their level 2. Granger causality is placed at level 0, but we assert that Granger causality may be an important tool for discovering mechanisms, as illustrated, for instance, by Brovelli et al. (2004), and so it is also related to level 2. Furthermore, the classification of Cox and Wermuth (2004) does not include time, and since Granger causality explicitly tries to sort out the effects of the past on the future, it is closer to causality than mere statistical association.

Cox and Wermuth (2004) also stress the provisional nature of much causal explanation, realizing, for example, that there would often be a more detailed explanation on a deeper level, which is presently not available. In general, there is a distinction between actions, where one wants to be fairly sure that the impact of the actions are as predicted, and understanding of mechanisms, which would often be much more provisional.

The concept of action, or intervention, has a special place in evaluation of causality. Often the aim of a causal understanding is precisely to intervene in a meaningful way to achieve a certain purpose. A medical doctor gives his patients medicine for them to get well, and so the medicines are supposed to have specific causal effects. In preventive medicine, measures like stopping to smoke are supposed to have a causal effect on health. In natural science, in general, large numbers of experiments are carried out inside and outside of laboratories where specific interventions are made with the aim of increasing the understanding of mechanisms governing physical, chemical, and biological systems. When human beings are involved as subjects of study, the possibility of experimental interventions will be strictly limited for ethical reasons, but one still has the aim of understanding what the effects of specific interventions might be. The problems of causality arise particularly in this latter area, which is important in medicine and the social sciences.

One should be aware, however, that ethical considerations are not the only reason that the possibilities of intervention and experimentation may be limited. For instance, since Newton, it has been assumed that the moon is the reason for the tides moving around the world, but one could hardly design a decisive experiment for this. Obviously, cosmology and other parts of science may be beyond experimentation, but one is still trying to understand mechanisms. So causal understanding is not necessarily tied to interventions.

A recent such example of great importance is the issue of global warming and climate change. It is now generally assumed that man-made carbon dioxide emissions have caused an increase in global temperature with great consequences in the future. The understanding behind this is mainly based on complex models with a mechanistic basis. Although it is assumed that human intervention is responsible for global warming, it is not realistic to test this in large-scale experiments. And the idea of a counterfactual world is not simple, since a world without large-scale carbon dioxide emissions would be a different place in many respects. Of course, the hope is that mankind will intervene and reduce emissions over time, but this will be more based on an understanding of mechanisms than on large-scale experiments.

9.1.3 Traditional approaches to causality in epidemiology

A common check for possible causal connections in epidemiology is the famous Bradford Hill criteria. These can be found in any textbook in epidemiology, see, for example, Rothman and Greenland (1998) or Rothman (2002). The criteria are as follows:

1. Strength of association: the stronger, the more likely to be causal
2. Consistency: the findings should be consistent over many different studies
3. Specificity in the causes: ideally only one cause
4. Temporality: cause must precede effect
5. Dose response relationship: the higher the dose, the stronger the response
6. Theoretical plausibility: there should be a theoretical basis
7. Coherence: consistent with other knowledge
8. Experimental evidence: evidence from experiments strengthens a hypothetical cause
9. Analogy: analogous to accepted principles in other areas

The criteria are simple and sometimes useful tools, but at the same time disputed. For instance, Höfler (2005) asserts that the criteria can easily be misapplied and argues for counterfactual thinking and sensitivity analysis. Some criteria are more accepted than others, for example, the important criterion of temporality. Also criteria 1, 2, 5, and 8 seem important. Criterion 6, concerning plausibility, is more difficult because, as pointed out by Höfler, one can often come up with plausible explanations of almost any outcome of a study. Höfler also points out that if the explanation has been suggested prior to the empirical study, instead of being proposed afterward, it will have a much stronger position. In any case, the Bradford Hill criteria will not be sufficient to prove causality, and they were never intended to do so.

Another classical model within epidemiology is the causal pie model of Rothman; see Rothman (2002). This promotes the important distinction between necessary and sufficient causes but lacks the time aspect. The model in itself is promising but should have been developed further. It may be viewed as a kind of counterfactual model.

9.1.4 The great theory still missing?

As has been shown here, the concept of causality shows up in many different and rather confusing shapes in empirical research. This could be an indication that "the great theory of causality" is still missing. However, this does not justify the "caution and avoidance" to which Pearl (2000) refers. Therefore, we shall discuss causality, focusing on some specific models. The idea is not necessarily to make definitive statements about causality, but more modestly to follow the idea that events and processes influence one another.

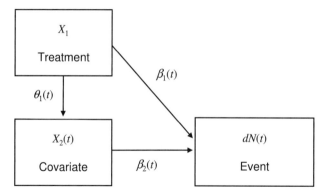

Fig. 9.1 *Example of a path diagram for a single jump in the counting process.*

9.2 Graphical models for event history analysis

Graphical models provide the basis of one of the major approaches in causal modeling. Traditionally, time has played little role in this approach; rather the aim has been to elucidate how variables influence one another. We believe that the thinking behind graphical models, as presented, for instance, in Pearl (2000) is really about processes developing over time, although this is unfortunately not reflected in the graphical formulation. Our view is that causality should primarily be perceived as a relationship between events and other developments in stochastic processes, not between fixed variables, and that it should be explicitly considered that cause has to precede effect. Notice that "events and other developments" are to be understood in a broad sense, as both specific occurrences like death, divorce, or myocardial infarction but also as changes in the measured level of some covariate or marker. We shall present a theory suited to analyze processes along these lines, but which at the same time uses the machinery of graphical models. Our graphical models are defined locally in time and change as time passes. We shall call our method *dynamic path analysis* (Fosen, Borgan et al., 2006; Fosen, Ferkingstad et al., 2006). An introduction for recurrent event data was given in Section 8.4, but here we shall present a more general point of view. The justification for a theory of the kind described here follows, for instance, from the difficulties of analyzing time-dependent covariates in survival analysis.

We start with a counting process. The idea is to define a graphical model for each jump in the counting process, that is, for each event occurring. For illustration we use a simple example with a fixed covariate X_1 (e.g., a treatment indicator), a possibly time-dependent covariate $X_2(t)$, and a counting process; see Figure 9.1. The graph in the figure is a *directed acyclic graph*, often abbreviated DAG. Note that we use infinitesimal graphs, that is, a whole series of graphical models indexed by time, and that the final dependent variable, on the right in the figure, is the differential of a counting process where, in practice, we would only consider the jumps. In order for this to be feasible, one must be able to carry out regression analysis on the occurrence of single jumps. The additive model is precisely the right tool for doing this.

In Figure 9.1 the regression coefficient in a standard linear model (ordinary linear regression) of $X_2(t)$ on X_1 is denoted $\theta_1(t)$. It is important to note that $\theta_1(t)$ may vary with t also when both covariates are fixed. This is because the risk set will change over time, and all analysis is carried out on the relevant risk set at any given time. In survival data, the correlation even between fixed covariates may change as time passes due to high-risk individuals leaving the risk set.

The remaining functions in Figure 9.1, $\beta_1(t)$ and $\beta_2(t)$, are regression functions in the additive model. In the path diagram they are the *direct* effects of X_1 and $X_2(t)$ respectively. Following the terminology of traditional path analysis, we denote the functions $\theta_1(t)$, $\beta_1(t)$, and $\beta_2(t)$ as path coefficients. The structural equations corresponding to the path diagram are the following:

$$dN(t) = (\beta_0(t) + \beta_1(t)X_1 + \beta_2(t)X_2(t))dt + dM(t);$$
$$X_2(t) = \theta_0(t) + \theta_1(t)X_1 + \varepsilon(t).$$

The first equation is the classical Doob-Meyer decomposition, meaning that $\beta_0(t) + \beta_1(t)X_1 + \beta_2(t)X_2(t)$ is the intensity process of $N(t)$. In the second equation, $\varepsilon(t)$ is an error term that is assumed independent of $M(t)$ and X_1. Inserting the second equation into the first yields the following marginal equation:

$$dN(t) = \{\beta_0(t) + \beta_2(t)\theta_0(t) + (\beta_1(t) + \theta_1(t)\beta_2(t))X_1 + \beta_2(t)\varepsilon(t)\}\,dt + dM(t).$$

The effect of treatment here, called the *total* effect, is $(\beta_1(t) + \theta_1(t)\beta_2(t))dt$. Subtracting the direct effect $\beta_1(t)dt$, we get the *indirect* effect of treatment given by $\theta_1(t)\beta_2(t)dt$. Hence, we may write

$$\text{total effect} = \text{direct effect} + \text{indirect effect} = (\beta_1(t) + \theta_1(t)\beta_2(t))dt \qquad (9.1)$$

as also explained in Section 8.4. In many cases, the total effect would coincide with what one terms the *causal* effect.

Since the estimates we use in the additive model with $dN(t)$ as a dependent variable are just least squares estimates like in standard path analysis, we conclude that the algebraic relationships between estimates of standard path analysis will be valid here too. This means that the marginal effect of treatment will exactly equal the total effect. In fact, one major advantage of the additive model is that it fits so well with standard linear regression analyses.

For estimation purposes, we consider n copies of the variables defined earlier, that is, counting processes $N_i(t)$ and covariates $X_{i,1}$ and $X_{i,2}(t)$ for $i = 1,...,n$. The estimation is carried out as in Section 4.2, where the design matrix $\mathbf{X}(t)$ has as its ith row the elements $\{Y_i(t), Y_i(t)X_{i,1}, Y_i(t)X_{i,2}(t)\}$.

The estimation in this model is done in the following way:

1. The regression functions $\beta_0(t)$, $\beta_1(t)$, and $\beta_2(t)$ are estimated by the additive hazards regression model, that is, in a cumulative fashion. The resulting cumulative
   estimates are denoted $\widehat{B}_0(t)$, $\widehat{B}_1(t)$, and $\widehat{B}_2(t)$.
2. The function $\theta_1(t)$ is estimated at each jump time of $N(t)$ by standard linear
   regression of $X_2(t)$ on X_1. Denote this estimate $\widehat{\theta}_1(t)$.
3. The integral $\int_0^t \theta_1(s)\beta_2(s)ds$ is estimated by multiplying each increment in the
   estimate of $\int_0^t \beta_2(s)ds$ by the relevant estimate of $\theta_1(s)$. Hence, the indirect effect
   of X_1 on the counting process is estimated by the cumulative regression function

$$\widehat{B}_{\text{indir}}(t) = \int_0^t \widehat{\theta}_1(s)d\widehat{B}_2(s)$$

and the total effect is $\widehat{B}_{\text{tot}}(t) = \widehat{B}_1(t) + \widehat{B}_{\text{indir}}(t)$.

9.2.1 Time-dependent covariates

The analysis of internal time-dependent covariates has been a thorny issue in survival analysis. The usual approach has been to include a time-dependent covariate
in a Cox model, where at each new event the contribution to the partial likelihood
uses the most recent value of the covariate. This is standard in several statistical program packages. From Figure 9.1, and imagining for a moment that $\beta_1(t)$ and $\beta_2(t)$
are coefficients in a Cox model, one sees that this corresponds to estimating the *direct* effects of the covariate and of treatment. However, when looking for the causal
effect for treatment, one often prefers to estimate the total effect (under the assumption that the treatment is randomized). This is again easily estimated for both the
additive and the Cox models by using a marginal model containing only the covariate X_1 (although, in a more complex situation the causal treatment effect would not
necessarily be a simple total effect). However, if one wants to split the total effect
in a sum of a direct effect and an indirect effect, thereby increasing the understanding of the process, one needs a decomposition formula like the one shown in (9.1).
This gives a combination of an event regression model, giving estimates of $\beta_1(t)$
and $\beta_2(t)$, and a model for how treatment influences the time-dependent covariate.
When both models are linear, there is no difficulty in this combination, but when the
event model is nonlinear, for instance, a Cox model, no decomposition is available.

We assert that the dynamic path analysis presented here throws a clear light on
the issue of time-dependent covariates. It explains why a standard analysis of time-
dependent covariates is often insufficient and how the problem should be mended.
Also, our approach gives an analysis of both direct and indirect effects. If interaction
is present, it may be possible to divide the sample into groups such that linearity
holds (approximately) in each group.

There are also other approaches for analyzing time-dependent covariates in a
causal manner. Several methods have been developed by Robins and coworkers,

Table 9.1 *Estimates with standard errors (SE) from a Cox analysis with baseline or current pro-thrombin.*

Covariates	Standard Cox model		Time-dependent model	
	Estimate	SE	Estimate	SE
Treatment	−0.28	0.14	−0.06	0.14
Sex	0.27	0.16	0.31	0.15
Age	0.041	0.008	0.043	0.008
Acetylcholinesterase	−0.0019	0.0007	−0.0015	0.0006
Inflammation	−0.47	0.15	−0.43	0.15
Baseline prothrombin	−0.014	0.007		
Current prothrombin			−0.054	0.004

and we shall discuss one of these, the marginal structural model, under the counter-factual approach; see Section 9.6.1.

Example 9.1. Survival of patients with liver cirrhosis. The liver cirrhosis data are described in Example 1.7. The covariates to be used are as follows:

- Receiving placebo or treatment with prednisone (0 = placebo, 1 = prednisone).
- Sex (0 = female, 1 = male).
- Age, ranging mostly from 44 to 77 years, but with four persons age 27.
- Acetylcholinesterase: an enzyme that breaks the neurotransmitter acetylcholine down after the transmission of a nerve impulse. This is a necessary breakdown process in order to enable rapid neurotransmissions (range: 42–566).
- Inflammation in liver connective tissues (0 = none, 1 = present).
- Baseline prothrombin [percentage of normal value of a blood test of no. II, VII, and X of blood coagulation factors produced in the liver]: prothrombin level mea-sured at time zero.
- Current prothrombin (time-dependent): prothrombin level measured most re-cently. (Time-dependent covariates were recorded at follow-up visits to a physi-cian. The visits were scheduled at 3, 6, and 12 months after randomization, and then once every year.)

Based on a preliminary analysis, the prothrombin values have been transformed so that if $v(t)$ is the untransformed prothrombin value (the last prothrombin value measured prior to time t for $t > 0$, and baseline prothrombin for $t = 0$), then we have

$$\text{Transformed prothrombin} = \begin{cases} v(t) - 70 & \text{if } v(t) < 70, \\ 0 & \text{if } v(t) \geq 70. \end{cases}$$

Note that a prothrombin value above 70 is considered "normal." Thus, the trans-formed prothrombin is zero for "normal" prothrombin levels and negative when the level is lower than "normal."

We start with a Cox analysis, as shown in Table 9.1. The standard Cox model presents results for an analysis with only fixed covariates, and one sees that there is a significant treatment effect. The right column in the table shows the results where

5358 9 Causality

Marginal model

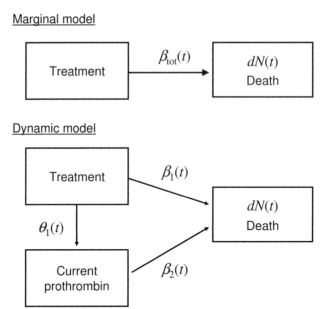

Dynamic model

Fig. 9.2 *The marginal and dynamic path models.*

prothrombin is introduced as a time-dependent covariate; in that case, the treatment effect almost disappears. This underestimation of the treatment effect is due to part of the treatment effect being picked up by the time-dependent covariate (current prothrombin). The question is how to get a clear picture of this relationship.

A detailed analysis based on dynamic path analysis is given by Fosen, Ferkingstad et al. (2006). We shall start with a simplified version, concentrating on treatment and current prothrombin. In the setting of the additive model, the marginal model and the dynamic path model are shown in Figure 9.2. The cumulative regression estimates of the β-functions are shown in Figure 9.3. The upper panel shows a significant treatment effect in the marginal model and how it is weakened in the dynamic model, while the lower panel shows a strong effect of current prothrombin on survival.

The solution to the problem consists in introducing the path from treatment to current prothrombin in the dynamic path model, see Figure 9.2. The strength of this path is analyzed by direct linear regression of current prothrombin on treatment at each failure time. One sees a strong initial effect, which tapers off. The marginal treatment effect at time t is now given as:

$$\beta_{\text{tot}}(t)dt = \beta_1(t)dt + \theta_1(t)\beta_2(t)dt.$$

Hence, the additive hazards model gives a nice resolution of the underestimation problem, partitioning the effect into a direct effect of treatment on survival and an indirect effect going through prothrombin. The estimate of $\theta_1(t)$ is given in Figure 9.4,

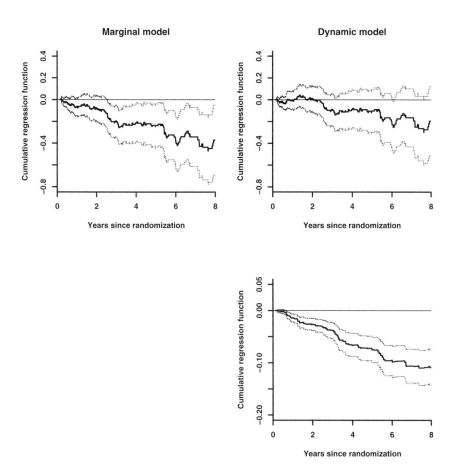

Fig. 9.3 *Cumulative regression estimates for the marginal and the dynamic model. Upper panel: treatment, with $\widehat{B}_0(t)$ to the left and $\widehat{B}_1(t)$ to the right. Lower panel: current prothrombin with estimated cumulative regression function $\widehat{B}_2(t)$.*

which clearly shows a strong initial effect of treatment on prothrombin level, with this effect disappearing quite soon. The conclusion is that there is no direct significant effect of treatment on survival, but there is a significant indirect effect through prothrombin, which however does not last.

When the other covariates are also included, a more complex picture emerges; see Figure 9.5. The figure is based on Fosen, Ferkingstad et al. (2006), which provides the details of the estimation. Main pathways for the various covariates are indicated. It appears that several covariates affect survival through prothrombin. Sex and age are exceptions; the influence of sex goes mainly through the covariate inflammation.

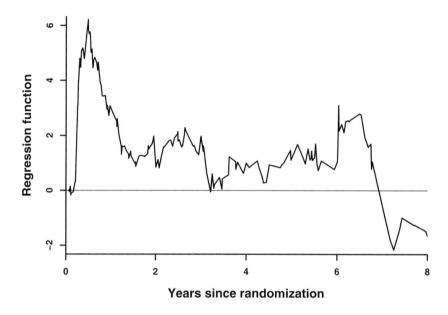

Fig. 9.4 *Least square regression estimates of current prothrombin on treatment.*

The idea behind this analysis is to approach a simple mechanistic model for how the treatment effect works. Within the data that are measured, it is clear that the treatment effect is to a large extent mediated through the current prothrombin level. This is a kind of causal thinking since we try to understand how effects are mediated through specific processes. Nevertheless, the depth of this analysis is not any greater than that permitted by the data. We just have a few measurements, and certainly we have no data on, say, the signaling and pathways within and between different cells. So, biologically, the analysis is quite superficial. If we did have more detailed data, we could possibly have used similar approaches to get a better analysis. But we have applied causal thinking to the data that are available. The resulting model will be highly provisional, just like any mechanistic model, and subject to various confounders. Since the prothrombin level is an indicator of liver function, we can make the provisional causal suggestion that treatment prolongs survival because it improves liver function. □

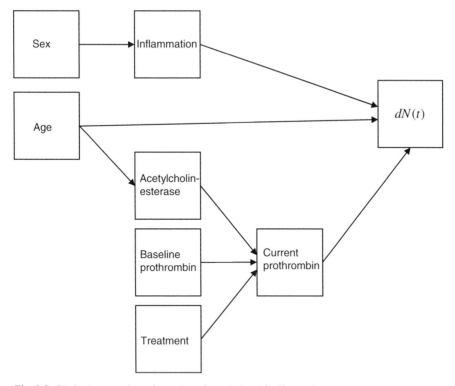

Fig. 9.5 *Cirrhosis example: a dynamic path analysis with all covariates.*

9.3 Local characteristics - dynamic model

Although the dependent variable in event history analysis is typically just the occurrence of an event, we shall briefly define a more general framework. It will often be artificial to distinguish event processes (counting processes) from other processes, and this section tells us how to assume a more general view.

We shall introduce a general type of statistical model that will be useful in this chapter. We recall the Doob decomposition of a time-discrete process (Section 2.1.4):

$$X_n = E(X_n \mid \mathscr{F}_{n-1}) + M_n - M_{n-1}. \tag{9.2}$$

The quantity $E(X_n \mid \mathscr{F}_{n-1})$ is a function of the past only, and we shall talk about a *dynamic statistical model* if $E(X_n \mid \mathscr{F}_{n-1})$ is parameterized in a suitable manner. Examples would be an auto-regressive model like

$$E(X_n \mid \mathscr{F}_{n-1}) = \alpha X_{n-1} + \beta X_{n-2}$$

or some model

$$E(X_n \mid \mathscr{F}_{n-1}) = \alpha U_{n-1} + \beta V_{n-1}, \tag{9.3}$$

where U_{n-1} and V_{n-1} are possibly complex functions of the past. Statistical parameters, like α and β, could be estimated by least squares or another estimating equation approach.

From equation (9.2), it is clear that the martingale differences, or innovations, are precisely the residuals. *Notice that no structure, like Markovian assumptions, need to be imposed on U_n and V_n. The key to statistical estimation is an explicit, not too complex, dependence on statistical parameters. The probabilistic dependence, on the other hand, can be almost arbitrarily complex.* This is due to the very general nature of the Doob decomposition and the asymptotic properties that follow from the martingale property of the cumulative residuals.

It will sometimes be useful to consider more general processes, and for this purpose we recall the time-continuous Doob-Meyer decomposition; see Section 2.2.3. A semimartingale can, according to the Doob-Meyer decomposition, be written as the sum of a predictable process and a martingale. Denote such a process $X(t)$; one can then write:

$$X(t) = X^*(t) + M(t). \tag{9.4}$$

The predictable part, $X^*(t)$, is denoted the compensator and is just the integral of the intensity process in the counting process case. As for the Doob decomposition in the time-discrete case, the idea of (9.4) is to decompose the process into what can be predicted from the past and what is the innovation or surprising element (the differential $dM(t)$ of the martingale).

Sometimes, the predictable part, $X^*(t)$, is differentiable, and we can write

$$X^*(t) = \int_0^t \mu(s)ds.$$

The stochastic process $\mu(t)$ is an example of a local characteristic because it determines the local development of the process. An intensity process is an example of a local characteristic. Another example is a diffusion process defined by the following stochastic differential equation:

$$X(t) = \int_0^t \mu(s)ds + \int_0^t \sigma(s)dW(s),$$

where both the drift coefficient $\mu(t)$ and the variance coefficient $\sigma^2(t)$ are denoted local characteristics.

A generalization of the idea of a dynamic model defined for discrete time [see (9.3)] may be introduced by putting a parametrization into, say, $\mu(t)$, for instance:

$$\mu(t) = \alpha(t)U(t) + \beta(t)V(t) \tag{9.5}$$

for predictable processes $U(t)$ and $V(t)$. We shall return to the use of such models in Section 9.4, in connection with Granger-Schweder causality.

When $X(t)$ is a counting process we are back to our usual framework, and the preceding model is a special case of the additive hazards regression model. Martinussen and Scheike (2006) have applied the additive model to general longitudinal data, thereby considerably extending the framework developed for counting processes. Extensions to binary longitudinal data have been made by Borgan et al. (2007).

In an interesting development, Farewell (2006) and Diggle et al. (2007) developed a linear, statistical analysis of the dynamic model. This is based on modeling the increments of a time-discrete longitudinal process and applying a linear model to the increments, in analogy with the additive hazards model. Their approach exploits the martingales efficiently, in an analogous way to counting processes for event histories. They also give an elegant way of handling missing data. These papers show the *basic unity of event history analysis and longitudinal data analysis*; to a large extent, similar tools can be used in both fields. This is also the idea in the present chapter, where in Examples 9.1 and 9.3 the processes of prothrombin and cortisol, respectively, are analyzed by linear models.

An interesting approach to likelihood analysis for mixed event history and longitudinal data is presented by Commenges and Gégout-Petit (2007).

9.3.1 Dynamic path analysis – a general view

Here we shall connect the dynamic statistical analysis with time-indexed graphs to produce dynamic path analysis. A special case has already been considered in Section 9.2, but we shall here put it into a more general framework.

A dynamic path diagram is a set of directed acyclic graphs (DAGs) $G(t) = (V(t), E(t))$ indexed by time t. Here $V(t)$ denotes the vertices and $E(t)$ the edges at time t. At any time t, the vertex set $V(t)$ is partitioned into a *covariate set* $V_c(t) = \{X_1(t), X_2(t), \ldots, X_p(t)\}$ and an *outcome process* $Y(t)$. We will assume that the vertices in the graph as well as the partition into covariates and outcome are time-invariant. However, the edge set $E(t)$ may vary with time. We shall assume that any set of edges respecting the DAG assumption is allowed, except edges pointing from the outcome to a covariate. The set of vertices that point to a vertex Z are called the parents of Z and are denoted $pa(Z)$, while the set of vertices to which there is an arrow originating in Z are called the children of Z and denoted $ch(Z)$. The outcome process $Y(t)$ has no children.

Estimation in dynamic path analysis is performed each time there is a change in the outcome process. By "change" we mean, for instance, an event in a counting process or the discrete changes of a discretized continuous process. As usual in standard path analysis, the estimation is done by recursive least squares regression, that is, each variable is regressed on its parents. In particular, the increment of the outcome process $Y(t)$ is regressed onto $pa(Y)$. If $Y(t)$ is a counting process we use the additive regression model, but a version of this can also be used when $Y(t)$ is a more general type of process; see Martinussen and Scheike (2000), who study human growth data as an example. In their case, $Y(t)$ is the speed of growth. Informally, the dynamic path diagram may be seen as continuously evolving over time, with edges possibly appearing or disappearing any time we collect new information. In the case of a counting process outcome, this happens at each event time.

As illustrated in Section 9.2, total effects may be decomposed into direct and indirect effects. A direct effect is an effect that is transmitted through a single edge in a graph, that is, from a covariate to one of its children, while an indirect effect is an effect of a covariate working through a directed path of length greater than one. In other words, an indirect effect is an effect that is *mediated* through one or more other covariates. Note that there may be several indirect effects of a covariate on an outcome. The decomposition into direct and indirect effects depends on the linearity of the system. In nonlinear systems, the notion of an indirect effect is problematic and cannot be defined simply as the difference between the total effect and the direct effect. According to Pearl (2001), such a "definition" would have no operational meaning, although a more complex definition would be possible.

In our model, the indirect effect along each path is simply the product of the (ordinary linear or additive hazard) regression functions along the path, and there is no problem of interpretation. The direct and indirect effects simply add up to the total effect. This is like in classical path analysis, where linearity gives this simple relationship between direct and indirect and total effects, with the multiplication rule for indirect effects; see Loehlin (2004). The new aspect here is that this is applied to the increment of a process. A special case was explicitly shown in formula (9.1) for the additive hazard model.

To formalize these notions, let $\varphi(X_h(t), X_j(t))$ denote the regression coefficient of $X_h(t)$ when $X_j(t)$ is regressed onto its parents (assuming that there is an edge from $X_h(t)$ to $X_j(t)$), and let $dB_{X_j}(t)$ be the additive hazards regression coefficient of $X_j(t)$ when $dY(t)$ is regressed onto its parents. We shall consider direct, indirect, and total effects of an arbitrary covariate $X_h(t)$ on the outcome $dY(t)$. Let $Z_1(t)$, $Z_2(t), \ldots, Z_k(t)$ be the ordered intermediate vertices along a directed path leading from $X_h(t)$ to $dY(t)$. The indirect effect of $X_h(t)$ on the infinitesimal change in $Y(t)$, $dY(t)$, is given by

$$\text{ind}(X_h(t) \rightarrow dY(t)) \tag{9.6}$$

$$= \sum_{\text{all paths}} \varphi(X_h(t), Z_1(t)) \left(\prod_{i=1}^{k-1} \varphi(Z_i(t), Z_{i+1}(t)) \right) dB_{Z_k}(t).$$

The direct effect, $\text{dir}(X_h(t) \rightarrow dY(t))$, is simply $dB_{X_h}(t)$ and the total effect is

$$\text{tot}(X_h(t) \rightarrow dY(t)) = \text{dir}(X_h(t) \rightarrow dY(t)) + \text{ind}(X_h(t) \rightarrow dY(t)). \tag{9.7}$$

Further, the cumulative indirect effect is given by

$$\text{cind}(X_h(t) \rightarrow Y(t)) \tag{9.8}$$

$$= \int_0^t \sum_{\text{all paths}} \varphi(X_h(t), Z_1(t)) \left(\prod_{i=1}^{k-1} \varphi(Z_i(t), Z_{i+1}(t)) \right) dB_{Z_k}(t),$$

while the cumulative direct effect is $\mathrm{cdir}(X_h(t) \rightarrow Y(t)) = B_{X_h}(t)$. The cumulative total effect, $\mathrm{ctot}(X_h(t) \rightarrow Y(t))$, is the sum of the cumulative direct and indirect effects.

9.3.2 Direct and indirect effects – a general concept

The concept of direct and indirect effects has a general validity, as may be illustrated by a couple of examples. Hesslow (1976) discusses causality in the context of the effect of birth control pills on thrombosis. It has been shown that certain pills increase the risk of thrombosis (direct effect). Pregnancy also increases the risk of thrombosis, while, clearly, birth control pills lower the chance of getting pregnant. Thus, birth control pills have a negative indirect effect on thrombosis, through preventing pregnancy. Another example may be found in a study on education as a risk factor for breast cancer. It is well known that women with higher education have a greater risk of breast cancer. Braaten et al. (2004) analyzed this in relation to a number of other risk factors and found that when one adjusted for parity, age at first birth, and other risk factors, the effect of education disappeared. Hence, education has no direct effect, only an indirect effect, for instance, by making women postpone having children.

Direct and indirect effects are connected to a mechanistic view of causality (Aalen and Frigessi, 2007; Machamer et al., 2000). This can be studied at many different levels. In genetics, one may decompose the associations between parent and child into genetic and environmental components. Certainly, at some level this yields a mechanistic understanding; we may realize that genetic elements are at work and that the disease, say, is likely to be inherited. A much more detailed understanding would be derived if one could pinpoint the specific genes that are responsible. Even greater understanding would be derived if one could specify how the proteins made by these genes influence specific pathways that generate disease. Hence, the concepts of direct and indirect effects are completely dependent on the level at which one is operating. A direct effect may dissolve into a number of indirect effects when more detailed knowledge is gained.

It is important to realize these different levels of understanding. In statistics, causality tends to be viewed as something absolute; either a factor has a causal effect or the apparent effect is due to confounding or bias. When attempting to understand underlying mechanisms, it is natural to think in less definite terms. A particular study may make a step toward a mechanistic understanding, but due to limitations in the data, the understanding will remain at a superficial level. Entering deeper into the issue may require much more elaborate measurements and data. Even at a rather fundamental biological level, for example, in transcriptional regulation networks, the distinction between direct and indirect effects is important; see Shen-Orr et al. (2002) and Wagner (2001). For instance, note the following interesting quotation from Wagner: "The reconstruction of genetic networks is the holy grail of functional genomics. Its core task is to identify the causal structure of a gene network, that is, to distinguish direct from indirect regulatory interactions among gene products."

One should note that the concepts of direct and indirect effects are not only statistical concepts, they may also correspond to a physical reality, like the directions that the signals in a cellular pathway are flowing. The hope, of course, is that statistical analysis shall lead to an insight to this underlying physical reality.

An interesting introduction to the use of direct and indirect effects in epidemiology is the paper by Ditlevsen et al. (2005) that uses a version of path analysis with emphasis on the so-called mediation proportion. A critique of this has been forthcoming from epidemiologists (Kaufman et al., 2004, 2005). Still, the approach advocated by Ditlevsen et al. (2005) is undoubtedly very useful, and it is important to overcome the resistance to using these methods in epidemiology, although one should be aware of the limitations. Other common concepts in this area are *causal pathway*, which indicates various directions of causal effects, and *mediator*, which denotes variables that mediate effects along a pathway; see Li et al. (2007) for a recent statistical treatment. The pathway concept has also become very popular in systems biology; see, for instance, Ulrich et al. (2006).

Cole and Hernán (2002) present a critical view of the possibility of estimating direct effects, pointing out that the causal status of such estimation may not be assured. However, as argued earlier estimating direct and indirect effects must not necessarily have a high level of causal ambition, but may still be useful. Furthermore, Cole and Hernán (2002) miss the time dimension of dynamic path analysis. A more optimistic view of the possibility of estimating direct and indirect effects is presented by Geneletti (2007), who explicitly distances herself from the counterfactual interpretation, embracing instead the "decision theoretic framework for causal inference." It is interesting to see an analysis from another point of view than the dominating counterfactual one, although again the important time dimension is left out.

Wermuth (2005) argues strongly for causal thinking, for instance, in the quotation: "There may also be important unobserved variables that are intermediate between treatment and the finally measured outcome variable. In this case, the overall treatment effect computed by omitting all intermediate variables may deviate strongly from the effect that describes pathways from the treatment to the response. ... Without mentioning such potential drawbacks and how they may possibly be corrected, statistics is likely to not appear as a trustworthy science to insightful collaborators from other fields." It appears that she is also thinking about issues like Simpson's paradox, which is related to the frailty artefacts discussed in Chapter 6.

Statistical estimation of direct and indirect effects are most easily done in the context of a linear model, as in dynamic path analysis. However, these concepts can also be defined in parametric models that explicitly allow for a causal structure; see Section 9.7.1.

9.4 Granger-Schweder causality and local dependence

In econometrics there has been considerable interest in what is termed "Granger causality." This was initiated by Granger (1969). Granger causality is focused on measurements taken over time and how they may influence one another. The idea is that causality is about the present and past influencing future occurrences. The idea behind Granger causality was also formulated by Schweder (1970) in a Markov chain setting, using the name local dependence, and this concept was later extended to more general stochastic processes by Mykland (1986) and Aalen (1987). The causality concept has a close relationship to dynamic stochastic processes defined through intensity functions, compensators, and martingales. One example of this is the counting process theory. The issues studied there are often about causality, although this term is not so much used in the stochastic process context.

The work of Schweder (1970) was independent of Granger's. The setting was also different: Schweder considered Markov chains while Granger studied time series; Schweder's model was at the individual or micro level, while Granger considered the macro level of econometric time series. However, the concept was fundamentally the same. To capture the breadth of the idea, we have chosen to name the concept "Granger-Schweder causality." We also note that the concept is dynamic in the sense used in this book, and one may therefore term it dynamic causality. This causality concept takes the time course into consideration in a much more direct fashion than the other approaches. Detailed data is a requirement for the analysis, but these are becoming increasingly available.

Note that Schweder (1970) avoids the term causality, instead using the concept local dependence, which we shall introduce in the following subsection. As in other settings, the use of the term causality here is controversial; see also some discussion in the Nobel Prize lecture of Granger (2003). We still follow the well-established tradition in economics and use the causality concept here, realizing that statistical analysis following these ideas will not necessarily result in establishing real causal connections.

9.4.1 Local dependence

Within the framework of a Markov chain, Schweder (1970) defines how one may think of the occurrence of one event influencing the occurrence of another, denoting the concept local dependence. For an illustration, see Figure 9.6, which shows a simple Markov chain for the occurrence of two events, A and B. The local dependence or independence is read off by comparing the transition intensities indicated on the arrows. Assume as an illustration that $\alpha(t) = \delta(t)$, while $\beta(t)$ and $\gamma(t)$ are different. If event A occurs first, then the intensity of event B is changed, hence A influences B. On the other hand, if event B occurs first, then the intensity for A is unchanged; hence B does not influence A. We say that B is locally dependent on A, while A is locally independent on B.

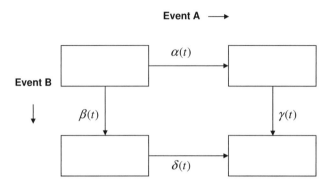

Fig. 9.6 *Graph illustrating the relationship between two events.*

A recent and very extensive study of local dependence has been given by Didelez in a series of papers (e.g., Didelez, 2006, 2007), where she also presents a novel type of graphical models for describing this kind of dependence. See also Pötter and Blossfeld (2001) for extensions and discussion of local dependence.

Example 9.2. Analysis of case series data. We shall briefly consider the analysis of case series data, by which we mean data on cases only, with no controls included. An analysis of such data using the local independence concept was done in Aalen et al. (1980). The interest in case series data has recently increased; see the papers by Whitaker et al. (2006) and Farrington and Whitaker (2006) for interesting new developments in this area. We shall use their data on the possible connection between the MMR vaccine and the occurrence of idiopathic thrombocytopenic purpura (ITP, a bleeding disorder) as an illustration of local dependence. The data concern 35 children who were admitted to hospital with ITP when between 1 and 2 years old. A few children experienced more than one event, so that there were 44 events in all. Following Whitaker et al. (2006), we shall analyze these as independent events; the authors show that accounting for multiple events has little effect on the results.

The aim of the analysis is to study the connection between the two types of events, getting the MMR vaccine (say event A) and being admitted with ITP (say event B). We shall do this by studying the four intensities, as given in Figure 9.6. We compute Nelson-Aalen estimators for the cumulative intensities and present the results in Figure 9.7.

The upper panel of Figure 9.7 indicates an increased rate of ITP admission after MMR vaccine compared to before the vaccine is given. The lower panel indicates a somewhat decreased rate of MMR vaccine after ITP admission compared to before. The main finding is a local dependence of ITP on MMR; after the MMR vaccine the risk of ITP is increased, confirming the results of Whitaker et al. (2006). Note, however, that since we only have cases available, the transition intensities will be conditional on occurrence of ITP between 1 and 2 years; see Aalen et al. (1980) for a discussion of this issue. In fact, this conditioning is a difficult issue that may create

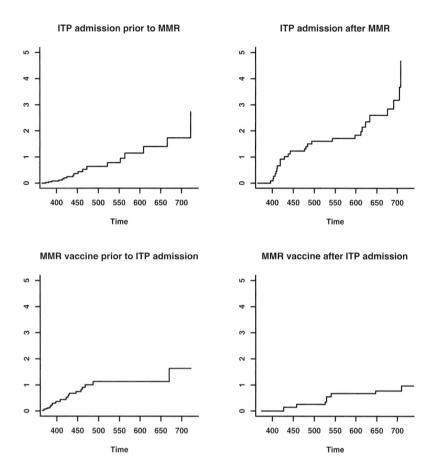

Fig. 9.7 *Nelson-Aalen estimators of cumulative transition intensities in the graph of Figure 9.6.*

bias in the comparison; it would therefore be better to have actual cohort data. So our analysis should be treated with caution and only be seen as an illustration of the concept of local dependence.

To give a little more detail we refer briefly to the Aalen and Didelez (2006) discussion of Farrington and Whitaker (2006). As stated there, we need to assume in our analysis, like Farrington and Whitaker (2006), that vaccination rates are unaffected by prior ITP to extract information from case series data. This is a crucial assumption that must be carefully scrutinized in every specific application, especially as it appears difficult to test empirically. Furthermore, unlike Farrington and Whitaker (2006), we cannot claim that our approach is self-controlled so that confounders can be ignored. This property of their method heavily relies on the multiplicativity assumption with no apparent counterpart in ours. □

9.4.2 A general definition of Granger-Schweder causality

Granger causality (Granger, 1969) is defined for two time series, $X(t)$ and $Y(t)$. One says that X does not Granger-cause Y if prediction of Y based on all predictors is no better than prediction based on all predictors with X omitted. This is essentially the same concept as local independence.

We shall give a more general formulation of Granger-Schweder causality, but first we extend the definition of local dependence. If there is a statistical association between two stochastic processes, say $X(t)$ and $Y(t)$, then this association may be of two major types. The first possibility is that the two processes are (partly) reflecting the same phenomenon. This is not a causal relationship. A medical example might be the presence of congested sinuses and of eye irritation in allergy, which might both express an underlying allergic process. When two processes are distinct, that is, not reflecting the same phenomenon in the sense discussed here, we call them *autonomous*; see Pötter and Blossfeld (2001).

The second type of association is where one process influences the likelihood of changes in the other process in a causal way. This may be a one-sided relationship with the influence going only in one direction, or it may be a two-sided relationship where both processes influence each other. In medicine, it is usually assumed that cholesterol level, which can be considered a stochastic process, causally influences the development of heart disease in a mainly one-sided fashion. This is expressed in the belief that lowering the cholesterol level reduces the risk of heart disease. (The reality may be more complex, however, with high cholesterol level, as well as high blood pressure, being not only a cause but also a symptom of heart disease.)

Let $\{\mathscr{F}_t^{X,Y}\}$ be the history generated by the processes $X(t)$ and $Y(t)$ and let $\mathscr{F}_0$ denote a σ-algebra encompassing other random events or influences. Under some regularity conditions on the two processes they can be represented by Doob-Meyer decompositions (see Section 2.2.3) defined with respect to the history $\{\mathscr{F}_0 \vee \mathscr{F}_t^{X,Y}\}$:

$$X(t) = X(0) + \int_0^t \mu_x(s)ds + M_x(t), \quad Y(t) = Y(0) + \int_0^t \mu_y(s)ds + M_y(t), \quad (9.9)$$

where the μs denote the local characteristics of the two processes and the Ms are martingales. Recall that the "differentials" of the martingales are often called innovations since they represent the new and "surprising" changes in the process.

One way of defining autonomous processes is in terms of the innovations. We say that processes X and Y are autonomous if the martingales are orthogonal, meaning that the product of the two processes is another martingale. If X and Y are counting processes, this reduces to the assumption that the two processes never jump at the same time. If the two processes are diffusion processes fed by Wiener processes orthogonality corresponds to stochastic independence of the Wiener processes. In practical terms autonomy means that the basic changes in the process, created by the innovations, are unrelated between the two processes. Autonomy, or orthogonality, is thus assumed to ensure that the two processes represent different phenomena. It seems reasonable to assume autonomy, before starting to talk about causal connections.

For orthogonal martingales, define $X(t)$ to be locally independent of Y at time t if $\mu_x(s)$ is only a function of X up to time t (and possibly of some extraneous events), but *not* a function of Y. If $X(t)$ is not locally independent of Y, it is locally dependent.

Let $\{\mathscr{F}_t^X\}$ be the history generated by the process $X(t)$. A formal definition of local independence may be given as follows:

Definition 9.1. Assume that $M_x(t)$ and $M_y(t)$ are orthogonal. Then $X(t)$ is locally independent of $Y(t)$ if $\mu_x(t)$ is adapted to the history $\{\mathscr{F}_0 \vee \mathscr{F}_t^X\}$.

Clearly, local dependence may be one-sided (only one process being dependent on the other) or two-sided. If there is only a one-sided local dependence over a time interval, it is tempting to talk about a causal connection of one process on the other. A practical medical application is presented in Aalen et al. (1980).

To define Granger-Schweder causality, we consider the ideal situation where all relevant confounders (that is, other variables influencing the relationship) are known and are a part of the history $\{\mathscr{F}_0 \vee \mathscr{F}_t^{X,Y}\}$. Then Definition 9.1 implies that the process Y *does not cause* X.

Consider a pair of counting processes $(N_1(t), N_2(t))$ with corresponding intensity processes $(\lambda_1(t), \lambda_2(t))$ and martingales $(M_1(t), M_2(t))$ defined with respect to the history $\{\mathscr{F}_0 \vee \mathscr{F}_t^{N_1, N_2}\}$. Then $N_1(t)$ is locally independent of $N_2(t)$ if $\lambda_1(t)$ does not depend on the development of N_2.

9.4.3 Statistical analysis of local dependence

To analyze statistically the presence of local dependence, and possibly Granger-Schweder causality one has to make suitable statistical models for the local characteristics. A simple possibility is to extend the additive hazards regression model to local characteristics. Local estimation of effects by least squares may be performed and then summed to informative cumulative functions.

Define the following models for the local characteristics in (9.9):

$$\mu_x(s) = \boldsymbol{\beta}_x^T \mathbf{z}_x(t), \qquad \mu_y(s) = \boldsymbol{\beta}_y^T \mathbf{z}_y(t)$$

for suitable parameter vectors $\boldsymbol{\beta}_x$ and $\boldsymbol{\beta}_y$, and covariate vectors $\mathbf{z}_x(t)$ and $\mathbf{z}_y(t)$. The covariate information contain various kinds of relevant information, including past information of the process itself and of the other process. Estimation may be performed by least squares regression.

Example 9.3. Sleep data. We shall continue the analysis of the sleep data (Example 1.14) from Section 8.5. We consider two processes. One is the event process registering transitions from the state "asleep" to the state "awake." Our counting process is then the process that counts the cumulative number of transitions of this kind after the first time the individual falls asleep. The number at risk at each time

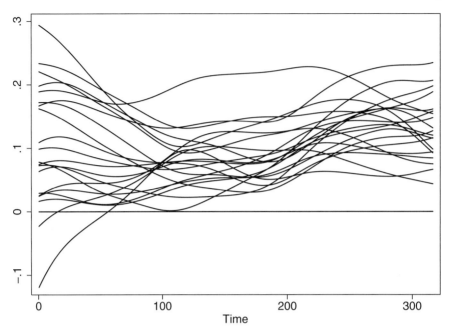

Fig. 9.8 *Cortisol change regressed on initial cortisol value, previous cortisol value, and number of previous awakenings; 20 bootstrap estimates of regression on initial cortisol value shown in figure.*

point is the number of people asleep just prior to this time. The counting process was analyzed in Section 8.5. Here we shall focus on a second process consisting of the cortisol measurements. Hence, the covariate process from Section 8.5 has become the main focus, and we want to estimate how the number of previous awakenings influence the cortisol level.

One must take into account that the cortisol measurements are not continuous but are taken every 20 minutes. The dependent variable in the analysis will simply be the difference between consecutive measurements (on a log scale). For a general approach to analyzing the increments of a longitudinal process, see Farewell (2006) and Diggle et al. (2007). A least squares analysis is performed at each point and then the estimates are summed up to cumulative functions. We include the following covariates:

- initial cortisol level (log scale)
- last measured cortisol level (log scale)
- cumulative number of times awake, divided by elapsed time

Twenty bootstrap estimates of the smoothed regression function for the covariates are shown in Figures 9.8, 9.9, and 9.10. Time is measured in hours from the first cortisol measurement after falling asleep. The two first figures show that those with a high initial cortisol level have a larger increase and that those with a high

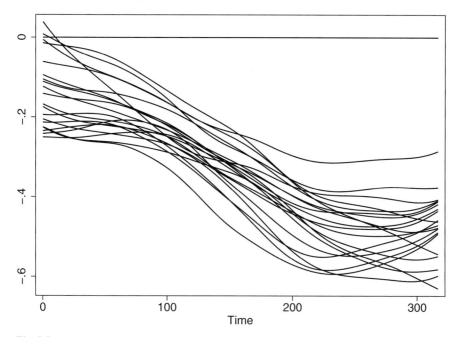

Fig. 9.9 *Cortisol change regressed on initial cortisol value, previous cortisol value, and number of previous awakenings; 20 bootstrap estimates of regression on previous cortisol value shown in figure.*

recent cortisol value have a smaller increase. Figure 9.10 shows no particular effect, hence the number of awakenings during the night do not appear to influence cortisol significantly. An effect in the opposite direction is found in Figure 8.5, where a high cortisol level in the morning would tend to increase the number of awakenings. This implies local dependence in only one direction, but clearly one should be cautious drawing a conclusion since cortisol is measured only every 20 minutes and this makes it difficult to pick up effects that happen on a shorter scale. The close relationship between awakening and cortisol does correspond to biological knowledge, see Edwards et al. (2001).

The analysis carried out can be seen as an extension of the path analysis described earlier. This is illustrated in Figure 9.11. Note that there are two final states in this model, as opposed to one in the previous path analysis. □

9.5 Counterfactual causality

The term "counterfactual" denotes one of the major philosophical approaches to causality (e.g., Menzies, 2001). In statistics, counterfactual ideas have been promoted especially by Donald Rubin and James Robins and their respective

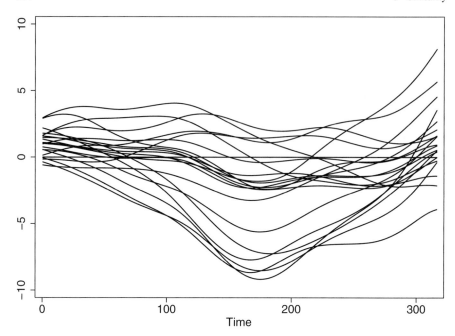

Fig. 9.10 *Cortisol change regressed on initial cortisol value, previous cortisol value and number of previous awakenings. 20 bootstrap estimates of regression on number of previous awakenings shown in figure.*

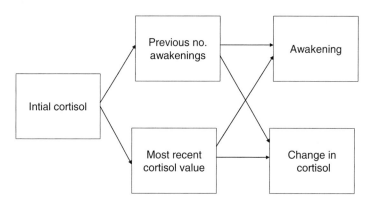

Fig. 9.11 *Path diagram corresponding to analysis of the relationship between the cortisol process and the process of awakening.*

coworkers; see, for instance, Rubin (1974) and Robins et al. (2000). The term "counterfactual" has been especially associated with the Robins school, while Rubin prefers the terminology "potential outcome," but the thinking seems closely related although the specific statistical methods differ somewhat between the schools. An introduction may be found in Pearl (2000), who also point out the relationship to classical graphical models.

The main idea is that one should consider what might happen if possibilities other than the actual one did occur, these other possibilities being "counter to fact." For instance, a smoker may develop certain diseases, and to judge the effect of smoking one considers what would have happened to the smoker had he not smoked. The obvious problem is that a complete analysis of this will depend on occurrences and measurements that cannot be observed, since any one person is in just one category. Nevertheless, statistical analysis of these issues can be done under certain assumptions, for example, the well-known one of "no unmeasured confounders," which is often associated with Robins but is really the classical requirement of epidemiological generalization. If assumptions of this kind are not valid, then one may make error bounds for the results or perform sensitivity analysis. The counterfactual thinking appears to be useful in many cases. The approach appears very pragmatic, attempting to secure correct conclusions in observational studies but is not so much directed toward gaining insight in the underlying processes.

A basic idea in the approach of Robins and coworkers is to weight the observed data in such a way that they mimic the data one would have if there were no censoring or no skewness due to correlation between treatment and time-dependent covariates. This strategy is similar to Rubin's use of the propensity score. One adjusts the data to mimic a complete and uncomplicated data set, a pseudo-population, which allows a simple analysis of the adjusted data. The approach can be seen as a generalization of the Horwitz-Thompson estimator from sampling theory. One important example is the marginal structural model, which we discuss in Section 9.6.1.

In the medical treatment of a patient it is typically the case that treatments change over time according to the clinical development of a patient. The evaluation of treatment strategies may therefore become complicated since different patients may undergo different treatment courses. Similar complex developments may be found in epidemiological studies and other contexts. Robins and others have introduced a method whereby one can nevertheless, under certain assumptions, compute probabilities for the outcome of various treatment strategies. One major result is the *G*-computation formula. The formula can be stated in many different ways. A special case can be formulated within our framework, where it turns out to be closely related to the empirical transition matrix.

An important aspect of the recent contributions of Robins and coworkers is the concept of a *treatment plan*. A treatment plan could be any series of actions or interventions, not necessarily medical treatments, of which we want to study the effect. Note that treatment changes may depend on previous outcomes for the patient, and this creates intricate problems that one has to solve.

The concept of counterfactual causality is attractive in many ways. It represents a very clear view of what is meant by a causal effect, and undoubtedly the introduction of these ideas into epidemiology and other fields has clarified thinking around causality. There are also controversies surrounding the concept, and it is quite clear that one should not expect one formulation to cover all notions of causal effect.

First we connect the idea of counterfactuals to ordinary survival and event history analysis.

9.5.1 Standard survival analysis and counterfactuals

Survival data are almost always subject to censoring, meaning that some of the survival times in most data sets will be censored. The aim of the estimation, however, is to make statements that would be valid in the absence of censoring. In fact, censoring is typically considered a nuisance, something one wants to make corrections for.

A Kaplan-Meier curve, for instance, is supposed to estimate the survival curve in a world where there is no censoring present. Hence, presenting a Kaplan-Meier curve is a simple example of a counterfactual statement. It is only valid when censoring is independent, see Sections 1.4 and 2.2.8 for a discussion of independent censoring. It is also important to note that censoring cannot even depend on observable covariates; if it does, then the standard Kaplan-Meier curve is not correct but must be adjusted (Section 4.2.7). This issue is often ignored in practice.

Counterfactual thinking in survival analysis has deep historical roots. In 1760, Daniel Bernoulli studied the consequences for mortality if smallpox could be erased. The method developed by him is today known as competing risks. Statements about what would happen under the hypothetical assumptions that a certain risk decreases or increases can be viewed as counterfactual. The basic question, obviously, is whether the remaining risks (e.g., diseases) are uninfluenced by the changes that take place. Eradicating cancer, for instance, may impact on other diseases. This also illustrates the basic fact that in order to make correct counterfactual statements, one will often need detailed information about the underlying biology, and simple statistical observations may not be sufficient.

A modern version of Bernoulli's approach has been carried out by Keiding and coworkers in several papers; see for instance Keiding et al. (2001). Instead of merely considering a set of competing risks, they look at more complex multistate models, usually formulated as Markov chains. After estimating the cumulative transition intensities by Nelson-Aalen estimators and then the transition probabilities by the empirical transition matrix (Section 3.4), they consider the effect of modifying certain intensities. An example they use is the study of the effect of treatment of graft-versus-host disease after bone marrow transplantation. The multistate model is shown in Figure 9.12, where the various relevant transitions are indicated.

A problem they consider is that graft-versus-host disease may have a positive effect in the sense of preventing relapse, by killing off leukemia cells, but a negative effect in the sense of increasing death risk in remission, and the question is to sort out the balance of these two effects. For this purpose, the analyses shown in Figures 9.13 and 9.14 have been made. By computing probabilities relating to

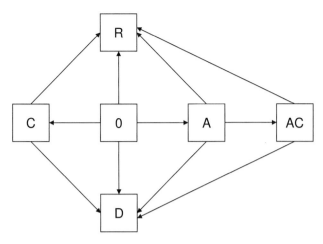

Fig. 9.12 *Simple multistate model for events after bone marrow transplantation. All patients start in state 0 at transplantation and ultimately either relapse (R) or die in remission (D). As intermediate states we include acute (A) or chronic (C) or both acute and chronic (AC) graft-versus-host disease. Figure redrawn from Keiding et al. (2001).*

hypothetical worlds where graft-versus-host disease is prevented, by setting intensities equal to zero, one can sort out the effect of such an action. These probabilities may be viewed as counterfactual. One sees that the prevention would slightly increase the relapse rate but considerably reduce the death rate in remission.

9.5.2 Censored and missing data

As discussed in the previous subsection, the treatment of censored data is based on estimating what would have been observed in the absence of censoring. This is a counterfactual point of view, since in most situations censoring cannot be avoided in the real world. When is such estimation possible? Missing data in general have been discussed by Rubin, who defined the concepts of "missing at random" and "missing completely at random." Under such assumptions, counterfactual estimation is feasible. Robins prefers to formulate a similar assumption as "no unmeasured confounders." In survival analysis, on the other hand, one prefer to talk about "independent censoring." As discussed in Section 2.2.8, there is also a close relationship to the concept of stopping time in martingale theory.

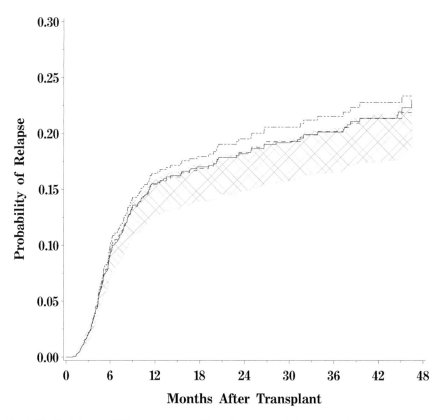

Fig. 9.13 *Shaded area: 95% pointwise confidence limits around the estimated probability of having relapsed by time t in this world. Predicted probability of having relapsed by time t in a hypothetical world where (—) acute, (– – –) chronic, and (– - - –) acute and chronic GVHD were prevented. Figure reproduced with permission from Keiding et al. (2001). Copyright John Wiley & Sons Limited.*

9.5.3 Dynamic treatment regimes

The ambition in causal analysis is to handle much more complex situations than one would have in an ordinary survival or event history analysis. In particular, one imagines treatments that can change dynamically over time. This is quite realistic as treatments in clinical practice will typically be changed and adapted to the development of the patient and his disease. Therefore the concept of a *treatment plan* is important. Ideally one would wish to evaluate the effect of possibly complex treatment plans. One then has to make assumptions as to how treatment decisions are made and how treatment is changing as a result of the clinical development. There

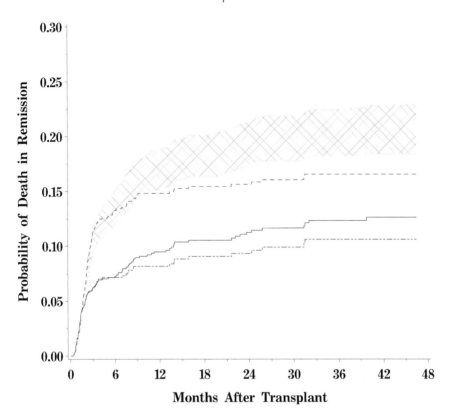

Fig. 9.14 *Shaded area: 95% pointwise confidence limits around the estimated probability of having died in remission by time t in this world. Predicted probability of having died in remission by time t in a hypothetical world where (—) acute, (– – –) chronic, and (– - - –) acute and chronic GVHD were prevented. Figure reproduced with permission from Keiding et al. (2001). Copyright John Wiley & Sons Limited.*

could be a random element in the choice, possibly by explicit randomization, or the treatment may be decided on the basis of the status of the patient. The treatment plan is a stochastic process, and we shall denote it by $G(t)$.

The aim of the statistical analysis of a clinical study will typically be to estimate the causal effect of treatment, by comparing an active treatment and a control. The causal effect is defined in a counterfactual manner: one considers a given individual and imagines that this individual could be given both treatments. What would then be the observed difference between the active treatment and the control? Note that the counterfactual situation may be purely hypothetical; in many cases, for instance, acute diseases, one can only imagine one of the treatments being given. The important question is when counterfactual effects can still be estimated. Again assumptions of the type discussed in the previous subsection for missing data must be made, like the "no unmeasured confounders" assumption. Basically, this amounts to assuming that given all the observed information, the treatment choices are made randomly. For this to be true in practice, sufficient information must be collected. If the doctor's treatment decisions depend on information that is not recorded in the file, one may be in trouble.

9.5.4 Marginal versus joint modeling

There are two fundamentally different ways of handling issues like dynamic treatment regimes, time-dependent covariates, and censoring. One is the marginal procedure; here the aim is to estimate, say, a treatment effect, and all other effects are considered nuisances to be adjusted for. Elaborate weighting procedures have been developed; these constitute the backbone of the work of Robins and coauthors. These procedures also focus mainly on estimating what would happen in a population with the same composition as the available sample, thereby limiting the generalizability.

The other approach is a joint one. Here, all or most components in the process are modeled in their own right. For example, the change of treatment as a function of previous observations is explicitly modeled, as is also time-dependent covariates. Even censoring may occasionally be explicitly modeled. In fact, censoring should more often be analyzed in its own right, for example, to check the independent censoring assumption required to estimate Kaplan-Meier curves.

9.6 Marginal modeling

We now present the analysis of marginal models, following the work of Robins and coauthors. We start with the marginal structural models, and then continue with the G-computation formula.

9.6.1 Marginal structural models

The aim of the marginal structural model, which was introduced by Robins (1986), is to estimate the effect of a treatment in the presence of time-dependent covariates, which may both influence and be influenced by the treatment. Essentially, the approach consists in weighting the observations appropriately before applying a Cox analysis. The ideas are most easily understood through an example, and we shall use the one applied in Robins et al. (2000). The authors study the effect of the medication zidovudine on the survival of HIV-positive men. A number of time-dependent covariates are measured, including the CD4 count. This is a time-dependent confounder because it is a risk factor for mortality and also a predictor of initiation of zidovudine therapy. Here, it is assumed that an individual stays on zidovudine therapy once it is started.

When including just treatment as a covariate (that is, on zidovudine or not), the hazard ratio is estimated to 3.55 (2.9–4.3), and when including baseline covariates this is adjusted to 2.32 (1.9–2.8). Hence, one may get the impression that zidovudine increases the risk of dying. The reason, of course, is that treatment is started when CD4 count is low, so one has to include information on the time-varying CD4 count. Doing that through a simple time-dependent Cox model would not be sufficient, because zidovudine therapy is influenced by and influences CD4 count. What is done in a marginal structural model is to weight the observations in an appropriate way.

Let $A(t)$ be 1 if the patient is on zidovudine treatment and 0 otherwise, and let $\overline{A}(t)$ denote an individual's treatment history up to, but not including, t. Furthermore, let $\overline{L}(t)$ denote the corresponding history of time-dependent covariates, and let V denote the baseline values of these covariates. It is common to define the weights based on a discretization of the time interval. Usually the weights are determined by logistic regression, so the intervals should not be too small. Let t be the end of time interval. The following weights are used, where the multiplication is performed over all time intervals up to time t:

$$sw_i(t) = \prod^{\text{int}(t)} \frac{P(A(k) = a_i(k) \mid \overline{A}(k-1) = \overline{a}_i(k-1), V = v)}{P(A(k) = a_i(k) \mid \overline{A}(k-1) = \overline{a}_i(k-1), \overline{L}(k-1) = \overline{l}_i(k-1))}. \quad (9.10)$$

The final step is to carry out a Cox analysis, where each observation in the risk set is weighted according to the above weights, and where treatment as well as V are included as covariates.

These weights are denoted inverse probability of treatment weights (IPTW). They are closely related to the propensity scores of Rosenbaum and Rubin (1983). The rationale behind them is as follows: the observations are weighted according to the probability of observed treatment given their covariates. Those who have an unusual treatment will be upweighted, and those who have a common treatment will be downweighted. Initially, the weights are the product formed from the denominators in (9.10). However, it has been found that more efficient estimators are achieved by stabilizing the weights, that is, multiplying with the numerator terms in (9.10). When there is no time-dependent confounding, the numerators will equal the denominators and give unweighted estimation. In the presence of confounding, the stabilized weights increase the efficiency.

When carrying out the weighted analysis, the hazard ratio for treatment is reduced to 0.74 (0.57–0.96), showing that there is an effect of treatment.

9.6.2 G-computation: A Markov modeling approach

The idea of *G*-computation was developed by Robins (1986); see also Lok et al. (2004). The aim is to compute the effect of a particular control policy *G*, for example, a sequence of treatments. The *G*-computation formula was originally expressed within a very general framework. The formulations of Robins are mostly for discrete outcomes and discrete time. An extension to the case of continuously distributed covariates and treatments was made by Gill and Robins (2001). Gill (2004) discussed extension to the time-continuous case, but so far this is incomplete. Here we give a special case of the *G*-computation formula for time-continuous Markov chains, where the result is closely related to the empirical transition matrix studied in Section 3.4. In this case, precise statements can be made, and the formula appears less "mystical" than it may be in the more general formulations. Our approach is similar to ideas in Keiding (1999); see also Section 9.5.1.

We consider individuals moving between different states. To get a reasonably simple situation, we shall assume that the development can be described as a Markov chain. The state space of the Markov chain contains the information of relevance for the decision regarding treatment, that is, whether to continue, to stop, or to change treatment. The state space shall also describe the clinical development that one wants to study. As in Section 3.4 we consider a Markov chain with a finite state space $\mathscr{I} = \{0, 1, \ldots, k\}$, and let $\alpha_{ij}(t)$ denote the transition intensity from state $i \in \mathscr{I}$ to state $j \in \mathscr{I}, i \neq j$. Further, for all $i, j \in \mathscr{I}$, we let $P_{ij}(s,t)$ denote the probability that an individual who is in state i at time s will be in state j at a later time t, and we write $\mathbf{P}(s,t)$ for the $(k+1) \times (k+1)$ matrix of these transition probabilities.

Suppose that we have a sample of n individuals from the population under study. The individuals may be followed over different periods of time, so our observations of their life histories may be subject to left-truncation and right censoring. A crucial assumption, however, is that truncation and censoring are independent so that the entry and censoring times do not carry any information on the risks of future transitions between the states. We assume that exact times for transitions between the states are recorded and denote by $T_1 < T_2 < \cdots$ the times when transitions between any two states are observed. Further, for $i, j \in \mathscr{I}, i \neq j$, we let $N_{ij}(t)$ count the number of individuals who are observed to experience a transition from state i to state j in $[0,t]$, and introduce $N_i(t) = \sum_{i \neq j} N_{ij}(t)$ for the number of transitions out of state i in this time interval. Finally, we let $Y_i(t)$ be the number of individuals in state i just prior to time t. Then the empirical transition matrix takes the form

$$\widehat{\mathbf{P}}(s,t) = \prod_{s < T_m \leq t} \left(\mathbf{I} + \widehat{\boldsymbol{\alpha}}_m \right);$$ (9.11)

cf. Section 3.4. Here $\mathbf{I}$ is the $(k+1) \times (k+1)$ identity matrix, $\widehat{\boldsymbol{\alpha}}_j$ is the $(k+1) \times (k+1)$ matrix with entry (i,j) equal to $\widehat{\alpha}_{ijm} = \triangle N_{ij}(T_m)/Y_i(T_m)$ for $i \neq j$ and entry (i,i) equal to $\widehat{\alpha}_{iim} = -\triangle N_i(T_m)/Y_i(T_m)$, and the matrix product is taken in the order of increasing T_m.

The estimator in (9.11) may be seen as a special case of the G-computation formula (Keiding, 1999). In particular, when evaluating treatment effects, there is the following correspondence: one considers a given treatment plan, and an individual is considered censored if he deviates from the treatment of interest. That is, an individual is considered "under observation" at a given time only if the treatment of interest at this time is being chosen. Otherwise the individual is "censored." This is termed artificial censoring (Hernán et al., 2006). The estimated transition matrix, which we shall denote $\widehat{\mathbf{P}}_G(s,t)$ will then be a valid estimate of the development under the given treatment sequence. For this to hold, the treatment decisions must satisfy the same requirement as the censoring. This is the "no unmeasured confounders" assumption of Robins.

The computation outlined here is called counterfactual because one estimates probabilities that are supposed to hold if all individuals are subjected to the particular treatment process. Since they were not, the computation is "counter to fact," but it may still be a valid statement.

Example 9.4. An illustration. Consider patients with angina pectoris on appropriate treatment. The patients may experience myocardial infarction (MI), whereupon the treatment is changed. This may be according to a strict protocol for treatment, or it may be part of a clinical trial where treatment is determined by randomization. Later, some patients may have a second infarction, or possibly heart failure, and then new treatments are instituted. The endpoint to be considered could be death.

Assume that at the first MI, the treatment is decided randomly according to two alternatives. Then, since the further developments in the treatment groups may be different, they may later get other treatments that differ between the groups. It may then be difficult to decide what the real effect of the original randomization was. A nonbeneficial effect of an original treatment may be partially offset by aggressive treatment of a later complication. In fact, it may be an advantage to receive an inferior treatment at an earlier stage, since this may bring out symptoms that lead to instituting a more effective treatment. Those who had the better treatment at the earlier stage may have to wait longer before receiving the more effective treatment because symptoms are suppressed. The methods discussed above may be of interest in such a case. $\square$

9.7 Joint modeling

When explicitly modeling treatment changes and other components of the process, a more explicit understanding is possible than in the marginal model. The weighting procedures in the marginal model are necessary because the effect is accumulated, but it will not be needed when a more detailed and explicit analysis is being performed.

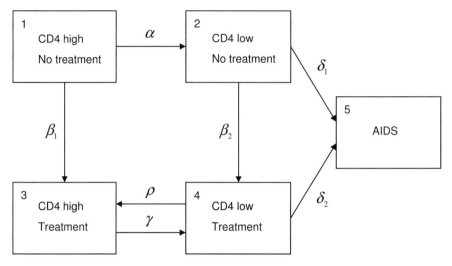

Fig. 9.15 *Development of HIV infection. A simple example of a dynamic treatment regime. The uptake of treatment is explicitly modeled.*

9.7.1 Joint modeling as an alternative to marginal structural models

One may alternatively include treatment as a state in the state space, as illustrated in Figure 9.15. Consider the development of AIDS. Treatment is instituted when the CD4 count falls below a certain limit. In actual fact, this will not be immediate; there will be some time, varying between individuals, before the treatment is started. Therefore, this can be seen as the transition to a new state. Transition intensities may be estimated. Probabilities of getting AIDS may then be computed dependent on various rates for starting treatment, for example, by comparing the hazard of AIDS with a realistic set of values for β_1 and β_2 with the hazard one gets when those parameters are zero. In this way, a treatment effect can be estimated.

In fact, Figure 9.15 illustrates the distinction between direct and indirect effects. The direct effect would correspond to the difference between δ_1 and δ_2, that is the effect on an individual who has a low CD4 count. The indirect effect is due to the treated group moving into a higher CD4 group (state 3). In this case, treatment works through changing the CD4 value of the individual. Hence, direct and indirect effects may be defined without reference to a linear model. The point of view indicated here is an extension of the local dependence concept and indicates that modeling through stochastic processes can give a detailed picture of how causality operates.

9.7.2 Modeling dynamic systems

Dissatisfaction with the present state of statistical approaches in epidemiology has led to calls for "dynamic systems modeling" (Ness et al., 2007; Joffe and Mindell, 2006). The idea is to try to understand the mechanisms or the causal structure of the phenomenon. The AIDS model in the previous subsection could be one very simple example of this, and dynamic path analysis and Granger-Schweder causality also fit into the general idea of understanding how systems work. Ness et al. (2007) criticize the counterfactual approach and also randomized clinical trials as yielding little insight into the processes behind the effect of a treatment or a risk factor. Although counterfactual approaches are useful, the construction of pseudopopulations by adjusting for imbalance in other variables or processes may limit the understanding of the process that actually occurs. There might be advantages in concentrating on the analysis of the process as it presents itself without the creation of pseudopopulations.

9.8 Exercises

9.1 Compute the transition probabilities for the Markov chain in Figure 9.6 when the transition rates are constant over time.

9.2 Consider three events A, B, and C instead of the two events in Figure 9.6. Make a similar diagram, but this time with three directions and eight states. Indicate the various local dependence relationships that you can have between these three events.

9.3 The incidence of a certain skin disease occurring, particularly in women, is influenced by the occurrence of menopause (cessation of menstruation), in the sense that the incidence increases after menopause. On the other hand, the occurrence of the disease does not influence the likelihood of menopause occurring. Neither disease nor menopause influences the likelihood of death. Explain the local dependence operating here. For details, see Aalen et al. (1980) or Didelez (2007).

Chapter 10
First passage time models: Understanding the shape of the hazard rate

In Chapter 6 we saw how the individual variability in hazard can be described in terms of a random frailty variable. For instance, frailty may enter the hazard in a multiplicative fashion, with the individual hazard h described as $h(t) = Z\alpha(t)$. Here, α is a "basic" rate and Z is a nonnegative random variable taking distinct values for each individual. This may be an appropriate way to account for missing (time-independent) covariates such as those genetic and other effects that may remain constant over time. A limitation of the standard frailty approach, however, is that Z is fixed at time zero. Once the level of frailty has been set for an individual, it is retained for the rest of the lifespan. Individuals may be endowed with a set of time-dependent (external) covariates encompassing changes in their environment, but conditional on the covariates the deviation of the individual hazard from the population baseline is completely determined. Thus, knowing the hazard ratio at one instant will completely determine the hazard ratio for the future. However, it is reasonable to believe that for each individual there will be a developing random process influencing the individual hazard and leading up to an event. This fact is usually ignored in the standard models, mostly because the process itself is usually poorly understood and may in fact not be observed at all. However, this does not imply that it should be ignored. A consideration of the underlying process, even in a speculative way, may improve our understanding of the hazard rate.

In particular, a model for the underlying process may suggest that different covariates influence the hazard in entirely different ways. In Cox regression, for instance, the covariates influence the hazard directly, through a (log-) linear combination of covariates. It may, however, be more fruitful to think that some covariates work on the underlying process rather than directly on the hazard. For example, driving your car recklessly will lead to an immediate increase in the accident hazard. On the other hand, poor maintenance of the car over a long period of time will lead to a gradual process of mechanical deterioration. To begin with, this will give an unnoticeable increase in the hazard, but eventually it will strongly increase the risk of a mechanical failure, which may lead to an accident. Thus, the covariates "reckless driving" and "poor maintenance" operate through different pathways and may be seen as fundamentally different in the ways they influence hazard.

Another type of covariate may not even effectively influence the underlying process but may rather be a measure of how far this process has advanced. Examples of such covariates are frequently found in clinical studies. For instance, bilirubin in liver disease, CD4 counts in HIV infection, and various staging measures in cancer may all be covariates of this type. Such covariates are often measured repeatedly over time and are then referred to as *markers*; see, for example, Nielsen and Linton (1995) and Jewell and Kalbfleisch (1996).

A classical example of a model that incorporates some of these aspects is the accelerated failure time model (Klein and Moeschberger, 2003). It is imagined that the underlying deterioration process can be accelerated or slowed according to one parameter. At the same time, another parameter controls the basic shape of the hazard. Since covariates influencing each of the two different parameters will influence the hazard in drastically different ways, the interpretation of the covariate effects will necessarily be different. Whereas one parameter measures the speed at which the event is approached, the other measures the "level of severity" of the condition of the individual.

A fairly general approach to this issue is to model a stochastic process for each individual, indicating the current level of severity. Once the process has developed far enough to hit a prespecified barrier, an event is said to occur. Thus, the event time is realized as the first passage time of a suitable stochastic process. The process may start at a low or high level, depending on the condition of the individual at time zero, and it may be accelerated or slowed depending on individual characteristics. In this chapter we will, somewhat informally, refer to such processes as *risk processes*. We consider hitting time models based on both continuous time Markov chains (discrete state space) and diffusion processes (continuous in time and space). The discrete and continuous space processes lead to models that enjoy much of the same essential features. Nevertheless, it is convenient to study both these classes since they offer somewhat different advantages in terms of explicit solutions and model flexibility. We will begin by looking at models based on the first passage time of finite state space Markov chains, so-called *phase type* models. An important practical example of this is the modeling of the progression of HIV infection by Longini and coauthors (see, e.g., Longini et al. (1989)). Aalen et al. (1997) consider an extended Markov chain model. We will also look at the infinite case, in particular birth and death processes.

As we shall see, the first passage time distributions tend to produce hazards that have a characteristic development over time. One particular feature frequently seen is a hazard that stabilizes at a positive level (plateau) over time. This indicates that the risk distribution (the distribution of the underlying process) for the survivors stabilizes over time. Such a limiting distribution is called a *quasi-stationary* distribution. It is not a stationary distribution for the full population but rather for those individuals that survive over time, hence the term *quasi*. We will discuss the phenomenon of quasi-stationary distributions to some extent in this and the following chapter. Just as for ordinary stationary distributions, if you start with a population whose underlying risk process at time zero matches the quasi-stationary distribution, this population will remain in the quasi-stationary distribution as time passes, and it will produce a constant hazard rate. If the starting distribution deviates from the quasi-stationary distribution, it would be expected that initially the hazard will not be constant, but after a while it will stabilize. The initial behavior is determined to a large extent by how the starting value relates to the quasi-stationary distribution. This will be seen in the examples in the following sections.

It should be remarked that the models expounded in this and the following chapter are somewhat tentative; in particular, it may not always be easy to derive an explicit parametric model even if the principles are clear. However, we give several examples of situations where this can be done. More important, we believe that this line of approach improves our basic understanding of how different hazard shapes may arise, including hazards seen in a number of practical situations.

In the following, we will assume some familiarity with standard results and terminology from Markov processes. We refer the reader to Appendix A for a brief introduction to Markov processes, including concepts such as stationarity and quasi-stationarity. More details about the topics covered in this chapter can be found, for instance, in O'Cinneide (1990), Aalen and Gjessing (2001, 2003, 2004), Lee and Whitmore (2006), and the review in Aalen (1995).

10.1 First hitting time; phase type distributions

10.1.1 Finite birth-death process with absorbing state

As our primary example, we will look at a finite state Markov chain with a single absorbing state. Consider the continuous time Markov chain X on the state space $\{0, 1, \ldots, 5\}$. In Figure 10.1, each box represents a state of the process, and the

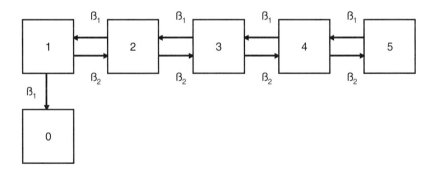

Fig. 10.1 *A six-state birth-death process with one absorbing state (0) and one reflecting (5).*

arrows indicate the possible transitions. The parameters β_1 and β_2 are the transition intensities for moving one step down or up, respectively. Properties of this process are discussed in some detail in Appendix A.2. Note that state 0 is absorbing and state 5 is reflecting. The infinitesimal generator of X is

$$
\alpha = \begin{pmatrix}
0 & 0 & 0 & 0 & 0 & 0 \\
\beta_1 & -(\beta_1 + \beta_2) & \beta_2 & 0 & 0 & 0 \\
0 & \beta_1 & -(\beta_1 + \beta_2) & \beta_2 & 0 & 0 \\
0 & 0 & \beta_1 & -(\beta_1 + \beta_2) & \beta_2 & 0 \\
0 & 0 & 0 & \beta_1 & -(\beta_1 + \beta_2) & \beta_2 \\
0 & 0 & 0 & 0 & \beta_1 & -\beta_1
\end{pmatrix},
$$

where the first row is only zeros since state 0 is absorbing. Since the process can only move up (a birth) or down (a death) one step at a time on a finite number of states, it is known as a finite *birth-death* process. The birth-death process has a reasonably simple structure which enables us to simplify some of the discussions, but many of the following illustrations will carry over to more general Markov chains.

10.1.2 First hitting time as the time to event

As discussed earlier, suppose that each individual in the population has its own "re-alization" of an underlying risk process modeled by a Markov chain $X(t)$ with a single absorbing state, that is, the risk process of an individual is just a sample path of X. The risk process is unobserved for the individual. When the individual risk process reaches the absorbing state, the individual experiences an event, and we de-note the time of the event by T. Thus, the event time $T = \inf_{t \geq 0}\{t : X(t) = 0\}$ is the first hitting time for the absorbing state 0. In the preceding model, it is clear that the process will only spend a finite amount of time in the other states before even-tually reaching state 0. Consequently, all individuals will eventually experience an event. We therefore refer to the nonabsorbing part of the state space as the *transient* space. Intuitively, it might be more natural to assume that a higher value of X means higher risk, and that the individual experiences an event when X reaches an upper boundary of the state space, rather than 0. Clearly, however, the two formulations are equivalent, and for notational simplicity we will assume that 0 is the absorbing state.

Notice that the only observable effect of X is through the individual event times. It is obvious, then, that as long as we do not observe time-dependent covariates closely related to the process X, any supposition about the structure of X will be mostly speculative. Nevertheless, modeling in terms of the underlying process is in many instances a natural way to represent the individual randomness, or frailty.

With the process X defined as before, we can start X according to a probability distribution on the transient space and wait for it to reach the absorbing state. The distribution of the event time T is known as a phase type distribution, and these have received a lot of attention in the literature; see e.g. O'Cinneide (1989, 1990, 1991) or Aalen (1995).

An interesting modeling feature of phase type distributions is that one may distinguish between two conceptually different ways to control the event time distribution. The first is to control the starting value of the process. Clearly, the further it starts from the absorbing state the longer it will take to reach it. The second possible control is to influence the transition intensities between the different states. For instance, in the example shown in Figure 10.1, when increasing β_1 the process is driven with greater force toward the endpoint, whereas increasing β_2 will increase the tendency of the process to "lump" around the reflecting upper barrier at state 5, thus keeping it away from the terminal state. To illustrate this, if X indicates the health condition of a patient enrolled in a randomized clinical trial, we may assume state 0 means cure and that the higher the state the more serious the condition of the patient. Since there is randomization, treatment does not influence the starting value of the patient, which reflects his or her condition at the time of entry into the study. Treatment *may*, however, influence the speed of recovery of the patient, which, in fact, is the intention of the treatment. A medication, for instance, will typically not result in instant recovery but rather an increased force toward a positive outcome. In this situation, it may be natural to think that the treatment influences the intensity parameters by increasing β_1 and/or decreasing β_2.

For a numerical example, put $\beta_1 = 1$ and $\beta_2 = 1.5$. We start by exploring the effect of starting in different states. Figure 10.2 compares the hazards obtained when starting in states 1 to 5. (The hazards can be computed from formula (10.1) using the probability distribution of the chain at time t.) When starting in state 1, the hazard starts at 1 and decreases. This is reasonable since at time zero all individuals in state 1 have an intensity $\beta_1 = 1$ of moving to state 0. Immediately after, the individuals starting in state 1 have been absorbed (and thus do not contribute any further to the hazard); or they remain in state 1 (with a hazard 1 of moving to state 0); or they have moved to state 2 or higher (in which they have zero hazard until they move back to state 1). For states 2 and higher, the closer one starts to the absorbing state the faster the hazard increases in the beginning, as expected. The hazard starts at zero, reaches a peak, and then levels off. Interestingly, regardless of starting state all hazards converge to a common limiting hazard, indicating a long-run stability of the system. In this example, the limiting value is 0.037. This is discussed in more detail later in this chapter. We notice that for the states closest to the absorbing barrier, the hazard is essentially falling toward the equilibrium value, whereas for the states farthest away it is essentially climbing toward it. In between, the peak effect is most pronounced.

We are also interested in understanding the effect of the parameters β_1 and β_2 on the shape of the hazard. Since a change of time scale is irrelevant for the actual hazard shape, it is only the balance between β_1 and β_2 that is relevant, not the actual sizes. We assume $\beta_1 = 1$ and look at the effect of varying β_2. Since β_2 is the force pulling the process away from zero, one would expect the hazard to diminish with increasing β_2 values. This is precisely what happens, as seen in Figure 10.3. We also observe that the essential shape of the hazard does not change much; all curves exhibit a peak; the larger the value of β_2 the earlier and sharper the peak.

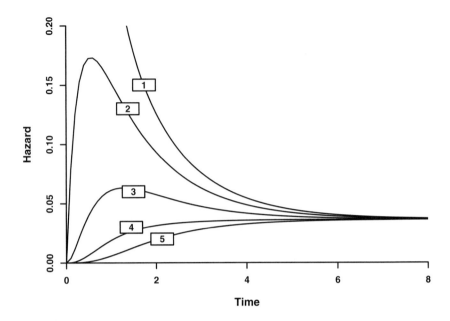

Fig. 10.2 *Hazard rates for time to absorption depending on starting state in phase type model. Starting state number is indicated in boxes.*

10.1.3 The risk distribution of survivors

From the model in Figure 10.1 we see that at any given time new events happen by individuals in state 1 passing to state 0. None of the other states can reach the absorbing state within very short time intervals. The intensity of new events thus depends on the parameter β_1, together with the proportion of individuals in state 1 at time t. This links the hazard of T to the distribution of risk for those not yet absorbed in state 0, that is, the risk distribution of the "survivors". To be precise, the hazard α is given by

$$\alpha(t) = \beta_1 P(X(t) = 1 | X(t) \geq 1) = \beta_1 \tilde{p}_1(t), \qquad (10.1)$$

where $\tilde{p}_1(t)$ is the first element of the vector $\tilde{\mathbf{p}}(t)$ containing the probability distribution *of those not yet absorbed in zero*. To illustrate, Figure 10.4 shows how the distribution $\tilde{\mathbf{p}}(t)$ changes with time for individuals starting in state 1, again using $\beta_1 = 1$ and $\beta_2 = 1.5$. To begin with (time zero), all mass is concentrated at state 1; when time increases, many of the individuals close to the absorbing state disappear and the remaining individuals are those who "escape" to higher states, away from the absorbing zero. Equation (10.1) tells us that the left endpoint of the distribution

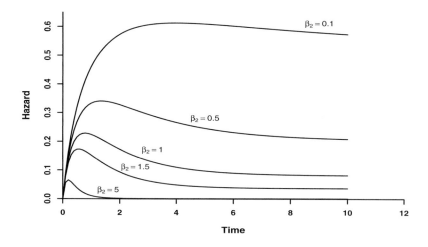

Fig. 10.3 *Hazards when starting in state 2, with $\beta_1 = 1$ and varying values of β_2.*

(state 1 in Figure 10.4) changes in accordance with the hazard graph (state 1 in Figure 10.2) as t increases. In particular, the long run probability of a nonabsorbed individual to be in state 1 approaches a constant value of 0.037, corresponding to the long-run level seen in Figure 10.2. In addition, the rest of the distribution converges to the distribution shown with time $t = \infty$ in Figure 10.4.

Note that the actual computation of the distribution of $X(t)$, and thus $\widetilde{\mathbf{p}}(t)$, can be done by solving the so-called *Kolmogorov equations*. The solution, giving the transition probabilities of X, is $\mathbf{P}(t) = \exp(\boldsymbol{\alpha} t)$, where exp is the matrix exponential function. The numerical value of the matrix exponential for a finite-dimensional $\boldsymbol{\alpha}$ is usually easy to compute in software such as Mathematica. For more details, see Appendix A.2.3 and Allen (2003).

10.1.4 Reversibility and progressive models

When studying stochastic processes for biological or social phenomena, a major distinction is made between reversible and irreversible processes. Some diseases, like nontreatable cancer, might be irreversible, while others are clearly reversible in the sense that the patient can be completely cured. Many diseases, like rheumatism or migraine, may be chronic but still go back and forth between good and bad periods, that is, being at least partially reversible. As an example of a social phenomenon, consider the incidence of divorce: it is clear that most marriages will have good and bad periods, undergoing reversible processes of deterioration and improvement,

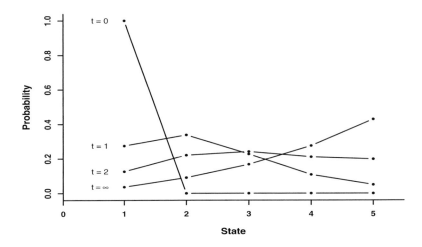

Fig. 10.4 *Probability distributions* $\tilde{\mathbf{p}}(t)$ *for individuals not yet absorbed in zero when starting in state 1. At time zero all probability is concentrated at state one; when time increases the distribution shifts toward the higher states, away from the absorbing state. At $t = \infty$ the quasi-stationary distribution appears. All distributions use $\beta_1 = 1$ and $\beta_2 = 1.5$.*

before some may end in divorce. By reversibility, we loosely mean that the transient states of the process make up a single class. This irreducibility of the transient state space is important when considering quasi-stationary distributions, as is done in the following sections. Typically, such distributions are well defined for Markov processes with absorption when the transient states constitute a single class.

Another distinction, which is related to the preceding one, is useful when studying hazard rates: by a progressive model, we mean that with a high probability the process starts out in an "extreme" state, which represents the natural starting point in the state space, and then moves toward the end state. Even though there may be temporary backward movements and some reversibility, there is still a clear direction in the development. For example, when studying the development of a disease, the individual starts out as completely disease-free, but the disease has a progression once it is established. A nonprogressive model, on the other hand, has no clear direction in the development. The process is not supposed to start in an extreme state, and it does not necessarily have any force of movement toward the absorbing state. Again, marriage can serve as an example; a marriage might improve or deteriorate after the wedding, but there is no law stating that it has to move in one direction. The distinction between progressive and nonprogressive models is tentative. Discussion and examples for phase type models are given in Aalen (1995).

Typically, progressive models tend to have increasing hazard rates. We see this in Figure 10.5, where the increase occurs for a wide range of values of β_2 when starting in state 5. As we have seen, for nonprogressive models the most typical shape will

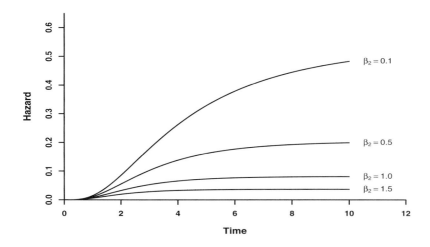

Fig. 10.5 *Hazards when starting in state 5, with $\beta_1 = 1$ and varying values of β_2.*

be a hazard rate that first increases and then decreases.

10.2 Quasi-stationary distributions

In order to understand the shapes of the hazard rates that may result from first passage time distributions, an essential element is the quasi-stationary distribution. In
Figure 10.2 it was seen that regardless of initial state the hazard stabilizes around
a limiting value. Through (10.1) this is linked to the behavior of the distribution of
survivors and suggests that this distribution may stabilize over time. It turns out that
there often exists a distribution on the transient state space such that if the process
starts according to this distribution it will in the future have a constant hazard of
transition to the absorbing state. This means that although probability mass is continuously being drained from the transient space, the remaining probability distribution on this space converges to a limiting distribution. The manifestation of such
a quasi-stationary distribution on the transient space in our birth-death example is
seen in Figure 10.4, for $t = \infty$. The quasi-stationary distribution does not depend on
the starting state but will typically depend on the process parameters. In the birth-
death example we see how the quasi-stationary distribution changes with changing
values of β_2 in Figure 10.6. As would be expected, the larger the drift away from
the absorbing state is, the more the quasi-stationary distribution is shifted away from
the absorbing state.

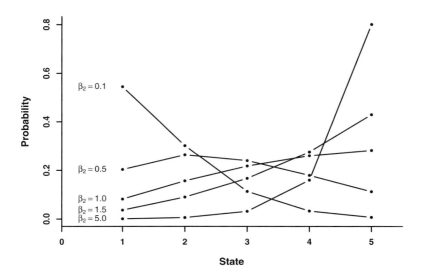

Fig. 10.6 *Quasi-stationary distributions for* $\beta_1 = 1$ *and varying* β_2.

In order to compute the quasi-stationary distribution for a finite-state Markov chain, consider the generator $\boldsymbol{\alpha}$ restricted to the transient space, that is, the intensity matrix

$$
\begin{pmatrix}
-(\beta_1 + \beta_2) & \beta_2 & 0 & 0 & 0 \\
\beta_1 & -(\beta_1 + \beta_2) & \beta_2 & 0 & 0 \\
0 & \beta_1 & -(\beta_1 + \beta_2) & \beta_2 & 0 \\
0 & 0 & \beta_1 & -(\beta_1 + \beta_2) & \beta_2 \\
0 & 0 & 0 & \beta_1 & -\beta_1
\end{pmatrix}.
$$

It can be shown that the quasi-stationary distribution is a normalized version of a left eigenvector of this reduced intensity matrix (see Appendix A.2.4). More precisely, it is the left eigenvector corresponding to the dominant eigenvalue (i.e., the one closest to zero). Whenever the transient state space constitutes a single finite class (which is the case here), the quasi-stationary distribution is unique and strictly positive on all states. The absolute value of the dominant eigenvalue is the constant hazard rate of absorption corresponding to a quasi-stationary starting distribution. In our numerical example, with $\beta_2 = 1.5$ and $\beta_1 = 1.0$, the quasi-stationary distribution on states 1 to 5 is given as 0.037, 0.090, 0.167, 0.276, 0.430. The absolute value of the dominant eigenvalue is 0.037, which equals the constant hazard rate of absorption under quasi-stationarity. It also equals the first probability in the quasi-stationary distribution (state 1), which follows from (10.1) since $\beta_1 = 1$. See Appendix A.3 and Aalen (1995) for examples and references.

10.2.1 Infinite birth-death process (infinite random walk)

Consider an infinite birth-death process $X(t)$ with absorbing state in 0, that is, an extension of the preceding model where instead of just five transient states there is an infinite number of states. The parameters β_2 and β_1 still denote the rates of moving up and down the state space. The advantage of considering this extension is that simple explicit formulas may be given for interesting quantities. In particular, explicit formulas for quasi-stationary distributions are given by Cavender (1978) for the case when such distributions exist, namely when $\beta_2 < \beta_1$, when the "drive" is toward the barrier. Somewhat surprisingly, there is a whole set of stationary distributions, but one distribution is "canonical" in the sense that it represents the limiting case for a process starting with probability one in a specific state, that is, when the starting distribution is concentrated in a single state. Thus,

$$q_j = \lim_{t \to \infty} P(X(t) = j | X(t) > 0, X(0) = i)$$

exists and is independent of i. The formula for this canonical distribution is

$$q_1 = \left(1 - \sqrt{\frac{\beta_2}{\beta_1}}\right)^2, \quad \text{and} \quad q_j = j q_1 \left(\frac{\beta_2}{\beta_1}\right)^{\frac{j-1}{2}} \quad \text{when} \quad j = 2, 3, \dots.$$

Note that this is a negative binomial distribution. From (10.1) the constant hazard rate of absorption that arises under quasi-stationarity equals

$$\alpha = \beta_1 q_1 = \left(\sqrt{\beta_1} - \sqrt{\beta_2}\right)^2.$$

Again, this is the limit of the hazard rate of the first passage time to 0 (starting in a single state) when time goes to infinity. Notice how the rate is created in the balance between the forces driving the process toward absorption and those removing it.

In the case under study, the probability density of the hitting time distribution can also be found, for instance from Gross and Harris (1985, page 134, formula (2.109), and page 143). Starting in state i the density of time to absorption is

$$i \left(\frac{\beta_1}{\beta_2}\right)^{\frac{i}{2}} t^{-1} \exp(-(\beta_1 + \beta_2)t) I_i(2\sqrt{\beta_2 \beta_1} t), \qquad t > 0,$$

where $I_i(t)$ is the modified Bessel function of order i.

A general result of Keilson (1979) says that the hazard rate of the first passage time to a neighboring state in a general birth death process is always decreasing. Hence, when starting in the state closest to absorption, namely state 1, one will necessarily have a decreasing hazard rate, as we also saw in the finite case. Otherwise, when starting in a single state with probability one, experience indicates that the hazard rate will first increase and then decrease, but for states far removed from the absorbing one, the hazard will be virtually only increasing.

10.2.2 Interpretation

The present theory may explain decreasing hazards seen frequently in practice. For instance, it is well known that after myocardial infarction the death rate is initially very high and then sharply falls. Data on this have been analyzed from a frailty point of view by Hougaard (1986), that is, when assuming that some individuals have a much higher risk of dying from the disease than others. The application of a conventional frailty model fits the data well. A somewhat different, but related, interpretation of the declining death rate can be given in the present framework. If the absorbing state 0 signifies death, then state 1 might correspond to a very critical illness, like myocardial infarction, from which some individuals will die, while others soon improve, that is, they move upwards in the state space. According to the present theory, individuals starting out in state 1 would experience a decreasing hazard rate of dying, as observed for the myocardial infarction patients.

In the biological literature, one may find attempts at interpreting shapes of hazard rates and drawing biological conclusions from these. In a paper in *Nature*, Clarke et al. (2000) discuss the development of diseases involving neuronal degeneration (e.g., Parkinson's disease). In such diseases, the onset of clinical symptoms may be delayed for years or decades after the premature neuronal death has started. There has been a discussion of whether this delay reflects a cumulative damage process, or whether the death of a neuron is basically a random event, the likelihood of which is constant over time. Presumably, the first hypothesis would lead to increasing hazard of cell death, while the second would give a constant hazard. The cumulative damage model implies a progressive process in the organism whereby previous cell death creates damage that increases the likelihood of new cell death. The random event model, also called a "one-hit" model, implies that there has been damage at one specific time, such as a mutation, which then increases the likelihood of cell death to a new fixed level.

For a number of diseases and animal models, Clarke et al. studied the survival of neurons. They found that the hazard rate is generally constant, or sometimes decreasing, but not increasing. From this, they concluded that the cumulative damage model is incorrect and that the one-hit random model is the true one. In a more recent paper, Clarke et al. (2001) further discuss their hypothesis. A similar study for a type of mutant mice is presented in Triarhou (1998), with the same conclusion of a constant rate of cell death.

Another setting where biological interpretations of hazard rates have been presented is in the understanding of sleep. Lo et al. (2002) study the changes between sleep and wake states throughout the night. They find that the duration of sleep periods have a more or less exponential distribution, while the duration of wake states has a distribution with a decreasing hazard rate. This is interpreted within a stochastic process context, but basically, the implication is that sleep represents a random walk with no drift.

What is common for these papers is that they draw biological conclusions from the shapes of hazard rates at an aggregated level. It is then important to understand how very different underlying models may lead to similar hazard rates. In particular, an approximately constant hazard rate will be a common phenomenon for many models due to convergence to quasi-stationarity. Hence the development of risk at the level of an individual neuron, say, is very hard to deduce, and it therefore appears difficult to draw conclusions about the underlying process with any degree of certainty.

This criticism does not necessarily imply that the conclusions in the mentioned papers are wrong. But it points to the necessity of understanding the hazard rate and how its various shapes can arise.

10.3 Wiener process models

In the remaining part of this chapter we focus on the shape of hazard functions derived from first passage time distributions of diffusion processes, in particular the Wiener process with drift and the Ornstein-Uhlenbeck process. We do not give general theorems concerning the shape of the hazard but will look at examples, special cases, and illustrations. As we have seen for discrete state space Markov chains, the quasi-stationary distribution will again turn out to be useful in understanding the shape of the hazard. Even though the Wiener process has a continuous state space, it is not very surprising that many of the basic features are parallel to the discrete space situation. Still, there are notable differences that warrant a separate study of the continuous case. An introduction to Wiener processes and related topics is given in Appendix A.4.

In addition to the basic results for the Wiener process, a randomized drift version of the Wiener process is studied in this section in connection with a practical example of analyzing survival data with covariates. We also briefly consider general diffusion processes on the positive real line with absorption in zero. Such diffusions will have a state distribution on the positive half-line, which, when normalized to mass 1, converges to a quasi-stationary distribution. In general, the hazard rate of the time to absorption is proportional to the derivative of the density of the normalized distribution at 0, a relationship that may give useful intuitive insight. We also study the hazard ratios resulting from comparing subgroups in some of the models.

A Wiener process $W(t)$ is a stochastic process continuous in time and space, taking values in $\mathbf{R}$. It has increments that are independent and normally distributed, with

$$E[W(s+t) - W(t)] = 0 \quad \text{and} \quad \text{Var}[W(s+t) - W(t)] = s.$$

The Wiener process is also frequently referred to as a *Brownian motion*. As a continuous random walk, the Wiener process will "diffuse" away from its starting point, that is, it will wander back and forth between increasingly larger positive and negative values in the state space.

Consider a Wiener process that starts in a positive value and moves freely until it is absorbed when reaching zero. As in the discrete case we think of the Wiener process as an underlying risk process for each individual; the time when the process hits zero is the event time for that individual. Thus, $T = \inf_{t \geq 0}\{t : W(t) = 0\}$. If the process starts in a specific point $W(0) = c$, the distribution of T is an inverse Gaussian distribution (see Section 10.3.1). Notice that because of the diffusive nature of W it will always reach zero in a finite time, that is, $P(T < \infty) = 1$. This changes if we consider a Wiener process with *drift*. The process $X(t) = c + mt + \sigma W(t)$ is called a Wiener process with initial value c, drift coefficient m, and diffusion coefficient σ. The parameters m and σ^2 are also referred to as the *(infinitesimal) mean and variance*, respectively, since

$$E[X(t)] = c + mt \quad \text{and} \quad \text{Var}[X(t)] = \sigma^2 t.$$

Clearly, for a Wiener process starting in $c > 0$ with drift $m = -\mu$, $\mu > 0$, the movement is markedly in the direction of zero, and if σ^2 is small in comparison to μ it will move in almost a straight line, $X(t) \approx c - \mu t$. The hitting time will then be nearly a deterministic function of c and μ, $T \approx c/\mu$. For a larger σ^2, the diffusion part is more dominant and the hitting time less predictable.

A particular situation occurs when the drift is *away* from zero, that is, $\mu < 0$. If the diffusion part is insignificant, the process will simply drift off to infinity without ever hitting the barrier. Increasing σ^2 will increase the likelihood that $X(t)$ will hit the barrier by chance before drifting away. It will always be the case that $0 < P(T < \infty) < 1$, so there is a positive probability that the individual will never experience an event. For some situations this may be just the appropriate behavior. If, for instance, the event of interest is time of relapse for a cancer patient receiving chemotherapy at time 0, there is a positive probability that the patient is completely cured and will never experience a relapse. This is an example of a *cure model*, which was discussed in Chapter 6.

There are a number of papers on applying Wiener processes in survival analysis; see e.g. Lancaster (1982), Whitmore (1986a,b, 1995), Whitmore et al. (1998), Eaton and Whitmore (1977), Doksum and Høyland (1992), Lee et al. (2000), Aalen and Gjessing (2001, 2003, 2004), Weitz and Fraser (2001), and Lee and Whitmore (2006). We will study Wiener processes as well as more general diffusion processes in the rest of this chapter.

10.3.1 The inverse Gaussian hitting time distribution

When the Wiener process with drift starts at a given point $c > 0$, the distribution of time T to absorption in zero is an inverse Gaussian one (see, e.g., Chhikara and Folks (1989) or Karatzas and Shreve (1991) and Exercise 10.2), with density

$$f(t) = \frac{c}{\sigma\sqrt{2\pi}} t^{-3/2} \exp\left[-\frac{(c - \mu t)^2}{2\sigma^2 t}\right]. \tag{10.2}$$

The associated survival function is

$$S(t) = \Phi\left(\frac{c - \mu t}{\sigma\sqrt{t}}\right) - \exp\left(\frac{2c\mu}{\sigma^2}\right)\Phi\left(\frac{-c - \mu t}{\sigma\sqrt{t}}\right),$$

where $\Phi(\cdot)$ is the cumulative standard normal distribution. The hazard rate is found, as usual, from $\alpha(t) = f(t)/S(t)$. There are three parameters in the inverse Gaussian distribution, namely c, μ, and σ, but the distribution only depends on these through the functions c/σ and μ/σ. Hence, from a statistical point of view, there are only two free parameters. This means, for instance, that we can put $\sigma = 1$ in a statistical analysis, without loss of generality. Note that this is only true when considering time to absorption, not when studying other aspects of the process, such as the risk distribution of survivors, as in Equation (10.3).

The shape of the hazard rate $\alpha(t)$ of this distribution is similar to the one observed for a phase type distribution (Figure 10.2). An illustration is shown in Figure 10.7, where the values 0.2, 1, and 3 are chosen for c. Like the phase type hazards, the

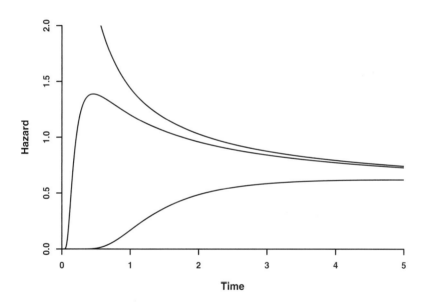

Fig. 10.7 *Hazard rates for time to absorption when process starts in $c = 0.2$ (upper curve), $c = 1$ (middle curve), and $c = 3$ (lower curve). In all cases $\mu = 1$ and $\sigma^2 = 1$.*

hazards for the Wiener process with drift exhibit the same stability phenomenon; regardless of initial value they all converge to the same limiting hazard. Indeed, for the Wiener process a quasi-stationary distribution also exists. If c is close to zero relative to the quasi-stationary distribution, one essentially gets a decreasing hazard rate; a value of c far from zero gives essentially an increasing hazard rate;

an intermediate value of c yields a hazard that first increases and then decreases. The continuous nature of the model and the noncompact state space yield hazard rates that, strictly speaking, always increase to a maximum and then decrease; see, for example, Seshadri (1998, proposition 5.1). For c small or large, however, for practical purposes the rates can be seen as just decreasing or just increasing.

10.3.2 Comparison of hazard rates

Much of survival analysis focuses on relative hazard rates or hazard ratios, in fact often assuming that the hazard rates are proportional, an assumption that despite its popularity is problematic on theoretical grounds; see Section 6.5. Two hazard rates, for different values of c, are shown in Figure 10.8. Computing the hazard ratio,

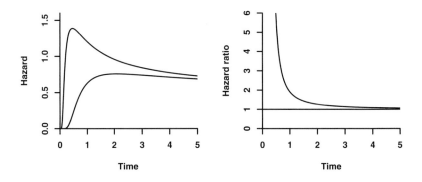

Fig. 10.8 *Left panel: hazard rates for time to absorption when the process starts in $c = 1$ (upper curve) and $c = 2$ (lower curve). Right panel: ratio of the two hazard rates. Parameters: $\mu = 1$ and $\sigma^2 = 1$.*

that is, one hazard divided by the other, reveals a strongly decreasing function, as shown in the figure. This feature, which is typical in these comparisons, is the same phenomenon as that observed in frailty models, where the relative hazards often decline (see, e.g., Aalen (1994) and Section 6.5). In the present setting it means that if a high-risk group is defined as being closer to the point of absorption than the low-risk group, then comparing the hazards in these two groups would give a declining hazard ratio. In fact, convergence toward a quasi-stationary distribution implies that the relative hazards decline toward a ratio of one.

It is also of interest to compare the hazard rates when the starting point c is the same, while the drift is different. The result of this is shown in Figure 10.9, where the picture that emerges is rather different from the previous one. In fact, the hazard rates seem more or less parallel after some time. The ratio of the two hazards has a "bathtub" shape and levels off at later times (Aalen, 1995). It should be kept in

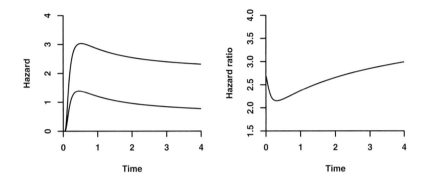

Fig. 10.9 *Left panel: hazard rates for time to absorption when process has drift parameters $\mu = 2$ (upper curve) and $\mu = 1$ (lower curve). Right panel: ratio of the two hazard rates. In both cases $c = 1$ and $\sigma^2 = 1$.*

mind, however, that proportionality eventually sets in in both cases; the difference lies in how the parameters control the limiting ratio and the early development of the hazards. In fact, the two curves ($\mu = 2$ and $\mu = 1$) converge to 2 and 1/2, respectively, and their ratio to 4. This can be computed from the simple formula for the limiting hazard given in (10.4).

As explained in the introduction to this chapter, covariates will often be indicators of *how far* or *how fast* the underlying process has advanced. Such covariates may influence the values of c and μ, respectively. The hazard comparisons given here thus clearly suggest that proportional hazards models would not provide an appropriate description of the covariate effects, particularly not at the earliest part of the time scale.

It is interesting to note the similarity between the rates in Figure 10.9 and the divorce rates in Figure 1.4; this might give a rough indication that increasing divorce rates is to a large extent due to cohort effects, with the drift toward divorce increasing in the more recent cohorts.

In conclusion, when comparing hazard functions, the result depends substantially on the causes of the differences between the groups. Difference in distance from point of absorption gives one type of result, while difference in the basic dynamics of the process, such as drift and variance of a Wiener process, gives another.

10.3.3 The distribution of survivors

As in 10.1.3 for the Markov chain models, we now look more closely at the "survivors" at time t. A well-known result, found in Cox and Miller (1965, page 221), gives the probability density at time t of the Wiener process with drift and absorption as

$$\psi_t(x) = \frac{1}{\sigma\sqrt{2\pi t}}\left\{\exp\left[-\frac{(x-c+\mu t)^2}{2\sigma^2 t}\right]\right.$$
$$\left. - \exp\left(\frac{2c\mu}{\sigma^2}\right)\exp\left[-\frac{(x+c+\mu t)^2}{2\sigma^2 t}\right]\right\}. \tag{10.3}$$

(See Exercise 10.3 for a derivation.) This density is a linear combination of two normal distributions. The integral of the density will be less than 1 and will decrease with time since there is an increasing probability of absorption at time 0, that is, when $x > 0$ we have

$$\psi_t(x) = \frac{1}{dx}P(X(t) \in (x,x+dx]) = \frac{1}{dx}P(X(t) \in (x,x+dx], T > t),$$

and thus

$$\int_0^\infty \psi_t(x)dx = S(t).$$

On the other hand, the conditional density along the positive axis *given that absorption has not taken place* is

$$\phi_t(x) = \psi_t(x)/S(t),$$

that is, the normalization of $\psi_t(x)$, so that $\int_0^\infty \phi_t(x)dx = 1$. The distribution ϕ_t is illustrated in Figure 10.10, with parameter values $c = 1$, $\mu = 1$, and $\sigma^2 = 1$. The distributions are shown at times 0.05, 0.2, 1, and 10, together with the quasi-stationary distribution. We see that the distribution of surviving individuals is rapidly spreading out from $c = 1$ (time zero) and approaches a quasi-stationary distribution, discussed in more detail in 10.3.4.

From general results for diffusion processes (see (10.12)), it follows that the hazard rate of absorption at time t equals the x-derivative at $x = 0$ of $\phi_t(x)$, multiplied by $\sigma^2/2$. This is illustrated in Figure 10.11, which is similar to Figure 10.10 with lines added showing the slope at $x = 0$. That is, in Figure 10.11 the hazard at time t is half the slope of the corresponding curve at zero. It is apparent from the figure that the distribution of survivors moves like a wave toward the barrier at $x = 0$ and is then "pushed back" toward the quasi-stationary distribution. The slope at $x = 0$, and hence the hazard, increases for small t to reach a maximum and then declines to the value for the quasi-stationary distribution. This pattern is seen in the middle curve of Figure 10.7. This development of an increase followed by a decrease is quite reasonable; the mass will rapidly approach the absorbing state at zero, resulting in

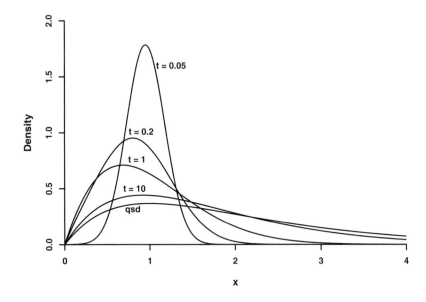

Fig. 10.10 *Normalized distribution of "survivors", $\phi_t(x)$, for $c = 1$, $\mu = 1$ and $\sigma^2 = 1$ at times $t = 0.05$, 0.2, 1 and 10. The quasi-stationary distribution (qsd) is added for comparison.*

an increasing slope, but later the slope declines as more mass moves out into the extreme right-hand part of the distribution.

10.3.4 Quasi-stationary distributions for the Wiener process with absorption

As earlier, let $X(t)$ be a Wiener process with drift $-\mu$ and diffusion coefficient σ. Quasi-stationary distributions for this case have been studied fairly recently (Martinez and San Martin, 1994), and it turns out that there is a whole family of such distributions (due to the infinite state space) in the case when $\mu > 0$ (i.e., the process has a drift toward zero). One of these distributions is "canonical" in the following sense: starting in a single given state with probability 1, the distribution of "survivors" will converge to the canonical one, that is, we have the convergence $\lim_{t\to\infty}\phi_t(x) = \phi(x)$, where ϕ is the canonical quasi-stationary distribution. The distribution ϕ is defined on the positive real line and satisfies $\int_0^\infty \phi(x)dx = 1$. The other quasi-stationary distributions are more heavy-tailed toward infinity and cannot be reached in this way. In fact, the canonical one is given by

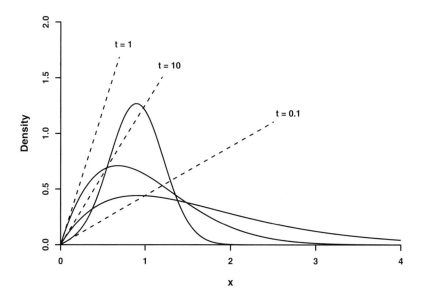

Fig. 10.11 *Normalized distribution of "survivors", $\phi_t(x)$, for $c = 1$, $\mu = 1$, and $\sigma^2 = 1$ at times $t = 0.1$, 1, and 10. The dashed lines show the slope at $x = 0$. The hazard of absorption at any time t is $\sigma^2/2$ times the slope at $x = 0$ at that time.*

$$\phi(x) = (\mu^2/\sigma^4) x \exp(-\mu x/\sigma^2),$$

which is a gamma distribution. We will show how this distribution is derived in 10.4.2.

The convergence to a quasi-stationary distribution is observed in Figure 10.10. If the process is initiated with this distribution on the positive real line, then the hazard rate of absorption in zero is constant and equal to the limiting hazard when starting in a specific state, that is,

$$\lim_{t \to \infty} \alpha(t) = \frac{1}{2}(\frac{\mu}{\sigma})^2. \tag{10.4}$$

It is of some interest to note that the (stationary) hazard rate depends on the *square* of μ.

Putting $\mu = 1$ and $\sigma^2 = 1$ results in the distribution shown in Figure 10.12. We will return to the derivation of the quasi-stationary distribution in Section 10.4.2, after first discussing more general diffusion processes.

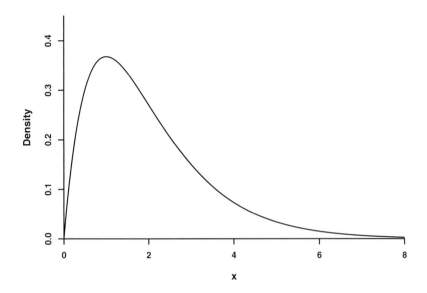

Fig. 10.12 *Quasi-stationary distribution for a Wiener process with absorption (parameters:* $\mu/\sigma^2 = 1$).

10.3.5 Wiener process with a random initial value

Instead of starting in a single state, one may initiate the process according to some distribution on the positive real line. The results become particularly simple if the starting distribution is related to the quasi-stationary one. For simplicity, assume that $\sigma = 1$, that is, the canonical quasi-stationary distribution equals $\mu^2 x \exp(-\mu x)$. Assume that the Wiener process with drift is initiated according to a distribution proportional to $x^k \exp(-\mu x)$ for some k. When $k < 1$, this will be a gamma distribution which is closer to the absorbing state 0 than the quasi-stationary one. If $k > 1$, the distribution is further removed from zero than the quasi-stationary one. By integrating the inverse Gaussian distribution with respect to the above distribution, one derives a new first passage distribution which is simply a gamma distribution proportional to

$$t^{(k-1)/2} \exp(-(\mu^2/2)t).$$

This gamma distribution is known to have a decreasing hazard rate for $k < 1$, while for $k > 1$ the rate is increasing. Hence, this example demonstrates how the shape of the hazard of the first passage distribution is related to how the starting distribution of the process compares with the quasi-stationary one.

One may also start with a distribution of the type $\phi_t(x)$, that is, the distribution of survivors at time t. One then gets the later part of a hazard rate (leaving out the beginning) and can hence achieve unimodal or decreasing hazard rates.

10.3.6 Wiener process with lower absorbing and upper reflecting barriers

When modeling a phenomenon by a Wiener process with an absorbing state at zero, it may be natural to think that there should be an upper bound to the risk process, a level that is never exceeded. The Wiener process with an absorbing lower barrier and a reflecting upper barrier provides such a model. We saw a discrete state version of this in Section 10.1, where state 5 was reflecting and thus an upper barrier. The continuous state (Wiener process) version of this is difficult to handle, and there is not much theory for this. A useful paper is Schwarz (1992), which presents formulas for first passage time distributions. A simple formula for the quasi-stationary distribution may be derived from his work in the case of a process without drift, an absorbing barrier at 0, and a reflecting barrier at B. Then the quasi–stationary distribution is

$$\frac{\pi}{2B}\sin\left(\frac{\pi x}{2B}\right),$$

and the corresponding hazard rate of absorption is

$$\frac{\pi^2\sigma^2}{8B^2}.$$

It is interesting to note that the quasi-stationary distribution is independent of σ. If there is drift, then more complex formulas may be derived from the work of Schwarz.

10.3.7 Wiener process with randomized drift

We now return to the Wiener process with absorption in zero but no reflection. A degree of individual "frailty" is allowed in this model through the diffusion effect: some individuals will have a risk process that diffuses away from zero for a long time in spite of the drift, whereas others will diffuse almost directly toward the barrier. It is often natural to think that some degree of "tracking" is taking place, that is, the drift itself may vary from individual to individual. This adds an extra level of frailty to the model and makes it more flexible. In fact, it adds a cure effect in that some individuals may have a drift *away* from the barrier and thus have a positive probability of never having an event.

To model the tracking explicitly, assume that the drift is randomized according to a normal distribution with expectation $-\mu$ and variance τ^2, independent of the Wiener process. We will set the variance coefficient σ^2 of the Wiener process equal to 1. As noted for the inverse Gaussian distribution in 10.3.1, this entails no loss of generality when considering time to absorption. By integrating over the distribution of the drift parameter in expression (10.3), one gets the density for the transition from c to x over a time period of length t. We use subscript R to denote that the drift has been randomized:

$$
\psi_{R,t}(x) = \frac{1}{\sqrt{2\pi(t^2\tau^2 + t)}} \left\{ \exp\left[-\frac{(x - c + \mu t)^2}{2(t^2\tau^2 + t)} \right] \right.
$$
$$
\left. - \exp\left(2c\mu + 2c^2\tau^2\right) \exp\left[-\frac{(x + c + 2ct\tau^2 + \mu t)^2}{2(t^2\tau^2 + t)} \right] \right\}. \tag{10.5}
$$

The computation of this distribution is basically a straightforward integration manipulating the quadratic forms in the exponents. The probability of not being absorbed by time t, which may be found e.g. in Aalen (1994), is

$$
S_R(t) = \Phi\left(\frac{c - \mu t}{\sqrt{t^2\tau^2 + t}} \right) - \exp\left(2c\mu + 2c^2\tau^2\right) \Phi\left(\frac{-c - 2ct\tau^2 - \mu t}{\sqrt{t^2\tau^2 + t}} \right). \tag{10.6}
$$

The probability density of the position of the process at time t conditioned on nonabsorption, that is, the distribution of survivors, is found by dividing (10.5) by (10.6), $\phi_{R,t}(x) = \psi_{R,t}(x)/S_R(t)$.

The probability density of time to absorption is the inverse Gaussian distribution with mixed drift parameter. By differentiating (10.6), we find that the density is given by

$$
f_R(t) = \frac{c}{\sqrt{2\pi}} \frac{1}{t\sqrt{t^2\tau^2 + t}} \exp\left[-\frac{(c - \mu t)^2}{2(t^2\tau^2 + t)} \right] \tag{10.7}
$$

(Whitmore, 1986b; Aalen, 1994). It should be noted that this is a defective survival distribution, because the entry into the absorbing state may never take place. This is due to the fact that only when the randomized drift is negative or zero will the process be certain to hit the barrier at zero. When the drift is away from zero, it may or may not reach zero before drifting to infinity. It is seen that

$$
\lim_{t \to \infty} S_R(t) = \Phi\left(-\frac{\mu}{\tau} \right) - \exp\left(2c\mu + 2c^2\tau^2\right) \Phi\left(-2c\tau - \frac{\mu}{\tau} \right) \tag{10.8}
$$

is the probability of never being absorbed. Defective survival distributions are important in practice; we already mentioned them under the name of "cure models" in the introduction to 10.3 and in Chapter 6.

The hazard rate for the time to absorption is now $\alpha_R(t) = f_R(t)/S_R(t)$. When t increases, the density and the hazard rate go to zero as $1/t^2$, and hence the hazard rates for different parameters will be asymptotically proportional. An example showing that approximate proportionality may be achieved very fast is given in Figure 10.13. The effect is similar to the nonrandomized model, except that the hazards themselves tend to zero over time.

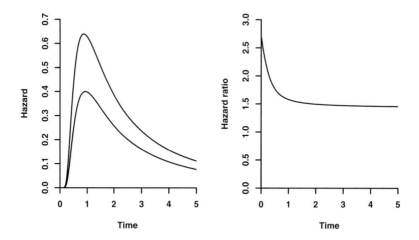

Fig. 10.13 *Wiener process with randomized drift. Left panel: hazard rates $\alpha_R(t)$ for time to absorption when the process has mean drift parameters $-\mu = -1$ (upper curve) and $-\mu = -0.5$ (lower curve). Right panel: ratio of the two hazard rates. Remaining parameters are $c = 2$ and $\tau^2 = 1$.*

10.3.8 Analyzing the effect of covariates for the randomized Wiener process

A test of the usefulness of the models presented here is whether they can be used in practice to analyze survival data with covariates, and how much information can be gleaned from the results. We will use the Wiener process with absorption and randomized drift, that is, a mixed inverse Gaussian distribution. We stress, however, that the model is meant only as an illustration of how the ideas in this chapter can be implemented for a real data set.

The model allows us to distinguish two different types of covariates, namely those that really only represent measures of how far the underlying process has advanced, and those that represent causal influences on the development. It is natural to model the first type as influencing the distance from absorption, that is, the parameter c, while the causal covariates influence the drift parameter μ. The distinction between two types of covariates is similar to the distinction between internal and external covariates, or between endogenous and exogenous effects. This was discussed in more detail in Chapter 9.

Using the explicit formulas (10.6) and (10.7) for the survival function and density, respectively, a full maximum likelihood approach is easily implemented; see Section 5.1. One or both of the parameters μ and c can be functions of covariates. The likelihood can be maximized by standard programs. The analysis presented here is carried out in R (R Development Core Team, 2007). The program is available from the book's Web site.

Example 10.1. Carcinoma of the oropharynx. Kalbfleisch and Prentice (2002) show a set of survival data concerning treatment of carcinoma of the oropharynx. Survival time is measured in days from diagnosis and some of the patients are censored. The following covariates, measured at diagnosis, will be considered here: x_1 = sex, x_2 = condition, x_3 = T-stage, and x_4 = N-stage. The data and variables are described in Example 1.4 (page 12). Note that we here use a different numbering of the covariates x_1, x_2, x_3, x_4. Also, survival time has in our analysis been recoded from days to years. In the data, case 141 has an invalid value (0) for the variable condition and is therefore excluded. However, Figure 10.14 demonstrates the effect of this outlier on one of the analyses. Some of the results from the analyses have previously been presented in Aalen and Gjessing (2001) and Halvorsen (2002).

With the exception of sex, the covariates in this example clearly measure how advanced the disease is at time of diagnosis. In light of the preceding discussion, it would be natural to try a model based on (10.6) and (10.7), where sex influences c and condition, T-stage and N-stage influence μ. Interestingly, since c and μ enter the functional form of (10.7) in two essentially different ways, we can model *both* parameters as functions of the same covariates. Specifically, we will assume that the parameters for an individual depend on the covariates as in

$$\mu = \boldsymbol{\beta}_\mu^T \mathbf{x},$$

$$g(c) = \boldsymbol{\beta}_c^T \mathbf{x},$$

where $\boldsymbol{\beta}_\mu = (\beta_{0\mu}, \ldots, \beta_{4\mu})^T$ and $\boldsymbol{\beta}_c = (\beta_{0c}, \ldots, \beta_{4c})^T$ are coefficient vectors and $\mathbf{x} = (1, x_1, x_2, x_3, x_4)^T$ is the vector of covariates for that individual. Following the standard terminology for generalized linear models (McCullagh and Nelder, 1989), the function g is a link function; if $g(x) = \log(x)$ then c is guaranteed to be positive, which is a requirement for the initial values. In addition to the log link, we test a model where $g(x) = x$. This model can be fit even if the resulting values of c are not constrained to be positive. The model is still acceptable since, given our data, none of the c-values are actually negative. A link function could also be applied to the μ-term, but we only use the identity link for μ here. We only explore linear (trend) effects of the ordinal variables.

We start by fitting models where μ is constant, that is, $\boldsymbol{\beta}_\mu = \beta_{0\mu}$ is just the intercept. Using Akaike's information criterion (AIC) (Pawitan, 2001), the fit is better (lower AIC) for the log link than for the identity link (Models 1 and 2 in Table 10.1). We then add all covariates also to the μ-term. Judging from the AIC, the increase in

Table 10.1 *AIC values for selected maximum likelihood fits. In Models 1–4,* **x** *is the full vector of covariates, in Model 5 sex is removed.*

Model	μ	c	AIC
1	$\beta_{0\mu}$	$\boldsymbol{\beta}_c^T \mathbf{x}$	436.7
2	$\beta_{0\mu}$	$\exp(\boldsymbol{\beta}_c^T \mathbf{x})$	428.0
3	$\boldsymbol{\beta}_\mu^T \mathbf{x}$	$\boldsymbol{\beta}_c^T \mathbf{x}$	442.6
4	$\boldsymbol{\beta}_\mu^T \mathbf{x}$	$\exp(\boldsymbol{\beta}_c^T \mathbf{x})$	436.7
5	$\beta_{0\mu}$	$\exp(\beta_c^T \mathbf{x})$	426.5

the number of parameters is not compensated by the improvement in the fit. With the full model, the log link (Model 4) still has a lower AIC value than the identity link (Model 3) but higher than the corresponding model with only the intercept value for μ (Model 2).

Even though the simpler models fit better according to the AIC, the full model tells us something more. The estimated parameters for Model 4 are shown in Table 10.2.

Table 10.2 *Estimated coefficients in regression Models 4 and 5.*

	Parameter	Model 4 Estimate	S.E.	Wald test	Model 5 Estimate	S.E.	Wald test
	τ	0.357	0.243	–	0.565	0.218	–
	Intercept	0.0684	0.557	0.123	0.748	0.162	4.61
	Sex	-0.106	0.228	-0.466			
μ	Condition	0.113	0.209	0.541			
	N-stage	0.0180	0.0837	0.215			
	T-stage	0.162	0.126	1.28			
	Intercept	1.42	0.373	3.81	2.04	0.250	8.13
	Sex	0.0958	0.155	0.619			
c	Condition	-0.638	0.127	-5.02	-0.685	0.100	-6.82
	N-stage	-0.0964	0.0531	-1.81	-0.134	0.0380	-3.52
	T-stage	-0.0976	0.0714	-1.37	-0.190	0.0498	-3.81

We see that the covariates condition and N-stage are clearly more influential when applied to c than to μ. This is consistent with the hypothesis that these covariates measure how far the disease has progressed at a given time rather than its progression rate. Sex has little influence on either of the parameters, whereas T-stage seems to have a possible (but not very strong) effect on both. Notice also that the signs of the parameters for condition, N-stage, and T-stage are all what you would expect: higher values of condition or stage give lower values of c and higher values of μ, both of which mean a shorter time to event. This effect is also seen in Figure 10.14 (right panel), which plots individual c-values against μ-values. Higher values of μ are associated with lower values of c.

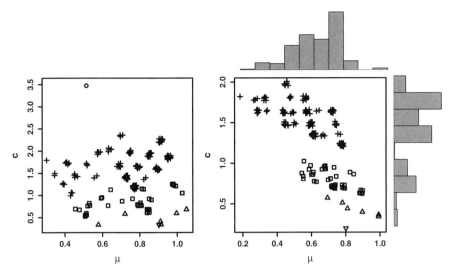

Fig. 10.14 *Scatterplot of estimated values of average drift and initial distance in Model 4, with histograms of marginal frequencies. Left scatterplot is before removal of outlier, right scatterplot is after removal. A "+" marks individuals with condition 1. Squares, triangles, and inverted triangles denote condition 2, 3, and 4, respectively. Points are slightly "jittered" in both directions to avoid overlap.*

From Figure 10.14 it is also seen that the pattern of average drift and initial values is largely dominated by the value of condition due to its large influence on the initial value. Another feature seen from Figure 10.14 is that all the estimated μ values are positive. Thus, all groups of individuals (defined by covariate combinations) have a predicted average drift *toward* the barrier, and consequently the majority of individuals will have zero "cure" probability. However, since the individual drift is assumed random with expectation μ, there will always be individuals who have a positive probability of no event. Notice that the original data contain a coding error resulting in an outlier, whose strong effect is easily seen in Figure 10.14 (left panel). The results presented in this example were computed after removal of this outlier.

Going through several submodels, Model 5 has the lowest AIC. This model has only an intercept for μ, and $c = \exp(\beta_{0c} + \beta_{2c}x_2 + \beta_{3c}x_3 + \beta_{4c}x_4)$, that is, sex has been removed from both parameters. The parameter estimates for this model are given in Table 10.2.

To check the goodness of fit of Model 5 we create a "prognostic index" from the initial distance computed in the model. Since Model 5 has no covariates in the drift term, it is not necessary to include drift in the index. We group the individuals in three groups: high risk (c less than 0.91), intermediate risk (c between 0.91 and 1.9), and low risk (c above 1.9). The three groups then comprise 20%, 60%, and 20% of the data, respectively. For each group we compute the Nelson-Aalen estimator and compare it to the model prediction. Since each group is somewhat heterogeneous, we predict the cumulative hazard in a group by first computing the average survival in each group, and then computing the negative logarithm of this. This ensures that the (cumulative) hazard at any moment is only computed among those who still survive, thus compensating for heterogeneity effects. The results are shown in Figure 10.15.

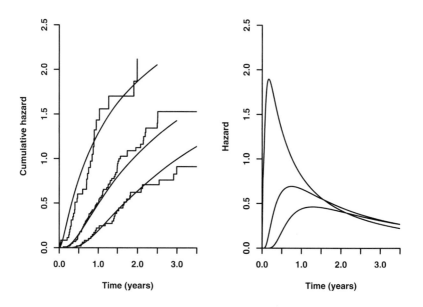

Fig. 10.15 *Cumulative hazards (left panel) according to estimated model (smooth curve) compared with the Nelson-Aalen estimate (step curve) within a high-risk, an intermediate-risk, and a low-risk group. The corresponding hazards (right panel) are clearly not proportional.*

Although this is a simple example, it shows that the idea of covariates measuring how far an underlying process has advanced might be useful. The estimated hazards in Figure 10.15 are clearly not proportional. Rather, the low-risk group shows a *delay* in the hazard, having almost zero hazard for a while, before it gradually catches up with the other groups. This delay is a natural consequence when regarding the low-risk group as having less advanced disease than the other groups. It is our experience that survival curves for groups with lower risk often show this feature of delay compared to groups with higher risk, which also implies nonproportionality. □

Example 10.2. Time between births. The data were introduced in Example 1.1. The data, taken from the Medical Birth Registry of Norway, record the time interval from the first birth to the second birth of a mother. In Example 3.1, it was seen that there was a large difference between mothers whose first child died and mothers who had a surviving first child. If the first child survived, the mother had a much lower chance of having a new child in the first two years than if it died. After two years, however, there was little difference between the two groups in the "hazard" of having a new birth. In the analysis, mothers without a second birth were censored.

We will now use the randomized Wiener process to evaluate the data on time between births. As in the previous example, Equations (10.6) and (10.7) define our model. With only the survival of the first child as our covariate, we do a completely stratified analysis, letting (c_0, μ_0, τ_0^2) and (c_1, μ_1, τ_1^2) be the initial value, mean drift, and drift variance of the Wiener process in the groups with a surviving and a dead first child, respectively. The results are seen in Table 10.3. The parameter estimates

Table 10.3 *Estimated coefficients in the Wiener process model with randomized drift, for time between births. Groups are 0 (first child survived) and 1 (first child died).*

Parameters	c_0	c_1	μ_0	μ_1	τ_0	τ_1
Estimate	34.1	14.8	0.0130	0.00926	0.00580	0.0265
Std. error	0.187	0.931	0.000326	0.00645	0.00159	0.0122

seem to have a natural explanation: the women whose first child survived start with a "wait and see" attitude, whereas the mothers whose first child died are more immediately interested in having a new child ($c_0 > c_1$). On the other hand, mothers with a live child seem more determined to have a second child a little later, having a (slightly) stronger drift toward a new birth ($\mu_0 > \mu_1$). Thus, if a mother with a dead first child does not have a new child soon, her attitude toward (or ability to) have a new one is less pronounced since μ_1 is less than μ_0 and the variability (τ_1) larger.

Such interpretations should, of course, be treated with healthy scepticism, particularly since μ_1 and τ_1 have large standard errors. Nevertheless, the model probably picks up some of the essential features of the phenomenon and provides insight not immediately available in the simpler Kaplan-Meier approach.

Comparing the Kaplan-Meier results with the survival functions predicted from the parametric model, the match is quite good for both groups (Figure 10.16), particularly for group 1. Figure 10.17 compares the model estimated hazards in the two groups. The picture reflects the intuitive discussion: mothers whose first child died will more quickly have a second child, but after three or more years the hazard is higher among those whose first child survived. It is abundantly clear that a proportional hazards model would be unrealistic in this example, at least until several years have passed (see Section 6.5).

From (10.8) we may compute an estimate of the probability of never having a second child, using the estimated values in Table 10.3. The estimates are 0.0015 and 0.13 for those with a live and a dead first child, respectively. However, as seen from Figure 10.16, the Kaplan-Meier curves, particularly for the live group, seem to flatten out more quickly than the model estimate. Thus, the extrapolation required to obtain these estimates is, at best, risky. Adding to this uncertainty is the fact that we have not included maternal age in the analysis. As a strong determinant of fertility,

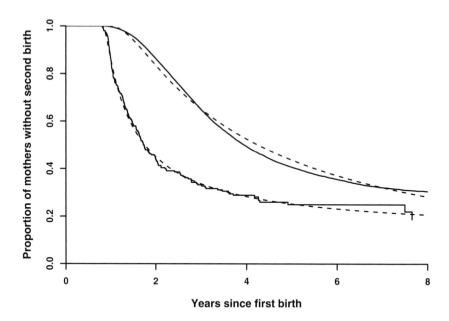

Fig. 10.16 *Survival curves for the time between first and second birth. Solid lines (step functions) are the Kaplan-Meier estimates. Dashed lines are fitted by the Wiener process model with randomized drift. Upper curves: first child survived one year. Lower curves: first child died within one year.*

maternal age is likely to seriously confound the analysis, particularly so over a long time range. □

10.4 Diffusion process models

We now briefly view the situation from a more general point of view, by considering a Markovian diffusion process on the positive half-line with zero as the absorbing state. While the Wiener process is the primary example of such a process, considering more general diffusions makes it possible to control the behavior of the underlying process in more detail. For instance, the Ornstein-Uhlenbeck process, described later, is a stationary Gaussian process, suitable for modeling systems that are more or less stable over time. A computational advantage of diffusion processes are the Kolmogorov equations, which allow one to derive expressions for transition probabilities and first passage time distributions under varying sets of conditions. More on this is also found in Appendix A.4.

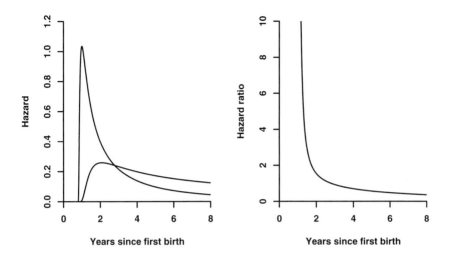

Fig. 10.17 *Hazards estimated from the Wiener process model with randomized drift, and their ratio. Left panel: hazard of second child for mothers whose first child died (sharply pointed curve) and for mothers whose first child survived first year (low curve). Right panel: hazard ratio (first child died over first child survived).*

Starting with a Wiener process $X(t) = c + mt + \sigma W(t)$ with drift m and diffusion σ, we see that, in differential notation,

$$dX(t) = mdt + \sigma dW(t)$$

is a simple example of a *stochastic differential equation*, which has X as its solution and where none of the coefficients depend on X. Note that if $\sigma = 0$, the equation is just a standard (trivial!) differential equation; the $dW(t)$ part makes it a stochastic differential equation. Since the $W(t)$ is not really a differentiable function of t, a special theory involving the so-called *Itô integral* is needed to make sense of a stochastic differential equation. More details about this topic are found in Appendix A.4, and in several excellent textbooks, such as Allen (2003) and Øksendal (1998). To define a *diffusion process* X, a more general stochastic differential equation can be set up, as in

$$dX(t) = \mu(X(t))dt + \sigma(X(t))dW(t), \tag{10.9}$$

with initial value $X(0) = c$. So for a diffusion process, both the drift and the diffusion coefficients may depend on the value of the process itself. The functions $\mu(x)$ and $\sigma(x)$ are real functions on the state space; for our purposes we assume they are nonrandom and do not depend on t. An important feature of stochastic differential equations is that the solution process X inherits the "wigglyness" of the driving stochastic force W. If, in contrast, $\sigma \equiv 0$, the equation reduces to a standard (possibly nonlinear) differential equation with smooth solutions. In that case, if c and μ were made stochastic, the solution process X would be random but still a smooth function of t. Thus, diffusion processes, as solutions to the full stochastic differential equation, have more flexible and interesting model features.

As we did for the Wiener process, we now assume that a diffusion X starts in $c > 0$ and let $T = \inf_{t \geq 0}\{t : X(t) = 0\}$ be the first hitting time at zero, where X is absorbed. Again, let $\psi_t(x)$, $x > 0$, be the probability density for the part of $X(t)$ that has not yet been absorbed, that is, at time t, the process is either in zero with probability $1 - S(t) = P(T \leq t)$ or positive with density $\psi_t(x)$, where $\int_0^\infty \psi_t(x)dx = S(t)$.

10.4.1 The Kolmogorov equations and a formula for the hazard rate

The celebrated Kolmogorov differential equations are among the primary tools for investigating properties of diffusion processes, in particular their transition densities. From the representation of X as the solution to a stochastic differential equation (10.9), the Kolmogorov equations for its transition density can be written down directly. Even though ψ_t is a "defective" density, in the sense that it holds only for the nonabsorbed part and does not integrate to one, it still satisfies the Kolmogorov differential equations. The absorption in zero is accounted for by choosing appropriate initial/boundary conditions. The forward equation for ψ_t is (see Appendix A.2 and Karlin and Taylor (1981, page 220))

$$\frac{\partial}{\partial t}\psi_t(x) = -\frac{\partial}{\partial x}\left(\mu(x)\psi_t(x)\right) + \frac{1}{2}\frac{\partial^2}{\partial x^2}\left(\sigma^2(x)\psi_t(x)\right), \qquad (10.10)$$

with the condition $\psi_0(x) = \delta_c(x)$, that is, a probability distribution with all mass placed in c at time 0, and the additional condition that $\psi_t(0) = 0$ for all $t \geq 0$. The last condition is appropriate for an absorbing boundary. Working with the Kolmogorov equations often involves finding solutions to second-order differential equations with appropriate initial/boundary conditions. A wealth of useful information can be found, for instance, in Abramowitz and Stegun (1964), Lebedev (1972), and Polyanin and Zaitsev (1995). It should also be kept in mind that modern computer algebra systems such as Mathematica (Wolfram, 1999) include a wide range of built-in solutions to differential equations.

Starting from the Kolmogorov equation, an expression of the hazard $\alpha(t)$ associated with the event time T can be found. A special case of this formula was mentioned in 10.3.3; here we derive a more general version. Substituting $\psi_t(x) = S(t)\phi_t(x)$ in (10.10), we have

$$-\alpha(t)\phi_t(x) + \frac{\partial}{\partial t}\phi_t(x) = -\frac{\partial}{\partial x}\left(\mu(x)\phi_t(x)\right) + \frac{1}{2}\frac{\partial^2}{\partial x^2}\left(\sigma^2(x)\phi_t(x)\right), \qquad (10.11)$$

and when integrating on both sides of this equation with respect to x, from 0 to ∞, we get

$$
\begin{aligned}
-\alpha(t) + 0 &= \left[-\mu(x)\phi_t(x) + \frac{1}{2}\frac{\partial}{\partial x}\left(\sigma^2(x)\phi_t(x) \right) \right]_0^\infty \\
&= \frac{1}{2}\left[2\sigma(x)\sigma'(x)\phi_t(x) + \sigma^2(x)\phi_t'(x) \right]_0^\infty \\
&= -\frac{1}{2}\sigma^2(0)\phi_t'(0),
\end{aligned}
$$

using boundary condition $\phi_t(0) = 0$, in addition to $\phi_t(\infty) = \phi_t'(\infty) = 0$ (which are reasonable requirements for a probability distribution). The prime denotes differentiation with respect to x. So α is expressed as

$$
\alpha(t) = \frac{1}{2}\sigma^2(0)\phi_t'(0), \tag{10.12}
$$

where the derivative of $\phi_t(x)$ is with respect to x. This formula is related to known results on first passage time distributions; see Goel and Richter-Dyn (1974, section 3.2, formula (24)). We gave an interpretation of this formula in 10.3.3, where we showed (Figure 10.11) that the distribution of survivors typically behaves like a wave: it starts out being focused at an initial value c. The "wave" then moves toward the barrier at zero. After building up mass close to zero, it typically starts receding, stabilizing at the quasi-stationary distribution. The slope at zero (and thus the hazard) follows this behavior by first increasing and then decreasing to the stable level. It is also interesting to observe that (10.12) is entirely parallel to formula (10.1) for the phase type distributions with discrete state space.

10.4.2 An equation for the quasi-stationary distribution

Just as for the Wiener process with drift, general diffusion processes may exhibit quasi-stationarity of the survivors when absorption takes place at a barrier. A discussion of this is found in Steinsaltz and Evans (2003). Assume that the process is in a quasi-stationary state. Then one can write $\psi_t(x) = e^{-\alpha t}\phi(x)$, where α is the constant hazard rate and $\phi(x)$ is the quasi-stationary distribution. Insertion into Equation (10.11) yields

$$
-\alpha\phi(x) = -\frac{\partial}{\partial x}[\mu(x)\phi(x)] + \frac{1}{2}\frac{\partial^2}{\partial x^2}\left[\sigma^2(x)\phi(x) \right]. \tag{10.13}
$$

Notice that the equation does not depend on t, and in general, this is an eigenvalue problem for the differential operator

$$
\alpha^\star = -\frac{\partial}{\partial x}\mu + \frac{1}{2}\frac{\partial^2}{\partial x^2}\sigma^2.
$$

Suitable boundary conditions for Equation (10.13) are

$$\phi(0) = 0 \quad \text{and} \quad \phi'(0) = \frac{2\alpha}{\sigma^2(0)}.$$

The first condition stems from the absorption at zero; the second ensures the correct limiting rate as derived in (10.12). In fact, the argument used to derive (10.12) shows that the second condition is equivalent to assuming $\int_0^\infty \phi(x)dx = 1$, but the condition on $\phi'(0)$ is sometimes more convenient to use.

The special case of a Wiener process with constant drift $m = -\mu$, $(\mu > 0)$ and variance coefficient σ^2 was introduced in Section 10.3, and the form of the quasi-stationary distribution in Section 10.3.4. We see that the eigenvalue problem for the quasi-stationary distribution in this situation reduces to

$$\frac{1}{2}\sigma^2\phi''(x) + \mu\phi'(x) + \alpha\phi(x) = 0.$$

This is a standard constant coefficients second-order differential equation (Polyanin and Zaitsev, 1995, page 132). The corresponding characteristic equation

$$\frac{1}{2}\sigma^2 r^2 + \mu r + \alpha = 0$$

has the roots

$$r_1 = \frac{-\mu - \sqrt{\mu^2 - 2\alpha\sigma^2}}{\sigma^2} \quad \text{and} \quad r_2 = \frac{-\mu + \sqrt{\mu^2 - 2\alpha\sigma^2}}{\sigma^2}.$$

When $\mu^2 - 2\alpha\sigma^2 < 0$, the oscillating nature of the solution is not compatible with a (nonnegative) probability distribution. So it is required that $\alpha \leq \mu^2/(2\sigma^2)$, that is, there is an upper bound to the possible limiting hazard. When $\alpha < \mu^2/(2\sigma^2)$, the solution is

$$\phi(x) = \frac{2\alpha}{\sigma^2(r_2 - r_1)}\left(e^{r_2 x} - e^{r_1 x}\right), \quad x \geq 0, \tag{10.14}$$

that is, an (opposite sign) mixture of two exponential distributions. Our assumption $\mu > 0$ guarantees that the process always moves toward the barrier at zero, creating a quasi-stationary distribution which is a valid probability distribution since $r_1, r_2 < 0$.

If we now let $\alpha \to \mu^2/(2\sigma^2)$ and notice that $\exp(r_2 x) - \exp(r_1 x) \approx \exp(r_1 x)(r_2 - r_1)x$ when $r_2 - r_1$ approaches zero, we obtain the limiting situation

$$\phi(x) = \frac{\mu^2}{\sigma^4}xe^{-\frac{\mu}{\sigma^2}x}, \quad x \geq 0, \tag{10.15}$$

when $\alpha = \mu^2/(2\sigma^2)$, that is, a gamma distribution. Indeed, as discussed in Section 10.3.4, this is the *canonical* solution, which is the one obtained when starting the Wiener process in a single value. The distributions in (10.14) have heavier tails (toward infinity) than the canonical distribution, resulting in lower limiting hazards, but they cannot be reached from a single starting point. For further details of the quasi-stationary distribution for the Wiener process, see Martinez and San Martin (1994).

10.4.3 The Ornstein-Uhlenbeck process

Another example of a diffusion process that is useful for modeling is the *Ornstein-Uhlenbeck* process. We define the Ornstein-Uhlenbeck process $X(t)$ by the stochastic differential equation

$$dX(t) = (a - bX(t)) \, dt + \sigma \, dW(t).$$

We typically assume $b > 0$. The intuitive interpretation of this equation is that the increment of X at time t has a deterministic part proportional to $a - bX(t)$ and a random part that is just a scaled increment of the driving Brownian motion. The solution X is a Gaussian process, provided $X(0)$ is Gaussian, which we assume. Taking expectations on both sides, we see that

$$g'(t) = a - bg(t),$$

where $g(t) = EX(t)$; thus the expectation of the process satisfies the deterministic part of the equation. Solving for $g(t)$, we obtain

$$g(t) = \frac{a}{b} + (g(0) - \frac{a}{b})e^{-bt}.$$

Clearly, as time passes the expectation moves from its starting value $g(0)$ and converges to the attraction point a/b. The process $X(t)$ itself will not converge since the noise $dW(t)$ is constantly added to it. Nevertheless, the term $a - bX(t)$ represents a force pulling X back toward a/b, with a pull that is stronger the further away it gets. This behavior is called *mean reversion*. The effect is that over time, X becomes stationary. The stationarity makes the Ornstein-Uhlenbeck process a suitable model for phenomena that remain fairly stable over time. More details about the Ornstein-Uhlenbeck process can be found in Appendix A.4. For applications of the Ornstein-Uhlenbeck and related processes in finance; see, for example, Musiela and Rutkowski (1997), Barndorff-Nielsen and Shephard (2001) and Aalen and Gjessing (2004).

As long as $a > 0$ (still assuming $b > 0$), the point of mean reversion a/b will be positive. If a/b is "large" the process will stay away from zero most of the time, and absorption in zero will take place only under extraordinary circumstances, when $X(t)$ gets far below what would be expected from its "safe" stationary state. If, on the other hand, $a < 0$, the process is attracted to negative values and is likely to cross the barrier at zero fairly soon. When $X(t)$ approaches the barrier, the term $-bX(t)$ contributes less and less to the drift, so that close to the barrier X will behave much like an ordinary Wiener process with drift.

In general, it is difficult to find explicit expressions for the distribution of time to absorption for diffusion processes. However, an example where the equations can be solved explicitly is the "symmetric" case of the Ornstein-Uhlenbeck process, that is, when $a = 0$, so that the process would be mean-reverting about zero but starts at a positive value x_0 and gets absorbed at zero. From Ricciardi and Sato (1988, page 46), we derive the formula (setting $a = 0$, and $b = 1$, $\sigma^2 = 2$ as an example)

$$f(t) = \sqrt{\frac{2}{\pi}} x_0 \frac{e^{2t}}{(e^{2t}-1)^{3/2}} \exp\left(-\frac{x_0^2}{2(e^{2t}-1)}\right) \qquad (10.16)$$

for the density of the hitting time T. The simple survival function belonging to this density is

$$S(t) = 2\Phi\left(\frac{x_0}{\sqrt{e^{2t}-1}}\right) - 1,$$

where $\Phi(\cdot)$ is the standard cumulative normal distribution function. As before, the hazard rate is computed as $\alpha(t) = g(t)/S(t)$. Perhaps not so surprisingly, the hazards resulting from this model show much similarity to the hazards found for the Wiener process with absorption, shown in Figure 10.7, with the position of the peak hazard depending on how close to the barrier the process starts, and a limiting hazard common to all starting values. For this example, the limiting hazard is 1.

In situations where explicit expressions for the survival are not available, one might use the Kolmogorov equations to derive differential equations for the Laplace transform of the hitting time distributions (Cox and Miller, 1965; Paulsen and Gjessing, 1997), and the Laplace transform may be a good starting point for either finding explicit solutions if such exist, or finding useful numerical approximations. We will not pursue that topic here. Rather, we will end this chapter by showing how the equations in 10.4.2 can be used to derive the limiting hazard properties of the general Ornstein-Uhlenbeck process.

The limiting absorption hazard α_0 for the Ornstein-Uhlenbeck process is the smallest positive zero of $H_{s/b}(-a/(\sigma\sqrt{b}))$, seen as a function of s. Here, $H_\nu(z)$ is the Hermite function (Abramowitz and Stegun, 1964). This value is easily computed numerically in software such as Mathematica (Wolfram, 1999). As seen for the Wiener process, the Ornstein-Uhlenbeck process also generates infinitely many quasi-stationary distributions, but α_0 belongs to the canonical one. In fact, all (canonical and noncanonical) quasi-stationary distributions can be found, but the canonical one has a simpler expression than the rest, given by the formula

$$\phi(x) = \frac{\sqrt{b}}{\sigma} e^{\frac{x(2a-bx)}{\sigma^2}} \frac{H_{\alpha_0/b}\left(\frac{-a+bx}{\sigma\sqrt{b}}\right)}{H_{(\alpha_0/b)-1}\left(-\frac{a}{\sigma\sqrt{b}}\right)}, \qquad x \geq 0.$$

The canonical distribution is in many instances fairly close to a normal distribution. The factor $\exp\left(x(2a-bx)/\sigma^2\right)$ defines a normal distribution with mean a/b (the mean-reverting value) and variance $\sigma^2/2b$, which are in fact the limiting mean and variance for the nonabsorbed Ornstein-Uhlenbeck process, see Appendix A.4.

The Hermite function skews the normal distribution and keeps it on the positive side of the barrier. It is also interesting to note that when $x \to \infty$ the Hermite part $H_{\alpha_0/b}((-a+bx)/\sigma\sqrt{b})$ increases to infinity proportionally with $x^{\alpha_0/b}$, so that the tail of $\phi(x)$ is not very different from a normal distribution. However, the noncanonical distributions have tails that approach zero proportionally to $x^{-1-\alpha/b}$, where α is the limiting hazard rate. Thus, the noncanonical quasi-stationary distributions have much heavier tails than the canonical one. As a particular case, when $a = b = \sigma^2 = 1$ we find, using Mathematica (Wolfram, 1999), $\alpha_0 \approx 0.23423$ as the smallest positive zero of $H_s(-1)$, and the canonical quasi-stationary distribution is

$$\phi(x) = e^{x(2-x)} \frac{H_{\alpha_0}(x-1)}{H_{\alpha_0-1}(-1)}.$$

In Figure 10.18, this distribution is compared to one of the noncanonical distributions ($\alpha = 0.05$). The heavy tail of the noncanonical distribution is evident; most of the mass of the distribution lies in the tail.

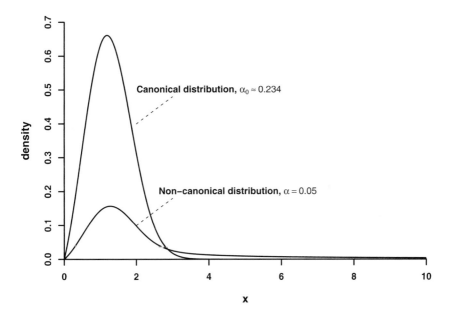

Fig. 10.18 *The canonical quasi-stationary distribution $\phi(x)$ for the Ornstein-Uhlenbeck process with an absorbing barrier at zero, compared to one of the noncanonical distributions. Both are shown for the special case $a = b = \sigma^2 = 1$.*

We will return to the Ornstein-Uhlenbeck process in the next chapter, then as a direct model for time-dependent stochastic hazard (as opposed to using absorption probabilities as here).

10.5 Exercises

10.1 Explain why the finite-dimensional distributions

$$(W(t_1), W(t_2), \ldots, W(t_k))$$

of the Wiener process are completely determined by the distribution of $W(t)$, that is, the one-dimensional distributions.

10.2 Prove that the hitting time distribution of a Wiener process (starting in $c > 0$, with $\mu = 0$ and $\sigma = 1$) is the inverse Gaussian distribution (10.2), using the following arguments. Recall that $W(t)$ has a Gaussian distribution with mean c and variance t. Let T be the first time $W(t)$ hits zero, but assume that W is not absorbed there. Note first that

$$P(T \le t) = P(T \le t, W(t) > 0) + P(T \le t, W(t) < 0).$$

Argue informally that $P(T \le t, W(t) < 0) = P(T \le t, W(t) > 0)$. Use this to prove that

$$P(T \le t) = 2P(W(t) < 0),$$

and derive the density of T. More details can be found in Karatzas and Shreve (1991).

10.3 Derive the expression (10.3) for the density (on the positive axis) of the survivors at time t, that is, find

$$\frac{1}{dx} P(X(t) \in (x, x+dx], T > t),$$

where X is a Wiener process with drift $\mu = 0$, diffusion $\sigma = 1$, and starting in $c > 0$. Hint: note that

$$P(X(t) \in (x, x+dx], T > t) = P(X(t) \in (x, x+dx]) - P(X(t) \in (x, x+dx], T \le t),$$

and use an argument like the one in the previous exercise.

10.4 Prove Equations (10.5) and (10.6).

Chapter 11
Diffusion and Lévy process models for dynamic frailty

In Chapter 10, we demonstrated how the concept of dynamically developing individual risk could be used to produce frailty-like effects on the population level. As soon as the risk process of an individual hits a barrier, that individual has had an event, leaving behind only those with a risk process not yet at the level of the barrier. The barrier hitting models can be seen as a way of relaxing the assumption of time-constant frailty found in the usual frailty models (Chapter 6). Another way of relaxing the constancy assumption is to allow dynamic covariates to capture a part of the variation experienced by an individual, as demonstrated in Chapter 8.

However, a natural direct extension of the frailty models is to let h be a "fully fledged" stochastic process, for instance, a diffusion process. Rather than letting a barrier determine an event, an event may occur at any time, governed by the individual hazard h. Just like the barrier-hitting models, covariates may be included to control the behavior of h. The stochastic process would then have a deterministic component which could be controlled through observed covariates, and a random component describing a level of uncertainty regarding unobserved covariates and a fundamental level of individual time-changing heterogeneity.

Introducing a frailty that develops as a stochastic process may be seen as a somewhat speculative approach. As with standard frailty models and barrier models, the problem is that we only observe an average (population) hazard, and that we frequently only have a single time of observation for each individual. As long as there is no specific information about the underlying process, and observations are only made once for each individual, there is little hope of identifying what kind of process is actually driving the development. On the other hand, all statistical models introduce noise to cope with the part of the observed phenomenon that cannot be described convincingly by deterministic structures. The success of a model derives from the interplay between the random and the deterministic part. For a random process frailty, we will see that the parameters can be given intuitive interpretations in terms of the unobserved development leading to events. It is too optimistic to believe that the model will be a precise description of this development. Rather, the results will be a distorted reflection of what actually happens at the lower levels. Nevertheless, interpreting the data in terms of the components of the model may increase our understanding of, for instance, what effect a patient treatment has, and at what levels the changes are occurring.

To connect the individual hazard with the population-level hazard we need a mathematical model that combines the flexibility of including relatively complex processes together with ease of computation. In the first part of this chapter, we will use diffusion processes for this purpose. As for the barrier models, diffusion processes allow relatively explicit solutions to be found in some instances, and they are flexible without being overly complicated. In the second part of the chapter, we will introduce the Lévy process as a model for dynamic frailty. Lévy models include, for instance, PVF frailty distributions as discussed in Chapter 6. It should be noted that there is a close connection between the barrier hitting time distribution approach and the direct modeling of h. We demonstrate this briefly in A.4.5.

Before turning to the explicit models, we begin by stating some general results relating the individual hazard h to the population-level (observable) hazard and survival functions.

11.1 Population versus individual survival

Let $h(t)$ be a nonnegative stochastic process on a continuous time and state space. The value $h(t)$ represents the hazard at time t of an individual randomly picked from the population. We could use the notation $h_i(t)$ to indicate the individual, but we will suppress the index and just say that a particular realization of the process h represents a randomly selected individual. In many applications, the event will be death, or any other "terminal" state. After such an event, an individual will not necessarily be under risk, and the process h will not really be defined. This is no different from other time to event models, and it poses only a conceptual problem, not a practical one. Also, since one may be interested in repeated (nondestructive) events, it is in many cases natural to have a hazard process h defined for all times.

Assume, for the moment, that h is given for a particular individual. Conditional on h, the (continuous) distribution of the time to event T for this individual is determined by the hazard h in the usual way, that is,

$$S(t|h) = P(T > t|h) = \exp\left(-\int_0^t h(s)\,ds\right), \qquad (11.1)$$

where $S(t|h)$ denotes the conditional (i.e., individual) survival function. Thus, T contains an additional amount of randomness; knowing the entire hazard history h of the individual does of course not tell exactly when the event will occur, and from the point of view of that specific individual the hazard h might be considered deterministic.

Then, let T be the event time of any (randomly sampled) individual from the population. Each individual is assumed to have its own "realization" of the stochastic process h as its hazard. Clearly,

$$S(t) = P(T > t) = E[S(t|h)] = E\exp\left(-\int_0^t h(s)\,ds\right). \qquad (11.2)$$

This corresponds to formula (6.2) for standard (constant) frailty.

There is also a direct connection to the population hazard $\mu(t)$. Let $f(t)$ be the (population) density of T. Again conditional on h, if $f(t|h) = -(d/dt)S(t|h)$ is the (conditional) density of T, we have

$$\begin{aligned} f(t) &= E[f(t|h)] = E[h(t)S(t|h)] \\ &= E[h(t)E[\mathbf{1}(T > t)|h]] = E[h(t)\mathbf{1}(T > t)] \\ &= E[h(t)|T > t]\,P(T > t) \end{aligned}$$

(using the formula $E[X|A] = E[X\mathbf{1}_A]/P(A)$ for a random variable X and an event A). It follows that the population survival $S(t)$ has the corresponding population hazard

$$\mu(t) = E[h(t)|T > t]. \qquad (11.3)$$

See also Yashin and Manton (1997).

The two relations (11.2) and (11.3) are fundamental connections between individual survival and what is observed at the population level. Equation (11.2) simply states that *the population survival is the average of the individual survivals, where the average is computed over the initial (time zero) distribution of individuals.* Equation (11.3), on the other hand, states that *the population hazard is an average of individual hazards, but where the average is computed only over the individuals remaining at time t.*

As a simple illustration, consider a population consisting of two homogeneous subgroups. Group 1 contains n_1 individuals at time 0, Group 2 contains n_2. In the first group all individuals have hazard $h_1(t)$, in the second all have $h_2(t)$, where h_1 and h_2 are deterministic functions. We assume no censoring. After time t, approximately $n_1 S_1(t)$ of the total population remain in Group 1, and $n_2 S_2(t)$ in Group 2. The total number of survivors in the population at large is thus $n_1 S_1(t) + n_2 S_2(t)$; relative to the original population $n = n_1 + n_2$ this is

$$S(t) = \frac{n_1}{n}S_1(t) + \frac{n_2}{n}S_2(t).$$

For the population hazard, observe that in a short time interval $[t, t+dt)$, approximately $n_1 S_1(t)h_1(t)dt$ individuals in Group 1 will have experienced an event, and $n_2 S_2(t)h_2(t)dt$ in Group 2. As a proportion of the total remaining population, this is

$$\begin{aligned} \mu(t) &= \frac{n_1 S_1(t)h_1(t) + n_2 S_2(t)h_2(t)}{n_1 S_1(t) + n_2 S_2(t)} \\ &= \frac{n_1 S_1(t)}{nS(t)}h_1(t) + \frac{n_2 S_2(t)}{nS(t)}h_2(t). \end{aligned}$$

The formulas (11.2) and (11.3) are direct extensions of this result. They correspond to the extreme case, where each individual represents its own stratum.

11.2 Diffusion models for the hazard

With formulas (11.2) and (11.3) as our basis, we will now explore relevant models
for a stochastic hazard. Our approach is to use a well-understood class of stochastic
processes, such as a diffusion process $X(t)$, and then define the hazard as a (nonnegative) function or functional of this process. One of the simplest models, and a
natural starting point, is the standard Wiener process.

11.2.1 A simple Wiener process model

Define $X(t) = m(0) + W(t)$, where W is a Brownian motion as before, and $m(0)$ is
a constant. Let the hazard be $h(t) = X^2(t)$, which ensures nonnegative values. Since
by its nature $W(t)$ will have larger and larger excursions away from zero (to both
the positive and negative sides) over time, the hazard $h(t)$ will exhibit increasingly
higher peaks but will always return to zero from time to time. Figure 11.1 illustrates

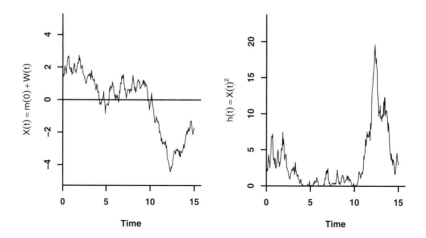

Fig. 11.1 *A simulated Wiener process with starting value $m(0)$ (left panel), and the resulting individual hazard $h(t)$ obtained by squaring the Wiener process (right panel).*

a typical path of the hazard process.

Recall that, from (11.3), the population hazard is determined as the mean hazard
of the individuals still under risk. It can be shown that (Yashin, 1985), conditional
on $T \geq t$, $X(t)$ is Gaussian with mean

$$m(t) = \frac{m(0)}{\cosh(\sqrt{2}t)}$$

and variance

$$\gamma(t) = \frac{1}{\sqrt{2}} \tanh(\sqrt{2}t).$$

We thus see that the population hazard is

$$\mu(t) = E[h(t)|T \geq t] = m^2(t) + \gamma(t).$$

Figure 11.2 shows the hazard, together with the mean and variance of $X(t)$ condi-

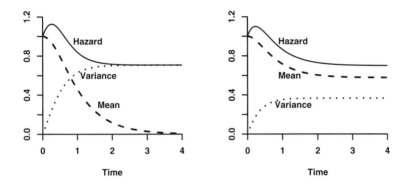

Fig. 11.2 *Population hazard $\mu(t)$, with mean and variance of the underlying stochastic process $X(t)$, conditional on no event before time t. Individual hazard is $h(t) = X(t)^2$. Left panel: $X(t) = m(0) + W(t)$, with $m(0) = 1$. Right panel: $X(t)$ is an Ornstein-Uhlenbeck process with $a = 1$, $b = 1$, and $\sigma^2 = 1$.*

tional on $T \geq t$. Let us consider for a moment what this implies for the conditional distribution of $X(t)$. At $t = 0$, X starts for all individuals in the point $X(0) = m(0)$, since the variance is zero at $t = 0$. Immediately after, the variance increases, and X has a nondegenerate Gaussian distribution. This also results in an overall hazard that is increasing. After a while, however, the mean value of $X(t)$ among the survivors drops since many of the highest-hazard individuals have had an event. Because X^2 is symmetric around zero as a function of X, the mean value does not drop below zero but stabilizes in an equilibrium at zero. The increasing diffusion effect of the underlying Brownian motion W tends to widen the distribution of X over time, but this is offset by the increasing hazard if X moves away from zero. The result is that X approaches a quasi-stationary distribution, which is Gaussian with mean 0 and variance $1/\sqrt{2}$.

What makes this model particularly tractable is the fact that the distribution of X conditional on survival is Gaussian. As one would expect, this property extends to the following more general situation.

11.2.2 The hazard rate as the square of an Ornstein-Uhlenbeck process

We introduced the Ornstein-Uhlenbeck process in 10.4.3 and A.4.3. Let $X(t)$ be an Ornstein-Uhlenbeck process defined from the equation

$$dX(t) = (a - bX(t))dt + \sigma dW(t),$$

and again define the hazard as a square of the underlying process, $h(t) = X^2(t)$ (Woodbury and Manton, 1977; Wenocur, 1990; Aalen and Gjessing, 2004). Figure 11.3 shows the Ornstein-Uhlenbeck process and its square. The strongly fluctuating,

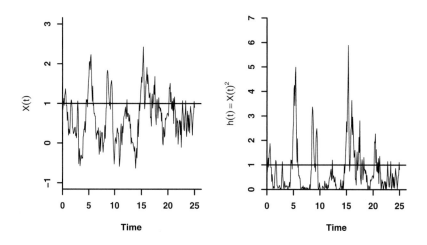

Fig. 11.3 *The path of an Ornstein-Uhlenbeck process $X(t)$ with $a = 1$, $b = 1$, $\sigma^2 = 1$, and initial value 1 (left panel). The square $h(t) = X(t)^2$ is used as a model for individual hazard (right panel). The horizontal line at 1 indicates the mean-reverting value a/b.*

but still stationary, behavior of X results in a hazard $h(t)$ that is stationary, with irregularly spaced sharp peaks. Contrary to the Wiener process example, the stationary Ornstein-Uhlenbeck process generates peaks that do not increase in size over time. It should also be noted that if the point of mean reversion of X, $a/b = 1$, is close to zero (relative to the diffusion), X frequently takes negative values. This will generate positive peaks in h, but peaks that are typically smaller than peaks resulting from a positive X. Consequently, h will have high peaks (when X is positive), interspersed with smaller peaks (when X is negative). This may or may not be a desirable modeling feature, but the effect is reduced if the mean reverting value is far from zero (or if it is zero).

Assume that the starting value of $X(0)$ is Gaussian with mean $m(0)$ and variance $\gamma(0)$. To analyze the resulting population hazard, it may again be shown (Yashin, 1985) that X is still Gaussian, conditional on $T \geq t$. The conditional moments, $m(t) = E[X(t)|T \geq t]$ and $\gamma(t) = \text{Var}[X(t)|T \geq t]$, solve the equations

$$m'(t) = a - bm(t) - 2m(t)\gamma(t) \tag{11.4}$$

$$\gamma'(t) = -2b\gamma(t) + \sigma^2 - 2\gamma^2(t). \tag{11.5}$$

Thus, it is possible to give a complete description of the hazard process conditional on survival. Equations (11.4) and (11.5) can be solved explicitly by solving first for γ and then for m. Define the constant $B = \sqrt{b^2 + 2\sigma^2}$ and let

$$l = \frac{1}{2}\log\left|\frac{B + (2\gamma(0) + b)}{B - (2\gamma(0) + b)}\right|.$$

Then

$$\gamma(t) = \frac{1}{2}B\tanh(Bt + l) - \frac{1}{2}b$$

and

$$m(t) = \frac{\cosh l}{\cosh(Bt + l)}\left(m(0) + \frac{a}{B\cosh l}(\sinh(Bt + l) - \sinh l)\right).$$

The population hazard is again $\mu(t) = m(t)^2 + \gamma(t)$. Figure 11.2 shows $m(t)$, $\gamma(t)$, and $\mu(t)$ when $a = 1$, $b = 1$, and $\sigma^2 = 1$. Interestingly, even though $m(t)$ and $\gamma(t)$ behave quite differently for the Ornstein-Uhlenbeck process than for the simple Wiener process example, the resulting population hazards in Figure 11.2 are seen to be almost identical. This again demonstrates the difficulty of separating different underlying mechanisms from observations of the population rates alone. In this example, the fluctuations of the Wiener process and the Ornstein-Uhlenbeck process are so similar to begin with that μ reaches a steady state before the difference becomes obvious. So the steadily increasing sizes of the peaks of the Wiener model do not become visible in the population hazard. Furthermore, already at time 4 there are only a few percent of the original population left.

11.2.3 More general diffusion processes

For a brief moment, consider a general diffusion process X defined by the equation

$$dX(t) = \mu(X(t))dt + \sigma(X(t))dW(t),$$

as we did in (10.9). In Section 10.4 we showed how to derive general results about X in the context of first passage times, that is, when time to event was the boundary hitting time of X. In particular, the density $\psi_t(x)$ of $X(t)$ is a solution to the forward Kolmogorov Equation (10.10). This is true even though ψ is a "defective" density in the sense that $\int_0^\infty \psi_t(x)dx < 1$ when $t > 0$, since there is a positive probability that it is already absorbed in zero at time $t > 0$, and the density only covers the part of the distribution where $x > 0$. The absorption is accounted for by the boundary condition $\psi_t(0) = 0$ for all $t \ge 0$.

We now assume that there is no absorbing boundary; rather, the events happen as described in (11.1). In Appendix A.4.5, we explain how the resulting survival function solves the Kolmogorov Equation (A.36), with an additional term to account for the event rates. Similar equations for ψ can be derived. Such equations may be used as starting points for analyzing properties of the model, in particular, equations for the quasi-stationary distributions may be derived. The squared Ornstein-Uhlenbeck process in 11.2.2 is an example in which particularly tractable solutions may be found. We will not delve further into diffusion process models here; refer to Allen (2003) and Øksendal (1998) for general information. In the following sections, we turn our attention to Lévy process models for dynamic hazard.

11.3 Models based on Lévy processes

Although the relation between individual and population survival is relatively simple, actually computing it in a specific model may not be. An implementation of frailty as a stochastic process needs to combine the flexibility of including relatively complex processes together with ease of computation. A framework that goes some way to answer both needs is in terms of Lévy processes, in particular nonnegative Lévy processes, known as subordinators. The use of subordinators as frailty models has been explored in Kebir (1991) and Gjessing et al. (2003). The topic is also discussed in Aalen and Hjort (2002), Hjort (2003), and Singpurwalla (1995). Closely related models are analyzed in Lee and Whitmore (1993), Singpurwalla (1995), and Hougaard et al. (1997). An interesting aspect of the Lévy approach is that many standard frailty models, such as PVF distribution models (see Section 6.2.3), are special cases in this framework. The Lévy process based models extend in a natural way the standard frailty distribution at time zero to a process developing over time. At any point in time, the frailty may be within the same distribution class, such as PVF distributions. However, the correlation between frailty at two different time points typically decreases with the distance between the points, at a rate partly determined by covariates in the model.

The models introduced in the following lead typically (but not necessarily) to parametric models, a collection of which can be given explicit formulations. A lot of work remains to be done, however, on how best to implement the models. We will give a few examples and suggestions.

We will start by reviewing some of the basic properties of Lévy processes. Rather than going into the more technical aspects of this material, we will focus on examples that will serve as a basis for our models. The structure of Lévy processes (independent and stationary increments) is fundamental and has been utilized in numerous applications of stochastic processes. General introductions to Lévy processes are found in Bertoin (1996) and Applebaum (2004). Appendix A.5 contains a brief introduction covering some of the same aspects as the following subsection.

11.4 Lévy processes and subordinators

The class of Lévy processes encompasses a wide range of stochastic processes. Still, the actual definition is simple. Let $Z(t)$ be right-continuous with $Z(0) = 0$, and consider the increments $Z(t) - Z(s)$ for $t > s \geq 0$. Assume these increments all have the same distribution as $Z(t - s)$ and that they are independent, that is, if $t_4 > t_3 > t_2 > t_1$ then $Z(t_4) - Z(t_3)$ is independent of $Z(t_2) - Z(t_1)$. We then call Z a *Lévy process*. One can thus say that Z is a stationary increment process. A *subordinator* is simply a nonnegative Lévy process, making it suitable as a model for frailty. A consequence of the nonnegativity is that the increments are nonnegative. Except for trivial cases, Z will have positive jumps. Many interesting applications stem from a Z with "large" jumps, such as compound Poisson processes, but other useful examples appear when the jumps are tiny but extremely frequent, obtainable as limits of compound Poisson processes.

Since our models will be based on nonnegativity, in the following we will assume that Z is a subordinator, although some of the introductory results are also true for Lévy processes in general. We start by introducing some fundamental tools and a collection of basic examples.

11.4.1 Laplace exponent

It follows immediately from the definition that a subordinator is a Markov process. The behavior of a stochastic process is determined by its finite-dimensional distributions, that is, the distribution of $\{Z(t_1), Z(t_2), \ldots, Z(t_k)\}$ for any selection of time points $t_1, \ldots, t_k$. In fact, all finite-dimensional distributions of Z can be computed from the the the one-dimensional distribution of $Z(t)$; see Exercise 11.1. In addition, in Chapter 6 we saw that the Laplace transform of the frailty variable played a central role in deriving expressions for hazard and survival functions. This is also the case for the more general Lévy process models. For this reason, it is extremely convenient to have a simple representation of the Laplace transform of $Z(t)$ at each time t. (For general Lévy processes, the characteristic function is often used, but for subordinators we can focus on the Laplace transform.) Denote the Laplace transform by $\mathscr{L}$. We then have

$$\mathscr{L}(c;t) = E\exp\{-cZ(t)\} = \exp\{-t\Phi(c)\},\qquad(11.6)$$

where $c \geq 0$ is the argument of the Laplace transform. Observe that the exponent in the representation is linear in t. This is a manifestation of the time-homogeneity of the increments of Z; see Exercise 11.2. The function $\Phi(c)$ is called the *Laplace exponent* of Z and is essential to the models in the remainder of this chapter. Various subordinators are defined by explicit formulas for Φ. In general, Φ is increasing and concave (its derivative Φ' is decreasing) and $\Phi(0) = 0$.

By differentiating inside the expectation with respect to c in (11.6), it is seen that

$$EZ(1) = \Phi'(0) \quad and \quad \text{Var}Z(1) = -\Phi''(0),$$

as long as $\Phi'(0)$ and $\Phi''(0)$ exist and are finite. Even though $\Phi'(c)$ and $\Phi''(c)$ always exist for $c > 0$, the derivatives may not exist when $c = 0$, and the expectation and variance of Z may thus be infinite. When the moments *do* exist, it is seen that

$$EZ(t) = tEZ(1),\qquad(11.7)$$
$$\text{Var}Z(t) = t\text{Var}Z(1),\qquad(11.8)$$

so that both the expectation and the variance increase as linear functions of t. For other details, see Exercise 11.4.

A completely trivial example of a subordinator is the deterministic process $Z(t) = \kappa t$, where $\kappa \geq 0$ is a constant. Obviously, $\Phi(c) = \kappa c$, and $EZ(t) = \kappa t$, $\text{Var}Z(t) = 0$ in accordance with the above formulas. (Note that if κ were random, the process would not have independent increments and would not be a subordinator.) However, the typical subordinator will exhibit jumps. The following subsections introduce some important examples of subordinators.

11.4.2 Compound Poisson processes and the PVF process

Perhaps the best known nontrivial example of a subordinator is a Poisson process N with intensity ρ. Since $N(t)$ has a Poisson distribution with parameter ρt, it follows that the Laplace transform is

$$\mathscr{L}_N(c;t) = \exp\{-t\rho(1-e^{-c})\},\qquad(11.9)$$

thus $\Phi(c) = \rho(1-e^{-c})$, $EN(t) = \rho t$, and $\text{Var}N(t) = \rho t$.

Now, let $X_1, X_2, \ldots$ be independent positive random variables with common probability distribution $P(X \in A) = p_X(A)$, $A \subset (0,\infty)$, not necessarily absolutely continuous. Define

$$Z(t) = \sum_{i=1}^{N(t)} X_i,$$

with the convention that the sum is zero when $N(t) = 0$. Thus, with the ith jump of $N(t)$ we associate the random variable X_i, and $Z(t)$ sums up all the X_is accumulated up to (and including) time t, or, in other words, $Z(t)$ is a *compound Poisson process* with jumps of sizes $X_1, X_2, \ldots$. The simple form (11.9) of the Laplace transform for the plain Poisson process carries over to the compound Poisson processes. In fact,

$$\mathscr{L}(c;t) = \exp\{-t\rho(1 - \mathscr{L}_X(c))\}, \tag{11.10}$$

where $\mathscr{L}_X(c)$ is the Laplace transform of X; see Exercise 11.3 for details. Consult Kingman (1992) for more information on compound Poisson processes. Thus, $\Phi(c) = \rho(1 - \mathscr{L}_X(c))$. It is readily seen that $EZ(t) = E[X]\rho t$ and $\mathrm{Var}Z(t) = (\mathrm{Var}X + (EX)^2)\rho t$, provided these quantities exist. Furthermore, its special form implies that Φ is bounded. In fact, *Φ is bounded if and only if Z is compound Poisson* (Bertoin, 1996). The compound Poisson distribution was introduced as a frailty distribution in Section 6.2.3.

A tractable special case of the compound Poisson process occurs when the jumps are gamma distributed with shape parameter $m > 0$ and scale parameter $v > 0$. In this case, $\mathscr{L}_X(c) = (v/(v+c))^m$, and

$$\mathscr{L}(c;t) = \exp\left\{-\rho t \left[1 - \left(\frac{v}{v+c}\right)^m\right]\right\}.$$

We will refer to this as the *standard compound Poisson process*. Figure 11.4 shows a typical sample path of the process.

In fact, the restriction $m > 0$ can be relaxed to $m > -1$, with the additional restriction $m\rho > 0$. The process then ceases to be compound Poisson (note that Φ is no longer bounded), but it is still a subordinator. We will refer to this as a *PVF (power variance function) process*, and at any time t the process $Z(t)$ has a PVF distribution. The PVF distribution was introduced as a frailty distribution in Section 6.2.3; see also Hougaard (2000) for more information. Apart from embracing the compound Poisson processes, the PVF class also includes the *inverse Gaussian process* (Applebaum, 2004; Barndorff-Nielsen, 1998) when $m = -1/2$; see Figure 11.5. Two borderline cases (described in more detail later) occur when $m = 0$ (the gamma process) and when $v = 0$ (the stable process), although this is not readily seen from the expression for $\mathscr{L}$. The PVF processes will figure prominently as examples in this chapter.

11.4.3 Other examples of subordinators

The *gamma process* is a subordinator with

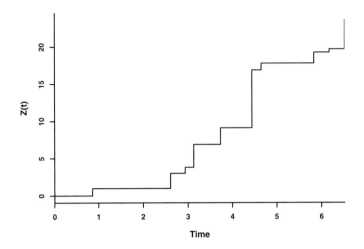

Fig. 11.4 *A sample path of the compound Poisson process with gamma distributed jumps. The intensity is $\rho = 1$; the gamma parameters are $m = 2$ and $v = 1$.*

$$\Phi(c) = \eta\{\log(v+c) - \log v\}, \quad \eta, v > 0.$$

It follows that $\mathcal{L}(c;t) = (v/(v+c))^{\eta t}$, and $Z(t)$ has a gamma distribution with shape parameter ηt and scale parameter v, hence the name. The gamma process can be obtained as a limit of the standard compound Poisson process by letting m (the shape parameter of the gamma jumps) go to zero and at the same time increase the rate ρ of the Poisson process in such a way that ρm converges to η; see Section 6.2.3.

Another important type of process is the *stable process*. The Laplace exponent is $\Phi(c) = ac^\beta$ with $a > 0$, $\beta \in (0,1)$. The corresponding Laplace transform is $\mathcal{L}(c;t) = \exp(-tac^\beta)$, which is the Laplace transform of a stable distribution; see Section 6.2.3. The stable processes can be obtained from the PVF processes by letting $v \to 0$ and $\rho \to -\infty$ in such a way that $-v^m \rho \to a$ (and keeping $m = -\beta$ constant).

11.4.4 Lévy measure

The general compound Poisson process is the archetypal subordinator. The jumps arrive according to a Poisson process, and the jump size follows a general distribution. This results in the simple relation between the Laplace transform $\mathcal{L}(c;t)$ of the process $Z(t)$ and the Laplace transform $\mathcal{L}_X(c)$ of the jump distribution as seen

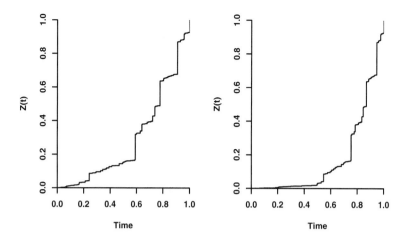

Fig. 11.5 *A sample path of the inverse Gaussian process with a constant rate (left) and with a rate that increases five-fold after time 0.5 (right).*

in (11.10). In fact, the Laplace exponent can be written

$$\Phi(c) = \rho(1 - \mathscr{L}_X(c)) = \rho \int_0^\infty (1 - e^{-cx}) p_X(dx).$$

This relation extends directly to *all* subordinators. For any Laplace exponent Φ, there is a corresponding measure Π on $(0, \infty)$ such that

$$\Phi(c) = \eth c + \int_0^\infty (1 - e^{-cx}) \Pi(dx). \qquad (11.11)$$

The measure Π is the *Lévy measure* of the subordinator Z. For the compound Poisson process, we see that $\Pi(dx) = \rho p_X(dx)$ is just the probability distribution of the jumps, multiplied by the intensity ρ of the Poisson process. The parameter $\eth$ is called the *drift coefficient* of the subordinator since it corresponds to the $\eth t$ part of a subordinator $Z(t) = \eth t + Z_0(t)$ when Z_0 is a subordinator with drift zero. For the compound Poisson process, $\eth = 0$. The crucial restriction of the compound Poisson process is that the measure Π is finite, that is, $\int_0^\infty \Pi(dx) < \infty$. This leads to a bounded Φ, which is equivalent to Z being compound Poisson. To cover all possible subordinators, Π is allowed to be infinite, with the restriction that $\int_0^\infty (1 \wedge x) \Pi(dx) < \infty$, thus the upper tail of Π must still be integrable, but the lower tail does not need to be. Note that this ensures that the integral on the right-hand side of (11.11) is finite for all c. From relations (11.7) and (11.8), it is seen that

$$EZ(t) = t(\eth + \int_0^\infty x \Pi(dx)),$$

$$VarZ(t) = t \int_0^\infty x^2 \Pi(dx),$$

so that the moments of Z are directly related to the moments of Π. For a general subordinator, the measure Π still has an interpretation as a jump size distribution.

For the standard compound Poisson process, Π has the gamma density

$$\Pi(dx) = \rho \frac{v^m}{\Gamma(m)} x^{m-1} e^{-vx} dx = m\rho \frac{v^m}{\Gamma(m+1)} x^{m-1} e^{-vx} dx, \tag{11.12}$$

where the last form extends naturally when $0 > m > -1$, that is, to the general PVF situation (recall the restriction $m\rho > 0$). For the gamma process, the Lévy measure is $\Pi(dx) = \eta e^{-vx}/x \, dx$, which is now clearly seen to be a "borderline" case for the compound Poisson when $m \to 0$, $m\rho \to \eta$. For the stable process, $\Pi(dx) = (\beta a/\Gamma(1-\beta)) x^{-1-\beta} dx$, and again this is easily obtained by letting $v \to 0$, $-v^m \rho \to a$, and $m = -\beta$ in the PVF process.

11.5 A Lévy process model for the hazard

A subordinator Z is in itself a valid model for the random development of the frailty of an individual. However, it has a limited flexibility because of its increasing nature and because of the stationarity of its increments. To allow for more realistic models we use the subordinator as a building block for a class of hazard process models. Let $r(t) \geq 0$ be a deterministic function and let $R(t) = \int_0^t r(u) \, du$. Then $(Z \circ R)(t) = Z(R(t))$ has independent and nonnegative, but no longer stationary, increments. The function $R(t)$ works as a deterministic time change, which enables us to speed up or slow down the process at will. For instance, we could choose $r(t) \equiv r_1$ when $t \leq t_0$ and $r(t) \equiv r_2$ when $t > t_0$, where t_0 is the time of an event thought to speed up or slow down the underlying hazard development for an individual. Figure 11.5 shows the effect of speeding up time in an inverse Gaussian process after a point in time. This effect would be a natural way to implement time-varying covariates in the model. Notice that the intensity parameter ρ now becomes redundant since it can be absorbed in r, but we will retain it for some applications.

The time-transformed process $Z \circ R$ is still somewhat restricted as a frailty model. If $Z \circ R$ were to describe the accumulated environmental exposure of an individual, for instance exposure to a toxic substance, one would think that exposures earlier in life might have a different influence on current hazard of developing a disease than more recent exposures. This could go either way. Early exposures could set off a process leading almost certainly to disease in later years, or in other cases the effect of early exposures could, over time, become negligible compared to recent exposures. To incorporate this in the model, we include a (deterministic) weighting function $g(t,v)$ and a deterministic hazard function $\alpha(t)$. Our model for the individual hazard is then introduced as

$$h(t) = \alpha(t) \int_0^t g(s, t - s) \, dZ(R(s)). \tag{11.13}$$

Clearly, the deterministic hazard $\alpha(t)$ is redundant since it could be absorbed in g, but we prefer to keep the direct dependence on t explicit. This model for h allows great flexibility, through choosing the time transformation R, the weighting g, and the deterministic part α, in addition to the type of subordinator Z.

The function g allows for two different time scales to be handled. In the integral, the process $Z \circ R$ contributes $dZ(R(s))$ at time s. This is weighted by $g(s, t - s)$, which depends on the *time of contribution* (first argument) and *time elapsed since the contribution* (second argument). For instance, a constant $g(s, t - s) \equiv g$ is the simplest example of dependence on past events. All contributions from $Z \circ R$ are weighted equally, regardless of when they occur, and the resulting hazard is $h(t) = g\alpha(t)Z(R(t))$. If, in addition, α is constant, $h(t) = g\alpha Z(R(t))$ is increasing (by jumps) for all individuals. This is the consequence of adding contributions to the hazard along the way, a cumulation of exposure. On the other hand, if, for instance, $g(s, t - s) = g(t - s)$ only depends on the second argument, the behavior may be different. Define $G(t) = \int_0^t g(s) \, ds$. We later show that the behavior of h depends crucially on the value of $G(\infty)$. When $G(\infty) < \infty$, the cumulated effect of previous contributions converges, and the hazard process will typically enter a stationary phase. When $G(\infty) = \infty$, the cumulation continues indefinitely, and the hazard will typically go to infinity. Yet another example would be if $g(s, t - s) = g1_{\{s \le t_0\}}$ for a fixed time t_0. Then only the actual time of the contribution matters, and only if it occurs before time t_0.

It is instructive to consider these cases in light of the general formulas for the expectation and variance of h,

$$E[h(t)] = \alpha(t)E[Z(1)] \int_0^t g(s, t - s) r(s) \, ds, \tag{11.14}$$

$$\text{Var}[h(t)] = \alpha(t)^2 \text{Var}[Z(1)] \int_0^t g(s, t - s)^2 r(s) \, ds. \tag{11.15}$$

(A corresponding formula for the correlation $\text{Cor}(h(t), h(t + u))$ is given in Exercise 11.4) When α and g are constant we have

$$E[h(t)] = \alpha g E[Z(1)]R(t) \quad \text{and} \quad \text{Var}[h(t)] = \alpha^2 g^2 \text{Var}[Z(1)]R(t),$$

again demonstrating the increasing nature of h.

If, for instance, $g(s, t - s) = g \exp(-\kappa(t - s))$, $\kappa > 0$, and r and α are constant, both $E[h(t)]$ and $\text{Var}[h(t)]$ converge to finite values as $t \to \infty$, suggesting the approach toward stationarity. In this case, applying the formula in Exercise 11.4 yields

$$\lim_{t \to \infty} \text{Cor}(h(t), h(t + u)) = e^{-\kappa u}, \quad u \ge 0.$$

This is analogous to the continuous Ornstein–Uhlenbeck process discussed in Section 10.4 and Appendix A.4, and is a typical example of a "moving average" process. Moving averages will appear frequently in the rest of this chapter.

The preceding discussion applies to the behavior of the individual hazard only if it could be observed indefinitely (regardless of whether an event occurred) but is still useful to keep in mind when choosing the appropriate model. As discussed in Section 11.1, the observed population hazard μ at any time will be computed only from the individuals still alive. We thus need to derive a formula for μ from model (11.13).

11.5.1 Population survival

Equations (11.2) and (11.3) can now be applied to our model (11.13) to derive the formulas for population survival and hazard. First define

$$b(u,t) = \int_u^t \alpha(s)g(u,s-u)\,ds.$$

Note that $\int_0^t h(s)\,ds = \int_0^t b(u,t)\,dZ(R(u))$ by changing the order of integration. Entering this into (11.2) and using the independent increments property, we obtain

$$S(t) = \exp\left\{-\int_0^t \Phi(b(u,t))r(u)\,du\right\}. \tag{11.16}$$

[For more details, see Exercise 11.5 or Gjessing et al. (2003).] The cumulative population hazard is thus $\int_0^t \Phi(b(u,t))r(u)\,du$, and by an appropriate differentiation of this we arrive at

$$\mu(t) = \alpha(t)\int_0^t \Phi'(b(u,t))g(u,t-u)r(u)\,du \tag{11.17}$$

for the population hazard.

11.5.2 The distribution of h conditional on no event

Until now, our focus has been on $\mu(t) = \mathrm{E}[h(t)|T > t]$, which is the expectation in the hazard distribution of survivors. A natural next step would be to find an expression for the variance of h, conditional on $T > t$. Even further, we could wish to find the entire distribution, for instance through computing the conditional Laplace transform of $h(t)$. Starting with the most general part, we can prove (see Exercise 11.6) that

$$\mathscr{L}^{\text{surv}}(c;t) \overset{\text{def}}{=} E\left[e^{-ch(t)}|T>t\right] \tag{11.18}$$

$$= \exp\left\{-\int_0^t [\Phi(c\alpha(t)g(s,t-s)+b(s,t)) - \Phi(b(s,t))]\, r(s)\, ds\right\}.$$

Indeed, this enables us in principle to compute the conditional distribution of $h(t)$ at any time by inverting (possibly numerically) the Laplace transform. By differentiating the Laplace transform, the variance of the hazard of survivors can be derived:

$$\text{Var}[h(t)|T>t] = -\alpha(t)^2 \int_0^t \Phi''(b(s,t))g(s,t-s)^2 r(s)\, ds.$$

We now present some cases that are of special interest.

11.5.3 Standard frailty models

The frailty models discussed in Chapter 6 can easily be expressed in the Lévy framework. Assume that $g(s,t-s) \equiv 1$, let $r(t) = 1_{(t\leq 1)}$, that is, it is equal to 1 until time 1 and equal to zero thereafter. Let $\alpha(t)$ be equal to zero before time 1. We see that $h(t) = \alpha(t)Z(R(t)) = \alpha(t)Z(1)$ for all $t \geq 0$. It follows that after time 1 (which can be chosen as the new starting point of the time axis), the model has all the features of a standard frailty model. As seen earlier, $Z(1)$ may follow most of the standard frailty distributions, including gamma, PVF, and compound Poisson.

Note that the formula for μ reduces to $\mu(t) = \alpha(t)\Phi'(A(t))$, which coincides with formula (6.4). It is also seen that since $b(u,t)$ depends on α, a proportionality in the baseline individual hazards $\alpha(t)$ will not lead to a corresponding proportionality in the population hazard $\mu(t)$, as discussed in Section 6.5 for the standard frailty case. For the basic process obtained when α, g, and r are all constant, there is the simple expression $\mu(t) = r\Phi(\alpha gt)$. This is a special case of the following tractable models.

11.5.4 Moving average

Assume $\alpha(t) \equiv r(t) \equiv 1$. We have seen that the second argument in the weight function $g(s,t-s)$ represents time elapsed since $dZ(s)$ made a contribution to the hazard $h(t)$. When $g(s,t-s) = g(t-s)$ depends only on this argument,

$$h(t) = \int_0^t g(t-s)\, dZ(s)$$

is a moving average of the increments of Z from time 0 to t. In the following we refer to this as the *moving average model*. For instance, if $Z(t) = \sum_{i=1}^{N(t)} X_i$ is compound

Poisson, then $h(t) = \sum_{i=1}^{N(t)} g(t - T_i)X_i$, where $T_1, T_2, \ldots$ are the jump times of $N(t)$. We also notice that if $g(v) = 0$ when $v \geq v_0$ then $h(t)$ becomes stationary when $t \geq v_0$, and more generally if $G(\infty) = \int_0^\infty g(s)\, ds < \infty$ then $h(t)$ approaches a stationary distribution as $t \to \infty$. Otherwise, $h(t)$ keeps accumulating contributions from dZ. Figure 11.6 shows the process h resulting from an exponentially decaying g combined with a compound Poisson process Z (this is related to "shot-noise" processes; see, for instance, Singpurwalla and Youngren (1993)). Similarly, Figure 11.7

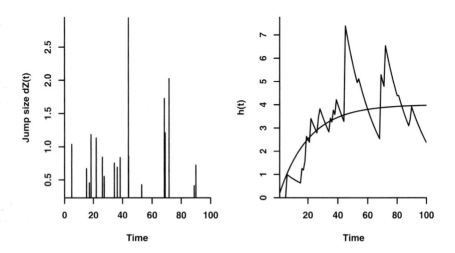

Fig. 11.6 *Left: a Poisson point process (jump times and jump sizes) of* Z(t). *Right: the resulting individual hazard* h(t) *(jagged curve), with its expected value* Eh(t) *(smooth curve).*

shows the same g but combined with the inverse Gaussian process shown in Figure 11.5.

For the moving average, formulas (11.16) and (11.17) simplify to

$$S(t) = \exp\left\{ -\int_0^t \Phi(G(u))\, du \right\}$$

and

$$\mu(t) = \Phi(G(t)).$$

It is seen that μ is increasing. In addition, a decreasing g will lead to a concave μ. Two natural examples would be the exponentially decaying influence, that is,

$$g_1(x) = \exp(-\kappa x), \quad G_1(t) = \frac{1}{\kappa}(1 - \exp(-\kappa t)),$$

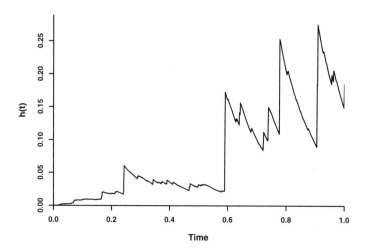

Fig. 11.7 *An individual hazard process h when g is exponentially decaying and Z is the inverse Gaussian process from Figure 11.5.*

and the more slowly decreasing

$$g_2(x) = \frac{\kappa x}{1 + (\kappa x)^2}, \quad G_2(t) = \frac{1}{2} \log(1 - (\kappa t)^2).$$

Figure 11.8 demonstrates some of the shapes that are possible for μ when these examples are used with the PVF process (11.19).

An important additional feature of μ is that it will frequently reach a limit, $\lim_{t \to \infty} \mu(t) = \Phi(G(\infty)) < \infty$. In our two examples, it is seen that if Φ is bounded this limit will always exist, whereas if Φ is unbounded this limit will still exist if G is bounded, that is, if g decreases fast enough. This will be discussed further in Section 11.8.

11.5.5 Accelerated failure times

Let S_0 and S_1 be the survival functions of two groups of individuals. When comparing the two groups, the accelerated failure time model specifies that $S_1(t) = S_0(t/w)$ for all t, for some constant w. In other words, $T_1 = wT_0$, where T_0 and T_1 are the survival times for individuals in group 0 and 1, respectively, and the equality is taken in the sense of distributions. The corresponding transformation is $\mu_1(t) = \mu_0(t/w)/w$ on the hazard level. Using model (11.16) for the survival, the accelerated failure

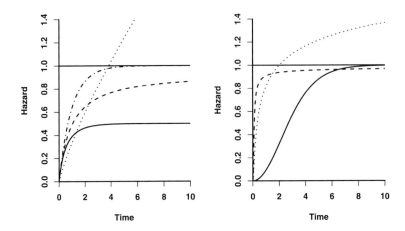

Fig. 11.8 *Population hazard $\mu(t)$ under different variants of a moving average of a PVF process. Left: when g_1 is combined with Φ_{PVF}, $\mu(t)$ will level off unless both $m < 0$ and $\kappa < 0$. Right: when g_2 is combined with Φ_{PVF}, the hazard will level off if and only if $m < 0$.*

time model is readily incorporated. Define $\alpha_w(t) = \alpha(t/w)/w$, $r_w(t) = r(t/w)/w$, and $g_w(t,v) = g(t/w,v/w)$. By change of variable,

$$S_1(t) = \exp\left\{ -\int_0^t \Phi(b_w(u,t))r_w(u)\,du \right\},$$

where b_w is computed as b, but using α_w and g_w instead of α and g. Thus, the constant w can be absorbed in the design functions of the Lévy model, not particularly surprising due to the overall flexibility of the model.

More generally, we may compare an individual i with covariate vector $\mathbf{x}_i$ to the reference group by writing $S_i(t) = S_0(t/w_i)$, where, for instance, $w_i = \exp(\beta^T \mathbf{x}_i)$, and again w_i can be incorporated in the model.

11.6 Results for the PVF processes

Recall that the PVF Lévy process is defined by the Laplace exponent

$$\Phi_{PVF}(c;\rho,v,m) = \rho\left\{ 1 - \left(\frac{v}{v+c} \right)^m \right\} \tag{11.19}$$

with the restrictions $v > 0$, $m > -1$, and $m\rho > 0$. Similarly to the standard compound Poisson process, many computations regarding the PVF processes simplify considerably due to the convenient form of Φ_{PVF}. Since the PVF processes extend the class of compound Poisson processes to include processes with densely spaced jumps, we will use PVF processes as a basis for some of our explicit models. Before specifying the model in more detail, note that since $v/(v+c) = 1/(1+c/v)$, we can conclude from formula (11.16) that the constant v can be absorbed in either α or g. For the PVF models we will assume $v = 1$.

Using the Weibull parametric model as our starting point, we let $\alpha(t) = t^{k-1}$ for some $k \geq 1$. Choosing $r(t) \equiv 1$ and $g(t, v) \equiv g$ constant, we can model

$$h(t) = gt^{k-1}Z(t)$$

as a random perturbation of the Weibull model. It is clear that h is increasing. According to formulas (11.14) and (11.15), we find that

$$Eh(t) = gMt^k \quad \text{and} \quad \text{Var}[h(t)] = g^2\tau^2 t^{2k-1}, \tag{11.20}$$

with $M = EZ(1) = m\rho$ and $\tau^2 = \text{Var}[Z(1)] = m(m+1)\rho$. Regardless of k, it is clear that $\text{Var}[h(t)] \to \infty$ as $t \to \infty$. These formulas are exemplified and interpreted in Example 11.1.

11.6.1 Distribution of survivors for the PVF processes

To study the resulting population hazard μ, we introduce the hypergeometric function $H(\alpha, \beta, \gamma; z)$. The hypergeometric function, often denoted as $_2F_1$, appears frequently in the literature as the solution to second-order differential equations (Lebedev, 1972; Polyanin and Zaitsev, 1995). Practical use of such functions has been made much simpler since they are now included in several software systems, in particular Mathematica (Wolfram, 1999). An integration of formula (11.17) using Mathematica leads to the expression

$$\mu(t) = m\rho k\gamma(t)^{-m}\left(\frac{\gamma(t)-1}{\gamma(t)}\right)H\left(\frac{1}{k}, m+1, \frac{k+1}{k}, \frac{\gamma(t)-1}{\gamma(t)}\right), \tag{11.21}$$

where

$$\gamma(t) = 1 + \frac{g}{k}t^k.$$

The corresponding survival function turns out to be

$$S(t) = \exp\left\{-\rho t\left(1 - \gamma(t)^{-m}H(\frac{1}{k}, m, \frac{k+1}{k}, \frac{\gamma(t)-1}{\gamma(t)})\right)\right\}. \tag{11.22}$$

For the conditional variance we get

$$\mathrm{Var}[h(t)|T > t] = m(m+1)\rho k^2 \frac{1}{t} \gamma(t)^{-m}$$

$$\times \left(\frac{\gamma(t)-1}{\gamma(t)} \right)^2 H \left(\frac{1}{k}, m+2, \frac{k+1}{k}, \frac{\gamma(t)-1}{\gamma(t)} \right),$$

and a computation of the conditional Laplace transform yields

$$\mathscr{L}^{\mathrm{surv}}(c;t) = \mathrm{E}[e^{-ch(t)}|T > t]$$

$$= \exp \left[-\rho t \left\{ \gamma(t)^{-m} H \left(\frac{1}{k}, m, \frac{k+1}{k}, \frac{\gamma(t)-1}{\gamma(t)} \right) \right. \right.$$

$$\left. \left. -(\gamma(t)+cgt^{k-1})^{-m} H \left(\frac{1}{k}, m, \frac{k+1}{k}, \frac{\gamma(t)-1}{\gamma(t)+cgt^{k-1}} \right) \right\} \right].$$

Figure 11.9 presents a typical selection of population hazard curves $\mu(t)$ (upper row), together with the resulting lifetime densities (lower row). The graphs are generated by individual hazards with $g = 1/2$ and $\alpha(t) = t^2$ but varying values of m and ρ, that is, $h(t) = 1/2t^2 Z(t)$.

The most prominent feature of the hazard is that the compound Poisson case ($m = 1$) leads to convergence, whereas the case $m = -1/2$ leads to a hazard going to infinity. However, there is an important distinction between the case where $|\rho| = 5$ and $|\rho| = 0.25$. In the former case, the hazard increases rapidly and reaches a high level already before time $t = 2$, and the consequence is that within time 2 almost all individuals have had an event. Since the hazards for $m = -1/2$ and $m = 1$ are fairly similar over this time interval, the corresponding event time densities are very close in shape (Figure 11.9). In the latter case, however, even though the shape of the hazard curves is the same, the overall level of hazard is much lower. The consequence is that the densities for $m = -1/2$ and $m = 1$ differ greatly, in particular $m = 1$ results in a much longer right-hand tail of the event time density, and this tail will resemble an exponential distribution since the hazard is nearly constant.

Seeing the behavior of $\mu(t)$ as $t \to \infty$, it is interesting to see how much can be learned about the distribution of $h(t)$ conditional on no events, that is, conditional on $T > t$, particularly when $t \to \infty$. Section 11.8 will discuss some general results on this, after which we will return to the PVF processes.

11.6.2 Moving average and the PVF process

Letting $k = 1$ in the PVF model leads to particularly simple results. The hazard is just $h(t) = gZ(t)$, and since g is constant this can also be treated as a special case of the moving average model. In particular,

$$\mu(t) = \Phi_{\mathrm{PVF}}(G(t);\rho,\nu,m) = \Phi_{\mathrm{PVF}}(gt;\rho,1,m) = \Phi_{\mathrm{PVF}}(t;\rho,1/g,m)$$

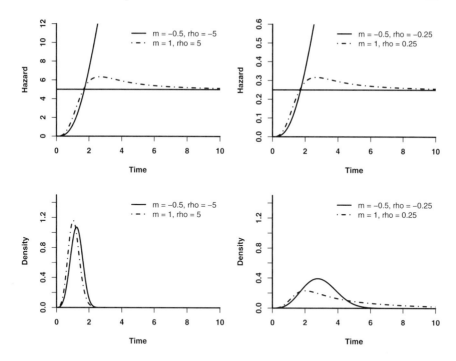

Fig. 11.9 *Population hazard rate (upper row) and their corresponding lifetime densities (lower row). The hazards are generated by the individual hazard $h(t) = \frac{1}{2}t^2 Z(t)$, where $Z(t)$ is PVF with $v = 1$ and varying values of m and ρ. Note the different hazard scales in the upper row.*

and

$$\text{Var}[h(t)|T > t] = g\Phi'_{\text{PVF}}(0; \rho, v, m) - \mu'(t) = \Phi_{\text{PVF}}(t; gm\rho, 1/g, m+1).$$

The expression for the Laplace transform becomes

$$\mathscr{L}^{\text{surv}}(c; t) = \exp\left\{-\frac{\rho}{g(m-1)}\left(1 + \left(\frac{1}{1+g(t+c)}\right)^{m-1}\right.\right.$$
$$\left.\left. -\left(\frac{1}{1+gc}\right)^{m-1} - \left(\frac{1}{1+gt}\right)^{m-1}\right)\right\}$$
$$= \frac{\exp\left\{-\Phi_{\text{PVF}}\left(c; \frac{\rho}{g(m-1)}, 1/g, m-1\right)\right\}}{\exp\left\{-\Phi_{\text{PVF}}\left(c; \frac{\rho}{g(m-1)}\left(\frac{1}{1+gt}\right)^{m-1}, 1/g+t, m-1\right)\right\}}, \tag{11.23}$$

which in the compound Poisson case $(m > 0)$ is a quotient of two PVF Laplace transforms.

11.7 Parameterization and estimation

The model for h introduced with (11.13) opens for a wide range of parameterizations and model specifications. Each of the functions α, g, and r have intuitive and natural interpretations in the interplay with the driving frailty process Z. However, in lieu of direct observation of the process Z, particularly without multiple observations per individual, the effects attributed to the different components of the model must be considered only reflections of the true underlying mechanisms causing events to occur. Nevertheless, when comparing groups, it may be valuable to identify what parts of the estimated model differ among groups. We will give an example of this below.

In 11.5.5 we gave an example of how covariates can be introduced in an accelerated failure time setting, and this fits nicely into the framework. The accelerated failure time property carries over naturally from the individual hazard level to the population hazard.

Considering the proportional hazard property, we see that proportionality on the individual level does not necessarily carry over to the population level. For instance, it might seem natural to let the baseline hazard $\alpha(t)$ depend on covariates in a multiplicative manner, using $\exp(\beta^T \mathbf{x})\alpha(t)$ for an individual with covariate $\mathbf{x}$. Since α is a part of b, formula (11.17) reveals that μ lacks the required proportionality. An interesting alternative may be to model the rate at which the frailty process $Z(R(t))$ progresses. Replacing $r(t)$ by $\exp(\beta^T \mathbf{x})r(t)$ leads to a frailty process $Z(\exp(\beta^T \mathbf{x})R(t))$, which is sped up or slowed down according to the values of $\mathbf{x}$. On the population level, this leads to proportional hazards. Note that the same effect could be achieved by substituting $\exp(\beta^T \mathbf{x})\Phi$ for Φ, since this also has the effect of changing the rate of Z.

For data with a single observation per individual, estimation proceeds by standard maximum likelihood. An event at time T_i for individual i contributes $\mu(T_i)S(T_i)$ to the likelihood, whereas a censoring at time $\widetilde{T}_j$ contributes $S(\widetilde{T}_j)$; see Section 5.1.

Example 11.1. Divorce in Norway. As an illustration of how Lévy process models can be applied, we use the PVF model to analyze the divorce rate data introduced in Example 1.2 and further discussed in Example 5.4. The data given in Table 5.2 contain a follow-up of marriages contracted in 1960, 1970 and 1980. For the 1960 and 1970 cohorts we used data for 26 years of follow-up, the 1980 cohort for 16 years. For each of the cohorts we know the number of individuals under risk at the beginning of each year and the number of divorces during that year. The data are given as closed cohorts, ignoring an (fairly moderate) amount of loss-to-follow-up due to emigration, death, etc. Comparing Figure 11.9 to the observed rates in Figure 1.4 suggests that we might try a model of the form

$$h(t) = gt^{k-1}Z(t)$$

where g is constant, $\alpha(t) = t^{k-1}$, and Z is a PVF process with parameters ρ, ν, and m. This is the model described in Section 11.6. We fit the model using parametric maximum likelihood. For simplicity, we assign an event time $T_i = i$ to all couples that divorce during year i. At the end of a follow-up period of κ years we assign a (censored) time of $\tilde{T} = \kappa + 1$ to all couples that "survive" through the last year κ. The parametric likelihood for a single cohort is thus (see Section 5.1)

$$\mathscr{L} = \prod_{i=1}^{\kappa} [\mu(T_i)S(T_i)]^{n_i} S(\tilde{T})^l,$$

where n_i is the number of divorces during year i, $i = 1, \ldots, \kappa$, and l is the number of couples still married at time $\kappa + 1$. The explicit expressions for μ and S given in (11.21) and (11.22) can be used for the numeric calculations if the hypergeometric functions are available in the software. Alternatively, numerical integration can be used in formulas (11.17) and (11.16).

We set $v = 1$ as before, and we estimate ρ, m, g, and k separately for each of the three years 1970, 1980, and 1990. The results are found in Table 11.1.

Table 11.1 *Estimated parameters for the divorce rates. The model for individual (couple) hazard is $h(t) = gt^{k-1}Z(t)$, where Z is a PVF process with parameters ρ, $v = 1$ and m.*

Parameter		1960	1970	1980
ρ	Est.	0.0064	0.011	0.017
	95%CI	(0.0062, 0.0067)	(0.011, 0.011)	(0.016, 0.018)
g	Est.	0.21	0.093	0.094
	95%CI	(0.15, 0.27)	(0.076, 0.11)	(0.075, 0.11)
m	Est.	0.92	1.0	0.63
	95%CI	(0.17, 1.7)	(0.52, 1.56)	(0.35, 0.92)
k	Est.	3.0	3.8	4.2
	95%CI	(2.3, 3.6)	(3.3, 4.3)	(3.6, 4.8)

The fitted rates are shown together with the observed rates in Figure 11.10. The fit appears excellent for the years 1970 and 1980. For 1960, although the fit is good, the slight increase after 15 years is not quite properly fitted. (In all fairness it should also be remarked that the 1960 rates start to decline at some point past 25 years, indicating that very long term marriages have their own dynamics.)

It is interesting to observe how the individual hazard processes develop. Indeed, assuming that we could observe all couples at all times (no events or censoring), the average hazard would be the polynomial function given in (11.20). As seen from the variance given in the same formula, the individual (couple) variability (standard deviation) in hazard of divorce also increases as a polynomial function over time. This is illustrated in Figure 11.11.

Thus, according to this model, most couples would have a fairly rapid increase in hazard of divorce over the years, but there would be much variability among couples, and over the years the marriages with the slowest increase in hazard would remain.

The example clearly illustrates the dangers of overinterpreting the shape of the population rates. The "obvious" interpretation of Figure 11.10 is that there is a

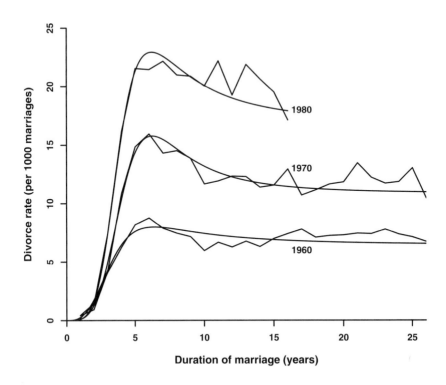

Fig. 11.10 *Divorce rates in Norway for marriages contracted in 1960, 1970 and 1980 (rough curves) and fitted population values from the model $h(t) = gt^{k-1}Z(t)$ (smooth curves). Based on data from Statistics Norway.*

seven-year crisis (Bergman et al., 1977); once this point in time has been passed, the rates start declining. Our model shows that this population rate could equally well be the visible consequence of hazard rates that are increasing for every couple and that show great couple-to-couple variability. However, in this specific example, there is a particular reason why the model produces a stable limiting hazard. We return to this in 11.8.1, after looking at more general limit results. □

11.8 Limit results and quasi-stationary distributions

As seen in Sections 11.5.4 and 11.6, the population hazard μ frequently approaches a limit as $t \to \infty$. It is natural to expect that this is related to a quasi-stationary distribution for $h(t)$ as $t \to \infty$. This would be similar to what is seen in Section 10.3. We

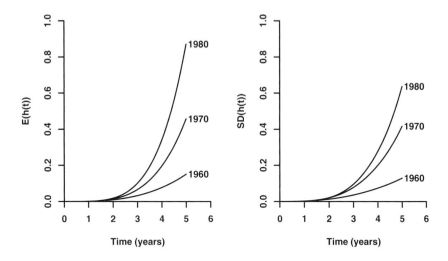

Fig. 11.11 *Average individual hazard of divorce for a fully observed (hypothetical) population cohort for the years 1960, 1970, and 1980 (left), and the corresponding standard deviations of the hazards (right).*

now look more closely at the reasons for this behavior in the context of the moving average model and closely related models. Recall that for the moving average model we have $\mu(t) = \Phi(G(t))$. Convergence of μ may happen for two reasons. First, if $G(\infty) < \infty$, h will approach a stationary state and it is intuitively clear that this also will lead to a quasi-stationary state for h, resulting in a converging $\mu(t)$ as $t \to \infty$. Second, on the other hand, when $G(t) \to \infty$ as $t \to \infty$, $\mu(t)$ will still stabilize if $\Phi(\infty) < \infty$. This is precisely the compound Poisson case, and the reason for the convergence is different. For any $t > 0$, the compound Poisson case admits a group of individuals for which $Z(t) = 0$, that is, the group for which no positive jump in risk has yet occurred. Since $h(s) = 0$ for all $s \leq t$, these individuals have not accumulated any risk and will thus still be alive. Using relation (11.6),

$$P(Z(t) = 0) = \lim_{c \to \infty} \mathrm{E}\exp(-cZ(t)) = \exp(-t\Phi(\infty)).$$

Hence, the group for which $Z(t) = 0$ will constitute a reservoir from which new individuals are "released" at a rate $\Phi(\infty)$. Once an individual is released, it will be under risk of an event. Thus, there is a "bottleneck" effect; eventually this release-rate will become the dominant force determining the limiting rate.

In addition to the convergence of μ, in both cases we will see a corresponding quasi-stationary distribution of h. By taking the limit in (11.18) we get

$$\lim_{t\to\infty} \mathscr{L}^{\text{surv}}(c;t) = \exp\left\{ -\int_0^\infty [\Phi(cg(v)+G(v)) - \Phi(G(v))]\, dv \right\}.$$

Alternatively, this can be written

$$\lim_{t\to\infty} \mathscr{L}^{\text{surv}}(c;t) = \exp\left\{ -(\eth G(\infty))c - \int_0^\infty \left(1 - e^{-cz}\right)\widetilde{\Pi}(dz) \right\}, \qquad (11.24)$$

where $\widetilde{\Pi}$ is the new Lévy measure

$$\widetilde{\Pi}(dz) = \int_{\{w:g(w)>0\}} \exp\left(-\frac{G(w)}{g(w)} z \right) \Pi\left(\frac{dz}{g(w)} \right) dw.$$

From the second expression, it is seen that the quasi-stationary distribution of h is the distribution of $\widetilde{Z}(1)$, where $\widetilde{Z}$ is a new subordinator with drift $\eth G(\infty)$ and Lévy measure $\widetilde{\Pi}$. The result can be extended to somewhat more general situations [see theorem 2 of Gjessing et al. (2003)].

11.8.1 Limits for the PVF process

Even though g is constant (and thus a function "only" of the second argument), our PVF example does not strictly fit into the general limit results discussed for moving average processes, since $\alpha(t) = t^{k-1}$ is not constant unless $k = 1$. In Section 11.6 we saw that this resulted in "exploding" hazard rates in the noncompound Poisson cases ($k = 3$, $m = -0.5$.) Nevertheless, in the compound Poisson case the bottleneck phenomenon again manifests itself, causing the hazard rate to converge even when $k > 1$. We will now look a bit more closely at our specific PVF example to see what happens when $t \to \infty$. By directly considering formula (11.21) it follows from the asymptotic properties of the hypergeometric functions (Lebedev, 1972) that in the compound Poisson case $m > 0$, $\mu(t) \to \rho$ as $t \to \infty$, regardless of k. Otherwise, when $0 > m > -1$, we have

$$\mu(t) \sim m\rho(\frac{k}{g})^m B(\frac{1}{k}, -m) t^{-mk} \to \infty \text{ as } t \to \infty,$$

where $B(x,y) = \int_0^1 t^{x-1}(1-t)^{y-1}\, dt$ is the beta function. *Thus, the parameter m alone determines whether μ converges.*

However, it is only in the case where $k = 1$ that the general results guarantee that a quasi-stationary *distribution* exists. Indeed, the variance behaves asymptotically as

$$\text{Var}[h(t)|T > t] \sim gm\rho\, t^{k-1}, \qquad t \to \infty,$$

indicating that an actual quasi-stationary distribution with finite variance is found only when $m > 0$ and $k = 1$. Taking the limit in (11.23) when $k = 1$ reveals, as expected, that a nondegenerate distribution

$$\lim_{t \to \infty} \mathscr{L}^{\text{surv}}(c;t) = \exp\left\{-\Phi_{\text{PVF}}(c; \frac{\rho}{g(m-1)}, 1/g, m-1)\right\}$$

occurs only when $m > 0$. In fact, it is the distribution of a PVF process with new parameters; in particular, m is replaced by $m-1$.

This last result can be verified very simply from the general Equation (11.24). Using expression (11.12) for the Lévy measure (density) of the PVF process h, the new Lévy measure is

$$\tilde{\Pi}(dz) = \int_0^\infty \exp\left(-\frac{gw}{g}z\right)\Pi\left(\frac{dz}{g}\right)dw$$

$$= m\rho \frac{1}{\Gamma(m+1)}\left(\frac{z}{g}\right)^{m-1}e^{-z/g}\frac{dz}{g}\int_0^\infty \exp(-wz)\,dw$$

$$= \frac{\rho}{g(m-1)}\frac{(1/g)^{m-1}}{\Gamma(m-1)}z^{m-2}e^{-z/g}\,dz.$$

Example 11.2. Divorce in Norway (continued). From Table 11.1 we see that for the divorce data, the estimated values of the PVF parameter m are positive for all three cohorts. As we have just argued, this leads to a converging μ, with limit ρ as $t \to \infty$. Comparing the estimates of ρ in Table 11.1 with the long-term hazard levels in Figure 11.10 this result is readily verified. If we are willing to accept the assumptions underlying our model, the conclusion is that there is a group of marriages that for a "long" time have zero hazard of breaking up, since $Z(t)$ has a positive probability of being zero at any time. After a while (some seven to ten years) almost all of the marriages that quickly developed a positive hazard have broken up. The remaining marriages are those that still have a (practically) zero probability of breaking up. Since $P(Z(t) = 0) = \exp(-t\Phi(\infty))$, these marriages enter a positive-risk group at a rate $\Phi(\infty)$. Since the Weibull part of the hazard, t^{k-1}, grows faster and faster, those marriages that get a positive risk will quickly dissolve, leaving only those that still have zero hazard, and resulting in the limiting rate of $\rho = \Phi(\infty)$.

As long as we do not observe any indicators of change in marriage quality over time, these interpretations are, of course, completely open to criticism. Still, they again illustrate how the same characteristic shapes of the population hazard arise from completely different assumptions on an individual level. $\square$

11.9 Exercises

11.1 Explain why the finite-dimensional distributions

$$(Z(t_1), Z(t_2), \ldots, Z(t_k))$$

of a Lévy process are completely determined by the distribution of $Z(t)$, that is, the one-dimensional distributions.

11.2 Let Z be a Lévy processes. Let

$$\mathscr{L}(c) = \mathrm{E}e^{-cZ(1)} = e^{-\Phi(c)}$$

be the Laplace transform of $Z(1)$. Prove that for any rational number $r \geq 0$, the Laplace transform of $Z(r)$ is $\exp(-r\Phi(c))$. Use this to prove the general expression for the Laplace transform of $Z(t)$ given in (11.6).

11.3 Let $N(t)$ be a Poisson process and let $Z(t) = \sum_{i=1}^{N(t)} X_i$ be a compound Poisson process, where $X_1, X_2, \ldots$ are *iid*. First, show that the conditional Laplace transform for Z given N (the number of jumps occurred up to time t) is

$$\mathrm{E}[e^{-cZ(t)}|N(t)] = (\mathscr{L}_X(c))^{N(t)},$$

where $\mathscr{L}_X$ is the Laplace transform of X. Use this to prove that the Laplace transform of Z is

$$\mathrm{E}e^{-cZ(t)} = \mathscr{L}_N(-\log(\mathscr{L}_X(c)); t),$$

where $\mathscr{L}_N(c; t)$ is the Laplace transform of $N(t)$, as given in (11.9).

11.4 Prove formulas (11.7) and (11.8). Prove that the autocorrelation of Z (the correlation of $Z(t)$ with $Z(t + u)$) is

$$\mathrm{Cor}(Z(t), Z(t+u)) = \sqrt{\frac{t}{t+u}}, \quad u \geq 0.$$

Explain (heuristically) why

$$\mathrm{E}[dZ(R(t))] = \mathrm{E}[Z(1)]r(t)dt$$

and

$$\mathrm{Var}[dZ(R(t))] = \mathrm{Var}[Z(1)]r(t)dt,$$

and use this to derive formulas (11.14) and (11.15). Furthermore, prove that

$$\mathrm{Cor}(h(t), h(t+u)) = \frac{\int_0^t g(s, t-s)g(s, t+u-s)r(s)\, ds}{\sqrt{\int_0^t g(s, t-s)^2 r(s)\, du \int_0^{t+u} g(s, t+u-s)^2 r(s)\, ds}}.$$

Note that this is independent of the Laplace exponent, that is, of what subordinator you use in the frailty formulation.

11.5 Argue that, for a suitably integrable function ψ,

$$E \exp\left(-\int_0^t \psi(s)\, dZ(s)\right) = \exp\left(-\int_0^t \Phi(\psi(s))\, ds\right),$$

using the independent increments property of Z. Extend this to an expression for $E \exp(-\int_0^t \psi(s)\, dZ(R(s)))$, and derive formula (11.16).

For a suitable function ψ, argue that

$$\frac{d}{dt}\int_0^t \psi(t,s)\, ds = \psi(t,t) + \int_0^t \frac{\partial}{\partial t}\psi(t,s)\, ds.$$

Use this to derive (11.17).

11.6 Prove formula (11.18).

Hint: Use the formula $E[X|A] = E[X 1_A]/P(A)$ for a random variable X with finite expectation and an event A with $P(A) > 0$, together with an argument similar to that of Exercise 11.5.

Appendix A
Markov processes and the product-integral

A main purpose of this book is to show how the theory of stochastic processes fits naturally into the more applied framework of event history analysis. In particular, the tools and ideas of martingale and Markov process theory are ubiquitous in an integrated approach. In spite of the close relationship, most presentations found in the literature do not acknowledge this; standard presentations of event history analysis rarely say much about Markov processes, and vice versa.

Our main text incorporates many applications of martingale theory and counting processes and this appendix will not say much about that. The fundamental connections between hazard, survival, Markov processes, the Kolmogorov equations, and the product-integral are also at the heart of many models covered in our book, but a full discussion is too much to include in the main text. Since these topics are hard to find as a comprehensive text, they are presented in this appendix. The results are particularly useful for understanding the connection between the Nelson-Aalen and Kaplan-Meier estimators (Chapter 3) and for multivariate survival data and competing risk (Section 3.4). This appendix deals mostly with the mathematical relationships; the parallel results used in estimation and asymptotics are in the main text. For extensions of multistate Markov models to semi-Markov models, the reader is referred to Huzurbazar (1999, 2005).

We also cover introductory ideas for diffusion processes, stochastic differential equations, and certain stationary processes with independent increments, known as Lévy processes. These models are used extensively in the final chapters of the book (Chapters 10 and 11).

The material in the appendix is meant to be self-contained, but a reader with no background in stochastic processes will find it useful (or even necessary!) to consult standard texts such as Karlin and Taylor (1975, 1981, 1998), Allen (2003), and Cox and Miller (1965).

A.1 Hazard, survival, and the product-integral

Let $T \geq 0$ be a random survival time with survival function $S(t) = P(T > t)$. It is common to assume that the survival function $S(t)$ is absolutely continuous, and let us do so for the moment. Let $f(t)$ be the density of T. The standard definition of the hazard rate $\alpha(t)$ of T is

$$\alpha(t) = \lim_{\Delta t \to 0} \frac{1}{\Delta t} P(t \leq T < t + \Delta t \mid T \geq t) = \frac{f(t)}{S(t)}, \tag{A.1}$$

where dt is "infinitesimally small", that is, the probability of something happening in the immediate future conditional on survival until time t. Then α is obtainable from S by

$$\alpha(t) = \frac{-S'(t)}{S(t)}. \tag{A.2}$$

Inversely, when α is available S solves the differential equation $S'(t) = -\alpha(t)S(t)$, which leads to

$$S(t) = \exp\left(-\int_0^t \alpha(s)ds\right) = \exp(-A(t)), \tag{A.3}$$

where $A(t) = \int_0^t \alpha(s)ds$ is the cumulative hazard rate. Note that T does not have to be finite; if $P(T = \infty) > 0$ then $\int_0^\infty f(s)ds < 1$ and $\int_0^\infty \alpha(s)ds < \infty$, which is sometimes referred to as a *defective* survival distribution.

We will now see how (A.2) and (A.3) may be generalized to arbitrary distributions, which need neither be absolutely continuous nor discrete. Such generalizations, in particular the *product-integral*, which generalizes (A.3), are useful for a number of reasons:

- It enables us to handle both discrete and continuous distributions within a unified framework.
- It is the key to understanding the intimate relation between the Kaplan-Meier and Nelson-Aalen estimators (Chapter 3).
- It is the basis for the martingale representation of the Kaplan-Meier estimator; cf. Section 3.2.6.
- It can readily be extended to Markov processes and used to derive a generalization of the Kaplan-Meier estimator to multivariate survival data; cf. Section 3.4 and Section A.2.4.

We now prepare the ground for the product-integral. When S is not absolutely continuous, it is still right-continuous with limits from the left, that is, a *cadlag* function. Informally, let $dS(t)$ be the increment of S over the small time interval $[t, t + dt)$, so that $-dS(t) = P(t \leq T < t + dt)$. If T has a density, we have $-dS(t) = f(t)dt$, and if T has point mass p_t in t then $-dS(t) = p_t$, still in a rather informal sense. If we let $S(t-)$ denote the left-hand limits of $S(t)$, the definition (A.1) gives

$$\alpha(t)dt = P(t \leq T < t + dt \mid T \geq t) = \frac{-dS(t)}{S(t-)},$$

and in this form, since S is monotone, it can be integrated on both sides to yield

$$A(t) = -\int_0^t \frac{dS(u)}{S(u-)}. \tag{A.4}$$

The integral on the right-hand side is now well defined as a Stiltjes integral (Rudin, 1976) regardless of whether S is absolutely continuous or not, and we take this as the definition of the cumulative hazard A in general, also when the hazard $\alpha(t) = dA(t)/dt$ itself does not exist. Note that for the absolutely continuous case $dS(u) = -f(u)du$ and $S(u-) = S(u)$, so in this case (A.4) specializes to $A(t) = \int_0^t f(u)/S(u)du$ in agreement with (A.2). On the other hand, for a purely discrete distribution, (A.4) takes the form $A(t) = \sum_{u \leq t} \alpha_u$, where the discrete hazard

$$\alpha_u = -\{S(u) - S(u-)\}/S(u-) = P(T = u \mid T \geq u)$$

is the conditional probability that the event occurs exactly at time u given that it has not occurred earlier.

Equation (A.4) expresses the cumulative hazard in terms of the survival function. We also need the inverse representation. In differential form (A.4) may be written

$$dS(t) = -S(t-)\,dA(t), \tag{A.5}$$

or more formally as an integral equation

$$S(t) = 1 - \int_0^t S(u-)dA(u) \tag{A.6}$$

for S. In the absolutely continuous case this is just an integrated version of (A.2), but the simple solution (A.3) is not valid in general. In fact, the much-studied integral equation (A.6) may serve as an implicit definition of S when given an increasing function A as the cumulative hazard.

We will sketch a (slightly simplified) argument of how an expression for the solution S can be found. Consider first the *conditional survival function* $S(v \mid u) = P(T > v \mid T > u) = S(v)/S(u)$, which is the probability of an event occurring later than time v given that it has not yet occurred at time u, $v > u$. The conditional survival function is an important concept that corresponds more generally to Markov transition probabilities when an individual can move between several different states, not only from "alive" to "dead". We will discuss the extension in A.2.4.

Partition the time interval $(0, t]$ into a number of subintervals $0 = t_0 < t_1 < t_2 < \cdots < t_K = t$. To survive from time 0 to time t, an individual needs to survive all the intermediate subintervals. By proper conditioning, we see that

$$S(t) = P(T_1 > t_1)P(T_2 > t_2 \mid T_1 > t_1) \cdots P(T > t_K \mid T > t_{K-1})$$

$$= \prod_{k=1}^K S(t_k \mid t_{k-1}). \tag{A.7}$$

Observe that, from (A.5), we have the approximation

$$S(t_k) - S(t_{k-1}) \approx -S(t_{k-1})(A(t_k) - A(t_{k-1})) \tag{A.8}$$

or

$$S(t_k \mid t_{k-1}) \approx 1 - (A(t_k) - A(t_{k-1})).$$

Entering this in (A.7), we get

$$S(t) \approx \prod_{k=1}^{K} \{1 - (A(t_k) - A(t_{k-1}))\}. \tag{A.9}$$

One would expect the approximation to improve with increasing number of subintervals, and if in this product we let the number K of time intervals increase while their lengths go to zero in a uniform way, the product on the right-hand side will indeed approach a limit, which is termed the product-integral. In fact, for an arbitrary cadlag function $B(t)$ (of locally bounded variation) the product-integral is defined as

$$\prod_{u \le t} \{1 + dB(u)\} \overset{\text{def}}{=} \lim_{M \to 0} \prod_{k=1}^{K} \{1 + (B(t_k) - B(t_{k-1}))\},$$

where $M = \max_k |t_k - t_{k-1}|$ is the length of the longest subinterval. Here the product-integral notation $\prod$ is used to suggest a limit of finite products $\prod$, just as the integral $\int$ is a limit of finite sums $\sum$.

Whereas (A.4) expresses A in terms of S the inverse relation can thus be expressed in terms of the product-integral

$$S(t) = \prod_{u \le t} \{1 - dA(u)\}. \tag{A.10}$$

For a purely discrete distribution, this relationship takes the form

$$S(t) = \prod_{u \le t} (1 - \alpha_u).$$

When A is absolutely continuous we write $dA(u) = \alpha(u)du$. Using the approximation $\exp(\alpha(u)du) \approx 1 - \alpha(u)du$, valid for small du, it is seen (informally) that

$$S(t) = \prod_{u \le t} \{1 - dA(u)\} = \prod_{u \le t} \{1 - \alpha(u)du\}$$
$$= \exp\left\{-\int_{u \le t} \alpha(u)du\right\} = \exp\{-A(t)\},$$

so the product-integral specializes to the well-known relation (A.3). More generally, if we decompose the cumulative hazard into a continuous and a discrete part, that is, $A(t) = A_c(t) + A_d(t)$, where $A_c(t)$ is continuous and $A_d(t)$ is a step function, then the product-integral may be factored as

$$\prod_{u \leq t} \{1 - dA(u)\} = e^{-A_c(t)} \prod_{u \leq t} \{1 - \triangle A_d(u)\}. \tag{A.11}$$

We have shown how the product-integral provides the natural link between the survival function and the cumulative hazard for all types of distributions. It unifies the exponential formulation in the continuous case with the geometric formulation in the discrete case. It provides a similar link between the nonparametric estimates of survival and cumulative hazard, that is, the Kaplan-Meier and Nelson-Aalen estimators. In A.2.4, we will show how the product-integral also extends naturally to expressing transition probabilities in time-inhomogeneous Markov chains.

One important application of the product-integral formulation is to show the "closeness" of two survival functions by using the closeness of their cumulative hazards. In particular, one can find the asymptotic distribution of the Kaplan-Meier estimator using the properties of the corresponding Nelson-Aalen estimator (see Chapter 3). Let $S_1(t)$, $S_2(t)$ be two survival functions with cumulative hazards $A_1(t)$ and $A_2(t)$. To compare S_1 and S_2 look at the ratio $S_1(t)/S_2(t)$. From the usual formula for differentiating ratios one would expect that $d\left(S_1/S_2\right) = \left(\left(dS_1\right)S_2 - S_1 dS_2\right)/S_2^2 = S_1 d\left(A_2 - A_1\right)/S_2$ using (A.5). Integrating this relationship on both sides, accounting for possible discontinuities in S_1 and S_2, we have

$$\frac{S_1(t)}{S_2(t)} = 1 + \int_0^t \frac{S_1(s-)}{S_2(s)} d\left(A_2 - A_1\right)(s), \tag{A.12}$$

which is *Duhamel's equation*. Since $S_1(t)/S_2(t) - 1 = \left(S_1(t) - S_2(t)\right)/S_2(t)$ this gives an expression for the relative difference between S_1 and S_2 in terms of $A_2 - A_1$. For further details, see Andersen et al. (1993) and Gill and Johansen (1990). The multivariate extension of Duhamel's equation is given in A.2.4.

A.2 Markov chains, transition intensities, and the Kolmogorov equations

In a standard analysis of survival data, Markov processes play a minor role. The standard setup involves only individuals in one of two possible states, either alive or dead, sick or cured, in working order or defective, etc., and individuals move only from the first state to the second. However, if $N(t)$ counts the number of individuals in a closed group (no loss or influx of new individuals, in particular no censoring) who have experienced an event by time t, and all individuals have a constant hazard α of experiencing an event, then N is one of the prime examples of a continuous time Markov chain, namely the so-called pure birth process. From this observation alone, it should be clear that a basic understanding of the behavior of Markov chains may be useful when applying the theory of counting processes to event history analysis.

When the standard event history models are extended to multistate models such as the competing risk model and the illness-death model, it is necessary to keep track of the states an individual moves through and of the transition intensities between states. Markov models are then essential tools for a detailed probability description, and counting process theory is used in the estimation of transition intensities for the Markov models.

In this book, even more extensive use of Markov models is made in Chapters 10 and 11, where Markov models and extensions of these will be used to describe unobserved dynamic frailty, that is, models that describe how the risk of a specific individual develops over time, until disease or other events eventually manifest themselves as a consequence of an elevated risk level. In such models, the states of the underlying Markov process are not always observed directly. The Markov model may then be seen as a convenient and intuitive tool for constructing parametric hazards. If the Markov model is successful in describing the underlying risk process, this will be reflected in corresponding hazard models whose properties reflect those of the observed hazard.

In the following we give a brief review of standard discrete and continuous time Markov chain models. We discuss both time-homogeneous and -inhomogeneous Markov chains, as both concepts will be of specific use in the probability models introduced in this book.

Together with the martingale concept, the Markov principle is among the most frequently used methods for simplifying the dependence structure of a stochastic process. Intuitively, a Markov process is defined through the property that once we know the current state of the process, any knowledge of the past gives no further information for predicting the future. That is, to describe the probability distribution of the process in the future it suffices to make use of its current state; there is no extra information to be extracted from past history.

The practical upshot of the Markov definition is a simplification of the transition probabilities that describe the probability for the process to move from one state to another within a specified time interval; these are the same regardless of the behavior of the Markov chain in the past. We now give a more formal definition. Let $X(t)$ be a stochastic process where the time index t can either be a nonnegative integer (discrete time) or a nonnegative real value (continuous time), and let $\mathscr{S}$ denote the set of possible values X can attain (the state space). We briefly discuss discrete time processes, but most of our applications will focus on continuous time. In the following introduction we assume X takes values on the integers, that is, $\mathscr{S} = \{0, \pm 1, \pm 2, \ldots\}$, in which case X is referred to as a Markov *chain*. Typically, X will be nonnegative, thus $\mathscr{S} = \{0, 1, 2, \ldots\}$. However, the general Markov process theory also covers processes that are continuous in both space and time, that is, the state space is $\mathbb{R}$ (or restrictions like $\mathbb{R}^+$). We use such processes, in particular the Wiener process and the Ornstein-Uhlenbeck process, as models for random risk development (Chapters 10 and 11) and as limit processes for point process martingales (see, for example, Section 2.3). They are discussed in more detail in Section A.4. The basic Markov principle is still the same, although the theory requires some technical modifications.

We say that $X(t)$ is Markov if

$$P(X(t) = x \,|\, X(t_k) = x_k, X(t_{k-1}) = x_{k-1}, \ldots, X(t_1) = x_1)$$
$$= P(X(t) = x \,|\, X(t_k) = x_k) \quad \text{(A.13)}$$

for any selection of time points $t, t_1, \ldots, t_k$ such that $t_1 < \cdots < t_{k-1} < t_k < t$, and integers $x, x_1, \ldots, x_k$. As mentioned, this intuitively means that as long as the value x_k of $X(t_k)$ is known the value of X at earlier times is uninformative when predicting future outcomes of X, or alternatively that the past and the future are independent given the present. If we let $\mathscr{F}_t = \sigma(X(s), s \leq t)$ denote the entire cumulated history (σ-algebra) generated by X until time t, the Markov property can be written

$$P(X(t) \in B \,|\, \mathscr{F}_s) = P(X(t) \in B \,|\, X(s)), \quad t \geq s,$$

for a subset B of the state space $\mathscr{S}$.

The Markov chain X is said to be *time-homogeneous* when

$$P(X(s+t) = y \,|\, X(s) = x) = P(X(t) = y \,|\, X(0) = x), \quad \text{(A.14)}$$

that is, the transition probabilities depend only on the separation in time (t), not on the time of starting (s). Thus, a time-homogeneous Markov chain completely "restarts" at time s if $X(s)$ is known. The past history of the process can be discarded, as can the current time s. We will consider both homogeneous and inhomogeneous chains in what follows.

A.2.1 Discrete time-homogeneous Markov chains

Discrete time-homogeneous Markov chains do not only have a discrete state space, the time axis is also discrete, with the process $X(n)$ defined for $n = 0, 1, 2, \ldots$. Due to the Markov property (A.13), the behavior of a discrete time-homogeneous Markov chain is completely described by the distribution of the starting value $X(0)$ and the one-step transition probability matrix $\mathbf{P}$, with elements

$$p_{ij} = P(X(n+1) = j \,|\, X(n) = i) = P(X(1) = j \,|\, X(0) = i).$$

For instance, the two-step transition probabilities can be computed by summing over all possible states at time 1, as in

$$P(X(2) = j \,|\, X(0) = i) = \sum_{k \in \mathscr{S}} P(X(2) = j \,|\, X(1) = k, X(0) = i)$$
$$\times P(X(1) = k \,|\, X(0) = i)$$
$$= \sum_{k \in \mathscr{S}} p_{kj} p_{ik},$$

which in fact correspond to the elements of the matrix $\mathbf{P}^2$. The principle of summing over an intermediate time point leads in general to equations of the form

$$P(X(n) = j \mid X(0) = i) = \sum_{k \in \mathscr{S}} P(X(n) = j \mid X(m) = s)P(X(m) = s \mid X(0) = i),$$

$n > m > 0$, which are known as the *Chapman-Kolmogorov equations*. Repeated applications of the Chapman-Kolmogorov equations lead to the general result that the n-step transition probabilities can be written as a matrix product, in that $P(X(n) = j \mid X(0) = i) = p_{ij}(n)$, where $p_{ij}(n)$ are the elements of $\mathbf{P}^n$. Similarly, if we let $\mathbf{p}(n)$ be the vector of probabilities $p_i(n) = P(X(n) = i)$, and $\mathbf{p}(0)$ the distribution of the starting value $X(0)$, then

$$\mathbf{p}(n) = (\mathbf{P}^T)^n \mathbf{p}(0). \tag{A.15}$$

It should by now be clear that all finite-dimensional distributions of X, that is, all probabilities of the form

$$P(X(t_k) = x_k, X(t_{k-1}) = x_{k-1}, \ldots, X(t_1) = x_1),$$

can be expressed in terms of the initial distribution $\mathbf{p}(0)$ and the transition matrix $\mathbf{P}$, thus completely describing the probability structure of X.

As a simple example, consider the one-step transition probability matrix

$$\mathbf{P} = \begin{pmatrix} 1 & 0 & 0 & 0 \\ 1/2 & 1/2 & 0 & 0 \\ 0 & 1/2 & 0 & 1/2 \\ 0 & 0 & 1 & 0 \end{pmatrix}.$$

(Note that the first row/column corresponds to state 0, the second to state 1, etc.) A numerical computation of matrix power shows, for instance, that

$$\mathbf{P}^5 = \begin{pmatrix} 1 & 0 & 0 & 0 \\ 0.969 & 0.031 & 0 & 0 \\ 0.656 & 0.219 & 0 & 0.125 \\ 0.563 & 0.188 & 0.25 & 0 \end{pmatrix}, \quad \mathbf{P}^{10} = \begin{pmatrix} 1 & 0 & 0 & 0 \\ 0.999 & 0.001 & 0 & 0 \\ 0.938 & 0.030 & 0.031 & 0 \\ 0.908 & 0.061 & 0 & 0.031 \end{pmatrix}.$$

Thus, all n-step transition probabilities are easily computed. It is evident, however, that as n grows, the transition probabilities in our example all go to zero, except the probability of being in state 0, which increases toward 1. This means that in the long run, $X(n)$ will be in state 0 with almost absolute certainty. The state 0 is thus an *absorbing state*. Since we are interested in survival models defined in terms of hitting times, that is, the time it takes X to reach state 0, the limiting behavior is of importance to understand how hazard develops. This is explored in more detail in A.3 and Chapter 10.

Meanwhile, we will look at Markov chains in continuous time and see how they relate to multistate models of event histories.

A.2.2 Continuous time-homogeneous Markov chains

We now assume that the time axis of the Markov chain is continuous, that is, the process $X(t)$ is defined for all $t \in [0, \infty)$. The state space is still assumed discrete. There is an obvious connection between a continuous and a discrete time-homogeneous Markov chain. If for the continuous chain we only keep track of what happens at each transition, that is, what states the chain moves between but not at what times, the resulting discrete time process is called the *embedded* chain. It is intuitively clear that the embedded chain is itself a discrete time Markov chain. In addition to the transitions, the time spent by the process in each state is random. The time-homogenous Markov property dictates that the time from one transition to the next follows an exponential distribution, where the expected time to transition may depend on the current state. The reason for this is that the exponential distribution has no "memory", since $P(T \geq t + s \mid T \geq s) = P(T \geq t)$ for all t, s when T has an exponential distribution, and that the hazard of a new transition at any time is constant, that is, $P(t \leq T < t + dt \mid T \geq t)/dt = \alpha$ does not depend on t. When this is the case, the only valuable information for future predictions of the chain is the current state of the chain, not how long the chain has been in that particular state. Hence, knowing the past states is useless, in agreement with the Markov property.

The continuous time-homogeneous Markov chain can be described by its transition probability matrix $\mathbf{P}(t)$ with elements $p_{ij}(t) = P(X(t) = j \mid X(0) = i)$, $t \geq 0$, together with the initial distribution. The transition matrix can be computed from a formula similar to the $\mathbf{P}^n$ used in the discrete time situation. Before doing this, however, it is useful to look at an alternative description of the behavior of the continuous chain, namely the *infinitesimal generator*. We will introduce this by way of an example, a finite, continuous time Markov chain X on the states $\{0, 1, \ldots, 5\}$. In Figure A.1, each box represents a state of the process, and the arrows indicate the possible transitions. The process is known as a *birth-death* process since it can move up (a birth) or down (a death) one step at a time. The embedded (time discrete) process known as a *random walk*. The parameters β_1 and β_2 are the *transition intensities*. If the process is in state 3, it can either move up to state 4 (intensity β_2) or down to state 2 (intensity β_1). A common way of picturing the process leading up to a jump is to imagine two independent random times T_1 and T_2, each exponentially distributed with parameters β_1 and β_2, such that T_1 measures the time X spends in a specific state until it moves down, and T_2 the time until it moves up. The two random times compete in the sense that if $T_1 < T_2$ the process moves down at time T_1, and up at time T_2 if $T_2 < T_1$. The time actually spent in a state is thus $T = T_1 \wedge T_2$, which is exponentially distributed with parameter $\beta_1 + \beta_2$, and the probabilities of moving down and up are $\beta_1/(\beta_1 + \beta_2)$ and $\beta_2/(\beta_1 + \beta_2)$, respectively, derived from the property of the exponential distribution. However, one usually does not wish

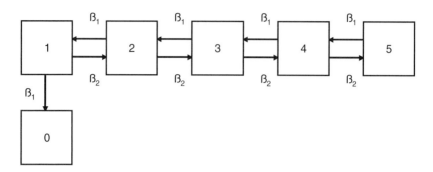

Fig. A.1 *A six-state birth-death process with one absorbing state (0) and one reflecting (5).*

to tie the interpretation of the transition intensities too closely to an explicit model of independent waiting times. Rather, it is customary to say that the process has an intensity of $\beta_1 + \beta_2$ of having an event, and that the ratio of the two transition probabilities, conditional on a transition actually occurring, is β_1/β_2.

We now see that the one-step transition probabilities for the embedded chain of our six-state chain are described by

$$P = \begin{pmatrix} 1 & 0 & 0 & 0 & 0 & 0 \\ l_1 & 0 & l_2 & 0 & 0 & 0 \\ 0 & l_1 & 0 & l_2 & 0 & 0 \\ 0 & 0 & l_1 & 0 & l_2 & 0 \\ 0 & 0 & 0 & l_1 & 0 & l_2 \\ 0 & 0 & 0 & 0 & 1 & 0 \end{pmatrix},$$

with $l_1 = \beta_2/(\beta_1 + \beta_2)$ and $l_2 = 1 - l_1$. Notice that whereas states 1 through 4 all behave as state 3, state 5 is *reflecting*. Whenever the process reaches state 5, it will eventually return to state 4, after a sojourn time in 5 of expected length $1/\beta_1$. State 0 is *absorbing*; once X has reached this state it will stay there forever. The infinitesimal generator of X is defined as

$$\alpha = \begin{pmatrix} 0 & 0 & 0 & 0 & 0 & 0 \\ \beta_1 & -(\beta_1+\beta_2) & \beta_2 & 0 & 0 & 0 \\ 0 & \beta_1 & -(\beta_1+\beta_2) & \beta_2 & 0 & 0 \\ 0 & 0 & \beta_1 & -(\beta_1+\beta_2) & \beta_2 & 0 \\ 0 & 0 & 0 & \beta_1 & -(\beta_1+\beta_2) & \beta_2 \\ 0 & 0 & 0 & 0 & \beta_1 & -\beta_1 \end{pmatrix}. \qquad (A.16)$$

The generator is just the matrix of transition intensities. The values along the diagonal are (with signs inverted) the total intensity of a process in that state, that is, if the process is in state i, $i = 1, \ldots, 4$, the intensity of leaving that state is $\beta_1 + \beta_2$. In this constant intensity example, the expected time spent in any of the states $1, \ldots, 4$

is thus $1/(\beta_1 + \beta_2)$. The process will, on average, spend a time $1/\beta_1$ in state 5 before being reflected to state 4, and the parameter 0 for state 0 illustrates the infinite amount of time spent in an absorbing state.

A.2.3 The Kolmogorov equations for homogeneous Markov chains

Let us formalize the definition of the generator of a time-homogeneous Markov chain. The generator α is defined as the time-derivative (elementwise) at time zero of the transition probability matrix $\mathbf{P}(t)$. That is,

$$\alpha = \mathbf{P}'(0) = \lim_{\Delta t \to 0^+} \frac{1}{\Delta t}(\mathbf{P}(\Delta t) - \mathbf{P}(0)). \tag{A.17}$$

For our birth-death generator example (A.16), this is seen from the following intuitive reasoning: if $X(t) = i$, $1 \le i \le 4$, then during the short interval $[0, t + \Delta t)$ the probability of X moving up one step is approximately equal to $\beta_2 \Delta t$, the probability of moving down is $\beta_1 \Delta t$, and of not moving at all is $1 - (\beta_1 + \beta_2)\Delta t$. The probability of a simultaneous up/down movement within a short interval is negligible, as are the probabilities of more than one upward or more than one downward movement. Similar arguments hold for the boundary states 0 and 5. Using this to express $\mathbf{P}(\Delta t)$ approximately for small Δt, and noting that $\mathbf{P}(0)$ is just the identity matrix, we see that the right hand side of (A.17) ends up as the generator matrix.

Note that each element of α is thus

$$\alpha_{ij} = \lim_{\Delta t \to 0^+} \frac{1}{\Delta t} P(X(t + \Delta t) = j \mid X(t) = i)$$

when $i \ne j$ (independent of t). Its similarity to definition (A.1) should be clear: the hazard measures the infinitesimal probability of going from "live" to "dead" at time t; the Markov intensity measures the more general infinitesimal probability of going from state i to state j at time t. This is a basic connection between event history analysis and Markov theory, and we develop it further in A.2.4.

Proceeding with the relationship between α and $\mathbf{P}$, we can now use the Chapman-Kolmogorov equations to write

$$\mathbf{P}(t + \Delta t) - \mathbf{P}(t) = \mathbf{P}(t)\mathbf{P}(\Delta t) - \mathbf{P}(t) \tag{A.18}$$
$$= \mathbf{P}(t)(\mathbf{P}(\Delta t) - \mathbf{I})$$
$$\approx \mathbf{P}(t)\alpha\Delta t,$$

and after using (A.17) this leads to the *Kolmogorov forward (differential) equation*

$$\mathbf{P}'(t) = \mathbf{P}(t)\alpha,$$

with the initial condition $\mathbf{P}(0) = \mathbf{I}$. Similarly, one can derive the *Kolmogorov backward differential equation*

$$\mathbf{P}'(t) = \boldsymbol{\alpha}\mathbf{P}(t)$$

with the same initial condition. In fact, the Kolmogorov equations are the matrix equivalents of (A.2), the differential equation for $S(t)$, and if $\mathbf{P}$ and $\boldsymbol{\alpha}$ were scalar, we would have the familiar solution

$$\mathbf{P}(t) = \exp(\boldsymbol{\alpha}t). \tag{A.19}$$

This formula also holds true in the matrix case. The exponential function must then be defined as the *matrix exponential*, where

$$\exp(\mathbf{B}) \overset{\text{def}}{=} \mathbf{I} + \sum_{n=1}^{\infty} \frac{1}{n!}\mathbf{B}^n.$$

It is not hard to verify that when (formally) differentiating $\exp(\boldsymbol{\alpha}t)$ with respect to t it satisfies the Kolmogorov equations. It should be noted that the above relations depend heavily on the chain being time homogeneous. For inhomogeneous chains, the generator $\boldsymbol{\alpha}$ will depend on t, which makes the situation more complex. The extension to inhomogeneous chains is discussed in the following section. For more information on the matrix exponential, see Moler and van Loan (1978, 2003).

Let us return for a moment to the time-discrete chains. The Kolmogorov forward equation simply corresponds to the difference equations $\Delta\mathbf{P}(n) = \mathbf{P}(n) - \mathbf{P}(n-1) = \mathbf{P}(n-1)\boldsymbol{\alpha}$ with $\boldsymbol{\alpha} = \mathbf{P} - \mathbf{I}$, and the solution is, as we have seen, $\mathbf{P}(n) = \mathbf{P}^n$. This can also be seen as a continuous time chain with jumps only at integer time points, and $\mathbf{P}(t) = \mathbf{P}(n)$ when $n+1 > t \geq n$ is a cadlag function. Then $d\mathbf{P}(t) = \mathbf{P}(t-)\boldsymbol{\alpha}$. So both time-discrete and time-continuous homogeneous processes behave in a similar fashion and the resulting transition probabilities behave like exponential (or geometric) functions of time. In the next subsection, we show how the product-integral serves to unify the two, as in the case of survival functions seen in the start of this appendix.

A.2.4 Inhomogeneous Markov chains and the product-integral

We now consider Markov chains where the transition intensities $\boldsymbol{\alpha} = \boldsymbol{\alpha}(t)$ depend on time, that is, time-inhomogeneous processes. To begin with we will consider the simplest example possible and show how a two-state Markov process relates to the concept of survival, or time-to-event. Let $X(t)$ be defined on the state space $\{0,1\}$ by the transition intensity matrix

$$\boldsymbol{\alpha}(t) = \begin{pmatrix} -\alpha(t) & \alpha(t) \\ 0 & 0 \end{pmatrix}. \tag{A.20}$$

State 1 is thus absorbing, and the intensity of leaving state 0 and entering state 1 is $\alpha(t)$ at time t. The process is illustrated in Figure A.2. When starting in state 0, we

Alive   Dead

$$\boxed{0} \longrightarrow \boxed{1}$$

Fig. A.2 *The standard survival model. From its initial state 0 ("alive") the process can move only to state 1 ("dead"). The hazard of moving to state 1 at time t is $\alpha(t)$.*

can define the event time $T = \min\{t \mid X(t) = 1\}$, that is, the time until absorption in state 1. Then T is a survival time with hazard $\alpha(t)$. Let $f(t)$ be the density of T, and $S(t) = P(T > t)$ the corresponding survival function, and $\mathbf{P}(t)$ the transition probabilities, with elements $P_{ij}(t)$, $i, j = 0, 1$. Notice that $P_{00}(t) = S(t)$ is the survival function (the probability of still being in state 0 after time t), and we know from (A.3) that $S(t) = \exp\left(-\int_0^t \alpha(s)ds\right)$. Since $P_{01}(t) = 1 - P_{00}(t)$, the full transition probability matrix can thus be written

$$\mathbf{P}(t) = \begin{pmatrix} S(t) & 1 - S(t) \\ 0 & 1 \end{pmatrix} = \begin{pmatrix} \exp\left(-\int_0^t \alpha(s)ds\right) & 1 - \exp\left(-\int_0^t \alpha(s)ds\right) \\ 0 & 1 \end{pmatrix}. \quad \text{(A.21)}$$

This reveals the simple connection between survival and Markov processes in this two-state example.

Note that when α is constant, formula (A.19) applies, and we find

$$(\boldsymbol{\alpha}t)^n = \begin{pmatrix} (-1)^n(\alpha t)^n & (-1)^{n+1}(\alpha t)^n \\ 0 & 0 \end{pmatrix}$$

and

$$\mathbf{P}(t) = \exp(\boldsymbol{\alpha}t) = \mathbf{I} + \sum_{n=1}^{\infty} \frac{1}{n!}(\boldsymbol{\alpha}t)^n$$

$$= \begin{pmatrix} \sum_{n=0}^{\infty} \frac{1}{n!}(-\alpha t)^n & 1 - \sum_{n=0}^{\infty} \frac{1}{n!}(-\alpha t)^n \\ 0 & 1 \end{pmatrix}$$

$$= \begin{pmatrix} \exp(-\alpha t) & 1 - \exp(-\alpha t) \\ 0 & 1 \end{pmatrix},$$

which verifies (A.21). In fact, in this simple example (A.19) also holds for the time-inhomogeneous case, in the sense that $\mathbf{P}(t) = \exp\left(-\int_0^t \boldsymbol{\alpha}(s)ds\right)$ where the integration is taken elementwise and the matrix exponential is defined as before. Unfortunately, this simple formula is not true in the general inhomogeneous case. To find a correct expression for the solution in the inhomogeneous case we need to use the product-integral. Recall from (A.11) that the product-integral provided a unification of discrete and absolutely continuous survival distributions. We will now see that with a little extra care it also extends in a natural way to the matrix situation resulting from a multistate Markov process with many possible transitions.

For inhomogeneous processes we cannot use the simplifying relation (A.14) to express the transition probabilities as the matrix $\mathbf{P}(t)$ with only the single time parameter. Rather, we need to write $\mathbf{P}(s,t)$ for the transition probabilities, where the matrix elements are $P_{ij}(s,t) = P(X(t) = j \mid X(s) = i)$. In the homogeneous case, this reduces to $\mathbf{P}(s,t) = \mathbf{P}(t-s)$.

When the transition probabilities are absolutely continuous, we can again use the Chapman-Kolmogorov equations as in (A.18) to see that the Kolmogorov forward equation

$$\frac{\partial}{\partial t}\mathbf{P}(s,t) = \mathbf{P}(s,t)\boldsymbol{\alpha}(t) \tag{A.22}$$

holds, with

$$\boldsymbol{\alpha}(t) = \lim_{\Delta t \to 0+} \frac{1}{\Delta t}(\mathbf{P}(t,t+\Delta t) - \mathbf{I}). \tag{A.23}$$

Indeed, in the general case (not necessarily absolutely continuous transition probabilities), the forward equation can be written

$$\mathbf{P}(s,t) = \mathbf{I} + \int_s^t \mathbf{P}(s,u-)d\mathbf{A}(u),$$

which is the equivalent of (A.6) for multistate processes. The matrix function $\mathbf{A}(t)$ is the matrix of cumulative transition intensities and as such generalizes the cumulative hazard. In the absolutely continuous case, it is just the elementwise integral of $\boldsymbol{\alpha}(t)$.

To find a solution in the general case we use the Chapman-Kolmogorov equations to write

$$\mathbf{P}(s,t) = \mathbf{P}(t_0,t_1)\mathbf{P}(t_1,t_2)\cdots\mathbf{P}(t_{K-1},t_K)$$

for a partition $s = t_0 < t_1 < t_2 < \cdots < t_K = t$. Notice that the matrix product must be taken carefully in sequence from left to right with increasing time index. This product formulation is the multistate equivalent of the product of conditional survival probabilities given in (A.7). Arguing as we did to derive (A.9), we can now write

$$\mathbf{P}(s,t) \approx \prod_{k=1}^{K} \{\mathbf{I} + (\mathbf{A}(t_k) - \mathbf{A}(t_{k-1}))\}$$

with the partition $s = t_0 < t_1 < t_2 < \cdots < t_K = t$. A crucial difference from the one-dimensional case, though, is that the product needs to be taken in the appropriate increasing order from left to right, that is, as

$$\{1 + (\mathbf{A}(t_1) - \mathbf{A}(t_0))\}\{1 + (\mathbf{A}(t_2) - \mathbf{A}(t_1))\}\cdots\{1 + (\mathbf{A}(t_K) - \mathbf{A}(t_{k-1}))\}.$$

The matrices $\mathbf{A}(t_k) - \mathbf{A}(t_{k-1})$ do not necessarily commute, and this is the reason (A.19) does not extend in general to the inhomogeneous case, not even in the absolutely continuous case. We now let the lengths of the subintervals go to zero and arrive at the solution as a *matrix product-integral*

$$\mathbf{P}(s,t) = \prod_{u\in(s,t]} \{\mathbf{I}+d\mathbf{A}(u)\}.$$

In the continuous case, we can write

$$\mathbf{P}(s,t) = \prod_{u\in(s,t]} \{\mathbf{I}+\boldsymbol{\alpha}(u)du\}. \tag{A.24}$$

It should be kept in mind, however, that the sequence of the product matters also in the continuous case. For a time discrete process, the increments of the cumulative transition intensities are simply $d\mathbf{A}(t_k) = \mathbf{P}_k - \mathbf{I}$, where $\mathbf{P}_k$ is the transition probability matrix at time t_k. The product-integral is thus simply a restatement of the Chapman-Kolmogorov equations $\mathbf{P}(t_k,t_{k+l}) = \mathbf{P}_k\mathbf{P}_{k+1}\cdots\mathbf{P}_{k+l}$.

The product-integral is a compact way of expressing the solution to the Kolmogorov forward equation. In general, it may not be possible to reduce the product-integral to "simpler" expressions even in the continuous case. In the next subsection we look at useful examples where relatively explicit solutions can be found.

For completeness, we end this subsection with the multivariate formulation of Duhamel's equation (A.12). For two different transition probability matrices $\mathbf{P}_1$ and $\mathbf{P}_2$ with cumulative transition intensity matrices $\mathbf{A}_1$ and $\mathbf{A}_2$, the formula is

$$\mathbf{P}_1(s,t)\mathbf{P}_2(s,t)^{-1} - \mathbf{I} = \int_0^t \mathbf{P}_1(s,u-)d\,(\mathbf{A}_1-\mathbf{A}_2)\,(u)\mathbf{P}_2(s,u)^{-1}, \tag{A.25}$$

which again is useful for deriving the asymptotic properties of the empirical transition probability matrices; cf. Section 3.4 and Gill and Johansen (1990).

A.2.5 Common multistate models

As a simple but important illustration extending (A.20), we consider the *competing risk* model. Let $X(t)$ describe the state of an individual at time t. Assume individuals in state 0 are those still alive, and that these individuals may die from one of r different causes. The intensities $\alpha_{0h}(t)$, $h = 1,2,\ldots,r$, represent the instantaneous risk of dying from cause h at time t given that the individual is still alive at this time. The model is illustrated in Figure A.3.

Since death is absorbing, we have the transition intensity matrix

$$\boldsymbol{\alpha}(t) = \begin{pmatrix} -\sum_{j=1}^r \alpha_{0j}(t) & \alpha_{01}(t) & \cdots & \alpha_{0r}(t) \\ 0 & 0 & \cdots & 0 \\ \vdots & \vdots & \ddots & \vdots \\ 0 & 0 & 0 & 0 \end{pmatrix}.$$

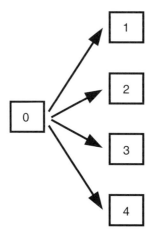

Fig. A.3 *The competing risk model. From its initial state 0 the process can move to any of a number of terminal stages. The hazard of moving to state j at time t is $\alpha_{0j}(t)$.*

The transition probability matrix is of the form

$$\mathbf{P}(s,t) = \begin{pmatrix} P_{00}(s,t) & P_{01}(s,t) & \cdots & P_{0r}(s,t) \\ 0 & 1 & \cdots & 0 \\ \vdots & \vdots & \cdots & \vdots \\ 0 & 0 & 0 & 1 \end{pmatrix}.$$

By the Kolmogorov forward equation (A.22), we obtain the one-dimensional differential equations

$$\frac{\partial}{\partial t}P_{00}(s,t) = -P_{00}(s,t)\sum_{j}\alpha_{0j}(t),$$

$$\frac{\partial}{\partial t}P_{0h}(s,t) = -P_{0h}(s,t)\alpha_{0h}(t).$$

The equations are solved by standard methods to get

$$P_{00}(s,t) = \exp\left(-\int_{s}^{t}\sum_{j}\alpha_{0j}(u)du\right),$$

$$P_{0h}(s,t) = \int_{s}^{t}\alpha_{0h}(u)P_{00}(s,u)du.$$

Seen from a survival perspective, $S(t\,|\,s) = P_{00}(s,t)$ is the probability of surviving from time s to time t conditional on not yet being dead at time s. The product $\alpha_{0h}(t)P_{00}(s,t) = \alpha_{0h}(t)S(t\,|\,s)$ is the cause-specific probability density, and $P_{0h}(s,t)$

is the cause-specific cumulative probability, that is, the probability of dying from cause h in the interval $(s,t]$ conditional on being alive at time s. Note that $1 - P_{0h}(s,t)$ is harder to interpret since it is the probability of *not* dying from cause h in the interval $(s,t]$, which means either surviving *or* dying from another cause. Clearly, the states in this example may signify things other than death, as long as the terminal states $1,\ldots,r$ are "terminal". For instance, the process may describe the waiting time in a queue, which ends the first time one of several operators becomes available for service. The competing risk situation is analyzed further in Section 3.4.

Another useful example is a model sometimes referred to as a *conditional model*, see, for instance, Therneau and Grambsch (2000, p. 187) and Prentice et al. (1981). In this model, a subject moves through successive stages in one direction, for each stage experiencing a new hazard of moving to the next stage. It might, for instance, describe the "hazard" of giving birth to a child, conditional on already having had none, one, or two or more children. We can use age of the mother as the time scale and let $\alpha_{01}(t)$, $\alpha_{12}(t)$, $\alpha_{23}(t),\ldots$ model the hazard of passing from 0 to 1 child, 1 to 2, etc., as functions of age t. Thus, $\alpha_{12}(33)$ would be the instantaneous "risk" of giving birth for a mother of age 33 with one child. It should be noted that a more detailed model should include information about time since last birth since there must be a minimum separation in time between successive births. Alternatively, a model using time since last birth as the main time scale could be used. With age as the time axis, our model can be illustrated as in Figure A.4.

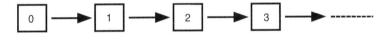

Fig. A.4 *The conditional model. The process moves along a series of stages, with the transition hazards $\alpha_{01}(t)$, $\alpha_{12}(t),\ldots$.*

The transition intensity matrix takes the form

$$\boldsymbol{\alpha}(t) = \begin{pmatrix} -\alpha_{01}(t) & \alpha_{01}(t) & 0 & 0 & \cdots \\ 0 & -\alpha_{12}(t) & \alpha_{12}(t) & 0 & \cdots \\ 0 & 0 & -\alpha_{23}(t) & \alpha_{23}(t) & \cdots \\ 0 & 0 & 0 & -\alpha_{34}(t) & \cdots \\ \vdots & \vdots & \vdots & \vdots & \ddots \end{pmatrix},$$

and we can still appeal to the Kolmogorov forward equation (A.22) to obtain the equations

$$\frac{\partial}{\partial t} P_{ii}(s,t) = -P_{ii}(s,t)\alpha_{i,i+1}(t),$$

$$\frac{\partial}{\partial t} P_{ij}(s,t) = P_{i,j-1}(s,t)\alpha_{j-1,j}(t) - P_{ij}(s,t)\alpha_{j,j+1}(t),$$

where $j \geq i+1$. The solutions can be obtained by first computing

$$P_{ii}(s,t) = \exp\left(-\int_s^t \alpha_{i,i+1}(u)du\right).$$

The remaining solutions can be obtained recursively in j. For instance, from

$$\frac{\partial}{\partial t}P_{0j}(s,t) = P_{0,j-1}(s,t)\alpha_{j-1,j}(t) - P_{0j}(s,t)\alpha_{j,j+1}(t),$$

we get

$$P_{0j}(s,t) = P_{jj}(s,t)\int_s^t P_{0,j-1}(s,u)\alpha_{j-1,j}(u)P_{jj}^{-1}(s,u)du,$$

and similar solutions can be obtained for $P_{ij}(s,t)$ when $j \geq i+1$.

The last example we consider is the three-state *illness-death* model. An individual in state 0 is considered healthy. A healthy individual can then either get a disease, which puts him/her in state 1, or he/she can die of other causes, ending in state 2. Diseased individuals can also die, that is, go from state 1 to state 2. State 2, by its nature, is absorbing. In our example, diseased individuals cannot recover, although the possibility of a transition from state 1 to state 0 might also be added to the model. The process is depicted in Figure A.5. The corresponding transition

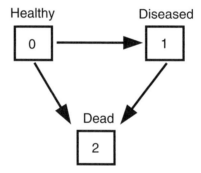

Fig. A.5 *The illness-death model. From its initial ("Healthy") state 0, an individual can either move to state 1 ("Diseased") or die (state 2). In this example, an individual in state 1 can only move to state 2, that is, there is no recovery.*

intensity matrix is

$$\boldsymbol{\alpha}(t) = \begin{pmatrix} -(\alpha_{01}(t) + \alpha_{02}(t)) & \alpha_{01}(t) & \alpha_{02}(t) \\ 0 & -\alpha_{12}(t) & \alpha_{12}(t) \\ 0 & 0 & 0 \end{pmatrix}.$$

It is seen that $P_{ij}(s,t) = 0$ when $i > j$, and $P_{22}(s,t) = 1$. Since $(\partial/\partial t)P_{00}(s,t) = -(\alpha_{01}(t) + \alpha_{02}(t))P_{00}(s,t)$ we get $P_{00}(s,t) = \exp\left(-\int_s^t \alpha_{01}(u) + \alpha_{02}(u)du\right)$. Similarly $P_{11}(s,t) = \exp\left(-\int_s^t \alpha_{12}(u)du\right)$, and clearly, $P_{12}(s,t) = 1 - P_{11}(s,t)$. Furthermore, we have $(\partial/\partial t)P_{01}(s,t) = \alpha_{01}(t)P_{00}(s,t) - \alpha_{12}(t)P_{01}(s,t)$, which is solved by

$$P_{01}(t) = \exp\left(-\int_s^t \alpha_{12}(u)du\right)\int_s^t \alpha_{01}(u)P_{00}(s,u)\exp\left(\int_s^u \alpha_{12}(v)dv\right)du$$

$$= \int_s^t P_{00}(s,u)\alpha_{01}(u)P_{11}(u,t)du.$$

Since $P_{02}(s,t) = 1 - P_{00}(s,t) + P_{01}(s,t)$, we have a complete set of solutions to the Kolmogorov equations.

A.3 Stationary and quasi-stationary distributions

We return for a while to the discussion at the end of A.2.1 and look at the long-run behavior of discrete time Markov chains. We saw that for the one-step transition probabilities

$$\mathbf{P} = \begin{pmatrix} 1 & 0 & 0 & 0 \\ 1/2 & 1/2 & 0 & 0 \\ 0 & 1/2 & 0 & 1/2 \\ 0 & 0 & 1 & 0 \end{pmatrix}, \tag{A.26}$$

the higher powers of $\mathbf{P}$ tended to concentrate mass at state 0, that is, $\lim_{n\to\infty} p_{ij}(n)$ is zero when $j \neq 0$ and 1 when $j = 0$, regardless of the value of i. We now discuss the limiting behavior in more detail. Additionally, one might ask if more can be said about the long-run probability distribution for those states that are *not* absorbing. The probabilities typically tend to zero, but conditionally on not yet having reached the absorbing state zero, what is the "balance" between the other states? We look more into this in the following.

A.3.1 The stationary distribution of a discrete Markov chain

To better understand the long-run behavior of a given Markov chain, it is important to distinguish between different types of states. Roughly speaking, there are the *recurrent* states, to which the process keeps returning over and over again, and there are the *transient* states, which the process may visit a finite number of times before leaving the state forever. Although we only need a basic understanding of this in our presentation, we include some details for completeness.

It is customary to divide the state space $\mathscr{S}$ into subgroups, or *communication classes,* where all elements of a subgroup *communicate.* We say that states i and j

communicate if X can reach j from i and reach i from j, that is, the element $p_{ij}(n)$ of $\mathbf{P}^n$ is positive for some n, and similarly $p_{ji}(n) > 0$ for some n. A Markov chain that contains only a single-state group is said to be *irreducible*; otherwise the state space can be reduced into a set of irreducible classes. If X cannot escape from a class, that is, there is zero probability of reaching any outside state from that class, the class is said to be *closed*. A class that is not closed is *transient*. This name is appropriate because if X starts in a transient part of the state space it will eventually leave this part and end up in a closed class, where it will remain forever.

Consider again the example given in (A.26). Recall that the first row/column corresponds to state 0, the second to state 1, etc. There are three classes, $C_1 = \{0\}$, $C_2 = \{1\}$, and $C_3 = \{2,3\}$. Only C_1 is closed, C_2 and C_3 are transient. If the process starts, say, in state 3, it will spend some time in class C_2, possibly going back and forth between states 2 and 3 for some time. It will, however, at some point leave this class and proceed to state 1. It may return to state 1 a few times, but eventually it will end up in state 0. State 0 makes up its own closed class and is an *absorbing* state. Because when entering state 3, the process will always return to state 2, we call state 3 a *reflecting* state.

In conjunction with formula (A.15), the class division of the transition probability matrix determines the long-run behavior of X. The numerical expressions for $\mathbf{P}^5$ and $\mathbf{P}^{10}$ in A.2.1 show the absorbing effect of state 0 but also suggest a *periodic* behavior of class C_3; transition probabilities alternate between zero and positive (albeit declining) values. One may then ask what happens with an irreducible chain, where the closed class consists of several states, not only a single absorbing one. It is clear that the process will never be absorbed in any one state, but it is still natural to consider the limiting value of the transition probabilities $\mathbf{P}^n$. A general result ensures the existence of $\lim_{n\to\infty}\mathbf{P}^n$ provided the Markov chain is *irreducible, aperiodic, and recurrent* (see Karlin and Taylor, 1975). Not only does the limit exist, but it is also independent of what state the process starts in, as one might expect. For a finite state space, a class is recurrent if and only if it is closed, so this result is true for all irreducible aperiodic finite Markov chains.

Since the limiting distribution is independent of starting state, it is clear that all rows of $\lim_{n\to\infty}\mathbf{P}^n$ are equal to the limiting distribution. Let $\boldsymbol{\pi}$ be a column vector denoting this limiting distribution. We can write $\boldsymbol{\pi} = (\lim_{n\to\infty}\mathbf{P}^n)^T = \mathbf{P}^T(\lim_{n\to\infty}\mathbf{P}^{n-1})^T = \mathbf{P}^T\boldsymbol{\pi}$, thus the equation

$$\boldsymbol{\pi} = \mathbf{P}^T\boldsymbol{\pi} \tag{A.27}$$

determines the limiting distribution. This is the same as saying that 1 is an eigenvalue of $\mathbf{P}^T$ and $\boldsymbol{\pi}$ is the corresponding eigenvector, normalized so that $\sum_{i\in\mathscr{S}}\pi_i = 1$, where π_i are the elements of $\boldsymbol{\pi}$. (Equivalently, $\boldsymbol{\pi}$ is the left eigenvector of $\mathbf{P}$.) This result guarantees the existence and uniqueness of the solution of (A.27). Any solution to (A.27) is known as a *stationary* distribution, since any Markov chain started in a stationary distribution will remain in that distribution in the future, as is easily verified from (A.15). It should be noted, however, that a stationary distribution is not necessarily a limiting distribution. For instance, if the Markov chain is periodic, a stationary distribution may exist, but in the strict sense the limiting distribution will not.

As another example, we slightly modify the transition matrix defined in (A.26) to

$$\mathbf{P} = \begin{pmatrix} 9/10 & 1/10 & 0 & 0 \\ 1/2 & 1/2 & 0 & 0 \\ 0 & 1/2 & 0 & 1/2 \\ 0 & 0 & 1 & 0 \end{pmatrix}.$$

The only difference is that the previously absorbing state 0 now communicates with state 1, having a relatively small probability of returning to state 1 after entering state 0. We then have

$$\mathbf{P}^5 = \begin{pmatrix} 0.835 & 0.165 & 0 & 0 \\ 0.825 & 0.175 & 0 & 0 \\ 0.581 & 0.294 & 0 & 0.125 \\ 0.515 & 0.235 & 0.25 & 0 \end{pmatrix} \quad \text{and} \quad \mathbf{P}^{10} = \begin{pmatrix} 0.833 & 0.167 & 0 & 0 \\ 0.833 & 0.167 & 0 & 0 \\ 0.792 & 0.177 & 0.031 & 0 \\ 0.769 & 0.200 & 0 & 0.031 \end{pmatrix}.$$

The eigenvalues of $\mathbf{P}^T$ are $\lambda_1 = 1$, $\lambda_2 = \sqrt{2}/2$, $\lambda_3 = 0.4$, and $\lambda_4 = -\sqrt{2}/2$, and the normalized eigenvector corresponding to the eigenvalue 1 is $\boldsymbol{\pi} = (5/6, 1/6, 0, 0)^T$, which is the limiting distribution. Note that the eigenvalue 1 is larger in absolute value than all the other eigenvalues. To understand why this dominant eigenvalue and the corresponding eigenvector determine the long-run behavior, assume we start the process in an arbitrary distribution $\mathbf{p}(0)$. Let $\mathbf{v}_1 = \boldsymbol{\pi}$; $\mathbf{v}_2$, $\mathbf{v}_3$, and $\mathbf{v}_4$ denote the four (linearly independent) eigenvectors of $\mathbf{P}^T$ belonging to the eigenvalues $\lambda_1 = 1$, λ_2, λ_3, and λ_4. Write the starting distribution vector as $\mathbf{p}(0) = \alpha_1 \mathbf{v}_1 + \alpha_2 \mathbf{v}_2 + \alpha_3 \mathbf{v}_3 + \alpha_4 \mathbf{v}_4$. We then apply the transitions $\mathbf{P}^T$ to this n times. Using (A.15) the result is

$$\mathbf{p}(n) = (\mathbf{P}^T)^n \mathbf{p}(0) = \alpha_1 \lambda_1^n \mathbf{v}_1 + \alpha_2 \lambda_2^n \mathbf{v}_2 + \alpha_3 \lambda_3^n \mathbf{v}_3 + \alpha_4 \lambda_4^n \mathbf{v}_4, \qquad \text{(A.28)}$$

and it is clear that $\alpha_1 \lambda_1^n \mathbf{v}_1 = \alpha_1 \mathbf{v}_1$ becomes the dominant part as $n \to \infty$. All other contributions converge at a geometric rate to zero. It is also clear why there cannot be any eigenvalues larger than 1; such eigenvalues would cause elements of $\mathbf{p}(n)$ to grow without restriction.

A.3.2 The quasi-stationary distribution of a Markov chain with an absorbing state

We now consider another form of stationarity that is important in the context of event history analysis. Assume state 0 is the only absorbing state for the Markov chain $X(n)$. Define the random time $T = \min\{n \mid X(n) = 0\}$, that is, the time until absorption in state 0. The time T may be considered a (discrete) time to event, where an event is defined as absorption. The process $X(n)$ will then serve as a model for individual "risk"; at time n, the hazard of getting absorbed for a specific individual is the probability that the Markov chain will be absorbed in zero in the next step given that it is currently in state $X(n)$. Models of this type will play a prominent role in Chapter 10, and here we look at some properties of the chain conditional on nonabsorption. Specifically, consider the conditional probability $P(X(n) = i \mid X(n) \neq 0) = P(X(n) = i \mid T > n)$. Since T is the time of absorption, this is the state space distribution for the "surviving" individuals.

Interestingly, in many situations (more general than the current setting) the distribution of survivors converges to a *quasi-stationary* distribution. In terms of transition probabilities this means that $\lim_{n \to \infty} P(X(n) = i \mid X(n) \neq 0)$ exists and defines a probability distribution on the state space without state 0. It is seen that if we block-partition the matrix $\mathbf{P}$ as

$$\mathbf{P} = \begin{pmatrix} 1 & \mathbf{0} \\ \mathbf{P}_0 & \mathbf{P}_{-0} \end{pmatrix},$$

where $\mathbf{0} = (0, 0 \dots)$, $\mathbf{P}_0 = (p_{10}(1), p_{20}(1), \dots)^T$ and $\mathbf{P}_{-0}$ is the transition matrix for all states except 0, then

$$\mathbf{P}^n = \begin{pmatrix} 1 & \mathbf{0} \\ \mathbf{P}_0(n) & \mathbf{P}_{-0}^n \end{pmatrix}$$

with $\mathbf{P}_0(n) = (p_{10}(n), p_{20}(n), \dots)^T$. We are interested in the behavior of $\mathbf{P}_{-0}^n$, but under the assumption of nonabsorption, that is, each row should be normalized to sum to one. As for the limiting distribution discussed earlier we look at the eigenvalues of $\mathbf{P}_{-0}^T$. In a typical situation, $\mathbf{P}_{-0}^T$ will have linearly independent eigenvectors and a dominant (largest absolute value) eigenvalue $0 < \lambda_1 < 1$. Using an expansion like (A.28) but for $\mathbf{P}_{-0}^T$ rather than $\mathbf{P}^T$ it becomes clear that when n is large the distribution $\mathbf{p}(n)$ (on the nonzero states) is dominated by the term $\alpha_1 \lambda_1^n \mathbf{v}_1$. Thus, the eigenvector corresponding to the dominant eigenvalue (normalized to sum to one) is the resulting quasi-stationary distribution.

We also notice that in the long run, the number of individuals still "alive" (not yet absorbed in zero) is reduced by a factor of λ_1 in each step. This means that, out of the live population, a proportion of $1 - \lambda_1$ die in each transition. The long run (discrete) hazard of dying is thus $1 - \lambda_1$ in this model.

To illustrate this, consider the following example of a discrete time birth-death process on the states $\{0, 1, \dots, 5\}$:

$$\mathbf{P} = \begin{pmatrix} 1 & 0 & 0 & 0 & 0 & 0 \\ 2/4 & 1/4 & 1/4 & 0 & 0 & 0 \\ 0 & 2/4 & 1/4 & 1/4 & 0 & 0 \\ 0 & 0 & 2/4 & 1/4 & 1/4 & 0 \\ 0 & 0 & 0 & 2/4 & 1/4 & 1/4 \\ 0 & 0 & 0 & 0 & 1 & 0 \end{pmatrix}.$$

State 0 is absorbing and state 5 is immediately reflecting. We are interested in seeing the quasi-stationary distribution on the states $\{1, 2, \dots, 5\}$ and finding the limiting rate at which individuals get absorbed in 0. Define as before $\mathbf{P}^T_{-0}$ the submatrix of $\mathbf{P}^T$ with first row and column removed. The approximate eigenvalues of $\mathbf{P}^T_{-0}$ are $-0.494, -0.214, 0.202, 0.619$, and 0.887, so the dominant eigenvalue is $\lambda_1 = 0.887$. The corresponding normalized eigenvector is $(0.227, 0.289, 0.254, 0.179, 0.051)^T$, and this is the quasi-stationary distribution. The rate at which individuals in the long run are absorbed into state 0 is $1 - \lambda_1 = 0.113$.

Notice that in this example only the individuals in state 1 can be absorbed at any time. Among those not yet absorbed at time n there is a proportion of about 0.227 in state 1 when n is big, the first element in the quasi-stationary distribution. Since the transition probability from 1 to 0 at any time is $1/2$, one should expect an influx of $(1/2) \times 0.227 = 0.114$, which matches the value $1 - \lambda_1$. In general, this relationship can be written

$$\text{Event rate} = 1 - \lambda_1 = \sum_{i \geq 1} \pi_i^{\text{q.s.}} p_{i0},$$

where $\pi_i^{\text{q.s.}}$ are the elements of the quasi-stationary distribution; it is a direct consequence of eigenvalue properties. The long run behavior is thus described in more detail by saying that the survivors follow an approximate quasi-stationary distribution and that the rate of events is determined by the flow from the quasi-stationary distribution into state 0, specifically a proportion of $\pi_i^{\text{q.s}} p_{i0}$ from state i. In other words, if $\boldsymbol{\pi}^{\text{q.s.}}$ is the quasi-stationary distribution on the states $\{1, 2, \ldots\}$ and the process is started in the distribution $\mathbf{p}(0) = \begin{pmatrix} 0 \\ \boldsymbol{\pi}^{\text{q.s.}} \end{pmatrix}$, then

$$\mathbf{p}(n) = (\mathbf{P}^T)^n \mathbf{p}(0) = \begin{pmatrix} 1 - \lambda_1^n \\ \lambda_1^n \boldsymbol{\pi}^{\text{q.s.}} \end{pmatrix}.$$

As one would expect, many of the discrete time results on stationarity and quasi-stationarity discussed here carry over more or less directly to continuous time Markov chains and also to the corresponding diffusion process models. Hitting time distributions and quasi-stationarity are central to Chapter 10 of the main text.

A.4 Diffusion processes and stochastic differential equations

As we have seen, Markov processes with discrete state space are natural models for multistate situations. The hazard of an event depends on the current state of the process, there is a well-defined set of possible classes, and the movements of the process between classes is reliably observed. On the other hand, in Chapter 10 we demonstrate that such models are also suited to describe an underlying unobserved hazard, where an event occurs at the time the process reaches a "terminal" state. For that purpose, the exact states of the process may be of less importance; it is only the resulting hazard at the terminal state that matters. In such situations, the underlying hazard development may just as well be seen as a continuously developing process.

Additionally, in situations where measurements are made on covariate processes, for instance, hormone levels or blood pressure in a hospital patient, it is natural to use stochastic process models with both continuous time *and* state space.

The most prominent of all continuous processes is the Wiener process. Its key position in the theory of stochastic processes is similar to that of the Gaussian distribution in statistical theory. It is the archetypal example of a diffusion process, and the driving force of the random part of stochastic differential equations. Frequently, continuous processes may be described using fairly explicit descriptions of the processes itself in terms of stochastic integrals, or of the transition probabilities as solutions to well-studied ordinary differential equations. Variants of the Wiener process also appear as limiting processes for counting process martingales and are as such indispensable in the asymptotic theory covered in this book, in particular in Chapters 2, 3, and 4.

This section starts by describing the Wiener process itself; we then describe extensions to more general diffusions as solutions to stochastic differential equations. In Section A.5 the Wiener process is extended to independent increment (Lévy) processes, which combines continuous processes with jump processes.

A.4.1 The Wiener process

Let $W(t)$ be a process defined for $t \in [0, \infty)$, taking values in $\mathbf{R}$, and with initial value $W(0) = 0$. If W has increments that are independent and normally distributed with

$$\mathrm{E}\{W(s+t) - W(t)\} = 0 \quad \text{and} \quad \mathrm{Var}\{W(s+t) - W(t)\} = s,$$

we call W a *Wiener process*. The Wiener process is also frequently referred to as a *Brownian motion*. Note that, in particular, $\mathrm{E}W(t) = 0$ and $\mathrm{Var}W(t) = t$. As a continuous random walk, the Wiener process will "diffuse" away from its starting point, that is, it will wander back and forth between increasingly larger positive and negative values in the state space; its position at time t is always a Gaussian distribution $N(0,t)$. To increase the flexibility of the Wiener process for applications, one can define the process $X(t) = x + mt + \sigma W(t)$, which is called a Wiener process with initial value x, drift coefficient m, and diffusion coefficient σ^2. The parameters m and σ^2 are also referred to as the *(infinitesimal) mean and variance*, respectively, since

$$\mathrm{E}\{X(t)\} = x + mt \quad \text{and} \quad \mathrm{Var}\{X(t)\} = \sigma^2 t$$

[which is also more generally true for diffusion processes when time t is replaced by the infinitesimal increment dt, as in (A.31) and (A.33)]. Excellent introductions to many aspects of Wiener processes are found in Cox and Miller (1965) and Allen (2003). Figure A.6 shows simulations of typical paths of the Wiener process. Note how a larger diffusion coefficient makes the variability of the process larger and partially masks the drift.

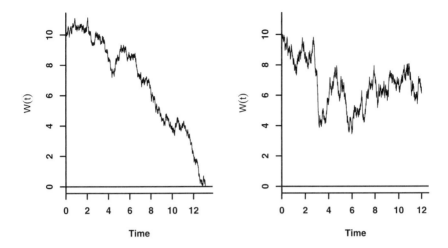

Fig. A.6 *Simulated paths of the Wiener process. Left panel:* $x = 10$, $\mu = -1$, *and* $\sigma^2 = 1$. *Right panel:* $x = 10$, $\mu = -1$, *and* $\sigma^2 = 3$. *In the left panel the process strikes a barrier at zero at about* $t = 13$.

Recall that a *martingale* is a stochastic process $X(t)$ such that $\mathrm{E}(X(t) \mid X(u), u \le s)$ $= X(s)$ (see Jacod and Protter, 2003; Billingsley, 1995). That is, a martingale is unbiased about the future in the sense that regardless of its history, the current value $X(s)$ is the "best guess" of the future value $X(t)$. (In the full definition, the history may include more events than those determined by X itself, as long as they do not create bias about the future of X.) Martingales play an important role in the estimation theory for point processes in Chapters 2, 3, and 4, and they are indeed central building blocks of the general theory of stochastic processes (Karatzas and Shreve, 1991; Kallenberg, 1997; Andersen et al., 1993). We do not present general results for martingales in this appendix, but we point out situations where they naturally occur in the context of diffusion processes.

A consequence of the independent increments of W is that it is a martingale and a time-homogeneous Markov process. The discrete-state transition probabilities for Markov chains are replaced by transition densities. For a stochastic variable X with density f, we use the intuitive notation $P(X \in (x, x+dx]) = P(X \in dx) = f(x)dx$ for an "infinitesimal" quantity dx. The transition probabilities for the Wiener process can be written as

$$P(W(s+t) \in dy \mid W(s) = x) = \phi(y - x; t)dy,$$

where $\phi(z; t) = \phi\left(z/\sqrt{t}\right)/\sqrt{t}$ and ϕ is the density of the standard normal distribution. Notice that the Wiener process is not only time-homogeneous but also homogeneous in space, since it only depends on the difference $y - x$, not on the actual values of x and y. The transition density for a Wiener process with general drift and diffusion can be expressed in a similar way.

A.4.2 Stochastic differential equations

As seen in Figure A.6, the path of the Wiener process is quite irregular, and in fact not differentiable as a function of time t. The "derivative" $dW(t)/dt$ is a "white noise" process, which is not a stochastic process in the ordinary sense. This problem is seen from the following argument: for a positive dt the differential $dW(t) = W(t+dt) - W(t)$ is normally distributed with mean zero and variance dt, and $dW(t)/\sqrt{dt}$ has a standard normal distribution. Since $dW(t)/dt = (dW(t)/\sqrt{dt})/\sqrt{dt}$, difficulties are to be expected when $dt \to 0$. Similarly, the sum of the absolute increments of W over an interval goes to infinity if the sum is taken over smaller and smaller subintervals, that is,

$$\lim_{\max |t_{i+1}-t_i| \to 0} \sum_i |W(t_{i+1}) - W(t_i)| = \infty$$

for partitions $0 < t_1 < \cdots < t_k < t$ of the interval $[0,t]$.

On the other hand, the square of the differential is more well behaved. It is easy to verify that

$$\mathrm{E}\left\{(dW(t))^2 - dt\right\}^2 = 2dt^2, \tag{A.29}$$

and from this (and the independence of increments) it follows that the sum of *squared* increments converges in mean square:

$$\lim_{\max |t_{i+1}-t_i| \to 0} \sum_i \left\{W(t_{i+1}) - W(t_i)\right\}^2 = t.$$

This is the *optional variation process* of W and is written $[W](t) = t$. (For obvious reasons, the optional variation process is sometimes called the *quadratic* variation process.) We discuss variation processes for other types of martingales in Chapter 2.

Relation (A.29) is often expressed in the simple identity

$$dW(t)^2 = dt, \tag{A.30}$$

which should be viewed as a formal identity to be used when computing with differentials. This observation is the starting point for a highly useful calculus based on the Wiener process. Let f be any sufficiently smooth function and consider the Taylor expansion

$$df(W(t)) = f'(W(t))dW(t) + \frac{1}{2}f''(W(t))dW(t)^2 + \cdots.$$

Using relation (A.30) and ignoring all higher-order terms of dt, we have

$$df(W(t)) = f'(W(t))dW(t) + \frac{1}{2}f''(W(t))dt.$$

This is the simplest version of the celebrated *Itô formula*. Notice that the first term in the Itô formula is what one would expect from a standard application of the chain rule, the second term is a peculiarity resulting from the path properties of the Wiener process. The Itô formula is easily extended to a more general process

$$dX(t) = \mu(t)dt + \sigma(t)dW(t) \tag{A.31}$$

by the formula

$$df(X(t)) = f'(X(t))dX(t) + \frac{1}{2}\sigma^2(t)f''(X(t))dt$$
$$= \left(\mu(t)f'(X(t)) + \frac{1}{2}\sigma^2(t)f''(X(t))\right)dt + \sigma(t)f'(X(t))dW(t). \tag{A.32}$$

Again, there is a dt-term that adds to the standard chain rule. Under the right conditions the Itô formula (A.32) is valid even when μ and σ are stochastic processes. Furthermore, if f also depend on t, the differential $df(t, X(t))$ is given as in (A.32) but with the additional term $f_1(t, X(t))dt$, where $f_1 = (\partial/\partial t)f$. Additional extensions to multivariate processes are relatively straightforward (Karatzas and Shreve, 1991).

It should be remarked that while the differential $dW(t)$ is not strictly defined, it makes sense in an integrated form, where, for instance, X in (A.31) is really defined by

$$X(t) = X(0) + \int_0^t \mu(s)ds + \int_0^t \sigma(s)dW(s).$$

The integral with respect to $dW(t)$ cannot be defined as an ordinary Stiltjes integral; the reason is that the usual approximating sums over subintervals used in the definition do not converge due to the unbounded variation of the Wiener process. Rather, the integral is defined as the *Itô integral*, which also extends to situations where σ is stochastic (Øksendal, 1998).

The process $\int_0^t \sigma(s)dW(s)$ is itself typically a mean zero martingale, and hence the Itô formula (A.32) decomposes the stochastic process $f(X(t))$ into an absolutely continuous (dt) part and a martingale ($dW(t)$) part. This is the Doob-Meyer decomposition (Karatzas and Shreve, 1991), which parallels the Doob decomposition of a counting process into a compensator and a martingale part, used to analyze the Nelson-Aalen estimator in Section 3.1.5. In fact, when σ is deterministic, the process $\int_0^t \sigma(s)dW(s)$ is a Gaussian martingale (i.e., the joint distribution at any finite selection of time points is multivariate normal), with distribution $N(0, \int_0^t \sigma^2(s)ds)$ at time t and covariance function

$$c(s,t) = \text{Cov}\left(\int_0^s \sigma(u)dW(u), \int_0^t \sigma(u)dW(u)\right) = \int_0^{s \wedge t} \sigma^2(u)du.$$

Such processes appear as the limit of compensated counting processes and are central to the asymptotic properties of hazard and survival estimators (Chapter 2).

The process X is a simple example of a *diffusion process* with drift μ and diffusion σ. Diffusion processes are continuous Markov processes whose sample paths resemble those of the Wiener process. A common way of introducing diffusion processes is to extend the differential equation for X to the more general equation

$$dX(t) = \mu(t, X(t))dt + \sigma(t, X(t))dW(t), \tag{A.33}$$

that is, by allowing the coefficients μ and σ also to depend on the process itself. Equation (A.33) is, for obvious reasons, known as a *stochastic differential equation*, and the solution X is a diffusion process. The initial value $X(0)$ may be random in many examples. In the context of diffusion processes, the coefficients are assumed nonrandom except for the dependence on X. In the special case $\sigma = 0$, Equation (A.33) is a standard first-order differential equation (possible nonlinear), and the solution $X(t)$ is random only through a possibly random initial value $X(0)$. In that case, the equation can be solved as a deterministic differential equation with smooth paths, using a random initial value. The more interesting case where σ is nonzero is entirely different, and the solution paths are usually not differentiable.

For some specific examples, an explicit solution of (A.33) in terms of the Itô integral can be found. Such explicit expressions enable us to derive properties of the solution directly from the properties of the stochastic integral. This complements the more traditional approach to stochastic differential equations through the Kolmogorov equations described in Section A.4.4. A standard example is the Ornstein-Uhlenbeck process introduced below. In general, all linear stochastic differential equations permit such solutions (Karatzas and Shreve, 1991; Kallenberg, 1997).

A.4.3 The Ornstein-Uhlenbeck process

One of the classical examples of a linear stochastic differential equation is the *Langevin equation*

$$dX(t) = (a - bX(t))dt + \sigma dW(t),$$

where a, b, and σ are constants. The solution X to this equation is known as the *Ornstein-Uhlenbeck process*. We assume here that $b > 0$. A look at the equation itself reveals that there is a self-containing mechanism in the process; if $X(t)$ exceeds a/b the dt-part of the increment becomes negative and pushes the process downward again. Similarly, if $X(t)$ goes below a/b, it is pulled up again by a positive increment. The effect of this is that $X(t)$ fluctuates around a/b, and the random component $dW(t)$ perturbs $X(t)$ sufficiently that it never converges toward a/b, as would the corresponding deterministic equation ($\sigma = 0$). This property of the process X is called the *mean-reverting* property. Using the Itô integral, the explicit solution to the Langevin equation is

$$X(t) = \frac{a}{b} + \left(X(0) - \frac{a}{b}\right)e^{-bt} + \sigma \int_0^t e^{-b(t-s)}dW(s).$$

From what was said in A.4.2 we see that $X(t)$ is Gaussian with mean

$$EX(t) = \frac{a}{b} + \left(EX(0) - \frac{a}{b}\right)e^{-bt}$$

and variance

$$Var(X(t)) = e^{-2bt}Var(X(0)) + \frac{\sigma^2}{2b}(1 - e^{-2bt}).$$

There is a "run-in" effect caused by an initial value $X(0)$ different from a/b, but apart from that, the initial value loses its influence over time, and $EX(t) \to a/b$ and $Var(X(t)) \to \sigma^2/2b$ as $t \to \infty$. Except for the initial effect, $X(t)$ is stationary. The autocorrelation $Cor(X(t+h), X(t))$ goes to zero exponentially fast (proportional to $\exp(-bh)$) as $h \to \infty$. Figure A.7 shows what two sample paths of the Ornstein-Uhlenbeck process might look like. The Ornstein-Uhlenbeck process is particularly

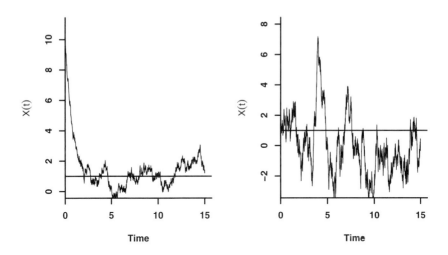

Fig. A.7 *Simulated paths of the Ornstein-Uhlenbeck process. Left panel: $X(0) = 10$, $a = 1$, $b = 1$, and $\sigma^2 = 1$. Right panel: $X(0) = 1$, $a = 1$, $b = 1$, and $\sigma^2 = 9$. The mean-reverting value $a/b = 1$ is marked by a horizontal line in both plots.*

useful as a model for variables that remain fairly stable over time, that is, they may fluctuate in a random fashion but tend to return to the same region over and over again. The Gaussian property and stationarity make it easy to work with. We present models based on the Ornstein-Uhlenbeck process in Chapters 10 and 11.

A.4.4 The infinitesimal generator and the Kolmogorov equations for a diffusion process

As we did for the continuous Markov chains in Section A.2.2, we now have a brief look at the infinitesimal generator for diffusion processes defined by a stochastic differential equation (A.33). The generator provides a direct link between the approach through differentials and stochastic differential equations and the more traditional approach through the Kolmogorov equations.

For inhomogeneous Markov chains we defined the generator as the derivative of the transition probabilities at time t, as in (A.23). A similar approach for diffusions is to define the generator as the derivative at time t of the expectation of suitable functions. When, as before, f is a sufficiently smooth function, we define the generator $\boldsymbol{\alpha}(t)$ as

$$(\boldsymbol{\alpha}(t)f)(x) = \lim_{\Delta t \to 0} \frac{1}{\Delta t} \mathrm{E}\left[f(X(t+\Delta t)) - f(X(t)) \mid X(t) = x\right]$$
$$= \frac{1}{dt} \mathrm{E}\left[df(X(t)) \mid X(t) = x\right].$$

The operator $\boldsymbol{\alpha}(t)$ transforms, for each t, the function f into a new function $\boldsymbol{\alpha}(t)f$ which takes the argument x. Using the Itô formula (A.32) on the differential $df(X(t))$ it can be seen (strictly speaking using a martingale argument) that the expectation of the dW-part is zero, thus

$$(\boldsymbol{\alpha}(t)f)(x) = \mu(t,x)f'(x) + \frac{1}{2}\sigma^2(t,x)f''(x),$$

or, written as a differential operator,

$$\boldsymbol{\alpha}(t) = \mu(t,x)\frac{d}{dx} + \frac{1}{2}\sigma^2(t,x)\frac{d^2}{dx^2}.$$

Notice the analogy to the finite-valued Markov chain situation: if the state space is $\mathscr{S} = \{0,1,\ldots,n\}$, then any vector $\mathbf{v}$ of length $n+1$ can be seen as a function mapping the state space into $\mathbf{R}$. The generator $\boldsymbol{\alpha}(t)$ is an $(n+1) \times (n+1)$ matrix, which, by matrix multiplication, maps $\mathbf{v}$ into the vector $\boldsymbol{\alpha}(t)\mathbf{v}$. For diffusions, $\boldsymbol{\alpha}(t)$ maps the function f into the function $\boldsymbol{\alpha}(t)f$. In the simplest case, when X is a standard Wiener process, $\boldsymbol{\alpha}(t) = \boldsymbol{\alpha} = (1/2)d^2/dx^2$, that is, just a simple second-order differential operator. In the corresponding discrete situation, where X is a symmetric random walk,

$$\boldsymbol{\alpha} = \begin{pmatrix} 0 & 0 & 0 & 0 & \cdots \\ \beta & -2\beta & \beta & 0 & \cdots \\ 0 & \beta & -2\beta & \beta & \cdots \\ \vdots & \vdots & \vdots & \vdots & \ddots \end{pmatrix},$$

and left-multiplying a vector $\mathbf{v}$ with $\boldsymbol{\alpha}$ yields

$$\beta v_{i+1} - 2\beta v_i + \beta v_{i-1} = \beta \left((v_{i+1} - v_i) - (v_i - v_{i-1}) \right),$$

which is a second-order *difference* operator (except at the boundary of the state space, where absorption occurs).

As a further illustration, we show how the Kolmogorov forward equation for the density $\psi_t(x) = P(X(t) \in dx)/dx$ of a diffusion process $X(t)$ can be derived, following along the lines of Cox and Miller (1965, page 217). Starting with a sufficiently smooth function f and using the Chapman-Kolmogorov equations, we see that

$$\frac{1}{dt} \mathrm{E} df(X(t)) = \int_{-\infty}^{\infty} \frac{1}{dt} \mathrm{E}(df(X(t)) \mid X(t) = x) \psi_t(x) dx$$

$$= \int_{-\infty}^{\infty} (\boldsymbol{\alpha}(t)f)(x) \psi_t(x) dx$$

$$= \int_{-\infty}^{\infty} \left(\mu(t,x) f'(x) + \frac{1}{2} \sigma^2(t,x) f''(x) \right) \psi_t(x) dx.$$

Using integration by parts, assuming that the integrand vanishes at $-\infty$ and $+\infty$, the differential operator can be "moved" from f to ψ_t:

$$\frac{1}{dt} \mathrm{E} df(X(t)) = \int_{-\infty}^{\infty} f(x) (\boldsymbol{\alpha}^\star(t) \psi_t)(x) dx,$$

where $\boldsymbol{\alpha}^\star(t)$ is the (adjoint) differential operator

$$(\boldsymbol{\alpha}^\star(t)g)(x) = -\frac{d}{dx} [\mu(t,x)g(x)] + \frac{1}{2} \frac{d^2}{dx^2} \left[\sigma^2(t,x)g(x) \right].$$

However, we can also compute the same expression more directly as

$$\frac{1}{dt} \mathrm{E} df(X(t)) = \frac{d}{dt} \int_{-\infty}^{\infty} f(x) \psi_t(x) dx = \int_{-\infty}^{\infty} f(x) \frac{d}{dt} \psi_t(x) dx.$$

We have thus established the equality

$$\int_{-\infty}^{\infty} f(x) \frac{d}{dt} \psi_t(x) dx = \int_{-\infty}^{\infty} f(x) (\boldsymbol{\alpha}^\star(t) \psi_t)(x) dx$$

for all sufficiently smooth f, in particular when $f(x) = \exp(-ux)$, which yields the Laplace transform on both sides. It follows that

$$\frac{d}{dt} \psi_t(x) = (\boldsymbol{\alpha}^\star(t) \psi_t)(x),$$

which is the Kolmogorov forward equation for ψ.

The Kolmogorov equations appear in many forms, one version (Karatzas and Shreve, 1991, page 369) is that for a function f (continuous and with compact support), the function

$$u(t,x) = \mathrm{E}\{f(X(t)) \mid X(0) = x\}$$

solves the backward equation

$$\frac{\partial}{\partial t} u(t,x) = \boldsymbol{\alpha}(t) u(t,x), \tag{A.34}$$

with initial condition

$$u(0,x) = f(x).$$

Note that, in particular, if $f(x) = I(x \leq z)$, then

$$\mathrm{E}\{f(X(t)) \mid X(0) = x\} = P(X(t) \leq z \mid X(0) = x),$$

which informally relates this result to the corresponding result for transition probabilities (although f is not continuous in this case).

The Kolmogorov equations are useful to derive expressions not only of the transition densities of a process, but also long-run (stationary) distributions and quasi-stationary distributions. The resulting differential equations are well understood, and sometimes explicit solutions can be found (Polyanin and Zaitsev, 1995).

A.4.5 The Feynman-Kac formula

As we saw in A.3.2, a model for event times can be constructed from a diffusion process, or more generally from a Markov process, by letting the event time T be defined as the first time a process hits a barrier. However, there may be an even more natural way of linking event histories to an underlying stochastic process: to use a nonnegative process directly as a model for a randomly varying hazard. This is the subject of Chapter 11. The basic idea is that if $X(t)$ is a stochastic process describing how the risk of an individual develops over time, we can assume that the hazard of this individual is $h(X(t))$ at time t, where h is a nonnegative real function. Thus, the individual survival function (for a given X) is

$$P(T > t \mid X) = \exp\left\{-\int_0^t h(X(s))ds\right\},$$

and the corresponding population survival is

$$S(t) = \mathrm{E}\exp\left\{-\int_0^t h(X(s))ds\right\}. \tag{A.35}$$

Let us assume that X is a diffusion process with generator $\boldsymbol{\alpha}(t)$ and recall the Kolmogorov backward equation (A.34). Interestingly, the Kolmogorov equation has a direct extension which applies to expression (A.35). Adding an extra term, consider the equation

$$\frac{\partial}{\partial t}v(t,x) = \boldsymbol{\alpha}(t)v(t,x) - h(x)v(t,x) \tag{A.36}$$

with the initial condition $v(0,x) = f(x)$ as before. The solution to this equation is given by the *Feynman-Kac* formula

$$v(t,x) = \mathrm{E}\left\{\exp\left(-\int_0^t h(X(s))ds\right) f(X(t)) \,|\, X(0) = x\right\},$$

which simply extends the definition of $u(t,x)$ to include the survival term in (A.35). In particular, the survival function

$$S(t,x) = \mathrm{E}\exp\left\{-\int_0^t h(X(s))ds \,|\, X(0) = x\right\}$$

solves Equation (A.36), with initial condition $S(0,x) \equiv 1$.

There is a rather suggestive connection between the random hazard formulation and the concept of "killing" a Markov process, a term frequently seen in Markov theory (Øksendal, 1998). Let $X(t)$ be a continuous time Markov process and let T be a random time. Let Ξ be an artificial extra state of the process, not already in the state space. Define a new process $\widetilde{X}$ by

$$\widetilde{X}(t) = \begin{cases} X(t) \text{ when } t \leq T, \\ \Xi \text{ when } t > T. \end{cases}$$

Then $\widetilde{X}$ is called a "killed" Markov process and T the killing time; the state Ξ is sometimes referred to as a "coffin state". One usually assumes that the distribution of T is determined from X by the relation

$$\frac{1}{dt}P(T \in [t,t+dt] \,|\, \widetilde{X}(t) = x) = h(x)$$

for some nonnegative function h for which $h(\Xi) = 0$. Thus

$$\frac{1}{dt}P(T \in [t,t+dt] \,|\, T > t, X(t)) = h(X(t)),$$

and so

$$P(X \text{ killed at } t \text{ or later}) = P(T \geq t) = \mathrm{E}\left\{\exp\left(-\int_0^t h(X(s))ds\right)\right\}.$$

Thus, this is entirely equivalent to a time to event situation where $h(X(t))$ is the individual hazard, as described earlier. It is not hard to see that $\widetilde{X}$ is also Markov on the extended state space. The generator of $\widetilde{X}$ is $\boldsymbol{\alpha}(t) - h$. For related discussions, see Singpurwalla (1995, 2006) and Cox (1999). For further information on killing in Markov process theory, see Øksendal (1998) and Karlin and Taylor (1981).

There is also a natural connection between treating the hazard directly as a stochastic process and viewing a random process as a risk process generating an event when it hits a barrier. Let R be an exponentially distributed random variable with $ER = 1$, and assume R is independent of the hazard process $h(X(t))$. Define

$$T = \inf\{t : \int_0^t h(X(s))ds = R\},$$

that is, the time when the cumulative hazard hits the random barrier R. Conditional on X,

$$P(T > t \mid X) = P(R > \int_0^t h(X(s))ds \mid X) = \exp\left\{-\int_0^t h(X(s))ds\right\}.$$

Thus, an individual has an event (or equivalently, the Markov process is "killed") when $H(t) = \int_0^t h(X(s))ds$ strikes the (randomized) barrier R.

A.5 Lévy processes and subordinators

Recall that the Wiener process has the independent increment property, that is, that nonoverlapping increments are independent and all increments of the same length have the same distribution. Both the martingale property and the Markov property follow from this. Another well-known example of an independent increments process is the *Poisson process* $N(t)$. The Poisson process takes values $\{0, 1, 2, \ldots\}$, it starts in zero and increases with jumps of one at random times. The distances between jumps are independently exponentially distributed with parameter (intensity) λ, that is, with mean $1/\lambda$. As a result, the increments $N(t) - N(s)$ of the Poisson process are Poisson distributed with parameter $\lambda(t - s)$. Note that the distribution of the increment depends only on the length of the time interval, $t - s$, not on s or t. The Poisson process is one of the simplest examples of a *counting process*; it counts the number of jumps up to time t. Figure A.8 shows a typical Poisson process path. Note that jump processes in our context are usually assumed to be right continuous, and that limits from the left exist at all points, i.e. it is cadlag.

There are two differences between the Wiener process and the Poisson process. First, the Poisson process is a pure jump process, whereas the Wiener process has only continuous paths. Second, even though the Poisson process is Markov, it is not really a martingale since its expectation is $EN(t) = \lambda t$, which is increasing. However, it is a submartingale, and subtracting λt makes $M(t) = N(t) - \lambda t$ a martingale (see Chapter 2). A natural question is whether the simplifying and highly useful property of independent and identically distributed increments (over same-length intervals) also holds true for other processes, and if so, what do they look like? A simple extension of the standard Wiener process is to introduce general (constant) drift and diffusion, as discussed in A.4.1. An extension of the Poisson process is to let the jump sizes be independent identically distributed random variables. This is

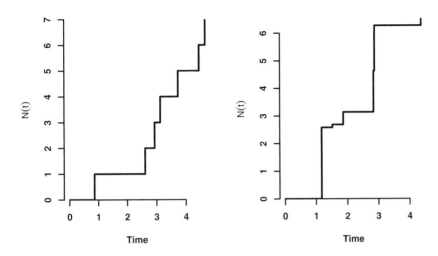

Fig. A.8 *A simulated path of a standard Poisson process with intensity* $\lambda = 1$ *(left), and a compound Poisson process with* $\lambda = 1$ *and exponentially distributed jumps with parameter 1 (right).*

known as a *compound Poisson process* and is a basis for many useful and easy-to-work-with models. Figure A.8 shows the path of a compound Poisson process with exponentially distributed jumps with expectation 1.

A.5.1 The Lévy process

In general, let Z be a right-continuous left-limit stochastic process. If the increments $Z(t) - Z(s)$ are independent (when nonoverlapping) and have the same distribution as $Z(t - s)$, we call Z a *Lévy process*. The increments of Z are said to be stationary since their distribution depends only on the length of the interval. By necessity, we must have $Z(0) = 0$. We see that both the compound Poisson process and the Wiener process $mt + \sigma W(t)$ described in A.4.1 are examples of Lévy processes. Due to the independent increments, Lévy processes are Markov. It is also clear that any linear combination of independent Lévy process is again a Lévy process; one such example would be the sum of a Wiener process and a Poisson process. There are many more examples of Lévy processes. In particular, the Wiener process has no jumps, and the jumps of the Poisson process are isolated, in the loose sense that you rarely have several jumps very close to one another. Lévy processes with densely spaced jumps are possible, provided the typical jump size is not too big. One such example is the *inverse Gaussian* process. Let W be a standard Wiener process and define the inverse Gaussian process $Z(t)$ by

$$Z(t) = \inf\{s \geq 0 \mid W(s) > t\},$$

that is, for each t, $Z(t)$ is the time W hits the barrier t. The construction is illustrated in Figure A.9. We note that the inverse Gaussian process is just the (pathwise) in-

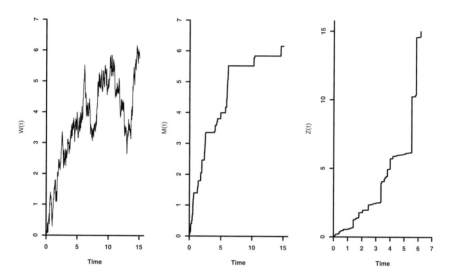

Fig. A.9 *The inverse Gaussian process. Starting with a standard Wiener process (left), compute the running maximum (middle). The inverse Gaussian process is obtained by inverting the axes of the running maximum process.*

verse of the running maximum $M(t) = \max_{s \leq t} W(s)$. The inverse Gaussian process can have a lot of small jumps before suddenly taking "large" jumps, as determined by the dips in the Wiener process. At any point in time, the inverse Gaussian process $Z(t)$ has an inverse Gaussian distribution (Applebaum, 2004).

Both the compound Poisson process and the inverse Gaussian share the property of being nondecreasing. In fact, any Lévy process that is nonnegative will have non-negative increments and thus be nondecreasing. Such a Lévy process is known as a *subordinator*. The nonnegativity of subordinators makes them suitable as models for time-changing frailty (Chapter 11). Although simple in definition, Lévy processes, in particular subordinators, may produce a wide range of models. In the following we mainly focus on subordinators, but much of the reasoning is valid for general Lévy processes.

Let $\{0, t/n, 2t/n, \ldots, t\}$ be a partition of the interval $(0, t]$ into equal length parts. Writing

$$Z(t) = \{Z(t) - Z((n-1)t/n)\} + \cdots + \{Z(2t/n) - Z(t/n)\} + Z(t/n), \qquad \text{(A.37)}$$

we see that for any n, $Z(t)$ can be written as a sum of independent and identically distributed variables. It seems reasonable from this that we can write the expected value and variance of a Lévy process $Z(t)$ as linear functions $EZ(t) = tEZ(1)$, and similarly for the variance. This is indeed the case, as seen in (A.39) and (A.40). By the same token, one would expect the characteristic function of $Z(t)$, $\chi(c;t) = E\{\exp(icZ(t))\}$, to be a power $\chi(c;1)^t$ of the characteristic function at time 1. This is also true; the corresponding result for the Laplace transform is given in A.5.2. The distribution of a random variable which can be written as a sum of n independent and identically distributed variables for any n is called an *infinitely divisible distribution*. One conceptually important property of infinitely divisible distributions is that they can be approximated by compound Poisson processes. In that sense, the compound Poisson process is the archetypal Lévy process (Feller, 1971; Bertoin, 1996).

A.5.2 The Laplace exponent

Knowing the distribution of $Z(t)$ at each time t tells us all we need to know about Z, since any finite-dimensional distribution $(Z(t_1), Z(t_2), \ldots, Z(t_k))$ can be computed using the independent stationary increments (Exercise 11.1). The distribution of $Z(t)$ is determined by its characteristic distribution, but since we are mostly working with nonnegative Lévy processes, the Laplace transform of the distribution of $Z(t)$ is a useful replacement. It is thus convenient to have a simple representation of the Laplace transform. Such a representation is given by

$$\mathcal{L}(c;t) \stackrel{\text{def}}{=} E\{\exp(-cZ(t))\} = \exp(-t\Phi(c)), \qquad \text{(A.38)}$$

where $c \geq 0$ is the argument of the Laplace transform. The function $\Phi(c)$ is called the *Laplace exponent* of Z and is central to many computations involving subordinators. It is seen that $\Phi(c) = -C(-c)$, where C is the cumulant generating function of $Z(1)$ (Billingsley, 1995). Observe that the exponent of the representation is linear in t, or alternatively $\mathcal{L}(c;t) = \mathcal{L}(c;1)^t$. As indicated, this is a manifestation of the time-homogeneity of the increments of Z; see Exercise 11.2. In addition, for many common applications, a simple expression is available for Φ. Several such examples are presented in Chapter 11.

In general, Φ is increasing and concave (its derivative Φ' decreasing), and $\Phi(0) = 0$. From the Laplace transform of Z, we easily derive

$$EZ(t) = tEZ(1), \qquad \text{(A.39)}$$

$$\mathrm{Var}Z(t) = t\,\mathrm{Var}Z(1), \qquad \text{(A.40)}$$

with

$$EZ(1) = \Phi'(0) \quad \text{and} \quad \mathrm{Var}Z(1) = -\Phi''(0),$$

demonstrating that both the expected value and the variance of $Z(t)$ increase linearly with time. Although $\Phi'(c)$ and $\Phi''(c)$ always exist for $c > 0$, these quantities may become infinite when $c = 0$, and thus $EZ(1)$ and $\text{Var}Z(1)$ may possibly be infinite. We also remark that the autocorrelation of Z, $\text{Cor}(Z(t), Z(t+u))$, decreases with u and does not depend on what subordinator Z we use (see Exercise 11.4).

Our main use of Lévy processes will be as models for random hazards, along the lines leading to (A.35). In Chapter 11 we show that for hazard models expressed as

$$h(t) = \alpha(t) \int_0^t g(s, t - s) dZ(R(s)),$$

where α, g, and R are deterministic, R is nondecreasing and Z is a subordinator, the expectation on the right-hand side of (A.35) can be given an explicit representation in terms of the Laplace exponent Φ. Such representations are good starting points for estimation.

Appendix B
Vector-valued counting processes, martingales and stochastic integrals

Chapter 2 reviews a number of results on counting processes, martingales, and stochastic integrals. In order to keep the presentation simple, we in Chapter 2 only consider scalar-valued processes. In some applications, however, we need results for vector-valued counting processes, martingales, and stochastic integrals. These results, which are "the obvious" extensions of the results of Chapter 2, are collected here for easy reference. For further details and results, the reader is referred to Andersen et al. (1993, chapter II).

B.1 Counting processes, intensity processes and martingales

A *multivariate counting process*

$$\mathbf{N}(t) = (N_1(t), \ldots, N_k(t))^T$$

is a vector of k counting processes $N_1(t), \ldots, N_k(t)$, with the additional assumption that no two components jump simultaneously (with probability one). The process is adapted to a filtration (or history) $\{\mathscr{F}_t\}$. The smallest filtration that makes $\mathbf{N}(t)$ adapted is the filtration generated by the process itself, denoted the self-exciting filtration. Typically, the filtration $\{\mathscr{F}_t\}$ is larger than the self-exiting filtration to accommodate information on covariates, censorings, late entries, etc.

The *intensity process* $\boldsymbol{\lambda}(t)$ of a multivariate counting process is the intensity process of each component collected together in a vector, that is,

$$\boldsymbol{\lambda}(t) = (\lambda_1(t), \ldots, \lambda_k(t))^T,$$

where $\lambda_j(t)$ is the intensity process of $N_j(t)$; $j = 1, \ldots, k$.

We write $\int_0^t \boldsymbol{\lambda}(u)du$ for the k-dimensional vector with elements $\int_0^t \lambda_j(u)du$; $j = 1, \ldots, k$. Then

$$\mathbf{M}(t) = \mathbf{N}(t) - \int_0^t \boldsymbol{\lambda}(u)du$$

is a vector-valued martingale.

The predictable variation process of this vector-valued martingale is the $k \times k$ matrix $\langle \mathbf{M} \rangle(t)$ with the (h, j)th entry equal to $\langle M_h, M_j \rangle(t)$. The optional variation process $[\mathbf{M}](t)$ is defined similarly. By (2.42)–(2.45), we have that

$$\langle \mathbf{M} \rangle(t) = \operatorname{diag} \int_0^t \boldsymbol{\lambda}(u) du, \tag{B.1}$$

$$[\mathbf{M}](t) = \operatorname{diag} \mathbf{N}(t), \tag{B.2}$$

where $\operatorname{diag} \mathbf{v}$ of a k-dimensional vector $\mathbf{v}$ is the $k \times k$ diagonal matrix with the elements of $\mathbf{v}$ on the diagonal.

Two or more components of $\mathbf{N}(t) = (N_1(t), \dots, N_k(t))^T$ do not jump at the same time. Therefore the *aggregated* process $N.(t) = \sum_{j=1}^k N_j(t)$ has jumps of size one, and hence it is a univariate counting process. Its intensity process is given by

$$\lambda.(t) = \sum_{i=1}^k \lambda_i(t), \tag{B.3}$$

and the corresponding martingale is $M.(t) = \sum_{j=1}^k M_j(t)$.

B.2 Stochastic integrals

We now consider vector-valued stochastic integrals. To this end, we introduce the $p \times k$ matrix $\mathbf{H}(t)$ whose (h, j)th entry is the predictable process $H_{hj}(t)$; $h = 1, \dots p$, $j = 1, \dots, k$. We denote by $\int_0^t \mathbf{H}(u) d\mathbf{M}(u)$ the p-dimensional vector whose hth element is the sum of stochastic integrals $\sum_{j=1}^k \int_0^t H_{hj}(u) dM_j(u)$. Then

$$\left\langle \int \mathbf{H} d\mathbf{M} \right\rangle(t) = \int_0^t \mathbf{H}(u) \operatorname{diag}\{\boldsymbol{\lambda}(u) du\} \mathbf{H}(u)^T, \tag{B.4}$$

$$\left[\int \mathbf{H} d\mathbf{M} \right](t) = \int_0^t \mathbf{H}(u) \operatorname{diag}\{d\mathbf{N}(u)\} \mathbf{H}(u)^T. \tag{B.5}$$

Written out componentwise, (B.4) and (B.5) take the form

$$\left\langle \sum_{j=1}^k \int H_{hj} dM_j, \sum_{j=1}^k \int H_{lj} dM_j \right\rangle(t) = \sum_{j=1}^k \int_0^t H_{hj}(u) H_{lj}(u) \lambda_j(u) du, \tag{B.6}$$

$$\left[\sum_{j=1}^k \int H_{hj} dM_j, \sum_{j=1}^k \int H_{lj} dM_j \right](t) = \sum_{j=1}^k \int_0^t H_{hj}(u) H_{lj}(u) dN_j(u); \tag{B.7}$$

cf. (2.48) and (2.49).

B.3 Martingale central limit theorem

Let

$$\mathbf{M}^{(n)}(t) = \mathbf{N}^{(n)}(t) - \int_0^t \boldsymbol{\lambda}^{(n)}(u)du; \qquad n = 1, 2, \ldots$$

be a sequence of vector-valued counting process martingales, and denote by k_n the dimension of $\mathbf{M}^{(n)}$. In some applications we will have $k_n = k$, the same for all n, while in other applications, k_n will depend on n, typically $k_n = n$. For each n we introduce the $p \times k_n$ matrix $\mathbf{H}^{(n)}(t)$ of predictable processes $H_{hj}^{(n)}(t)$; $h = 1, \ldots, p$; $j = 1, \ldots, k_n$.

We will consider the limiting behavior of the sequence of vector-valued stochastic integrals

$$\int_0^t \mathbf{H}^{(n)}(u) \, d\mathbf{M}^{(n)}(u); \qquad n = 1, 2, \ldots.$$

Specifically, we will state conditions for this sequence to converge in distribution to a p-variate mean zero Gaussian martingale. The distribution of a p-variate mean zero Gaussian martingale $\mathbf{U}(t)$ is determined by its covariance function $\mathbf{V}(t) = \mathrm{E}\{\mathbf{U}(t)\mathbf{U}(t)^T\}$, a continuous deterministic $p \times p$ matrix-valued function, zero at time zero, and with positive definite increments $\mathbf{V}(t) - \mathbf{V}(s); t > s$.

One requirement for the sequence of vector-valued stochastic integrals to converge to a mean zero Gaussian martingale is that the predictable variation processes converge in probability to the covariance function of the limiting Gaussian martingale; cf. requirement (i) in Section 2.3.3. By (B.4) this requirement now takes the form

$$\int_0^t \mathbf{H}^{(n)}(u) \operatorname{diag}\{\boldsymbol{\lambda}^{(n)}(u)du\} \mathbf{H}^{(n)}(u)^T \xrightarrow{\mathrm{P}} \mathbf{V}(t) \text{ for all } t \in [0, \tau], \text{ as } n \to \infty, \quad \text{(B.8)}$$

which is similar to (2.58).

In addition, we need a requirement corresponding to (ii) in Section 2.3.3, which ensures that the sample paths of the stochastic integrals become continuous in the limit. Here this requirement may be written

$$\sum_{j=1}^{k_n} \int_0^t (H_{hj}^{(n)}(u))^2 I\{|H_{hj}^{(n)}(u)| > \varepsilon\} \lambda_j^{(n)}(u)du \xrightarrow{\mathrm{P}} 0 \qquad \text{(B.9)}$$

for all $t \in [0, \tau]$, $h = 1, \ldots, p$, and $\varepsilon > 0$, as $n \to \infty$. This is similar to (2.59).

References

Aalen, O. O. (1975). *Statistical inference for a family of counting processes.* Ph. D. thesis, Univ. of California, Berkeley.

Aalen, O. O. (1976). Nonparametric inference in connection with multiple decrement models. *Scandinavian Journal of Statistics 3*, 15–27.

Aalen, O. O. (1978a). Nonparametric estimation of partial transition probabilities in multiple decrement models. *Annals of Statistics 6*, 534–545.

Aalen, O. O. (1978b). Nonparametric inference for a family of counting processes. *Annals of Statistics 6*, 701–726.

Aalen, O. O. (1980). A model for non-parametric regression analysis of life times. In W. Klonecki, A. Kozek, and J. Rosinski (Eds.), *Mathematical Statistics and Probability Theory*, Volume 2 of *Lecture Notes in Statistics*, pp. 1–25. New York: Springer-Verlag.

Aalen, O. O. (1987). Dynamic modelling and causality. *Scandinavian Actuarial Journal*, 177–190.

Aalen, O. O. (1988a). Dynamic description of a Markov chain with random time scale. *The Mathematical Scientist 13*, 90–103.

Aalen, O. O. (1988b). Heterogeneity in survival analysis. *Statistics in Medicine 7*, 1121–1137.

Aalen, O. O. (1989). A linear regression model for the analysis of life times. *Statistics in Medicine 8*, 907–925.

Aalen, O. O. (1991). Modelling the influence of risk-factors on familial aggregation of disease. *Biometrics 47*, 933–945.

Aalen, O. O. (1992). Modelling heterogeneity in survival analysis by the compound Poisson distribution. *The Annals of Applied Probability 2*, 951–972.

Aalen, O. O. (1993). Further results on the non-parametric linear regression model in survival analysis. *Statistics in Medicine 12*, 1569–1588.

Aalen, O. O. (1994). Effects of frailty in survival analysis. *Statistical Methods in Medical Research 3*, 227–243.

Aalen, O. O. (1995). Phase type distributions in survival analysis. *Scandinavian Journal of Statistics 22*, 447–463.

Aalen, O. O., E. Bjertness, and T. Sønju (1995). Analysis of dependent survival data applied to lifetimes of amalgam fillings. *Statistics in Medicine 14*, 1819–1829.

Aalen, O. O., Ø. Borgan, and H. Fekjær (2001). Covariate adjustment of event histories estimated from Markov chains: The additive approach. *Biometrics 57*, 993–1001. Correction *59*, 452–453.

Aalen, O. O., Ø. Borgan, N. Keiding, and J. Thormann (1980). Interaction between life history events: Nonparametric analysis of prospective and retrospective data in the presence of censoring. *Scandinavian Journal of Statistics 7*, 161–171.

Aalen, O. O. and V. Didelez (2006). Discussion on Farrington, C. P. and Whitaker, H. J.: Semiparametric analysis of case series data. *Journal of the Royal Statistical Society: Series C (Applied Statistics) 55*, 587.

Aalen, O. O., V. T. Farewell, D. De Angelis, N. E. Day, and O. N. Gill (1997). A Markov model for HIV disease progression including the effect of HIV diagnosis and treatment: Application to AIDS prediction in England and Wales. *Statistics in Medicine 16*, 2191–2210.

Aalen, O. O., V. T. Farewell, D. De Angelis, N. E. Day, and O. N. Gill (1999). New therapy explains the fall in AIDS incidence with a substantial rise in number of persons on treatment expected. *AIDS 13*, 103–108.

Aalen, O. O., J. Fosen, H. Weedon-Fekjær, Ø. Borgan, and E. Husebye (2004). Dynamic analysis of multivariate failure time data. *Biometrics 60*, 764–773.

Aalen, O. O. and A. Frigessi (2007). What can statistics contribute to a causal understanding? *Scandinavian Journal of Statistics 34*, 155–168.

Aalen, O. O. and H. K. Gjessing (2001). Understanding the shape of the hazard rate: A process point of view (with discussion). *Statistical Science 16*, 1–22.

Aalen, O. O. and H. K. Gjessing (2003). A look behind survival data: Underlying processes and quasi-stationarity. In B. H. Lindqvist and K. A. Doksum (Eds.), *Mathematical and Statistical Methods in Reliability*, pp. 221–234. Singapore: World Scientific Publishing.

Aalen, O. O. and H. K. Gjessing (2004). Survival models based on the Ornstein-Uhlenbeck process. *Lifetime Data Analysis 10*, 407–423.

Aalen, O. O. and H. K. Gjessing (2007). Stochastic processes in survival analysis. In V. Nair (Ed.), *Advances in Statistical Modeling and Inference. Essays in Honor of Kjell A Doksum*, pp. 23–44. Singapore: World Scientific Publishing.

Aalen, O. O. and N. L. Hjort (2002). Frailty models that yield proportional hazards. *Statistics & Probability Letters 58*, 335–342.

Aalen, O. O. and E. Husebye (1991). Statistical analysis of repeated events forming renewal processes. *Statistics in Medicine 10*, 1227–1240.

Aalen, O. O. and S. Johansen (1978). An empirical transition matrix for nonhomogeneous Markov chains based on censored observations. *Scandinavian Journal of Statistics 5*, 141–150.

Aalen, O. O. and S. Tretli (1999). Analyzing incidence of testis cancer by means of a frailty model. *Cancer Causes and Control 10*, 285–292.

Abate, J. and W. Whitt (1996). An operational calculus for probability distributions via Laplace transforms. *Advances in Applied Probability 28*, 75–113.

Abbring, J. H. and G. J. van den Berg (2007). The unobserved heterogeneity distribution in duration analysis. *Biometrika 94*, 87–99.

Abramowitz, M. and I. A. Stegun (1964). *Handbook of Mathematical Functions.* New York: Dover Publications.

Allen, L. J. S. (2003). *Stochastic Processes with Applications to Biology.* Upper Saddle River, NJ: Prentice Hall.

Almeida, O. P., G. K. Hulse, D. Lawrence, and L. Flicker (2002). Smoking as a risk factor for Alzheimer's disease: Contrasting evidence from a systematic review of case–control and cohort studies. *Addiction 97*, 15–28.

Altman, D. G. (1991). *Practical Statistics for Medical Research.* London: Chapman & Hall.

Altshuler, B. (1970). Theory for the measurement of competing risks in animal expreiments. *Mathematical Biosciences 6*, 1–11.

Andersen, P. K., Ø. Borgan, R. D. Gill, and N. Keiding (1993). *Statistical Models based on Counting Processes.* New York: Springer-Verlag.

Andersen, P. K. and R. D. Gill (1982). Cox's regression model for counting processes: A large sample study. *Annals of Statistics 10*, 1100–1120.

Andersen, P. K., L. S. Hansen, and N. Keiding (1991). Non- and semi-parametric estimation of transition probabilities from censored observations of a non-homogeneous Markov process. *Scandinavian Journal of Statistics 18*, 153–167.

Andersen, P. K. and M. Væth (1989). Simple parametric and non-parametric models for excess and relative mortality. *Biometrics 45*, 523–535.

Andersen, R. S., S. A. Husain, and R. J. Loy (2004). The Kohlrausch function: Properties and applications. *Australian & New Zealand Industrial and Applied Mathematics Journal 45 (E)*, c800–c816.

Antoniou, A. C. and D. F. Easton (2006). Risk prediction models for familial breast cancer. *Future Oncology 2*, 257–274.

Applebaum, D. (2004). *Lévy Processes and Stochastic Calculus.* Cambridge: Cambridge University Press.

Arjas, E. (1988). A graphical method for assessing goodness of fit in Cox's proportional hazards model. *Journal of the American Statistical Association 83*, 204–212.

Arjas, E. and M. Eerola (1993). On predictive causality in longitudinal studies. *Journal of Statistical Planning and Inference 34*, 361–386.

Armero, C. and M. J. Bayarri (1997). A Bayesian analysis of queueing system with unlimited service. *Journal of Statistical Planning and Inference 58*, 241–261.

Armitage, P. and R. Doll (1954). The age distribution of cancer and a multi-stage theory of carcinogenesis. *British Journal of Cancer 8*, 1–12.

Bagdonavicius, V., M. A. Hafdi, and M. Nikulin (2004). Analysis of survival data with cross-effects of survival functions. *Biostatistics 5*, 415–425.

Barinaga, M. (1992). Mortality: Overturning received wisdom. *Science 258*, 398–399.

Barlow, R. E., D. J. Bartholomew, J. M. Bremner, and H. D. Brunk (1972). *Statistical Inference Under Order Restrictions: The Theory and Applications of Isotonic Regression.* New York: Wiley.

Barlow, W. E. and R. L. Prentice (1988). Residuals for relative risk regression. *Biometrika 75*, 65–74.

Barndorff-Nielsen, O. E. (1998). Processes of normal inverse Gaussian type. *Finance and Stochastics 2*, 41–68.

Barndorff-Nielsen, O. E. and N. Shephard (2001). Non-Gaussian Ornstein-Uhlenbeck-based models and some of their uses in financial economics (with discussion). *Journal of the Royal Statistical Society: Series B (Statistical Methodology) 63*, 167–241.

Becker, N. G. (1989). *Analysis of Infectious Disease Data*. London: Chapman and Hall.

Bergman, E. M., W. R. Miller, N. Vines, and H. I. Lief (1977). The age 30 crisis and the 7-year-itch. *Journal of Sex and Marital Therapy 3*, 197–204.

Bertoin, J. (1996). *Lévy Processes*. Cambridge: Cambridge University Press.

Bhattacharyya, M. and J. P. Klein (2005). A note on testing in Aalen's additive hazards regression models. *Statistics in Medicine 24*, 2235–2240.

Bie, O., Ø. Borgan, and K. Liestøl (1987). Confidence intervals and confidence bands for the cumulative hazard rate function and their small sample properties. *Scandinavian Journal of Statistics 14*, 221–233.

Billingsley, P. (1995). *Probability and Measure* (3rd ed.). New York: Wiley-Interscience.

Birkhoff, G. and G.-C. Rota (1989). *Ordinary Differential Equations* (4th ed.). New York: Wiley.

Blumen, I., M. Kogan, and P. McCarthy (1955). *The Industrial Mobility of Labor as a Probability Process*. Ithaca, N.Y.: Cornell University Press.

Böhmer, P. E. (1912). Theorie der unabhängigen Warscheinlichkeiten. *Rapports, Mém. et Procés - verbaux 7e Congrès Internat. Act., Amsterdam 2*, 327–343.

Borgan, Ø. (2002). Estimation of covariate-dependent Markov transition probabilities from nested case-control data. *Statistical Methods in Medical Research 11*, 183–202. Correction *12*, 124.

Borgan, Ø., R. L. Fiaccone, R. Henderson, and M. L. Barreto (2007). Dynamic analysis of recurrent event data with missing observations, with application to infant diarrhoea in Brazil. *Scandinavian Journal of Statistics 34*, 53–69.

Borgan, Ø., L. Goldstein, and B. Langholz (1995). Methods for the analysis of sampled cohort data in the Cox proportional hazards model. *Annals of Statistics 23*, 1749–1778.

Borgan, Ø. and B. Langholz (1993). Non-parametric estimation of relative mortality from nested case-control studies. *Biometrics 49*, 593–602.

Borgan, Ø. and B. Langholz (1995). Estimation of excess risk from case-control data using Aalen's linear regression model. Statistical Research Report no 5, Department of Mathematics, University of Oslo.

Borgan, Ø. and B. Langholz (1997). Estimation of excess risk from case-control data using Aalen's linear regression model. *Biometrics 53*, 690–697.

Borgan, Ø. and B. Langholz (2007). Using martingale residuals to assess goodness-of-fit for sampled risk set data. In V. Nair (Ed.), *Advances in Statistical Modeling and Inference. Essays in Honor of Kjell A Doksum*, pp. 65–90. Singapore: World Scientific Publishing.

Borgan, Ø. and K. Liestøl (1990). A note on confidence intervals and bands for the survival curve based on transformations. *Scandinavian Journal of Statistics 17*, 35–41.

Borgan, Ø. and E. F. Olsen (1999). The efficiency of simple and counter-matched nested case-control sampling. *Scandinavian Journal of Statistics 26*, 493–509.

Boshuizen, H. C., G. J. Izaks, S. van Buuren, and G. Ligthart (1998). Blood pressure and mortality in elderly people aged 85 and older: community based study. *British Medical Journal 316*, 1780–1784.

Braaten, T., E. Weiderpass, M. Kumle, H.-O. Adami, and E. Lund (2004). Education and risk of breast cancer in the Norwegian-Swedish women's lifestyle and health cohort study. *International Journal of Cancer 110*, 579–583.

Brennan, P. (2002). Gene-environment interaction and aetiology of cancer: what does it mean and how can we measure it? *Carcinogenesis 23*, 381–387.

Breslow, N. E. (1970). A generalized Kruskal-Wallis test for comparing K samples subject to unequal patterns of censorship. *Biometrika 57*, 579–594.

Breslow, N. E. and N. E. Day (1987). *Statistical Methods in Cancer Research. Volume II – The Design and Analysis of Cohort Studies*, Volume 82 of *IARC Scientific Publications*. Lyon: International Agency for Research on Cancer.

Brookmeyer, R. and J. J. Crowley (1982). A k-sample median test for censored data. *Journal of the American Statistical Association 77*, 433–440.

Brovelli, A., M. Ding, A. Ledberg, Y. Chen, R. Nakamura, and S. L. Bressler (2004). Beta oscillations in a large-scale sensorimotor cortical network: Directional influences revealed by Granger causality. *Proceedings of the National Academy of Sciences of the United States of America 101*, 9849–9854.

Bulmer, M. G. (1985). *The Mathematical Theory of Quantitative Genetics*. Clarendon Press.

Bunea, F. and I. W. McKeague (2005). Covariate selection for semiparametric hazard function regression models. *Journal of Multivariate Analysis 92*, 186–204.

Byar, D. P. (1980). The Veterans Administration study of chemoprophylaxis for recurrent stage I bladder tumors: Comparisons of placebo, pyridoxine, and topical thiotepa. In M. Pavone-Macaluso, P. H. Smith, and F. Edsmyn (Eds.), *Bladder Tumors and Other Topics in Urological Oncology*, pp. 363–370. New York: Plenum.

Carey, J. R., P. Liedo, D. Orozco, and J. W. Vaupel (1992). Slowing of mortality rates at older ages in large medfly cohorts. *Science 258*, 457–461.

Cavender, J. A. (1978). Quasi-stationary distributions of birth-and-death processes. *Advances in Applied Probability 10*, 570–586.

Chhikara, R. S. and J. L. Folks (1989). *The Inverse Gaussian Distribution: Theory, Methodology and Applications*. New York: Marcel Dekker.

Chiang, C. L. (1968). *Introduction to Stochastic Processes in Biostatistics*. New York: Wiley.

Chiang, C.-T., M.-C. Wang, and C.-Y. Huang (2005). Kernel estimation of rate function for recurrent event data. *Scandinavian Journal of Statistics 32*, 77–91.

Clarke, C., R. A. Collins, B. R. Leavitt, D. F. Andrews, M. R. Hayden, C. J. Lumsden, and R. R. McInnes (2000). A one-hit model of cell death in inherited neuronal degenerations. *Nature 406*, 195–199.

Clarke, C., C. J. Lumsden, and R. R. McInnes (2001). Inherited neurodegenerative diseases: The one-hit model of neurodegeneration. *Human Molecular Genetics 10*, 2269–2275.

Coggon, D. I. W. and C. N. Martyn (2005). Time and chance: The stochastic nature of disease causation. *Lancet 365*, 1434–1437.

Cole, S. R. and M. A. Hernán (2002). Fallibility in estimating direct effects. *International Journal of Epidemiology 31*, 163–165.

Commenges, D. and A. Gégout-Petit (2007). Likelihood for generally coarsened observations from multistate or counting process models. *Scandinavian Journal of Statistics 34*, 432–450.

Cook, R. J. and J. F. Lawless (2007). *The Statistical Analysis of Recurrent Events*. New York: Springer-Verlag.

Cox, D. R. (1972). Regression models and life-tables (with discussion). *Journal of the Royal Statistical Society: Series B (Statistical Methodology) 34*, 187–220.

Cox, D. R. (1975). Partial likelihood. *Biometrika 62*, 269–276.

Cox, D. R. (1999). Some remarks on failure-times, surrogate markers, degradation, wear, and the quality of life. *Lifetime Data Analysis 5*, 307–314.

Cox, D. R. and H. D. Miller (1965). *The Theory of Stochastic Processes*. London: Methuen.

Cox, D. R. and D. Oakes (1984). *Analysis of Survival Data*. London: Chapman and Hall.

Cox, D. R. and N. Wermuth (2004). Causality: A statistical view. *International Statistical Review 72*, 285–305.

Datta, S. and G. A. Satten (2001). Validity of the Aalen-Johansen estimators of stage occupation probabilities and Nelson-Aalen estimators of integrated transition hazards for non-Markov models. *Statistics & Probability Letters 55*, 403–411.

Datta, S. and G. A. Satten (2002). Estimation of integrated transition hazards and stage occupation probabilities for non-Markov systems under dependent censoring. *Biometrics 58*, 792–802.

de Bruijne, M. H. J., S. le Cessie, H. C. Kluin-Nelemans, and H. C. van Houwelingen (2001). On the use of Cox regression in the presence of an irregularly observed time-dependent covariate. *Statistics in Medicine 20*, 3817–3829.

Di Serio, C. (1997). The protective impact of a covariate on competing failures with an example from a bone marrow transplantation study. *Lifetime Data Analysis 3*, 99–122.

Didelez, V. (2006). Asymmetric separation for local independence graphs. In R. Dechter and T. S. Richardson (Eds.), *Proceedings of the 22nd Annual Conference on Uncertainty in Artificial Intelligence*, pp. 130–137. Boston: AUAI Press.

Didelez, V. (2007). Graphical models for composable finite Markov processes. *Scandinavian Journal of Statistics 34*, 169–185.

Diggle, P., D. M. Farewell, and R. Henderson (2007). Analysis of longitudinal data with drop-out: Objectives, assumptions and a proposal. *Journal of the Royal Statistical Society: Series C (Applied Statistics) 56*, 499–550.

Ditlevsen, S., U. Christensen, J. Lynch, M. T. Damsgaard, and N. Keiding (2005). The mediation proportion: A structural equation approach for estimating the proportion of exposure effect on outcome explained by an intermediate variable. *Epidemiology 16*, 114–120.

Doksum, K. A. and A. Høyland (1992). Models for variable-stress accelerated life testing experiments based on Wiener processes and the inverse Gaussian distribution. *Technometrics 34*, 74–82.

Dominicus, A., A. Skrondal, H. K. Gjessing, N. L. Pedersen, and J. Palmgren (2006). Likelihood ratio tests in behavioral genetics: Problems and solutions. *Behavior Genetics 36*, 331–340.

Eaton, W. W. and G. A. Whitmore (1977). Length of stay as a stochastic process: A general approach and application to hospitalization for schizophrenia. *Journal of Mathematical Sociology 5*, 273–292.

Ederer, F., L. M. Axtell, and S. Cutler (1961). The relative survival rate: A statistical methodology. *National Cancer Institute Monograph 6*, 101–121.

Editors (2000). Looking back on the millennium in medicine. *The New England Journal of Medicine 342*, 42–49.

Edwards, S., P. Evans, F. Hucklebridge, and A. Clow (2001). Association between time of awakening and diurnal cortisol secretory activity. *Psychoneuroendocrinology 26*, 613–622.

Efron, B. (1967). The two sample problem with censored data. In *Proceedings of the Fifth Berkeley Symposium on Mathematical Statistics and Probability. Vol. 4*, pp. 831–853.

Fahrmeir, L. and A. Klinger (1998). A nonparametric multiplicative hazard model for event history analysis. *Biometrika 85*, 581–592.

Farewell, D. M. (2006). *Linear models for censored data*. Ph. D. thesis, Lancaster University, Lancaster, UK.

Farewell, V. T. (1977). A model for a binary variable with time-censored observations. *Biometrika 64*, 43–46.

Farrington, C. P. and H. J. Whitaker (2006). Semiparametric analysis of case series data. *Journal of the Royal Statistical Society: Series C (Applied Statistics) 55*, 553–594.

Feller, W. (1971). *An Introduction to Probability Theory and Its Applications*, Volume 2. New York: Wiley.

Fleming, T. R. (1978a). Asymptotic distribution results in competing risks estimation. *Annals of Statistics 6*, 1071–1079.

Fleming, T. R. (1978b). Nonparametric estimation for nonhomogeneous Markov processes in the problem of competing risks. *Annals of Statistics 6*, 1057–1070.

Fleming, T. R. and D. P. Harrington (1991). *Counting Processes and Survival Analysis*. New York: Wiley.

Fosen, J., Ø. Borgan, H. Weedon-Fekjær, and O. O. Aalen (2006). Dynamic analysis of recurrent event data using the additive hazard model. *Biometrical Journal 48*, 381–398.

Fosen, J., E. Ferkingstad, Ø. Borgan, and O. O. Aalen (2006). Dynamic path analysis – A new approach to analyzing time-dependent covariates. *Lifetime Data Analysis 12*, 143–167.

Freireich, E. J., E. Gehan, E. Frei, L. R. Schroeder, I. J. Wolman, R. Anbari, E. O. Burgert, S. D. Mills, D. Pinkel, O. S. Selawry, J. H. Moon, B. R. Gendel, C. L. Spurr, R. Storrs, F. Haurani, B. Hoogstraten, and S. Lee (1963). The effect of 6-mercaptopurine on the duration of steroid-induced remissions in acute leukemia: A model for evaluation of other potentially useful therapy. *Blood 21*, 699–716.

Gaastra, W. and A. Svennerholm (1996). Colonization factors of human enterotoxigenic *Escherichia coli* (ETEC). *Trends in Microbiology 4*, 444–452.

Gandy, A. and U. Jensen (2004). A nonparametric approach to software reliability. *Applied Stochastic Models in Business and Industry 20*, 3–15.

Gandy, A. and U. Jensen (2005a). Checking a semiparametric additive risk model. *Lifetime Data Analysis 11*, 451–472.

Gandy, A. and U. Jensen (2005b). On goodness-of-fit tests for Aalen's additive risk model. *Scandinavian Journal of Statistics 32*, 425–445.

Gandy, A., T. M. Therneau, and O. O. Aalen (2008). Global tests in the additive hazards regression model. *Statistics in Medicine 27*, 831–844.

Gasser, T. and H.-G. Müller (1979). Kernel estimation of regression functions. In T. Gasser and M. Rosenblatt (Eds.), *Smoothing Techniques for Curve Estimation*, Volume 757 of *Lecture Notes in Mathematics*, pp. 23–68. Berlin: Springer-Verlag.

Gehan, E. (1965). A generalized Wilcoxon test for comparing arbitrarily singly censored samples. *Biometrika 52*, 203–223.

Geneletti, S. (2007). Identifying direct and indirect effects in a non-counterfactual framework. *Journal of the Royal Statistical Society: Series B (Statistical Methodology) 69*, 199–215.

Ghali, W. A., H. Quan, R. Brant, G. van Melle, C. M. Norris, P. D. Faris, P. D. Galbraith, and M. L. Knudtson (2001). Comparison of 2 methods for calculating adjusted survival curves from proportional hazards models. *Journal of the American Medical Association 286*, 1494–1497.

Ghosh, D. and D. Y. Lin (2000). Nonparametric analysis of recurrent events and death. *Biometrics 56*, 554–562.

Gill, R. D. (1983). Large sample behavior of the product-limit estimator on the whole line. *Annals of Statistics 11*, 49–58.

Gill, R. D. (2004). Causal inference for complex longitudinal data: The continuous time G-computation formula. http://uk.arxiv.org/PS_cache/math/pdf/0409/0409436v1.pdf.

Gill, R. D. and S. Johansen (1990). A survey of product-integration with a view towards application in survival analysis. *Annals of Statistics 18*, 1501–1555.

Gill, R. D. and J. M. Robins (2001). Causal inference for complex longitudinal data: The continuous case. *Annals of Statistics 29*, 1785–1811.

Ginsberg, R. B. (1971). Semi-Markov processes and mobility. *Journal of Mathematical Sociology 1*, 233–262.

Gjessing, H. K., O. O. Aalen, and N. L. Hjort (2003). Frailty models based on Lévy processes. *Advances in Applied Probability 35*, 532–550.

Gjessing, H. K. and R. T. Lie (2008). Biometrical modelling in genetics: Are complex traits too complex? *Statistical Methods in Medical Research 17*, 75–96.

Goel, N. S. and N. Richter-Dyn (1974). *Stochastic Models in Biology*. New York: Academic Press.

Goldstein, L. and B. Langholz (1992). Asymptotic theory for nested case-control sampling in the Cox regression model. *Annals of Statistics 20*, 1903–1928.

Gorard, S. (2002). The role of causal models in evidence-informed policy making and practice. *Evaluation and Research in Education 16*, 51–65.

Granger, C. W. J. (1969). Investigating causal relations by econometric models and cross-spectral methods. *Econometrica 37*, 424–438.

Granger, C. W. J. (2003). Time series analysis, cointegration, and applications (Nobel Prize lecture). http://nobelprize.org/economics/laureates/2003/.

Grønnesby, J. K. (1997). Testing covariate effects in Aalen's linear hazard model. *Scandinavian Journal of Statistics 24*, 125–135.

Grønnesby, J. K. and Ø. Borgan (1996). A method for checking regression models in survival analysis based on the risk score. *Lifetime Data Analysis 2*, 315–328.

Gross, D. and C. M. Harris (1985). *Fundamentals of Queuing Theory* (2nd ed.). New York: Wiley.

Gunnes, N., Ø. Borgan, and O. O. Aalen (2007). Estimating stage occupation probabilities in non-Markov models. *Lifetime Data Analysis 13*, 211–240.

Hall, P. and H.-G. Müller (2003). Order-preserving nonparametric regression, with applications to conditional distribution and quantile function estimation. *Journal of the American Statistical Association 98*, 598–608.

Halvorsen, H. (2002). Application of survival time modelled as first passage time in a Wiener process. Master's thesis, University of Oslo, Norway.

Harrington, D. P. and T. R. Fleming (1982). A class of rank test procedures for censored survival data. *Biometrika 69*, 133–143.

Heckman, J. J. (2000). Microdata, heterogeneity and the evaluation of public policy (Nobel prize lecture). http://www.nobel.se/economics/laureates/2000/.

Heckman, J. J. and B. Singer (1982). *Population Heterogeneity in Demographic Models*. Orlando: Academic Press.

Heimdal, K., H. Olsson, S. Tretli, S. D. Fosså, A.-L. Børresen, and D. T. Bishop (1997). A segregation analysis of testicular cancer based on Norwegian and Swedish families. *British Journal of Cancer 75*, 1084–1087.

Helland, I. S. (1982). Central limit-theorems for martingales with discrete or continuous-time. *Scandinavian Journal of Statistics 9*, 79–94.

Henderson, R. and A. Milner (1991). Aalen plots under proportional hazards. *Journal of the Royal Statistical Society: Series C (Applied Statistics) 40*, 401–409.

Hernán, M. A., E. Lanoy, D. Costagliola, and J. M. Robins (2006). Comparison of dynamic treatment regimes via inverse probability weighting. *Basic & Clinical Pharmacology & Toxicology 98*, 237–242.

Hesslow, G. (1976). Discussion: two notes on the probabilistic approach to causality. *Philosophy of Science 43*, 290–292.

Hjort, N. L. (1990). Nonparametric Bayes estimators based on beta processes in models for life history data. *Annals of Statistics 18*, 1259–1294.

Hjort, N. L. (2003). Topics in non-parametric Bayesian statistics (with discussion). In P. J. Green, N. L. Hjort, and S. Richardson (Eds.), *Highly Structured Stochastic Systems*, pp. 455–487. Oxford: Oxford University Press.

Höfler, M. (2005). The Bradford Hill considerations on causality: a counterfactual perspective. *Emerging Themes in Epidemiology 2*. DOI:10.1186/1742-7622-2-11.

Hogan, J. W., J. Roy, and C. Korkontzelou (2004). Handling drop-out in longitudinal studies. *Statistics in Medicine 23*, 1455–1497.

Hornung, R. and T. Meinhardt (1987). Quantitative risk assessment of lung cancer in U.S. uranium miners. *Health Physics 52*, 417–430.

Hosmer, D. W. and S. Lemeshow (1999). *Applied Survival Analysis. Regression Modelling of Time to Event Data*. New York: Wiley.

Hougaard, P. (1984). Life table methods for heterogeneous populations: Distributions describing the heterogeneity. *Biometrika 71*, 75–83.

Hougaard, P. (1986). Survival models for heterogeneous populations derived from stable distributions. *Biometrika 73*, 387–396. Correction *7*, 395.

Hougaard, P. (2000). *Analysis of Multivariate Survival Data*. New York: Springer-Verlag.

Hougaard, P., M. L. T. Lee, and G. A. Whitmore (1997). Analysis of overdispersed count data by mixtures of Poisson variables and Poisson processes. *Biometrics 53*, 1225–1238.

Huffer, F. W. and I. W. McKeague (1991). Weighted least squares estimation for Aalen's additive risk model. *Journal of the American Statistical Association 86*, 114–129.

Hume, D. (1965). *An Abstract of a Treatise of Human Nature, 1740*. Hamden, Ct.: Archon Books. Introduction by J. M. Keynes and P. Sraffa. (Original work written in 1740; abstract published in 1938).

Hunter, K. W. (2004). Host genetics and tumor metastasis. *British Journal of Cancer 90*, 752–755.

Huzurbazar, A. V. (1999). Flowgraph models for generalized phase type distributions with non-exponential waiting times. *Scandinavian Journal of Statistics 26*, 145–157.

Huzurbazar, A. V. (2005). *Flowgraph Models for Multistate Time-To-Event Data*. New York: Wiley.

Jacobsen, S. J., D. S. Freedman, R. G. Hoffman, H. W. Gruchow, A. J. Anderson, and J. J. Barboriak (1992). Cholesterol and coronary artery disease: Age as an effect modifier. *Journal of Clinical Epidemiology 45*, 1053–1059.

Jacod, J. and P. Protter (2003). *Probability essentials* (2nd ed.). Berlin: Springer-Verlag.

Jewell, N. P. and J. D. Kalbfleisch (1996). Marker processes in survival analysis. *Lifetime Data Analysis 2*, 15–29.

Joffe, M. and J. Mindell (2006). Complex causal process diagrams for analyzing the health impacts of policy interventions. *American Journal of Public Health 96*, 473–479.

Johnson, N. L., S. Kotz, and N. Balakrishnan (1995). *Continuous Univariate Distributions* (2nd ed.), Volume 2. New York: Wiley.

Johnson, R. A. and D. W. Wichern (1982). *Applied Multivariate Statistical Analysis*. Englewood Cliffs, NJ: Prentice-Hall.

Jørgensen, B. (1982). *Statistical Properties of the Generalized Inverse Gaussian Distribution*. Lecture Notes in Statistics 9. New York: Springer-Verlag.

Jørgensen, B. (1987). Exponential dispersion models. *Journal of the Royal Statistical Society: Series B (Statistical Methodology) 49*, 127–162.

Kalbfleisch, J. D. and R. L. Prentice (2002). *The Statistical Analysis of Failure Time Data* (2nd ed.). Hoboken, N.J.: Wiley.

Kallenberg, O. (1997). *Foundations of Modern Probability*. New York: Springer-Verlag.

Kaplan, E. L. and P. Meier (1958). Non-parametric estimation from incomplete observations. *Journal of the American Statistical Association 53*, 457–481.

Karatzas, I. and S. E. Shreve (1991). *Brownian Motion and Stochastic Calculus* (2nd ed.). New York: Springer-Verlag.

Karlin, S. and H. M. Taylor (1975). *A First Course in Stochastic Processes* (2nd ed.). New York: Academic Press.

Karlin, S. and H. M. Taylor (1981). *A Second Course in Stochastic Processes*. New York: Academic Press.

Karlin, S. and H. M. Taylor (1998). *An Introduction to Stochastic Modeling* (3rd ed.). New York: Academic Press.

Kaufman, J. S., R. F. MacLehose, and S. Kaufman (2004). A further critique of the analytic strategy of adjusting for covariates to identify biologic mediation. *Epidemiologic Perspectives & Innovations 1*. DOI:10.1186/1742-5573-1-4.

Kaufman, J. S., R. F. MacLehose, S. Kaufman, and S. Greenland (2005). The mediation proportion. *Epidemiology 16*, 710.

Kebir, Y. (1991). On hazard rate processes. *Naval Research Logistics 38*, 865–876.

Keiding, N. (1999). Event history analysis and inference from observational epidemiology. *Statistics in Medicine 18*, 2353–2363.

Keiding, N. (2005). Expected survival curve. In P. Armitage and T. Colton (Eds.), *Encyclopedia of Biostatistics* (2nd ed.), Volume 3, pp. 1832–1835. Chichester: Wiley.

Keiding, N. and P. K. Andersen (1989). Nonparametric estimation of transition intensities and transition probabilities: A case study of a two-state Markov process. *Journal of the Royal Statistical Society: Series C (Applied Statistics) 38*, 319–329.

Keiding, N., J. P. Klein, and M. M. Horowitz (2001). Multi-state models and outcome prediction in bone marrow transplantation. *Statistics in Medicine 20*, 1871–1885.

Keilson, J. (1979). *Markov Chain Models – Rarity and Exponentiality*. New York: Springer-Verlag.

Kellow, J. E., T. J. Borody, S. F. Phillips, R. L. Tucker, and A. C. Haddad (1986). Human interdigestive motility: Variations in patterns from esophagus to colon. *Gastroenterology 91*, 386–395.

Kendall, M. and A. Stuart (1977). *The Advanced Theory of Statistics*. London: Charles Griffin & Company.

Kessing, L. V. and P. K. Andersen (2005). Predictive effects of previous episodes on the risk of recurrence in depressive and bipolar disorders. *Current Psychiatry Reports 7*, 413–420.

Khoury, J. M., T. H. Beaty, and K. Y. Liang (1988). Can familial aggregation of disease be explained by familial aggregation of environmental risk factors? *American* Journal of Epidemiology *127*, 674–683.

Kingman, J. F. C. (1992). *Poisson Processes*. Oxford: Clarendon Press.

Kirkwood, T. B. L. and C. E. Finch (2002). The old worm turns more slowly. *Nature 419*, 794–795.

Klein, J. P. and M. L. Moeschberger (2003). *Survival Analysis. Techniques for Censored and Truncated Data* (2nd ed.). New York: Springer-Verlag.

Klotz, L. H. (1999). Why is the rate of testicular cancer increasing? *Canadian Medical Association Journal 160*, 213–214.

Kopans, D. B., E. Rafferty, D. Georgian-Smith, E. Yeh, H. D'Alessandro, R. Moore, K. Hughes, and E. Halpern (2003). A simple model of breast carcinoma growth may provide explanations for observations of apparently complex phenomena. *Cancer 97*, 2951–2959.

Kopp-Schneider, A. (1997). Carcinogenesis models for risk assessment. *Statistical Methods in Medical Research 6*, 317–340.

Kosek, M., C. Bern, and R. L. Guerrant (2003). The global burden of diarrhoeal disease, as estimated from studies published between 1992 and 2000. *Bulletin of the World Health Organization 81*, 197–204.

Lambert, P. C., L. K. Smith, D. R. Jones, and J. L. Botha (2005). Additive and multiplicative covariate regression models for relative survival incorporating fractional polynomials for time-dependent effects. *Statistics in Medicine 24*, 3871–3885.

Lancaster, D. J. (1982). *Stochastic Models for Social Processes* (3rd ed.). Chichester: Wiley.

Langer, R. D., T. G. Ganiats, and E. Barrett-Connor (1989). Paradoxical survival of elderly men with high blood pressure. *British Medical Journal 298*, 1356–1357.

Langer, R. D., T. G. Ganiats, and E. Barrett-Connor (1991). Factors associated with paradoxical survival at higher blood pressures in the very old. *American Journal of Epidemiology 134*, 29–38.

Langholz, B. (2005). Case cohort study. In P. Armitage and T. Colton (Eds.), *Encyclopedia of Biostatistics* (2nd ed.), Volume 1, pp. 628–635. Chichester: Wiley.

Langholz, B. (2007). Use of cohort information in the design and analysis of case-control studies. *Scandinavian Journal of Statistics 34*, 120–136.

Langholz, B. and Ø. Borgan (1995). Counter-matching: A stratified nested case-control sampling method. *Biometrika 82*, 69–79.

Langholz, B. and Ø. Borgan (1997). Estimation of absolute risk from nested case-control data. *Biometrics 53*, 767–774. Correction *59*, 451.

Langholz, B. and L. Goldstein (1996). Risk set sampling in epidemiologic cohort studies. *Statistical Science 11*, 35–53.

Lawless, J. F. and C. Nadeau (1995). Some simple robust methods for the analysis of recurrent events. *Technometrics 37*, 158–168.

Lebedev, N. N. (1972). *Special Functions and Their Applications*. New York: Dover Publications.

Lee, M. L. T., V. DeGruttola, and D. Schoenfeld (2000). A model for markers and latent health status. *Journal of the Royal Statistical Society: Series B (Statistical Methodology) 62*, 747–762.

Lee, M. L. T. and G. A. Whitmore (1993). Stochastic-processes directed by randomized time. *Journal of Applied Probability 30*, 302–314.

Lee, M. L. T. and G. A. Whitmore (2006). Threshold regression for survival analysis: Modeling event times by a stochastic process reaching a boundary. *Statistical Science 21*, 501–513.

Lee, P. M. (1989). *Bayesian Statistics: An Introduction*. New York: Oxford University Press.

Levine, M. M., C. Ferreccio, V. Prado, M. Cayazzo, P. Abrego, J. Martinez, L. Maggi, M. M. Baldini, W. Martin, and D. Maneval (1993). Epidemiologic studies of *Escherichia coli* diarrheal infections in a low socioeconomic level periurban community in Santiago, Chile. *American Journal of Epidemiology 138*, 849–869.

Li, Y., J. A. Schneider, and D. A. Bennett (2007). Estimation of the mediation effect with a binary mediator. *Statistics in Medicine 26*, 3398–3414.

Lie, S. A., L. B. Engesæter, L. I. Havelin, H. K. Gjessing, and S. E. Vollset (2000). Mortality after total hip replacement. 0-10 year follow-up of 39,543 patients in the Norwegian Arthroplasty Register. *Acta Orthopaedica Scandinavica 71*, 19–27.

Lin, D. Y., L. J. Wei, I. Yang, and Z. Ying (2000). Semiparametric regression for the mean and rate functions of recurrent events. *Journal of the Royal Statistical Society: Series B (Statistical Methodology) 62*, 711–730.

Lin, D. Y. and Z. Ying (1994). Semiparametric analysis of the additive risk model. *Biometrika 81*, 61–71.

Lindqvist, B. H. (2006). On the statistical modeling and analysis of repairable systems. *Statistical Science 21*, 532–551.

Lindsey, C. P. and G. D. Patterson (1980). Detailed comparison of the Williams-Watts and Cole-Davidson functions. *Journal of Chemical Physics 73*, 3348–3357.

Lo, C. C., L. A. Nunes Amaral, S. Havlin, P. C. Ivanov, T. Penzel, J. H. Peter, and H. E. Stanley (2002). Dynamics of sleep-wake transitions during sleep. *Europhysics Letters 57*, 625–631.

Loehlin, J. C. (2004). *Latent Variable Models: An Introduction to Factor, Path, and Structural Equation Analysis* (4th ed.). Hillsdale, NJ: Lawrence Erlbaum Associates.

Lok, J., R. D. Gill, A. van der Vaart, and J. Robins (2004). Estimating the causal effect of a time-varying treatment on time-to-event using structural nested failure time models. *Statistica Neerlandica 58*, 271–295.

Longini, I. M., W. S. Clark, R. H. Byers, et al. (1989). Statistical analysis of the stages of HIV-infections using a Markov model. *Statistics in Medicine 8*, 831–843.

Longini, I. M. and M. E. Halloran (1996). A frailty mixture model for estimating vaccine efficacy. *Journal of the Royal Statistical Society: Series C (Applied Statistics) 45*, 165–173.

Lunde, A., K. K. Melve, H. K. Gjessing, R. Skjærven, and L. M. Irgens (2007). Genetic and environmental influences on birth weight, birth length, head circumference, and gestational age by use of population-based parent-offspring data. *American Journal of Epidemiology 165*, 734–741.

Machamer, P., L. Darden, and C. F. Craver (2000). Thinking about mechanisms. *Philosophy of Science 67*, 1–25.

Magnus, P., H. K. Gjessing, A. Skrondal, and R. Skjærven (2001). Paternal contributions to birth weight. *Journal of Epidemiology and Community Health 55*, 873–877.

Maller, R. A. and X. Zhou (1996). *Survival Analysis with Long-Term Survivors*. New York: Wiley.

Mamelund, S.-E., H. Brunborg, and T. Noack (1997). Divorce in Norway 1886-1995 by calendar year and marriage cohort. Technical Report 97/19, Statistics Norway.

Mantel, N. (1966). Evaluation of survival data and two new rank order statistics arising in its consideration. *Cancer Chemotherapy Reports 50*, 163–170.

Manton, K. G. and E. Stallard (1998). *Chronic Disease Modelling: Measurement and Evaluation of the Risks of Chronic Disease Processes*. London: Charles Griffin & Company.

Manton, K. G., E. Stallard, and J. W. Vaupel (1981). Methods for comparing the mortality experience of heterogeneous populations. *Demography 18*, 389–410.

Mardia, K. V., J. T. Kent, and J. M. Bibby (1979). *Multivariate Analysis*. London: Academic Press.

Martinez, S. and J. San Martin (1994). Quasi-stationary distributions for a Brownian motion with drift and associated limit laws. *Journal of Applied Probability 31*, 911–920.

Martinussen, T., O. O. Aalen, and T. H. Scheike (2008). The Mizon-Richard encompassing test for the Cox and Aalen additive hazards models. *Biometrics 64*, 164–171.

Martinussen, T. and T. H. Scheike (2000). A nonparametric dynamic additive regression model for longitudinal data. *Annals of Statistics 28*, 1000–1025.

Martinussen, T. and T. H. Scheike (2006). *Dynamic Regression Models for Survival Data*. New York: Springer-Verlag.

May, S. and D. W. Hosmer (1998). A simplified method for calculating a goodness-of-fit test for the proportional hazards model. *Lifetime Data Analysis 4*, 109–120.

McCullagh, P. and J. A. Nelder (1989). *Generalized linear models* (2nd ed.). London: Chapman and Hall.

McKeague, I. W. and P. D. Sasieni (1994). A partly parametric additive risk model. *Biometrika 81*, 501–514.

Menzies, P. (2001). Counterfactual theories of causation. http://plato.stanford.edu/archives/spr2001/entries/causation-counterfactual/. Stanford Encyclopedia of Philosophy, Edward N. Zalta (ed.).

Metcalfe, C. and S. G. Thompson (2007). Wei, Lin and Weissfeld's marginal analysis of multivariate failure time data: should it be applied to a recurrent events outcome? *Statistical Methods in Medical Research 16*, 103–122.

Miloslavsky, M., S. Keles, M. J. van der Laan, and S. Butler (2004). Recurrent events analysis in the presence of time-dependent covariates and dependent censoring. *Journal of the Royal Statistical Society: Series B (Statistical Methodology) 66*, 239–257.

Moger, T. A., O. O. Aalen, K. Heimdal, and H. K. Gjessing (2004). Analysis of testicular cancer data by means of a frailty model with familial dependence. *Statistics in Medicine 23*, 617–632.

Moger, T. A., M. Haugen, Y. Pawitan, H. K. Gjessing, and Ø. Borgan (2008). A frailty analysis of two-generation melanoma data from the Swedish Multi-Generation Register. In preparation.

Mølbak, K. (2000). The epidemiology of diarrhoeal diseases in early childhood: A review of community studies in Guinea-Bissau. *Danish Medical Bulletin 47*, 340–358.

Moler, C. and C. van Loan (1978). Nineteen dubious ways to compute the exponential of a matrix. *SIAM Review 20*, 801–836.

Moler, C. and C. van Loan (2003). Nineteen dubious ways to compute the exponential of a matrix, twenty-five years later. *SIAM Review 45*, 3–49.

Morens, D. M., A. Grandinetti, J. W. Davis, G. W. Ross, L. R. White, and D. Reed (1996). Evidence against the operation of selective mortality in explaining the association between cigarette smoking and reduced occurrence of idiopathic Parkinson disease. *American Journal of Epidemiology 144*, 400–404.

Moulton, L. H. and M. J. Dibley (1997). Multivariate time-to-event models for studies of recurrent childhood diseases. *International Journal of Epidemiology 26*, 1334–1339.

Moulton, L. H., M. A. Staat, M. Santosham, and R. L. Ward (1998). The protective effectiveness of natural rotavirus infection in an American Indian population. *The Journal of Infectious Diseases 178*, 1562–1566.

Müller, H.-G. and J.-L. Wang (1994). Hazard rate estimation under random censoring with varying kernels and bandwidths *Biometrics 50*, 61–76.

Murphy, S. A. (1995). Asymptotic theory for the frailty model. *Annals of Statistics 23*, 182–198.

Musiela, M. and M. Rutkowski (1997). *Martingale Methods in Financial Modelling*. New York: Springer-Verlag.

Myers, L. E. (1981). Survival functions induced by stochastic covariate processes. *Journal of Applied Probability 18*, 523–529.

Mykland, P. A. (1986). Statistical causality. Statistical report no. 14, Dept. of Mathematics, University of Bergen, Norway.

Nelson, W. (1969). Hazard plotting for incomplete failure data. *Journal of Quality Technology 1*, 27–52.

Nelson, W. (1972). Theory and applications of hazard plotting for censored failure data. *Technometrics 14*, 945–965.

Ness, R. B., J. S. Koopman, and M. S. Roberts (2007). Causal system modeling in chronic disease epidemiology: A proposal. *Annals of Epidemiology 17*, 564–568.

Nielsen, G. G., R. D. Gill, P. K. Andersen, and T. I. A. Sørensen (1992). A counting process approach to maximum likelihood estimation in frailty models. *Scandinavian Journal of Statistics 19*, 25–43.

Nielsen, J. P. and O. B. Linton (1995). Kernel estimation in a nonparametric marker dependent hazard model. *Annals of Statistics 23*, 1735–1748.

Nielsen, N. R., T. S. Kristensen, E. Prescott, K. S. Larsen, P. Schnohr, and M. Grønbæk (2006). Perceived stress and risk of ischemic heart disease: Causation or bias? *Epidemiology 17*, 391–397.

Oakes, D. (1981). Survival times: Aspects of partial likelihood (with discussion). *International Statistical Review 49*, 235–264.

O'Cinneide, C. A. (1989). On non-uniqueness of representations of phase type distributions. *Communications in Statistics – Stochastic Models 5*, 247–259.

O'Cinneide, C. A. (1990). Characterization of phase type distributions. *Communications in Statistics – Stochastic Models 6*, 1–57.

O'Cinneide, C. A. (1991). Phase type distributions and majorization. *The Annals of Applied Probability 2*, 219–227.

Øksendal, B. (1998). *Stochastic Differential Equations* (5th ed.). New York: Springer-Verlag.

Osnes, K. and O. O. Aalen (1999). Spatial smoothing of cancer survival: a Bayesian approach. *Statistics in Medicine 18*, 2087–2099.

Ouyang, A., A. G. Sunshine, and J. C. Reynolds (1989). Caloric content of a meal affects duration but not contractile pattern of duodenal motility in man. *Digestive Diseases and Sciences 34*, 528–536.

Parner, E. (1998). Asymptotic theory for the correlated gamma-frailty model. *Annals of Statistics 26*, 183–214.

Paulsen, J. and H. K. Gjessing (1997). Ruin theory with stochastic return on investments. *Advances in Applied Probability 29*, 965–985.

Pawitan, Y. (2001). *In All Likelihood: Statistical Modelling and Inference Using Likelihood*. Oxford: Clarendon press.

Pearl, J. (2000). *Causality: Models, Reasoning and Inference*. New York: Cambridge University Press.

Pearl, J. (2001). Direct and indirect effects. In J. Breese and D. Koller (Eds.), *Proceedings of the Seventeenth Conference on Uncertainty in Artificial Intelligence*, pp. 411–420. San Francisco: Morgan Kaufmann Publishers.

Peña, E. A. (2006). Dynamic modeling and statistical analysis of event times. *Statistical Science 21*, 487–500.

Peña, E. A., E. H. Slate, and J. R. González (2007). Semiparametric inference for a general class of models for recurrent events. *Journal of Statistical Planning and Inference 137*, 1727–1747.

Peña, E. A., R. L. Strawderman, and M. Hollander (2001). Nonparametric estimation with recurrent event data. *Journal of the American Statistical Association 96*, 1299–1315.

Perneger, T. V. (1998). Smoking and risk of myocardial infarction. *British Medical Journal 317*, 1017.

Petersen, J. H. (1998). An additive frailty model for correlated life times. *Biometrics 54*, 646–661.

Peterson, A. V. (1975). Nonparametric estimation in the competing risks problem. Technical report 73, Department of Statistics, Stanford University, USA.

Peto, R. and J. Peto (1972). Asymptotically efficient rank invariant test procedures (with discussion). *Journal of the Royal Statistical Society: Series A (General) 135*, 185–206.

Pfeifer, D. and U. Heller (1987). A martingale characterization of mixed Poisson processes. *Journal of Applied Probability 24*, 246–251.

Polyanin, A. D. and V. F. Zaitsev (1995). *Handbook of Exact Solutions for Ordinary Differential Equations*. Boca Raton, Fla.: CRC Press.

Porat, N., A. Levy, D. Fraser, R. J. Deckelbaum, and R. Dagan (1998). Prevalence of intestinal infections caused by diarrheagenic *Escherichia coli* in Bedouin infants and young children in Southern Israel. *The Pediatric Infectious Disease Journal 17*, 482–488.

Portier, C. J., C. D. Sherman, and A. Kopp-Schneider (2000). Multistage, stochastic models of the cancer process: A general theory for calculating tumor incidence. *Stochastic Environmental Research and Risk Assessment 14*, 173–179.

Pötter, U. and H.-P. Blossfeld (2001). Causal inference from series of events. *European Sociological Review 17*, 21–32.

Prentice, R. L. (1978). Linear rank tests with right censored data. *Biometrika 65*, 167–179. Correction *70*, 304 (1983).

Prentice, R. L., B. J. Williams, and A. V. Peterson (1981). On the regression analysis of multivariate failure time data. *Biometrika 68*, 373–379.

Protter, P. (1990). *Stochastic Integration and Differential Equations: A New Approach*. Berlin: Springer-Verlag.

Putter, H., M. Fiocco, and R. B. Geskus (2007). Tutorial in biostatistics: Competing risks and multi-state models. *Statistics in Medicine 26*, 2389–2430.

R Development Core Team (2007). *R: A Language and Environment for Statistical Computing*. Vienna, Austria: R Foundation for Statistical Computing. ISBN 3-900051-07-0, http://www.R-project.org.

Ramlau-Hansen, H. (1983). Smoothing counting process intensities by means of kernel functions. *Annals of Statistics 11*, 453–466.

Rastas, S., T. Pirttilä, P. Viramo, A. Verkkoniemi, P. Halonen, K. Juva, L. Niinistö, K. Mattila, E. Länsimies, and R. Sulkava (2006). Association between blood pressure and survival over 9 years in a general population aged 85 and older. *Journal of the American Geriatric Society 54*, 912–918.

Rebolledo, R. (1980). Central limit theorems for local martingales. *Zeitschrift für Wahrscheinlichkeitstheorie und verwandte Gebiete 51*, 269–286.

Ricciardi, L. M. and S. Sato (1988). First-passage-time density and moments for the Ornstein-Uhlenbeck process. *Journal of Applied Probability 25*, 43–57.

Robins, J. M. (1986). A new approach to causal inference in mortality studies with a sustained exposure period – application to control of the healthy worker survivor effect. *Mathematical Modeling 7*, 1393–1512.

Robins, J. M. and D. M. Finkelstein (2000). Correcting for noncompliance and dependent censoring in an AIDS clinical trial with inverse probability of censoring weighted (IPCW) log-rank tests. *Biometrics 56*, 779–788.

Robins, J. M., M. A. Hernán, and B. Brumback (2000). Marginal structural models and causal inference in epidemiology. *Epidemiology 11*, 550–560.

Rosenbaum, P. R. and D. B. Rubin (1983). The central role of the propensity score in observational studies for causal effects. *Biometrika 70*, 41–55.

Rothman, K. J. (2002). *Epidemiology: An introduction*. New York: Oxford University Press.

Rothman, K. J. and S. Greenland (1998). *Modern Epidemiology* (2nd ed.). Philadelphia: Lippincott-Raven.

Rothwell, P. M. and C. P. Warlow (1999). Prediction of benefit from carotid endarterectomy in individual patients: a risk-modelling study. *Lancet 353*, 2105–2110.

Rubin, D. B. (1974). Estimating causal effects of treatments in randomized and nonrandomized studies. *Journal of Educational Psychology 66*, 688–701.

Rudin, W. (1976). *Principles of Mathematical Analysis* (3rd ed.). London: McGraw-Hill.

Samuelsen, S. O. (1997). A pseudolikelihood approach to analysis of nested case-control studies. *Biometrika 84*, 379–394.

Satten, G. A., S. Datta, and J. M. Robins (2001). Estimating the marginal survival function in the presence of time dependent covariates. *Statistics & Probability Letters 54*, 397–403.

Savarino, S., R. Abu-Elyazeed, M. Rao, R. Frenck, I. Abdel-Messih, E. Hall, S. Putnam, H. El-Mohamady, T. Wierzba, B. Pittner, K. Kamal, P. Moyer, B. Morsy, A. Svennerholm, Y. Lee, and J. Clemens (2003). Efficacy of an oral, inactivated whole-cell enterotoxigenic *E. coli*/cholera toxin B subunit vaccine in Egyptian infants. The 6th Annual Conference on Vaccine Research. Arlington, Va.

Scheike, T. H. (2002). The additive nonparametric and semiparametric Aalen model as the rate function for a counting process. *Lifetime Data Analysis 8*, 247–262.

Scheike, T. H. and A. Juul (2004). Maximum likelihood estimation for Cox's regression model under nested case-control sampling. *Biostatistics 5*, 193–206.

Scheike, T. H. and M.-J. Zhang (2003). Extensions and applications of the Cox-Aalen survival model. *Biometrics 59*, 1036–1045.

Schou, G. and M. Væth (1980). A small sample study of occurrence/exposure rates for rare events. *Scandinavian Actuarial Journal*, 209–225.

Schrödinger, E. (1915). Zur Theorie der Fall-und Steigversuche an Teilchen mit Brownsche Bewegung. *Physikalische Zeitschrift 16*, 289–295.

Schwarz, W. (1992). The Wiener process between a reflecting and an absorbing barrier. *Journal of Applied Probability 29*, 597–604.

Schweder, T. (1970). Composable Markov processes. *Journal of Applied Probability 7*, 400–410.

Seshadri, V. (1998). *The Inverse Gaussian Distribution: Statistical Theory and Applications*. New York: Springer-Verlag.

Shen-Orr, S. S., R. Milo, S. Mangan, and U. Alon (2002). Network motifs in the transcriptional regulation network of Eschericia coli. *Nature Genetics 31*, 64–68.

Singpurwalla, N. D. (1995). Survival in dynamic environments. *Statistical Science 10*, 86–103.

Singpurwalla, N. D. (2006). The hazard potential: introduction and overview. *Journal of the American Statistical Association 101*, 1705–1717.

Singpurwalla, N. D. and M. A. Youngren (1993). Multivariate distributions induce by dynamic environments. *Scandinavian Journal of Statistics 20*, 251–261.

Skårderud, F., P. Nygren, and B. Edlund (2005). 'Bad Boys" bodies: The embodiment of troubled lives. Body image and disordered eating among adolescents in residential childcare institutions. *Clinical Child Psychology and Psychiatry 10*, 395–411.

Skrondal, A. and S. Rabe-Hesketh (2004). *Generalized Latent Variable Modeling. Multilevel, Longitudinal and Structural Equation Models*. Boca Raton, Fla.: Chapman & Hall/CRC.

S-PLUS® (2002). Insightful Corporation, Seattle, Washington. http://www.insightful.com.

Steenland, K. and J. A. Deddens (1997). Increased precision using countermatching in nested case-control studies. *Epidemiology 8*, 238–242.

Steinsaltz, D. and S. N. Evans (2003). Markov mortality models: Implications of quasistationarity and varying initial distributions. *Theoretical Population Biology 65*, 319–337.

Steinsland, H. (2003). *The epidemiology of enterotoxigenic* Escherichia coli *infections and diarrhea in early childhood*. Ph. D. thesis, Centre for International Health, University of Bergen, Norway.

Steinsland, H., P. Valentiner-Branth, H. K. Gjessing, P. Aaby, K. Mølbak, and H. Sommerfelt (2003). Protection from natural infections with enterotoxigenic *Escherichia coli*: Longitudinal study. *Lancet 362*, 286–291.

Steinsland, H., P. Valentiner-Branth, M. Perch, F. Dias, T. K. Fischer, P. Aaby, K. Mølbak, and H. Sommerfelt (2002). Enterotoxigenic *Escherichia coli* infections and diarrhea in a cohort of young children in Guinea Bissau. *The Journal of Infectious Diseases 186*, 1740–1747.

Strehler, B. L. and A. S. Mildvan (1960). General theory of mortality and aging. *Science 132*, 14–21.

Suppes, P. (1970). *A Probabilistic Theory of Causality*. Amsterdam: North-Holland.

Sweeting, M. J., D. De Angelis, and O. O. Aalen (2005). Bayesian back-calculation using a multi-state model with application to HIV. *Statistics in Medicine 24*, 3991–4007.

Tang, J.-L. and J. A. Dickinson (1998). Smoking and risk of myocardial infarction. Studying relative risk is not enough. *British Medical Journal 317*, 1018.

Tarone, R. E. and J. Ware (1977). On distribution-free tests for equality of survival distributions. *Biometrika 64*, 156–160.

Temkin, N. R. (1978). An analysis for transient states with application to tumor shrinkage. *Biometrics 34*, 571–580.

Therneau, T. M. and P. M. Grambsch (2000). *Modeling Survival Data: Extending the Cox Model*. New York: Springer-Verlag.

Thomas, D. C. (1977). Addendum to: Methods of cohort analysis: Appraisal by application to asbestos mining By Liddell, F. D. K., McDonald, J. C., and Thomas, D. C. *Journal of the Royal Statistical Society: Series A (General) 140*, 469–491.

Thomas, D. R. and G. L. Grunkemeier (1975). Confidence interval estimation of survival probabilities for censored data. *Journal of the American Statistical Association 70*, 865–871.

Triarhou, L. C. (1998). Rate of neuronal fallout in a transsynaptic cerebellar model. *Brain Research Bulletin 47*, 219–222.

Tsiatis, A. A. (1981). A large sample study of Cox's regression model. *Annals of Statistics 9*, 93–108.

Tsiatis, A. A., V. DeGruttola, and M. Wulfsohn (1995). Modeling the relationship of survival to longitudinal data measured with error. Applications to survival and CD4 counts in patients with AIDS. *Journal of the American Statistical Association 90*, 27–37.

Ulrich, C. M., H. F. Nijhout, and M. C. Reed (2006). Mathematical modeling: Epidemiology meets systems biology. *Cancer Epidemiology Biomarkers & Prevention 15*, 827–829.

van Houwelingen, H. C. (2007). Dynamic prediction by landmarking in event history analysis. *Scandinavian Journal of Statistics 34*, 70–85.

Vaupel, J. W. (2005). Lifesaving, lifetimes and lifetables. *Demographic Research 13*, 597–614.

Vaupel, J. W., K. G. Manton, and E. Stallard (1979). The impact of heterogeneity in individual frailty on the dynamics of mortality. *Demography 16*, 439–454.

Vaupel, J. W. and A. I. Yashin (1985). The deviant dynamics of death in heterogeneous populations. *Sociological Methodology 15*, 179–211.

Vaupel, J. W. and A. I. Yashin (1999). Cancer rates over age, time, and place: insights from stochastic models of heterogeneous populations. Max Planck Institute for Demographic Research, Rostock. Working Paper.

Vollset, S. E., A. Tverdal, and H. K. Gjessing (2006). Smoking and deaths between 40 and 70 years of age in women and men. *Annals of Internal Medicine 144*, 381–389.

Wagner, A. (2001). How to reconstruct a large genetic network from n gene perturbations in fewer than n^2 easy steps. *Bioinformatics 17*, 1183–1197.

Wang, J.-L. (2005). Smoothing hazard rate. In P. Armitage and T. Colton (Eds.), *Encyclopedia of Biostatistics* (2nd ed.), Volume 7, pp. 4986–4997. Chichester: Wiley.

Weedon-Fekjær, H., B. H. Lindqvist, O. O. Aalen, L. J. Vatten, and S. Tretli (2007). Breast cancer tumor growth estimated through mammography screening data. Unpublished manuscript.

Weisstein, E. W. (2006a). Bessel function of the first kind. *MathWorld* – a Wolfram Web resource. http://mathworld.wolfram.com/BesselFunctionoftheFirst Kind.html.

Weisstein, E. W. (2006b). Beta function. *MathWorld* – a Wolfram Web resource. http://mathworld.wolfram.com/BetaFunction.html.

Weitz, J. S. and H. B. Fraser (2001). Explaining mortality rate plateaus. *Proceedings of the National Academy of Sciences of the United States of America 98*, 15383–15386.

Wenocur, M. L. (1990). Ornstein-Uhlenbeck processes with quadratic killing. *Journal of Applied Probability 27*, 707–712.

Wermuth, N. (2005). Statistics for studies of human welfare. *International Statistical Review 73*, 259–262.

Wetherill, G. B. (1986). *Regression Analysis with Applications*. London: Chapman and Hall.

Whitaker, H. J., C. P. Farrington, B. Spiessens, and P. Musonda (2006). Tutorial in biostatistics: The self-controlled case series method. *Statistics in Medicine 25*, 1768–1797.

Whitmore, G. A. (1986a). First-passage-time models for duration data: Regression structures and competing risks. *The Statistician 35*, 207–219.

Whitmore, G. A. (1986b). Normal-gamma mixtures of inverse Gaussian distributions. *Scandinavian Journal of Statistics 13*, 211–220.

Whitmore, G. A. (1995). Estimating degradation by a Wiener diffusion process. *Lifetime Data Analysis 1*, 307–319.

Whitmore, G. A., M. J. Crowder, and J. F. Lawless (1998). Failure inference from a marker process based on a bivariate Wiener model. *Lifetime Data Analysis 4*, 229–251.

Wienke, A. (2007). *Frailty models in survival analysis*. Ph. D. thesis, Martin-Luther-Universität Halle-Wittenberg, Germany. http://sundoc.bibliothek.uni-halle.de/habil-online/07/07H056/habil.pdf.

Williams, D. and K. Baverstock (2006). Too soon for a final diagnosis. *Nature 440*, 993–994.

Wolfram, S. (1999). *The Mathematica Book*. Champaign, Ill: Wolfram Media/Cambridge University Press.

Woodbury, M. A. and K. G. Manton (1977). A random walk model of human mortality and aging. *Theoretical Population Biology 11*, 37–48.

World Health Organization (1998). *Life in the 21st Century: A Vision for All*. The World health report, 1998. Geneva, Switzerland: WHO.

Wright, S. (1921). Correlation and causation. *Journal of Agricultural Research 20*, 557–585.

Yashin, A. I. (1985). Dynamics in survival analysis: Conditional Gaussian property versus Cameron-Martin formula. In N. V. Krylov, R. S. Lipster, and A. A. Novikov (Eds.), *Statistics and Control of Stochastic Processes*, pp. 466–475. New York: Springer.

Yashin, A. I. and I. A. Iachine (1995). Genetic analysis of durations: Correlated frailty model applied to survival of Danish twins. *Genetic Epidemiology 12*, 529–538.

Yashin, A. I. and K. G. Manton (1997). Effects of unobserved and partially observed covariate processes on system failure: A review of models and estimation strategies. *Statistical Science 12*, 20–34.

Yassouridis, A., A. Steiger, A. Klinger, and L. Fahrmeir (1999). Modelling and exploring human sleep with event history analysis. *Journal of Sleep Research 8*, 25–36.

Zahl, P. H. (1995). A proportional regression model for 20 year survival of colon cancer in Norway. *Statistics in Medicine 14*, 1249–1261.

Zahl, P. H. (1996). A linear regression model for the excess intensity. *Scandinavian Journal of Statistics 23*, 353–364.

Zahl, P. H. (1997). Frailty modelling for the excess hazard. *Statistics in Medicine 16*, 1573–1585.

Zahl, P. H. (2003). Regression analysis with multiplicative and time-varying additive regression coefficients with examples from breast and colon cancer. *Statistics in Medicine 22*, 1113–1127.

Zahl, P. H. and O. O. Aalen (1998). Adjusting and comparing survival curves by means of an additive risk model. *Lifetime Data Analysis 4*, 149–168.

Zahl, P. H. and S. Tretli (1997). Long-term survival of breast cancer in Norway by age and clinical stage. *Statistics in Medicine 16*, 1435–1449.

Zeilinger, A. (2005). The message of the quantum. *Nature 438*, 743.

Zeng, D. and D. Y. Lin (2007). Maximum likelihood estimation in semiparametric regression models with censored data. *Journal of the Royal Statistical Society: Series B (Statistical Methodology) 69*, 507–564.

Zhang, J. and Ø. Borgan (1999). Aalen's linear model for sampled risk set data: A large sample study. *Lifetime Data Analysis 5*, 351–369.

Zijlstra, F., J. C. A. Hoorntje, M.-J. de Boer, S. Reiffers, K. Miedema, J. P. Ottervanger, A. W. J. van 't Hof, and H. Suryapranata (1999). Long-term benefit of primary angioplasty as compared with thrombolytic therapy for acute myocardial infarction. *The New England Journal of Medicine 341*, 1413–1419.

Author index

Index

Springer
the language of science

springer.com

Proportional Hazards Regression

John O'Quigley

This book focuses on the theory and applications of a very broad class of models—proportional hazards and non-proportional hazards models, the former being viewed as a special case of the latter—which underlie modern survival analysis. Unlike other books in this area the emphasis is not on measure theoretic arguments for stochastic integrals and martingales. Instead, while inference based on counting processes and the theory of martingales is covered, much greater weight is placed on more traditional results such as the functional central limit theorem.

2008. 430p. (Statistics for Biology and Health) Hardcover
ISBN 0-387-25148-6

The Frailty Model

Luc Duchateau and Paul Janssen

This book provides an in-depth discussion and explanation of the basics of frailty model methodology for such readers. The discussion includes parametric and semiparametric frailty models and accelerated failure time models. Common techniques to fit frailty models include the EM-algorithm, penalised likelihood techniques, Laplacian integration and Bayesian techniques. More advanced frailty models for hierarchical data are also included.

2008. 335 pp. (Statistics for Biology and Health) Hardcover
ISBN 978-0-387-72834-6

The Statistics of Gene Mapping

David Siegmund and Benjamin Yakir

This book details the statistical concepts used in gene mapping, first in the experimental context of crosses of inbred lines and then in outbred populations, primarily humans. It presents elementary principles of probability and statistics, which are implemented by computational tools based on the R programming language to simulate genetic experiments and evaluate statistical analyses. Each chapter contains exercises, both theoretical and computational, some routine and others that are more challenging. The R programming language is developed in the text.

2007. 331 pp. (Statistics for Biology and Health) Hardcover
ISBN 978-0-387-43684-9

Easy Ways to Order▶ Call: Toll-Free 1-800-SPRINGER • E-mail: orders-ny@springer.com • Write: Springer, Dept. S8113, PO Box 2485, Secaucus, NJ 07096-2485 • Visit: Your local scientific bookstore or urge your librarian to order.

Printed in the United States of America